BAYESIAN AND LIKELIHOOD METHODS IN STATISTICS AND ECONOMETRICS

Essays in Honor of George A. Barnard

STUDIES IN BAYESIAN ECONOMETRICS AND STATISTICS

Editors

ARNOLD ZELLNER

JOSEPH B. KADANE

Volume 7

NORTH-HOLLAND – AMSTERDAM • NEW YORK • OXFORD • TOKYO

BAYESIAN AND LIKELIHOOD METHODS IN STATISTICS AND ECONOMETRICS

Essays in Honor of George A. Barnard

Edited by

SEYMOUR GEISSER
University of Minnesota

JAMES S. HODGES
The RAND Corporation

S. JAMES PRESS
University of California, Riverside

ARNOLD ZELLNER
University of Chicago

1990

NORTH-HOLLAND – AMSTERDAM • NEW YORK • OXFORD • TOKYO

ELSEVIER SCIENCE PUBLISHERS B.V.
Sara Burgerhartstraat 25
P.O. Box 211, 1000 AE Amsterdam, The Netherlands

Distributors for the U.S.A. and Canada:
ELSEVIER SCIENCE PUBLISHING COMPANY, INC.
655 Avenue of the Americas
New York, N.Y. 10010, U.S.A.

Library of Congress Cataloging-in-Publication Data

Bayesian and likelihood methods in statistics and econometrics : essays in honor of George A. Barnard / edited by Seymour Geisser ... [et al.].
p. cm. -- (Studies in Bayesian econometrics and statistics ; v. 7)
Includes bibliographical references.
ISBN 0-444-88376-2 (Elsevier Science)
1. Econometrics. 2. Mathematical statistics. 3. Probabilities. 4. Bayesian statistical decision theory. 5. Barnard, George A. (George Alfred) I. Barnard, George A. (George Alfred) II. Geisser, Seymour. III. Series.
HB139.B394 1990
330'.01'5195--dc20 89-28088
CIP

ISBN: 0 444 88376 2

Printed in The Netherlands

CONTENTS

V. SELECTED INFERENCE PROBLEMS

VI. PREDICTIVE METHODS AND APPLICATIONS

VII. INCOMPLETELY SPECIFIED PRIOR DISTRIBUTIONS

VIII. APPROXIMATIONS AND COMPUTATIONAL METHODS

ACKNOWLEDGMENTS

Many people have helped us produce this book. We would like to thank the referees, who must remain nameless; Gerald Wanrooy of North-Holland for his many efforts; Dan Relles, director of the Statistical Research and Consulting Group at the RAND Corporation; Pamela Eckert of the University of Chicago, and Rosalie Heacock and Sharon Koga of the RAND Corporation for their excellent secretarial help. Sharon Koga prepared and corrected the papers in TeX and prepared the final camera-ready copy.

The Editors

I. INTRODUCTION

INTRODUCTION

This book honors Professor George A. Barnard, an internationally acclaimed statistician. He is renowned not only for his statistical expertise but also for his broad knowledge of the history of statistics and its relation to the philosophy of science and applications in many fields of natural, biological and social science. The breadth and depth of his research contributions to statistical inference, including likelihood theory, estimation, testing, information theory, decision theory, and other areas of statistics, are recognized world-wide. It is thus appropriate that this volume is dedicated to him.

In addition to his many technical research contributions, Barnard is unique in his deep understanding of and writings on various approaches to statistical inference. While most have classified him as a "likelihood advocate," this label does not do justice to his sophisticated understanding and appreciation of the various issues involved in drawing meaningful inferences from data. For example, he is aware of the difficulties in formulating appropriate likelihood functions for complex problems. Also, the delicate issues involved in comparing and testing hypotheses have been of great concern to him. Similarly, the quality and appropriate uses of prior information in statistical analyses have been key issues in his thinking and writings for many years. That he has exerted a considerable influence on the development of statistical inference over the years is recognized by all.

The present volume is titled "Bayesian and Likelihood Inference in Statistics and Econometrics" since we wish to take this opportunity to pursue a theme which is of concern to Barnard. He takes the position that when solid prior information is available and well represented, it can be used to good advantage in statistical analyses. Note that he insists that prior information be of high quality and securely founded before it is introduced in serious analyses of scientific data. In the present volume, authors employ prior information, not only in the form of prior distributions but also in the formulation of models and likelihood functions. The reader can judge whether Barnard's high standards for the introduction of such information have been met. Of course, if they have not been met, it is probably the case that thorough diagnostic checking and poor predictive performance will reveal inadequacies of model specifications and prior information. That prior information can be insecurely based is obvious to Barnard and perhaps explains his caution in using prior distributions in statistical inference. However, Barnard has used prior distributions to good advantage in applied decision problems.

In part II of this volume, Dennis Lindley's paper, "A Survey of George Barnard's Statistical Work" is a deep and perceptive account of Barnard's research and approach to applied and theoretical statistics. Lindley emphasizes that Barnard is exceptional in that he is outstanding in getting to the heart of thereotical issues and relating them to applied problems. Also, he has been very successful in performing analyses of important, applied problems with results that have had a substantial impact.

Topics included in Lindley's review include (1) the relationship of Barnard's work to that of R.A. Fisher, (2) Barnard's pioneering research on the likelihood principle and likelihood ratios for testing problems, (3) pivotals and fiducial inference, (4) Barnard and the Bayesian viewpoint, and (5) Barnard's applied statistical work. He also reviews Barnard's three presidential addresses, which he gave to the Operations Research Society, the Royal Statistical Society and the Institute of Mathematics and its Application. Along with an extensive bibliography of Barnard's publications, Lindley's paper provides the reader with a cogent review and discussion of Barnard and his work.

Four papers are included in Section III, Foundations. The first, "Bayes, Fisher, and Belief Functions" by Arthur Dempster, discusses Bayesian inference, Fisher's fiducial inference and the theory of belief functions. He begins by noting that Barnard in a 1987 article characterized Fisher as "a true Bayesian." In Dempster's paper, he presents an elementary exposition of the belief function approach, discusses its relation to Bayesian and fiducial inference approaches and points out an underlying logical unity among the three approaches. There is no doubt but that Dempster's examination of the common elements and independence assumptions of the three approaches will do much to further understanding of their prescriptions for reasoning from empirical data to reach posterior conclusions.

In Bruce Hill's paper, "A Theory of Bayesian Data Analysis," he addresses "some of the basic philosophical and practical issues that arise in attempting to integrate the Bayesian theory with data analysis." He argues that both conventional data analysis and conventional pre-data Bayesian theory can benefit from one another. To formalize informal data analysis techniques—widely used in practice—is a great challenge. Hill in this paper and his earlier work has been a pioneer in this area of research. His current paper provides readers with a lucid discussion of Bayesian data analysis and some new results. He notes that Barnard in 1947 apparently first understood the inadequacies of "pre-data theories" in connection with a sequential analysis problem.

The third paper in this section is "Can/May Bayesians Do Pure Tests of Significance?" by James Hodges. As is well known, testing theory has been a controversial issues for many years. Hodges takes up Barnard's argument that the likelihood approach cannot be used to test hypotheses when alternative hypotheses are not available and so that there is a need for classical pure significance tests. He proceeds to consider Bayesian testing procedures relative to *a very broad* class of alternative hypotheses and points out that this is unnecessary. Then he refutes Bayesian arguments against use of exploratory significance tests "by constructing an appealing alternative distribution and prior probabilities on the null and alternative that yield (to arbitrary accuracy) the P-value as the posterior probability of the null." He concludes that Bayesians can and should use P-values in an exploratory role, a conclusion that is significant in view of the many years of controversy regarding this issue.

In the last paper of this section, "Conglomerability, Coherence and Countable Additivity," David Lane offers answers to two foundational questions relating to the Law of Total Probability, which all use. The first question considered is, "Is the Law always available?" and the second, "When the Law is available, must it be respected?" Also, he analyzes the consequences of violating the Law. As is evident, these are very basic considerations and Lane's analyses of them are impressive and illuminating.

Section IV, Selected Applications, illustrates uses of Bayesian analysis. Siddhartha Chib, Ram Tiwari and S. Rao Jammalamadaka in their paper, "On Bayes Smoothing in a Time Varying Regression Model," analyze a time-varying parameter regression model with unknown noise variance parameters. They derive the exact distribution of the regression coefficients and noise variance parameters. Since regression coefficients are often time-varying rather than fixed, the results in this paper will be useful in applications.

In the Denby-Gnanadesikan-Kettenring-Suzansky paper, "An Analysis of Questionnaire Data on Work Design and Job Satisfaction: A Case Study in the Use of Simple Graphical Displays," it is argued that simple graphical tools "are more appropriate for these data then more traditional and more complex tools, such as factor analysis and multivariate analysis of variance, that are often applied in such situations." This is an example of a situation in which it is difficult to formulate an appropriate likelihood function for the observations, a problem which Barnard appreciates as the authors note. Also, their fruitful use of clever graphical techniques indicates a need to extend formal inference methods to cover such analyses.

Stephen Fienberg's paper, "Legal Likelihoods and A Priori Assessments: What Goes Where?" investigates the use of the Bayesian approach as a basis for legal decision-making in courtroom settings. He emphasizes difficulties in formulating prior distributions and likelihood functions in legal problems. After analyzing two paradoxes, he concludes with a discussion of critical forensic evidence in a recent criminal case and alternative ways to use that evidence in specifying likelihoods and priors. His analysis illustrates well problems that, as pointed out above, are of great concern to Barnard.

In "Seemingly Unrelated Regression with Linear Hierarchical Structure," Yoel Haitovsky extends the Bayesian analysis of the "seemingly unrelated regression" (SUR) model to take account of hierarchical structures. Since the SUR model is widely used, as Haitovsky notes, and hierarchical structures have been found to be very useful in many analyses, his results are valuable, theoretically and practically.

Hiroki Tsurumi's paper, "Comparing Bayesian and Non-Bayesian Limited Information Estimators" takes up a central problem, the estimation of parameters in an equation of a linear structural econometric model. After deriving two Bayesian estimators, he reports the results of Monte Carlo experiments designed to appraise the performance of Bayesian and non-Bayesian estimators, since establishing the "operating characteristics" of statistical procedures is of great importance. His results will be of interest to both theoretical and applied researchers.

Section V is devoted to the analysis of selected inference problems. M. Bayarri and Morris DeGroot in their paper, "Selection Models and Selection Mechanisms," consider problems in which experimenters are not able to draw a random sample from a population of interest, a frequently encountered circumstance. They provide Bayesian methods for analyzing observations obtained from selected portions of a population by use of various selection or truncation models. Several examples are provided to indicate when selection effects can be ignored and when they cannot be ignored, very important issues in the analysis of sample data.

The Broemeling-Son-Hamdy paper, "A Bayesian Solution to the Behrens-Fisher Problem" treats a classic problem, which Barnard analyzed earlier, from the Bayesian point of view. Using a Jeffreys' diffuse prior distribution, they obtain an approximate posterior distribution for $k-1$ independent contrasts for k means of a multivariate normal distribution and derive a HPD region for the contrasts which is used to test for the equality of the means. Their approximate normal posterior distribution is compared to the exact posterior distribution and two examples are analyzed.

In the Chen-Dickey paper, "Extending Families of Densities by Special-Function Factors," they generalize a number of parametric univariate and vector-variate distributions by multiplying the usual density by a generalized hypergeometric function or a multiple hypergeometric function with an appropriately transformed argument variable, which results in generalized densities with normalizing constants which have a closed functional form. These generalized densities have attractive properties and permit investigators to entertain broader assumptions in formulating models with which to analyze data.

George Duncan in his paper, "Bayes Estimation in the Markov Renewal Process," develops Bayesian estimators for parameters of a finite state Markov renewal process. Using a matrix beta prior distribution, he presents explicit approximate Bayes estimators for transition probabilities which will be useful in analyses based on this widely employed model.

Ghosh, Low and Meeden discuss a hierarchical Bayes approach to certain cross-classification problems in their paper, "A Hierarchical Bayes Analysis of Cross-Classification Models." Their approach provides a unified treatment of the fixed, random, and mixed models. As they point out, the Bayes approach can also handle certain nested model situations as special cases.

In Winkler and Gaba's paper, "Inference with Imperfect Sampling from a Bernoulli Process," they consider situations in which various sources of noise may cause some observations to be classified incorrectly. On introducing a noise parameter with unknown value, they encounter an identification problem in a likelihood analysis. Using a Bayesian approach with a joint prior distribution on the binomial parameter and the noise parameter, they show how to analyze the problem and provide an example to illustrate their approach. Since misclassification occurs in many situations, the results of this paper should be very useful in practice.

Section VI, Predictive Methods and Applications, contains four papers relating to prediction and forecasting, topics of great interest to George Barnard. Seymour Geisser's paper, "Predictive Approaches to Discordancy Testing" begins with a consideration of "pure significance tests," P-values, critical regions and alternative hypotheses. In his paper, Geisser develops "a Bayesian framework for predictive approaches to testing for discordant observations." He further notes that, "Hence this [his] approach is one that a Bayesian can comfortably use and will turn out to be considerably more flexible and more widely applicable than the standard frequentist approach." The operational methods that Geisser develops are illustrated using several simple parametric models. The predictive approach, which Geisser utilizes, can be fruitfully considered in relation to the issues raised in Hodges's paper relating to Barnard's views on uses of "pure significance tests."

Ruben Klein and James Press in their paper, "Bayesian Contextual Classification with Neighbors Correlated with Training Data," use a variety of examples to illustrate their methods for classifying an observation vector $\underset{\sim}{z}_0$, into one of K normal populations when past data, or training sample data, are available with a spatial correlation structure. These methods and applications will be of great interest to those analyzing satellite data and other problems in which a spatial structure is present.

Peter Lenk's paper, "Bayesian Predictive Distributions Under Multinomial Sampling," presents a general method for constructing families of distributions of parameters of the multinomial distribution such that these families are closed in prior-to-posterior Bayesian analysis. The special case of the generalized logistic normal distribution is analyzed and its predictive distribution is approximated with a cubic spline. These basic results are valuable for investigators using multinomial models for prediction.

The Zellner-Hong-Gulati paper, "Turning Points in Economic Time Series, Loss Structures and Bayesian Forecasting" presents a methodology for making turning point forecasts based on Bayesian decision theory. Also, the problem of choosing an optimal turning point forecast and a compatible optimal point forecast is analyzed. Using a relatively simple autoregressive leading indicator model, turning points in 18 countries' rates of growth of real output are forecasted for the period 1974-1984. The relatively good forecasting performance achieved suggests that the model and methods employed in the paper may be useful to private and public forecasters and decision-makers.

Section VII, Incompletely Specified Priors, includes papers in which "partial" prior information is employed, a topic of great interest to Barnard, who is vitally concerned about the nature and representation of prior information. The papers in this Section provide flexible and less restrictive prior assumptions which may help in certain cases to allay Barnard's concerns.

Mark Berliner and Prem Goel's paper, "Incorporating Partial Prior Information: Ranges of Posterior Probabilities," shows how to incorporate partial prior information about the value of a parameter in statistical and decision-making problems. Their approach yields ranges of posterior probabilities concerning the values of parameters of interest. An explicit example is analyzed thoroughly to provide an application of their methods.

Wojciech Bijak's paper, "Estimation of the Linear Regression Model in Conditions of Incomplete Prior Information" is another example of an approach in which the representation of prior information is relaxed. That is, instead of using a particular prior distribution for the regression model's parameters, Bijak introduces a class of prior distributions and shows that Bayesian estimators with respect to natural conjugate prior distributions are minimax estimators with respect to the class of prior distributions considered and the class of all estimators of the regression model's parameters. Bijak's results establish important operating characteristics of Bayesian estimation procedures when prior information is incomplete.

Albert Madansky analyzes situations in which only partial prior information is available in his paper, "Bayesian Analysis With Incompletely Specified Prior Distributions." He presents a Bayesian approach in which it is possible to obtain posterior distributions when only knowledge about some quantiles of the prior distribution is available or just when "one bit of prior knowledge, namely that the prior density is monotonically nondecreasing" is available. As with other papers in this Section, this one permits analysts to represent partial prior information quite flexibly and to use just partial prior information in making inferences and decisions.

The last Section, Approximations and Computational Methods, includes three papers with results which will be of value in operationalizing Bayesian methods.

Jorge Achcar and Adrian Smith have investigated the effects of reparametrization on the accuracy of Laplacian and related methods of approximating posterior densities and moments in their paper, "Aspects of Reparametrization in Approximate Bayesian Inference." They illustrate application of their methods with selected examples and show that reparametrization matters.

The Larkin-Kadane paper, "A Method of Maximizing Likelihood Functions" presents a useful numerical algorithm for computing maximum likelihood or maximum posterior distribution estimates. They compare their "unrestricted quadratics" approach to various local gradient and grid search methods. Their procedure is applied to an interesting problem in developmental psychology. They suggest that their ideas and methods are useful for studying complex likelihoods in multidimensional spaces.

In the Kass-Tierney-Kadane paper, "The Validity of Posterior Expansions Based on Laplace's Method," methods are presented for justifying heuristic derivations of asymptotic expansions for predictive and posterior densities, odds factors, and posterior moments given by various authors. By providing a formal framework within which it is possible to evaluate asymptotic techniques, the authors have made this area of analysis much more rigorous.

The papers in this volume treat a broad spectrum of topics ranging from theoretical inference to applied forecasting and computation. In this respect, the volume reflects George Barnard's broad-ranging interest in and work on theoretical and applied statistical problems. As has been recognized by many, theory without application and application without theory are unsatisfactory. George Barnard's work reflects a wholesome combination of theory and application and it was with this in mind that the current volume honoring him was conceived and completed. We are all very grateful to George Barnard, not only for his many research contributions but also for his constructive and beneficial influence on statistics and statisticians.

The Editors

II. A SURVEY OF GEORGE BARNARD'S STATISTICAL WORK

Bayesian and Likelihood Methods in Statistics and Econometrics
S. Geisser, J.S. Hodges, S.J. Press and A. Zellner (Editors)

A SURVEY OF GEORGE BARNARD'S STATISTICAL WORK

D. V. Lindley

1. Introduction

George Barnard[1] did his graduate work in mathematical logic, then in the Second World War he did statistical work for the government, with an emphasis on industrial applications. These two attitudes, of concern for basic principles and skill in practical applications, have been the basic themes in his statistical life. Rarely, in one statistician, do we find such a spread of ideas from abstract contemplation to material application. He can discuss quality control of condoms [36] and axiomatic theories of inference [11]. In justifying his choice of topic he says (in the reply to the discussion of [13]) that is "was governed partly...by my own predilection for the foundations of any subject". Yet this is the same person who can deliver a brilliant and critical lecture on the problems of manufacturing industry in Britain [96]. The concern with foundations never leads to technicalities obscuring the basic issues. He explains this nicely in [17]: Wolfowitz had criticized the argument in [16] because an inversion of limits was unjustified, to which the response is "undue attention to pure rigour is out of place [in statistics]" and then proceeds to the justification. Or in [8] he says "undue emphasis on niceties is a disease to which persons with mathematical training are specially prone".

I first met Barnard as a junior to him in the government's service during the war. (With Plackett, he has documented these, and other wartime, statistical activities in [94].) I have always found him to be the most stimulating statistician to interact with, either through the printed word, or in conversation. He has a manner of presenting ideas that makes them interesting. One cannot just say in response, that is right, or that is wrong: the views lead forward to consider other aspects, to special cases, or to generalities. The only real difficulty in this pleasurable task of writing about his work, has been to restrain myself from commenting on his writing: "Yes, George, but what about nuisance parameters...". He makes one think. The temptation has been resisted except in some comments at the end in summarizing his work. I have been enormously helped in the preparation of this essay through the provision by Barnard

[1] Reference to Barnard's papers are in square brackets [ ] and refer to the bibliography of his work following this paper.

of an almost complete set of reprints; thus easily enabling me to read work almost forgotten or never seen before.

Apart from logic and quality control, another dominating influence has been R.A. Fisher. Every statistician who has even a glimmer of the history of our subject appreciates the genius of Fisher and the revolution in our thinking that he created. But Barnard's respect goes beyond that. He feels, with justification, that it is worth attending carefully to everything the master said and trying to unravel the obscurity. In [38], after quoting Fisher, he says "I shall try to expound what I take this to mean". Even over what I.J. Good has called "Fisher's mistake" in transferring a probability from data to parameter (because of a defective notation), Barnard has sought to understand. He is certainly one of the few people who could discuss critically with Fisher on statistical issues. Perhaps this is because "I gained much credit in Fisher's eyes as being...the only statistician who had ever admitted he had been mistaken" [50]. The mistake, over 2×2 tables, is discussed below.

This ability to change his view is a marked feature of Barnard's writing. This is nowhere more evident that in discussion of the likelihood principle. A few years ago Barnard and I were at a dinner of the Research Section of the Royal Statistical Society discussing some statistical topic. After he had made a remark, a younger person said "That is not what you said ten years ago". "What did I say then?" replied Barnard. He was told, thought for a moment, and then said "Good, that shows I have learnt something in ten years". My memory does not tell me whether I asked him if his updating of his opinion had been done by Bayes' theorem, or what his reply was. This evolving nature of his thought means that it can be incorrect to say "Barnard's view is...". The context, and date, need to be included.

2. Likelihood

It is natural to begin any discussion of his work with the likelihood principle. This concept, of great foundational interest and enormous practical relevance, keeps recurring in his papers. He was perhaps the first to put forward the idea, in a thoughtful review of Wald's (1947) book on *Sequential Analysis* [6]. (Sequential ideas had been explored by him whilst in war service.) "If we make a comparison between two hypotheses, H and H', on the basis of observed results R, this can be done only by comparing the chances of, getting R, if H were true, with those of getting R, if H' were true." He then goes on to show, by introducing an independent, ancillary statistic, that the comparison must be by the ratio. At the end of the review he describes the large sphere of applicability of the "likelihood ratio method of analysis" not merely in sequential sampling but as a tool for the analysis of results obtained more generally. The idea is expressed more crisply in [11]: "if one is prepared to accept the validity of the standard probability ratio sequential test, applied to data obtained by the prescribed sequential procedure, then one must be prepared to accept the validity of the test as applied to data not so obtained". Then in [19], in 1953, he says "the likelihood function 'contains all the information in the data', and hence...the problem of inference reduces to the problem of describing the likelihood function".

In one of the papers that he read at a Royal Statistical Society (RSS) meeting, and which therefore has a discussion, he put his ideas on the foundations of statistics onto an axiomatic basis [11]. His own summary of the main points appears in [53]. The ideas have been used by Barndorff-Nielsen (1978). Here we have the concepts of "forwards odds"—the ratio of densities for the same parameter value—and "backwards odds"—the ratio for the same data. Forwards odds look forward to the experiment: backwards odds look back, once the results of the experiment are available. "*Before* the experiment is carried out...probabilities are relevant, *after* the experiment likelihoods are relevant" (his italics, in [30]). This is an elaboration of the distinction made by Fisher between probability and likelihood. His axioms lead to an odds form of Bayes' theorem which elegantly eliminates the prior. For two data sets, A, A', and two parameter values α, α', with l and λ for the logarithms of forwards and backwards odds

$$\lambda(A \mid \alpha, \alpha') - \lambda(A' \mid \alpha, \alpha') = l(A, A' \mid \alpha) - l(A, A' \mid \alpha').$$

The prior cannot be dispensed with altogether, and the related idea of a neutral result, that does not change the odds, is introduced. The quadratic approximation to the log-odds is discussed and an elegant illustration provided in connection with the sequential t-test. The sequential t-test is considered further in [16] and the hypotheses formulated in an ingenious way to eliminate the nuisance parameter. Here he also introduces group ideas and the notion of invariance, connecting the latter with Jeffreys' ideas on the topic.

The most important of his papers on likelihood is, in my view, [30] with Jenkins and Winsten, also read to the RSS. The title is "Likelihood Inference and Time Series" and much of the value of the paper lies in the fact that it produces examples of likelihood methods in practice, with surprising results, especially in time series. The discussants were puzzled by the strange ideas. The paper begins with the now-familiar discussion of binomial sampling and the irrelevance of the reference set. An example, which appears repeatedly in his writings, effectively concerns a distribution uniform over a finite number of consecutive integers of known range but unknown location. It brings out the point that the precision of an inference can vary with the sample, unlike the familiar normal mean case where it is always $n\sigma^{-2}$, so emphasizing the irrelevance of averaging over the reference set. Another most interesting idea, which is attributed to Fisher, is that the power function is essentially the likelihood function, not for the data as observed, but for the data and more extreme values (that is, in the tail area).

The paper is almost contemporary with that of Birnbaum (1962) and similarly develops the likelihood principle from more fundamental notions. (The memorial article with Godambe [79] admirably describes Birnbaum's work.) But its most interesting part is the discussion of time series and, in particular the contrast between ordinary regression and autoregression. This involves an excursion into economic modelling and feedback, and provides a striking contrast between likelihood and sampling theories. Markov chains are considered and there is a brief treatment of the difficult problem (to an adherent of likelihood) of nuisance parameters. The discussion, even by RSS standards, is full of interest and Stein produces a good counter-example to the likelihood principle. This he enlarged on later, Stein (1962), and Barnard replied [31].

In understanding Barnard's views on statistical inference, it is necessary to appreciate that, although he is an enthusiastic advocate of the likelihood principle, he does not consider that it applies to all post-data problems of inference, as do the Bayesians. He recognizes several types of inference problems, and likelihood applies when different numerical values for a parameter are being compared. An example he has often used is the comparison of θ-values in the density $(1+\theta x)/2$ for $|x| \leq 1$, a problem that arises in physics [38]. Another type of inference arises when a scientist wishes to test a null hypothesis and the alternatives are at most vaguely formulated. Then the likelihood function either does not exist, or its creation by the statistician is artificial and alien to the scientist's mode of thought. Here significance tests and tail-area probabilities are relevant. In [75] four models are discussed: the significance test model, the Bayes model, the likelihood model and the pivotal model.

One important consequence of the likelihood principle is the stopping-rule principle which effectively says that in a sequential experiment the analysis should depend only on the final data and not on the rule of stopping that led to that data. This principle was, of course, recognized by Barnard as early as 1947 [6] when he expressed the likelihood principle in terms of sequential ideas. Savage makes the following comment in Savage et. al., (1962).

> I learned the stopping-rule principle from Professor Barnard in conversation in the summer of 1952. Frankly, I then thought it a scandal that

> anyone in the profession could advance an idea so patently wrong, even as today I can scarcely believe that some people resist an idea so patently right.

This remark serves to remind us of the originality and depth of Barnard's thinking. The likelihood and stopping-rule principles are now incorporated into statistical thinking and have a fine exposition in Berger and Wolpert (1984), yet Barnard has been aware of the ideas for over forty years. Their general adoption, even in the wide, but limited, area advocated by Barnard would have a substantial effect on statistical practice. Unfortunately papers are still being written proposing methods that violate the principles, the authors appearing to be unaware of Barnard's work or recent expositions of it. Only recently he and I listened to a paper on sequential methods read to the Royal Statistical Society that demonstrated such ignorance. One puzzles why the referees accepted the paper without, at least, asking for an explanation and for references to be included.

3. 2×2 Tables

A problem that Barnard has repeatedly turned to is that of a test of independence in the 2×2 table. Others may work on multidimensional tables but Barnard, quite sensibly, still finds this "elementary" problem of interest because it exposes a very simple situation that still has not got a generally-accepted, small-sample solution. Three early papers [1, 3, 6] begin the tale and are the source of his "mistake" mentioned above. Three types of tables are distinguished, (i) neither margin fixed, (ii) one margin fixed, and (iii) both fixed. In consideration of (ii), a rejection region is found having properties of convexity (C), symmetry (S) and containing the maximum (M) number of points. This yields the CSM-test, very much along the lines of Neyman and Pearson. It is shown that the numerical values of significance can be very different from those given by Fisher's "exact" test which suppose (iii), but, since the margins are held to "contain no information" about independence, also covers (i) and (ii). The fundamental reason for the difference in significance levels is that the two tests, CSM and Fisher's, use different sample spaces, or as they would prefer to say, reference sets. CSM refers to all tables with one margin fixed: Fisher, to the subset with both fixed at the values observed. Barnard's "mistake" was to use the wrong reference set.

Much later, in [84] and [97], an analysis is presented in which the independence is expressed as a test of the hypothesis that the log-odds, $\log(p_1q_2/q_1p_2)$, is zero with $\log(p_1p_2/q_1q_2)$ as a nuisance parameter. (Here p_1 and p_2 are the chance probabilities for the two rows of the table, the row totals being fixed.) By considering the conditional distribution of the data given the column totals also fixed, the nuisance parameter is eliminated and the conditional analysis is free to deal with the parameter of interest. Asymptotically, of course, all is well, but the case of very small samples is not totally resolved. In [84, 97] the table,

3	0
0	3

is considered. It had been used in [1] and discussed by Fisher (1945) in his reply. An argument which is almost Bayesian is produced. The point being that the experimental set-up with owls in the first row and hawks in the second, is such that the scientist had expected hawks to have the property and owls not, as did the data. Notice again this lovely blend in Barnard's writing of concern for fundamental issues and involvement in practical issues, here ornithology.

4. Pivotals

The concern over the reference set is not confined to 2×2 tables. The point arises in the Fisher-Behrens problem where it is required to compare the means of two normal distributions, the variances not being assumed the same (as they are in Student's test). In [12], using ideas of Stein and Rubin, he shows that a sensible reference set enables the levels of the Fisher-Behrens' test to have the meaning ordinarily attached to them within the Neyman-Pearson theory. This paper also contains the lovely example of the Haldanes' cats with its interesting sampling procedure. Reference sets arise in connection with ancillary statistics, these providing a common method of reducing the set by conditioning on the observed value of the ancillary. To a Bayesian, for whom the reference set contains only one point, that observed, the whole exercise looks foolish; but he has to admit the discussion is by no means sterile and provides genuine illumination. The famous counter-examples of ancillary statistics due to Basu (1964) threaten to demolish conditional arguments. Barnard, with Sprott, [49] discuss these: Basu, in reply, is puzzled.

Another situation where the pure likelihood approach is not adopted is one in which there is a group structure. It is argued that the group adds information to that provided by the likelihood function, and that this extra information should be used in inference. The classic example is due to Fisher with location and scale in the normal distribution. This easily leads to the notion of a pivot, a function of the data and the parameters whose distribution is known. In the normal case, with sample mean $\overline{x}$ and standard deviation s for a sample of size n, $(\overline{x} - \theta)/(s/\sqrt{n})$ is a pivot having Student's distribution with $(n-1)$ degrees of freedom, θ being the population mean. This, in turn leads to the fiducial argument in which the known probability distribution of the pivot is used to pass from a probability distribution over a sample space to one over parameter space. The simplest example of a pivot is the case where there are two quantities, x and θ, not necessarily random, about which nothing is known except that the difference $e = \theta - x$ has a known distribution. The pivotal argument says that, on observing the numerical values of x, $\theta = x + e$ has the corresponding distribution shifted by x. This procedure cannot be justified within the usual probability calculus but, according to its supporters, can be defended as logically sensible. In [82] he says, "We cannot apply to 'fiducial probabilities' the same rules of conditioning which we apply to the probabilities we apply to observations." They echo a common view that probability is not the only possible explicatum for uncertainty. In [35] he connects the idea with the principle of sufficient reason.

Barnard explores the logic of the fiducial argument in [32]. He strikingly begins by admitting "The domain of applicability of the fiducial argument is not a large one" but defends it partly for the contribution it makes to our understanding of uncertainty—a claim which is surely correct. This paper also discusses the famous inadmissibility result for the multivariate, normal, sample mean; and explains its irrelevance to the fiducial argument. In [86], with Sprott, the scope of the pivotal argument is considerably widened by the observation that an extension of the probability integral transformation enables any likelihood model involving continuous distributions to be reformulated as a pivotal model. Another important property claimed for pivots is their ability to handle robustness [87]. Barnard has always been clear that one should discuss robust samples, rather than robust procedures. Most samples from a Cauchy distribution look like normal samples and the sample mean is sensible. But occasionally the sample has extreme observations and care is needed. The point is related to the consideration of the reference set: in the non-normal-like samples, the reference set includes only such samples. It is also related to what Barnard has called the "serendipity principle", according to which you should analyze the data you actually have, without bothering with the possibility that you might have gotten a set which would have been much more difficult to analyse. Readers wanting a good introduction to pivotal ideas should read the encyclopaedia article [92].

5. The Bayesian View

Since this paper is appearing in a volume that is predominantly Bayesian, it is especially appropriate to consider Barnard's attitude towards that position. First, it should be made clear that he was advocating the Bayesian position in connection with decision-making, and particularly in industrial inspection, before almost everyone else. (He cites Molina as preceding him.) From an early stage, a clear distinction had been made by him between decision and inference, a distinction that Fisher (1956, IV.1) adopted from Barnard. In the latter it is desired to express opinions: in the former action is required. The contrast is between the "laboratory" and the "market place". In decision-making the Bayesian argument is correct. The dispute lies in its role in inference. Even here an advocate of the likelihood principle comes close to the Bayesian position. Only in significance testing is the contrast marked and here, for reasons already discussed, Barnard denies the appropriateness of the coherent view. Of the many criticisms of this view, Barnard's is perhaps the most cogent because it does centre on a real, practical difficulty, namely the need to describe the alternatives rather carefully. He often (see, for example, [58]) refers to his agreement with the distinction made by de Finetti (1974) between techniques and the Bayesian standpoint. The form is just mathematics (as in admissibility results), the latter relates the mathematics to reality and is a way of thinking about real-world situations involving uncertainty.

In [99] he asks the question "Was Fisher a Bayesian?". The unsurprising reply is "No" but the reasoning involves a point in Bayes' original paper previously discussed by Shafer (1982). This concerns the distinction between the probability of an event that has not happened and one which has but whose happening is unknown to you: in the latter case Bayes refers to "the probability that I guess right if I guess". Conditional probability is tricky (compare the discussion of the Borel-Kolmogorov paradox in connection with the fiducial argument in [32]).

6. Miscellaneous Works

There are several papers that lie outside the fields of his main interest. "A theory of mathematical statistics independent of the calculus of probability" [14] concerns studies of populations and expectation operators. Essentially we have a second-order, linear theory akin to that developed from different premises in least squares; a topic that is considered in [34]. There the notion of unbiasedness is criticized and it is shown that this requirement can be replaced by the mere condition that the range of the parameters to be estimated is unbounded. An alternative formulation is provided in terms of sufficiency. The geometry of least-squares is considered in [64]. Another read paper to the RSS [13] served to introduce us in Britain to the ideas of Shannon on information. After a description of Shannon's results in connection with coding (also discussed in [22]) Barnard goes on to develop an axiomatic basis for the results and then connects them with Fisher's notion of information, arguing for the unexpected second derivative of the log-likelihood rather than Fisher's form with the expectation included. There is a further reference to information in [51], where, in discussing an example of Edwards, he points out that there is a real difference between being uninformative in a particular context and being absolutely uninformative.

Mention has been made of the papers of Barnard's read at RSS meetings with subsequent discussion. Equally he has been extremely active in joining in the discussion of other papers. Often he has made penetrating remarks with original ideas that he has never followed up. He publishes his "half-baked ideas" (to use Good's phrase) in the RSS. A separate paper could be written about these.

7. Applied Statistics

An extremely important aspect of Barnard's research has been his early work on industrial statistics. In [5] he is concerned with sequential aspects and in [15] with assembly processes; whereas in [20] the close link with decision theory is explored, though the prior seems to be thought of in a frequency connection. Sensitivity to the prior is discussed and in attribute sampling the suggestion is explored of using a mixed binomial for the "prior" distribution of the number of defectives in the batch. This corresponds to a two-point prior on the parameter of the binomial likelihood and, in modern parlance, has three hyperparameters. Barnard seems to have been amongst the first to use conjugate distributions, though the more sensible description of "closed under sampling" was used. They appear in special cases in [20] but in generality in a paper by Wetherill (1961), then a student of Barnard's. The discussion ends with the unequivocal statement: "my view is that we must specify the prior distribution explicitly, recognizing that is we only know it approximately, then our scheme may be only approximate". In [24] the emphasis is on a control chart as an estimation procedure, rather than as a significance test.

Once a chart is considered in this light, it is natural to look at the sums (as being sufficient in binomial and normal cases) and plot the cumulative sum (CUSUM) over successive batches. The CUSUM chart and its associated V-mask proved to be a useful device. In the discussion an important ideas is raised that does not seem to have been explored by him though others have usefully employed it. The idea is not to compare maxima of likelihoods but to take (say) four values of the parameter reasonably covering the range of interest and average the likelihood for these: effectively using a simple type of prior. He refers to this as the OAAAA method—obviously arbitrary and always admissible.

Another industrial inspection problem considered later in [44] and [47] concerned the testing of condoms, and led to the British Standard [36]. (By an interesting coincidence on the day this is being written, the British consumer magazine "Which" reports, using the test, in a survey of condoms.) More mathematically, the idea was to use a (sequential) test of power one, the law of the iterated logarithm being employed to establish the result. This is possible again by using the cumulative sum which will be sure to exceed imposed limits if the proportion of defectives exceeds a prescribed value.

One project with which he was much involved concerned the factors that determine a young person's choice of subject at a university. A rather full survey revealed that the choice was made essentially at school and at an early age. This provided the basis of another RSS paper [45] with McCreath. Statisticians are often to be heard saying that much debate that goes on in democratic societies today is deficient because the facts are not known; with the result that the emphasis is on what they might be rather than what might be done. Barnard has devoted a lot of his energies to attempts to rectify this, and this paper is a notable and successful example.

8. Presidential Addresses

Barnard has given three presidential addresses, to the Operations Research Society (ORS), the Royal Statistical Society, and the Institute of Mathematics and its Application (IMA). A presidential address is often rather dull because it represents an occasion calling for an idea, instead of a bright idea needing an occasion for expression. But Barnard's trio is exciting. In the OR address [37] he discusses a problem that had arisen during his membership of the University Grants Committee: how should students be allocated to universities, taking due account of the universities' preference for students and the students' for universities. He relates it to the stable marriage problem and discusses Gale's algorithmic solution. There are interesting comments on communication within British society that hindered the connection between the committee and the mathematicians over this problem. The IMA address [47] has already been referred to in connection with contraceptives but, as befits the audience, also discusses Arrow's result on democratic preferences and Robbins on empirical Bayes. To the RSS [52] he argues that "the meetings of our Society constitute the best forum of the discussion of the factual bases of public policy that exists in this country". Never one to admire the "Establishment," he hastens to disclaim membership. He argues that statistics' role is essentially that of midwife. With little data, there is little to discuss: with masses of data, the theory is established. The midwife brings the idea from conception, through the days where some data exists, to birth. Geology is cited as an example. After these applied topics he "cannot resist the temptation to relate this to current controversies on the foundations of statistical inference". There again appears this tolerance towards most views, and this is reflected in the title "The unity of statistics". Even here there is a digression to consider decision-making in the context of the site for an additional airport for London.

9. Conclusion

My own position lies rather firmly, but hopefully not rigidly, within the Bayesian paradigm, and it may be of some interest to view Barnard's work from that perspective. In decision theory we find him a full member and, indeed, one of the founders. In emphasizing and using prior probabilities and loss functions in industrial problems his influence has been strong. As mentioned above, it is in inference problems that the membership lapses. But even there, in the likelihood principle, he comes close to the coherent view. It should not be forgotten that he was advocating forms of that principle well before Bayesians were aware of it. Savage does not mention it in his book (1954) and heard about the idea in connection with the Bayesian view from Barnard and adopts it in print by the time of Birnbaum's (1962) paper. The likelihood principle is not just theoretical: it vitally effects statistical practice and, indeed, denies the validity of almost every tool in the statistician's kit.

According to Barnard, there are inference situations not covered by Bayes or likelihood, and it is here that we part company. Pivots and the whole fiducial edifice are "out" for the Bayesian purist because, although they are typically Bayesian, they only become so through the use of improper priors and these lead to difficulties. Barnard's defense of them is powerful but ultimately has to admit an inferential procedure that does not obey the rules of classical probability. A Bayesian sees no need for a separate type of argument: for him probability embraces all uncertainty. The dispute here is not usually of great practical importance because the fiducial and posterior statements are usually similar, though the multivariate case with many parameters can show substantial differences.

It is with tests of hypotheses that the real Bayes/Barnard differences emerge. With its need for a reference set (and with Basu around to heckle, how hard this is to find) and resulting determination of "more extreme" values, the significance test violates the likelihood principle and gives results quite different from its Bayesian counterpart. There are at least two points in its favour: it corresponds to a real practical need (with null well-defined and alternative not), and it has been, and continues to be, popular with scientists.

What Barnard has done is to prod us all into thinking hard about these foundational issues without at any time forgetting their practical relevance. The practitioner who is scornful of "theoretical niceties" cannot afford to ignore Barnard's work. If he argues a point, it is partly because it interests him, but, more importantly, because the point matters to the practitioner.

A Bayesian can always feel happy with Barnard's fine applied work: in quality control or university admissions he has not only used his statistical skills but has seen to it that the results are implemented. He has stimulated both theoreticians and users, and both have learned something from the encounter with an original mind that spreads its talents. He is certainly not one of those mathematical statisticians who are all too prone to hive off into abstract mathematics without taking proper care to ensure that their mathematical model is relevant to the scientific or practical situation.

References

Barndorff-Nielsen, O. (1978). *Information and Exponential Families in Statistical Theory.* Chichester: Wiley.

Basu, D. (1964). "Recovery of Ancillary Information," *Sankhya, A*, Vol. 26, 3-16.

Berger, J.O. and R.L. Wolpert (1984). *The Likelihood Principle*, Lecture notes, monograph series, Vol. 6. Hayward: IMS.

Birnbaum, A. (1962). "On the Foundations of Statistical Inference" (with discussion), *Journal of the American Statistical Association*, Vol. 57, 269-326.

De Finetti, B. (1974). "Bayesianism: Its Unifying Role for Both the Foundations and Applications of Statistics," *International Statistical Review*, Vol. 42, 117-130.

Fisher, R.A. (1945). "A New Test for 2×2 Tables," *Nature*, Vol. 156, 388.

Fisher, R.A. (1956). *Statistical Methods and Scientific Inference.* Edinburgh: Oliver & Boyd.

Savage, L.J. (1954). *The Foundations of Statistics.* New York: Wiley.

Savage, L.J., et al. (1962). *The Foundations of Statistical Inference.* London: Methuen.

Shafer, G. (1982). "Bayes's Two Arguments for the Rule of Conditioning," *Annals of Statistics*, Vol. 10, 1075-1089.

Stein, C. (1962). "A Remark on the Likelihood Principle," *Journal of the Royal Statistical Society, Series A*, Vol. 125, 565-568.

Wald, A. (1947). *Sequential Analysis.* New York: Wiley.

Wetherill, G.B. (1961). "Bayesian Sequential Analysis," *Biometrika*, Vol. 48, 281-292.

References of Publications of G.A. Barnard

[1] (1945). "A New Test for 2×2 Tables," *Nature*, Vol. 156, 177.

[2] (1945). "Economy in Sampling," *Nature*, Vol. 156, 208.

[3] (1945). "A New Test of 2×2 Tables," *Nature*, Vol. 156, 783.

[4] (1945). "Target-Handicap Charts for Sampling Inspection," Director General of Scientific Research and Development, Ministry of Supply, London.

[5] (1946). "Sequential Tests in Industrial Statistics," *Journal of the Royal Statistical Society, Supp. VIII*, 1-26.

[6] (1947). "Sequential Analysis" (Review), *Journal of the American Statistical Association*, Vol. 42, 658-664.

[7] (1947). "Significance Tests for 2×2 Tables," *Biometrika*, Vol. 34, 123-128.

[8] (1947). "A Comment on E.S. Pearson's Paper," *Biometrika*, Vol. 34, 123-128.

[9] (1947). "The Meaning of a Significance Level," *Biometrika*, Vol. 34, 134.

[10] (1948). "The Cost of Inspection," (Address to the British Association) reported in *The Engineer*, Vol. 166, 329.

[11] (1949). "Statistical Inference," *Journal of the Royal Statistical Society, Series B*, Vol. 11, 115-149.

[12] (1950). "On the Fisher-Behrens Test," *Biometrika*, Vol. 37, 203-207.

[13] (1951). "The Theory of Information," *Journal of the Royal Statistical Society, Series B*, Vol. 13, 46-54.

[14] (1951). "A Theory of Mathematical Statistics Independent of the Calculus of Probability," Congres International de Philosophie des Sciences, Paris, Hermann et Cie., 115-124.

[15] (1951). "Statistics Applied to Assembly Processes," Sheffield Conference, Royal Statistical Society Publications.

[16] (1952). "The Frequency Justification of Certain Sequential Tests," *Biometrika*, Vol. 39, 144-150.

[17] (1953). "The Frequency Justification of Certain Sequential Tests—Addendum," *Biometrika*, Vol. 40, 468-469.

[18] (1953). "Time Intervals Between Accidents - A Note," *Biometrika*, Vol. 40, 212-213.

[19] (1953). "Statistical Decision Functions" (Review), *Biometrika*, Vol. 40, 475-477.

[20] (1954). "Sampling Inspection and Statistical Decisions," *Journal of the Royal Statistical Society, Series B*, Vol. 16, 151-174.

[21] (1954). "Simplified Decision Functions," *Biometrika*, Vol. 41, 241-251.

[22] (1955). "Simple Proofs in Simple Cases of the Coding Theorem," *Proceeding of the 3rd London Symposium on Information Theory*, London: Butterworth's.

[23] (1955). "Technometrics and the Mathematical Needs of Modern Technology," Inaugural Lecture, Imperial College, University of London.

[24] (1959). "Control Charts and Stochastic Processes," *Journal of the Royal Statistical Society, Series B*, Vol. 21, 239-271.

[25] (1961). (with P. Linstead and M. McCreath). "1800 Students of Science and Technology," *Universities Quarterly*, Vol. 16, 36-54.

[26] (1962). *The Foundations of Statistical Inference*, G.A. Barnard and D.R. Cox, eds., London: Methuen.

[27] (1962). "Prepared Contribution" and "Discussion," in *The Foundations of Statistical Inference*, G.A. Barnard and D.R. Cox, eds., London: Methuen, 39-49 and 62-93.

[28] (1962). "An Analogue of Tchebycheff's Inequality in Terms of Range," *Technometrics*, Vol. 4, 133-134.

[29] (1962). "Distributions of Products of Random Variables," *Technometrics*, Vol. 4, 277-278.

[30] (1962). (with G.M. Jenkins and C.B. Winsten). "Likelihood Inference and Time Series," *Journal of the Royal Statistical Society, Series A*, Vol. 125. 321-372.

[31] (1962). "Comment on Stein's 'A Remark on the Likelihood Principle'," *Journal of the Royal Statistical Society, Series A*, Vol. 125, 569-573.

[32] (1963). "Logical Aspects of the Fiducial Argument," *ISI Bulletin*, 34th Session, Ottawa, 870-883.

[33] (1963). "Fisher's Contributions to Mathematical Statistics," *Journal of the Royal Statistical Society, Series A*, Vol. 126, 162-166.

[34] (1963). "The Logic of Least Squares," *Journal of the Royal Statistical Society, Series B*, 124-127.

[35] (1963). "Some Logical Aspects of the Fiducial Argument," *Journal of the Royal Statistical Society, Series B*, Vol. 25, 111-114.

[36] (1964). "Condoms (Otherwise Known as Protectives and Sheaths)—3.a. Sampling," British Standard 3704, (UDC 615, 477, 86: 678, 4-46).

[37] (1964). "The Future of Operational Research—Presidential Address," *Operational Research Quarterly*, Vol. 15, 63-69.

[38] (1965). "The Use of the Likelihood Function in Statistical Practice," *Proceedings of the 5th Berkeley Symposium on Mathematical Statistics and Probability*, University of California Press, 27-40.

[39] (1967). "The Bayesian Controversy in Statistical Inference," *Journal of the Institute of Actuaries*, Vol. 93, 229-269.

[40] (1968). "Enquiry into the Flow of Candidates in Science and Technology into Higher Education (Dainton Report)," HM Stat. Off. Cmd. 3541, esp. Annex G.

[41] (1968). "Computers, Statistics, and Politics," in *The Future of Statistics*, D.G. Watts, ed., New York: Academic Press, 39-43.

[42] (1968). "Measurement in the Social Sciences," in *The Future of Statistics*, D.G. Watts, ed., New York: Academic Press, 291-307.

[43] (1969). *New Developments in Survey Sampling – Summary Remarks*, New York: Wiley Interscience, 696-711.

[44] (1970). "Practical Applications of Tests with Power One," *ISI Bulletin*, Vol. 43, 389-393.

[45] (1970). (with M.D. McCreath). "Subject Commitments and the Demand for Higher Education," *Journal of the Royal Statistical Society, Series, A*, Vol. 133, 358-408.

[46] (1971). *Factors Influencing Choice of Higher Education.* London: Department of Education and Science.

[47] (1971). "Mathematics and Our Social Problems—Presidential Address," *Bulletin of the IMA*, Vol. 7, 3-6.

[48] (1971). "Higher Education in the 1970's," *Bulletin, Institute of Mathematics and Applications*, Vol. 7, 3-6.

[49] (1971). (with D.A. Sprott). "A Note on Basu's Examples of Anomalous Ancillary Statistics," in *Foundations of Statistical Inference*, V.P. Godambe and D.A. Sprott, eds.

[50] (1971). "Scientific Inference and Day to Day Decisions." Ibid., 289-300.

[51] (1972). "Two Points in the Theory of Statistical Inference," *British Journal of Philosophy and Science*, Vol. 23 329-346.

[52] (1972). "The Unity of Statistics—Presidential Address," *Journal of the Royal Statistical Society, Series A*, Vol. 136, 1-14.

[53] (1972). "The Logic of Statistical Inference" (Review Article), *British Journal of Philosophy and Science*, Vol. 23, 123-190.

[54] (1973). "Maximum Likelihood and Nuisance Parameters," *Sankhya, Series A*, Vol. 35, 133-138.

[55] (1973). *Conditionality, Pivotals and Robust Estimation*, Conference on Foundational Questions in Statistical Inference, O. Barndorff-Neilson, P. Blaesild, G. Schow, eds., Aarhus, 61-80.

[56] (1974). "Statistics and Numerical Analysis," *Bulletin, Institute of Mathematics and Applications*, Vol. 10, 344-347.

[57] (1974). "Comment on M. Stone's 'Cross-validatory Choice and Assessment of Statistical Predictions'," *Journal of the Royal Statistical Society, Series B*, Vol. 34, No. 2, 133-135.

[58] (1974). "The Foundations of Statistics, 1964-1974," *Bulletin of the Institute of Mathematics and Applications*, Vol. 10, 344-347.

[59] (1974). "Can We All Agree on What We Mean by Estimation?" *Utilitas Mathematica*, Vol. 6, 3-22.

[60] (1974). "The Two Aspects of Statistical Estimation," *Mathematics Gazette*, Vol. LVIII, 116-123.

[61] (1974). "On Singularities in the Likelihood Function," *The American Statistician*, Vol. 28, No. 4, November, 162.

[62] (1975). "Comments on J.D. Kalbfleisch's ' Sufficiency and Conditionality'," *Biometrika*, Vol. 62, 260-261.

[63] (1975). "Computers, Privacy and Probabilities," *Mathematics in School*, Vol. 4, 24-26.

[64] (1975). "On the Geometry of Estimation," in *Perspectives in Probability and Statistics*, J. Gani, ed., New York: Academic Press, 41-46.

[65] (1976). "Conditional Inference is Not Inefficient," *Scandinavian Journal of Statistics*, Vol. 3, 132-134.

[66] (1976). "Comments on M. Stone's 'Strong Inconsistency from Uniform Priors'," *Journal of the American Statistical Association*, Vol. 7, No. 353, March, 121-122.

[67] (1977). Comment on Stigler's "Do Robust Estimators Work with Real Data?" *Annals of Statistics*, Vol. 5, 1080-1081.

[68] (1977). "Problems of Statistical Inference in the Biological Sciences," in *Mathematics and the Life Sciences*, D.E. Mathews, ed., Springer-Verlag, 1-65.

[69] (1977). "How the University Grants Committee Determines Allocations of University Grants: A Case Study in Statistical Fallacies," *Journal of the Royal Statistical Society, Series A*, Vol. 140, 200-209.

[70] (1977). "On Ridge Regression and the General Principles of Estimation," *Utilitas Mathematica*, Vol. 11, 299-311.

[71] (1977). "Obituary: Maurice Henry Quenouille," *Journal of the Royal Statistical Society, Series A*, Vol. 140, 568-569.

[72] (1977). "Comments on Kiefer's 'Conditional Confidence Statements and Confidence Estimators'," *Journal of the American Statistical Association*, Vol. 72, No. 360, 808-810.

[73] (1979). (with D.A. Sprott). "Schätztheorie," in *Handwörterbuch der Mathematischen Wirtschaftwissenschaften 2: Ökonometrie und Statistik*, G. Menges, ed., Wiesbaden: Gabler.

[74] (1979) "In Contradiction to J. Berkson's Dispraise, Conditional Tests Can Be More Efficient," *Journal of Statistical Planning and Inference*, Vol. 3, 181-187.

[75] (1980). "Pivotal Inference and the Bayesian Controversy," *Trab. Estadist.*, Vol. 31, 295-318.

[76] (1980). "Note on an Especially Simple Partitioning of Chi-Square from Questionnaries," *Psychological Bulletin*, Vol. 87, 58.

[77] (1981). "The Conditional Approach to Robustness," in *Statistics and Related Topics*, M. Csörgö, D.A. Dawson, J.N.K. Rao, A.K.Md.E. Saleh, eds., Amsterdam: North-Holland, 235-241.

[78] (1982). "Bayesian Risk Ratio Analysis," *The American Statistician*, Vol. 36, 394.

[79] (1982). (with V.P. Godambe). "Allan Birnbaum – Memorial Article," *Annals of Statistics*, Vol. 10, 1033-1039.

[80] (1982). "2×2 Tables," *Journal of the Royal Statistical Society, Series C*, Vol. 31, No. 3, 304-305.

[81] (1982). "A New Approach to the Behrens-Fisher Problem," *Utilitas Mathematica*, Vol. 21B, 261-271.

[82] (1982). "Discussion on A.P Dawid and M. Stone's 'The Functional Model Basis of Fiducial Inference'," *Annals of Statistics*, Vol. 10, 1068-1069.

[83] (1982). Discussion of Lindley's "Scoring Rules and the Inevitability of Probability," *ISI Review*, Vol. 50, 11-14.

[84] (1982). "Conditionality Versus Similarity in the Analysis of 2×1 Tables," in *Statistics and Probability: Essays in Honor of C.R. Rao*, G. Kallianpur, P.R. Krishnaiah and J.K. Ghosh, eds., Amsterdam: North-Holland, 59-65.

[85] (1982). "Causation," in *Encyclopedia of Statistical Sciences*, Vol. I, N.L. Johnson and S. Kotz, eds., New York: J. Wiley and Sons.

[86] (1983). (with D.A. Sprott). "The Generalised Problem of the Nile: Robust Confidence Sets for Parametric Functions," *Annals of Statistics*, Vol. 11, 104-113.

[87] (1983). "Pivotal Inference and the Conditional View of Robustness," in *Scientific Inference, Data Analysis and Robustness*, G.E.P. Box, T. Leonard, C.-F. Wu, eds., Academic Press, 1-8.

[88] (1983). (with D.A. Sprott). "Likelihood," in *Encyclopedia of Statistical Sciences*, Vol. IV, N.L. Johnson and S. Kotz, eds., New York: J. Wiley and Sons.

[89] (1984). "Comparing the Means of Two Independent Samples," *Journal of the Royal Statistical Society, Series C*, Vol. 33, 266-271.

[90] (1985). "Discussion of Bather's 'On the Allocation of Treatments in Sequential Medical Trials', and Armitage's 'The Search for Optimality in Clinical Trials'," *International Statistical Review*, Vol. 53, No. 1, 25-26.

[91] (1985). "A Coherent View of Statistical Inference," University of Waterloo Technical Report.

[92] (1985). "Pivotal Inference," in *Encyclopedia of Statistical Sciences*, Vol. VI, N.L. Johnson and S. Kotz, eds., New York: Wiley, 743-747.

[93] (1985). "Darwin's Data on Growth Rates of Plants," in *Data*, D.F. Andrews and A.M. Herzberg, eds., New York: Springer-Verlag, 9-12.

[94] (1985). (with R.L. Plackett). "Statistics in the United Kingdom 1939-45," in *A Celebration of Statistics*, A.C. Atkinson and S.E. Fienberg, eds., Springer-Verlag, 31-55.

[95] (1986). Discussion of Johnstone's "Tests of Significance in Theory and Practice," *The Statistician*, Vol. 35, 499-502.

[96] (1986). "Rescuing Our Manufacturing Industry—Some of the Statistical Problems—Kendall Memorial Lecture," *The Statistician*, Vol. 35, 3-16.

[97] (1986). "Should All Statistical Inference Problems be Tackled Via Bayes' Theorem?", *The Professional Statistician*, Vol. 5, No. 8, 6-7.

[98] (1987). "What Influences Influence?" *Journal of Applied Statistics*, Vol. 14, 107-116.

[99] (1987). "R.A. Fisher—A True Bayesian?", *ISI Review*, Vol. 55, 183-189.

[100] (1988). "Two-by-Two Tables," in *Encyclopedia of Statistical Sciences*, Vol. IX, N.L. Johnson and S. Kotz, eds., New York: J. Wiley and Sons.

[101] (1988). "A Conversation with George Barnard," *Statistical Science*, Vol. 3, 196-212.

[102] (1988). "The Future of Statistics: Teaching and Research," in *Bayesian Statistics, Volume 3*, J.M. Bernardo, M.H. DeGroot, D.V. Lindley, and A.F.M. Smith, eds., Oxford University Press, 17-24.

[103] (1988). "Discussion of W.R. Rice's 'A New Probability Model for Determining Exact P Values for 2×2 Contingency Tables When Comparing Binomial Proportions'," *Biometrics*, Vol. 44, 16-18.

[104] (1988). "Discussion of I.J. Good's 'The Interface Between Statistics and the Philosophy of Science'," *Statistical Science*, Vol. 3, No. 4, 401-403.

[105] (1989). "Sophisticated Theory and Practice in Quality Improvement," *Phil. Trans. Roy. Soc. Lond.*, Vol. A 327, 581-589.

[106] (1989). "Must Clinical Trials Be Large?", *Statistics in Medicine*, forthcoming.

[107] (1989). "On Alleged Gains in Power from Lower P Values," *Statistics in Medicine*, forthcoming.

[108] (1989). "Linear Pivotals and a Bayes-Non-Bayes Compromise," Proceedings of the 47th Session of the International Statistical Institute, Paris.

[109] (1989). "Discussion of S. Zabell's 'R.A Fisher on the History of Inverse Probability'," *Statistical Science*, Vol. 4, No. 3, August 1989, to appear.

III. FOUNDATIONS

Bayesian and Likelihood Methods in Statistics and Econometrics
S. Geisser, J.S. Hodges, S.J. Press and A. Zellner (Editors)

BAYES, FISHER, AND BELIEF FUNCTIONS

A. P. Dempster
Harvard University

Abstract

Bayesian inference, Fisher's fiducial inference, and the theory of belief functions each offer prescriptions for reasoning from empirical data to posterior probability assessments. Here I offer an elementary exposition of the belief function approach that emphasizes the underlying logical unity among the three systems. All three are best understood in terms of an 'objectivist subjectivist' attitude to probability, and all three require nontrivial 'independence' assumptions, especially independence of observations from other inputs of deterministic and probabilistic knowledge.

1. Introduction

The three inference systems of my title offer paths from empirical data to posterior probability assessments. In some applications, illustrated by the familiar normal measurement error model discussed below, the separate formulations can yield identical numerical answers, but more typically there are differences. I argue that the differences are more in the choices of specific models to represent a common situation than in differences of logic.

I begin by reacting to a stimulating article (Barnard, 1987) characterizing Fisher as "a true Bayesian". As Barnard shows, there are examples where Fisher found Bayesian inference to be compelling. But Fisher disagreed on Bayes's original example. Evidently, however, Barnard expresses an important insight when he suggests that Fisher shared certain thought patterns with Bayes that are largely absent, not only among the frequentist followers of Kolmogorov and Neyman, but even among many modern Bayesians. Nowadays Bayesian reasoning is widely accepted, while Fisher's alternative is almost forgotten. I agree with Barnard that Fisher deserves more credit for his contributions.

What are the common elements? One key is that both Bayes and Fisher were 'objectivist' 'subjectivists'. The latter term refers to the universal (pre-frequentist) semantics that perceives probability as a formal quantitative representation of an individual's assessment of uncertainty given current evidence, while the former term conveys the desideratum that only knowledge and principles external to any individual scientist, and therefore shared or at least sharable among scientists, be used. For me, belief function models are formal vehicles for reasoning with probabilities, motivated by a similar commitment to objective subjectivism.

I contend that a full understanding of the role of probability in contemporary statistical practice can be obtained only by explicit recognition of objective subjectivist probability as the central concept of statistical inference. By contrast, objective frequentism is a very different concept, valid and important, and indeed consistent with objective subjectivism, when the mathematics of probability theory is used to represent variation in mass phenomena, but misused, as Fisher clearly thought, when seen as the basic, or even only, concept of probability needed for statistical inference, because it is limited to studying only long run properties of 'procedures'.

While Fisher advocated different forms of probabilistic inference as appropriate in different circumstances, namely, fiducial inference, likelihood inference, and tests of significance, and while each of Fisher's types have close connections with widely taught frequentist methodology (they motivated much of it), careful examination indicates that Fisher interpreted all three forms in subjectivist terms. For example, he insisted that fiducial probability is no different from ordinary "Mathematical Probability" understood as a precise measure of uncertainty (Fisher, 1956, 1958a, 1958b). His concept of likelihood (Fisher, 1922) as the probability of the observed sample, but treated as a sample statistic, suggests early awareness that the sampling density requires a different interpretation after sample values are known. Other basic examples are (a) Fisher's famous 'disjunction' interpretation of a significance test (Fisher, 1956), that makes obvious sense in a subjectivist, but not in a frequentist, framework, and (b) his direct subjectivist interpretation of likelihood-based confidence intervals (Fisher, 1956) and Neyman-style confidence limits (Fisher, 1962).

Barnard (1987) perceptively notes that Bayes, like Fisher, carefully distinguished between the probability of an event interpreted before the event has occurred, as a prospective measure of uncertainty, and the same probability interpreted after the event has occurred. Unlike the above examples of Fisher, however, Bayes is addressing the inferential circumstances after the event has occurred but before the outcome is known, using language such as the "probability that I guess right if I guess that the event has occurred". In context, the probability in this quote makes sense only as subjectivist probability. I suggest, moreover, that Barnard should not in his Addendum have conceded to a referee that the quote sounds very much like a Neyman-style confidence interpretation, since there is no evidence that the early innovators such as Bernoulli and Bayes were concerned about long run properties of procedures. Frequentist probability per se, as distinguished from Bernoulli's sampling probability argument for using sample frequencies to draw inferences about population frequencies (Bernoulli, 1713), did not exist in Bayes's time. As noted above, Fisher went beyond Bayes in providing semantics for probability after both the event has occurred and the outcome is known, specifically in his interpretation of likelihoods and p-values. These distinctions about timing are irrelevant to frequentists, but are understood, necessary, and important parts of subjectivist thinking.

A fundamental advantage of subjectivist inference over frequentist inference is its direct linkage with everyday logic and reasoning, as is seen most clearly in the belief function formulation of the theory. By contrast, modern frequentists (e.g., Wald, 1950) and some Bayesians (e.g., Savage, 1954), are strongly behaviorist or decision-oriented in outlook and theory. Both of these contemporary schools have contributed to skewing statistical practice towards a heavy emphasis on choosing and carrying out methods or 'procedures', in contrast to the 'inductive logic' emphasized by Fisher. Although 'reasoners' can be found if one looks for them, the weight of orthodoxy in contemporary statistics is on the side of 'proceduralists'. I find this unfortunate. In my opinion, statistics can remain fully engaged in scientific uncertainty assessment only if it encourages a much expanded role for subjectivist probabilistic reasoning about specific uncertain phenomena.

2. Three Prescriptions for Posterior Inference

Measurement error and sampling error are historically (Todhunter, 1865, Porter, 1986, and Stigler, 1986) the two main application areas that originally stimulated the development of ideas and prescriptions about statistical inference. Measurement error models appear to be less burdened with ambiguity and nonuniqueness, and so are used here as illustrations.

The simplest case of a single measurement illustrates the modelling strategies of Gauss and Fisher, which may then be compared to the general belief function approach. Suppose by definition that

$$X = U + E \tag{1}$$

where X denotes a measurement, U denotes the unknown true value being measured, and E denotes measurement error. Suppose that prior uncertainty about E is characterized by a $N(0, \sigma^2)$ probability distribution with known σ^2. Historically, the normal distribution for measurement error was postulated by Gauss (1809) on axiomatic grounds, the chief axiom being that the arithmetic mean of repeated independent measurements should be the correct "mean" of the posterior distribution.

A more acceptable modern view assumes that an empirical basis exists for the error distribution, in that repeated trials judged to produce errors exchangeable with the particular E in equation (1) have been observed to follow an approximate $N(0,\sigma^2)$ distribution for prospective subjective uncertainty about E. Such a basis would be equally meaningful and acceptable under any of the three prescriptions.

Having the normal error distribution in hand, Gauss in effect assumed a prior distribution for U which was uniform on $(-\infty,+\infty)$ and independent of E. He then followed the prescription laid down by his predecessors Bayes and Laplace, as though it were simple common sense to do so. He thus concluded that after observing X the uncertainty about U is characterized, in modern notation, by an $N(X,\sigma^2)$ distribution.

This Bayes-Laplace-Gauss prescription differs from what Barnard might wish to call Bayes-Fisher logic. The difference is that, while Gauss followed Bayes in explicitly introducing a uniform prior probability element for the unknown parameter, Fisher explicitly rejected the correctness of introducing such a prior probability element in the absence of specific evidence to back it up (c.f., Lane, 1980), and so proceeded to ask what could be inferred without it. In the normal measurement model he reasoned that the $N(0,\sigma^2)$ distribution for the measurement error E was adequate to justify a $N(X,\sigma^2)$ posterior or 'fiducial' distribution for U given X, on grounds that the 'pivotal quantity' $X-U=E$ should continue to have its uncertainty characterized by $N(0,\sigma^2)$ after X is fixed by observation. That is, to Fisher there seemed to be no reason to alter the prior distribution for E as the relevant subjectivist representation of uncertainty. In effect, Gauss reached the same conclusion that the posterior distribution of E should be the same as the prior, and to Fisher it must have seemed that the artificial uniform prior for U was unsupported, and therefore an undesirable and unnecessary crutch. Following Barnard (1987), one could speculate that Fisher felt Bayes would have agreed with him.

An explicit belief function formulation of the fiducial argument, leading to the same model and the same inference in the case of a single observation, was introduced by Dempster (1966, 1968), preceding the important general structure and terminology introduced by Shafer (1976, 1982). In particular, the terms 'belief', 'plausibility', 'frame of discernment', 'focal element', 'basic probability assignment', 'minimal extension', and 'Dempster's rule of combination' were introduced by Shafer, along with the name 'theory of belief functions'.

These ideas require careful explanation, in order to expose the formal logic of belief functions as applied to the normal measurement error model. For example, the customary terminology of 'probability measure' over a 'sample space' is no longer appropriate in belief function theory. Instead, two different pairs of corresponding terms are appropriate, one for capturing the semantics of real world application, and the other for thinking about mathematical syntax.

In the semantics of subjectivism, the 'sample space' is understood to mean the set of possible states of the 'idealized real world' represented in the model. The belief function analog of the 'sample space' in this sense is termed the 'frame of discernment', often abbreviated to 'frame'. Similarly, the semantic analog of the 'probability measure' is the 'belief function'. Whereas a probability measure determines a 'probability' $P(A)$ for every 'event' A defined as a subset of the sample space, a belief function determines a pair of numbers for each A, namely, the 'belief' or 'lower probability' of A, denoted $P_*(A)$, and the 'plausibility' or 'upper probability' of A, denoted $P^*(A)$. Mathematical definitions are given below. In an applied context, $P_*(A)$ is interpreted to be the minimum probability that the subjectivist commits to A, while $P^*(A)$ is the maximum probability so committed.

Mathematically, the belief function equivalents of 'probability measure' and 'sample space' are a different pair of concepts, called, in Shafer's terms, 'basic probability assignment' and 'space of focal elements'. A 'focal element' is formally a subset of the 'frame', and the 'basic probability assignment' is formally an ordinary probability measure over the 'space of focal elements'. Thus, ignoring issues of measurability that are unimportant in our examples, a real-valued $m(A)$ is specified for each subset A of the 'frame', where the $m(A)$ sum or integrate to unity over all A. There are several equivalent ways to describe how the technical quantities $m(A)$ determine the epistemically meaningful measures $P_*(A)$ and $P^*(A)$. One way is to imagine, purely formally since no random process is postulated to exist, that the $m(A)$ define the distribution of a random subset B of the 'frame', and that $P_*(A)$ for any A is defined to be the probability that the random B is contained in A, while $P^*(A)$ is the probability that B has a nonempty intersection with A. Heuristically, $P_*(A)$ is the accumulation of all of the 'basic' probabilities $m(B)$ constrained to A, while $P^*(A)$ is the accumulation of all $m(B)$ possibly associated with A. As an exercise, the reader may now show that

$$0 \leq P_*(A) \leq P^*(A) \leq 1, \tag{2}$$

for any A. More details can be found in Dempster (1967) or Shafer (1976, 1982).

While a 'focal element' A is formally a subset of the 'frame', which in turn is semantically identical to a set of possible states of the world, the 'focal element' A has no interpretation as an event or proposition, and hence the probability $m(A)$ assigned to A by the 'basic probability assignment' has no interpretation as the subjective probability of some assertion about the world. Only the 'belief' $P_*(A)$ and 'plausibility' $P^*(A)$ bear subjectivist interpretation. The case of an ordinary probability measure over the frame can be recovered within the belief function framework by arranging that $P_*(A) = P^*(A)$ for all A, which happens if and only if the 'focal elements' are all singleton subsets of the 'frame'. This special case is semantically important because it shows that ordinary subjectivist models can be specified, albeit indirectly, within the class of belief function models. Despite disadvantages of unfamiliarity and complexity, the extra layer of terminology is necessary to maintain the separation of semantic and syntactic components of belief function theory that happen to coincide in the simpler model of ordinary subjective probability.

The remaining concepts of 'minimal extension' and 'Dempster's rule of combination' are fundamental operations that drive the calculus of belief functions. To explain and motivate these operations, and also to illustrate the heavy dose of terminology already introduced, I return to the normal measurement error example. The frame for the single measurement case may be taken to be the ordinary 3-space of possible values of the triple (X, U, E). A first reaction of many statisticians might be to object that equation (1) restricts the possible values of (X, U, E) to a 2-plane in 3-space, so that the frame should be redefined to be the 2-plane. The natural belief function approach, however, is to retain 3-space as the frame and to regard equation (1) as defining a 'logical' belief function. That is, equation (1) specifies a belief function whose basic probability assignment is completely defined by assigning $m(A) = 1$ where A is the 2-space defined by (1), and $m(A) = 0$ for all other A.

It might seem an anticlimax that the first example of a belief function involves no probabilities at all, and in fact is a simple statement of Boolean logic, but in fact the easy capability of belief function calculus to incorporate Boolean statements is a critically important property, not only in providing a formal tie between probabilistic and propositional logic that is missing in the ordinary theory of subjective probability, but also leading in complex examples to novel algorithms that combine information sources in orders that would not otherwise seem natural.

How then to represent the $N(0, \sigma^2)$ probability assumption for measurement error E? The answer may be given in either of two equivalent forms, one heuristic, and the other involving the formal concept of 'minimal extension'. The heuristic answer seeks to render plausible the idea that one might wish to assign a probability measure to a class of subsets of a frame rather than to the elements of the frame itself. The argument is that the empirically based probability measure over the possible values of E can equivalently be expressed as a distribution over a one-to-one transform of E, which in this situation is naturally taken to be the family of parallel 2-planes in 3-space whose members are characterized as the planes on which E takes a constant value. The latter is evidently a belief function whose focal elements are the planes.

More abstractly, given only an initial belief function over the elements of a partition of a given frame, the basic probability assignment of the initial belief function can be used to define a semantically equivalent belief function over the frame itself, where the focal elements of the minimal extension so defined are found by extending each initial focal element to a consistent focal element of the extended frame, while the same $m(A)$ is used for initial and extended focal elements. In the measurement error example, the initial belief function is defined by the $N(0,\sigma^2)$ distribution over the singleton subsets of the 1-space of possible values of E, which is then minimally extended to a distribution over the corresponding 2-planes of points (X,U,E) associated with each value of E. A point to understand here is that, in contrast to theory of probability, the theory of belief functions provides a semantically equivalent belief function over any frame given only a marginal belief function. Again the reader is directed to Shafer (1976, 1982) for more details on projecting belief functions to finer or coarser spaces.

The process of model construction over the 3-space associated with (X,U,E) has now specified two different belief functions, the first a logical belief function associated with equation (1), and the second a probabilistic belief function based on the knowledge of the $N(0,\sigma^2)$ measurement error distribution. The next step in applying the theory of belief functions is to consider combining the two belief functions using 'Dempster's rule of combination'. Henceforth in this paper, however, I will use the substantively more descriptive term 'product-intersection rule' in place of Shafer's term 'Dempster's rule'.

The key concept underlying the rule is 'independence'. In mathematical statistics, 'independence' is widely invoked as an 'assumption', with emphasis on the formal permission to invoke in the mathematics a product rule which is a special case of the 'product-intersection rule' of belief function theory. When a model is adopted for specific real world application, however, perceptions and judgements about real world appropriateness are essential to the validation of formal mathematical, hence computational and syntactic aspects. In practice, independence means that learning the outcome of certain events whose uncertainty is characterized by one belief function will not compromise the uncertainty concerning another set of events described by the second independent belief function. For example, if the two belief functions refer to physically unrelated processes, then 'independence' generally makes sense. Such judgements are essential if knowledge is to accumulate. Validity is never cast in concrete, however, for part of the human condition, scientific and otherwise, is to be open to arguments and data which lead to altered and hopefully improved problem formulations.

The 'rule of combination' of Dempster (1966), which I here propose to call the 'product-intersection rule', joins the basic probability device of multiplying independent probabilities with the familiar Boolean device of representing the conjunction of assertions by intersecting the sets representing the assertions. Given a pair of belief functions associated with the same frame, each with its own set of focal elements and its own basic probability assignment, the rule asserts: Make a combined belief function over the same frame by forming new focal elements and new basic probabilities. Each new focal element consists of the intersection of a pair of focal elements, with one element of each pair coming from the sets of focal elements associated with the two given belief functions. The associated new basic probability is found by multiplying the original basic probabilities associated with the intersected focal elements, and since the same new focal element can arise from different pairs, the product probabilities of the pairs are accumulated, and normalized to sum to unity. Formulas may be found in Dempster (1966,1967) or Shafer (1976,1982).

If the logical belief function of equation (1) is combined with the $N(0,\sigma^2)$ belief function over the 2-planes of (X,U,E) indexed by E, the result is a belief function specified by a $N(0,\sigma^2)$ distribution over the family of parallel lines indexed by E in the 2-plane defined by equation (1). The next step in developing the example is to introduce a third belief function, specifically, a logical belief function assigning basic probability one to the 2-plane in (X,U,E) defined by fixing X at its observed value, thus representing knowledge of the observed value of X. The net effect of combining this belief function with the previous pair is a belief function assigning a $N(0,\sigma^2)$ distribution to the singleton subsets of U. That is, the belief function argument yields the same $N(0,\sigma^2)$ posterior distribution as the Gauss or Fisher arguments.

While there is no reason to question the independence of the original pair of belief functions, many statisticians might question the independence of that pair with the observation of X. Specifically, the question is whether the fact of X becoming known should be considered irrelevant to continuing to accept the $N(0,\sigma^2)$ distribution of E. I argue that the question must in principle be answered anew by each thinking user of the model in each application. For practical model construction, it is necessary to adopt many independence assumptions in order to have any analysis at all, and resource limitations may force routine use of 'assumptions' leading to standardized 'methods'. But such assumptions deserve review and possible modification reflecting improved knowledge and understanding of circumstances.

In particular, the routine Bayesian procedure of 'conditioning' and Fisher's corresponding 'continuing to trust' the $N(0,\sigma^2)$ distribution of the error are both subject to scrutiny for independence of the observation process from the other knowledge sources. Fisher often stressed the need to incorporate all relevant information, so surely was open to evaluating information that might compromise the independence assumption. Careful Bayesian applied statisticians are also alert to phenomena that might compromise reliance on formal combination and conditioning, although the point is surely not much stressed in Bayesian writing. The main difference in the belief function viewpoint is that the questioning attitude includes checking independence among all forms of Boolean and probabilistic information sources.

If, as in the preceding example, the useful quantities calculated from a belief function turn out to be ordinary subjectivist probabilities calculated from an ordinary distribution of measurement error, was it really necessary to invoke the heavy machinery of belief functions? I believe the answer is yes, because otherwise there is no unified perspective from which to compare the different models and reasoning processes of Fisher and Gauss. The big difficulty with Gauss's model is the assumed prior with infinite variance for U. Since this prior makes no subjectivist sense, it seems to have been introduced for purely technical reasons to facilitate staying within the paradigm. Fisher's approach of introducing a prior distribution only when objectively justifiable seems clearly preferable.

Extension of the example to the case of repeated independent measurements X_1 and X_2 is straightforward, and illustrates how the belief function scheme renders unnecessary an ad hoc principle invoked by Fisher. The single equation (1) is replaced by the pair of equations

$$\begin{aligned} X_1 &= U + E_1 \\ X_2 &= U + E_2 \end{aligned} \tag{3}$$

where E_1 and E_2 are independently distributed *a priori*, each with a $N(0, \sigma^2)$ distribution. As is well known, the Bayes-Laplace-Gauss inference in effect multiplies the likelihood function by a uniform prior density for U and so arrives at a $N(\bar{X}, \sigma^2/2)$ posterior distribution for U, where $\bar{X} = (X_1 + X_2)/2$. Fisher proceeds by invoking his principle of sufficiency to reduce the pair of observations X_1 and X_2 to the single observation $\bar{X}$ which is treated as a single measurement of U with error $\bar{E} = (E_1 + E_2)/2$ whose prior distribution is $N(0, \sigma^2/2)$, and thence uses his single measurement fiducial reasoning to obtain the same posterior inference as Gauss. A more natural belief function modelling strategy would adopt a 5-dimensional frame of possible values of (X_1, X_2, U, E_1, E_2) which is restricted first to a 3-dimensional hyperplane by the pair of equations (2), and then further restricted to a 1-dimensional hyperplane, or line, by the observations X_1 and X_2 after the observations are in hand. When the joint prior density of E_1 and E_2 is restricted to the line, it yields once again the Gaussian or Fisherian answer upon indexing the points of the line by their associated values of U.

Although the preceding description of the belief function argument did not explicitly refer to the product-intersection rule of combination, the rule was in fact used repeatedly in ways that statisticians have come to regard as automatic. First, the combination of the two equations (3) amounts to intersecting the 4-spaces defined by the separate equations to produce a 3-space, illustrating a case where in the absence of probability the rule is purely Boolean intersection. Second, the multiplication of the individual densities of E_1 and E_2 to produce the joint density of the pair illustrates a case where in the absence of Boolean restrictions the rule involves pure multiplication of probabilities. Thus before observing X_1 or X_2 the combined information is represented by a belief function describable as a normal distribution across a 2-dimensional family of parallel lines in a 3-space. Further intersecting with the pair of 4-dimensional hyperplanes determined by the observed X_1 and X_2, and multiplying probabilities of intersected focal elements, leads to the final belief function inference, which of course turns out to be an ordinary probability distribution.

Extension from 2 to n measurements introduces no new features. An important point is that the belief function argument does not make use of the special principles of sufficiency and ancillarity required by Fisher to reduce dimensions. Only the product-intersection rule of combination is required. Whether or not one finds Fisher's principles plausible, it surely avoids ambiguities to rely on fewer principles. Note also that the Bayesian argument can be interpreted as resulting from repeated application of the rule, the difference being only that one more independent prior component belief function is combined into the system. If there is independent evidence to support such a prior, then the Bayesian answer becomes the special belief function inference of choice. Otherwise, I suggest, the belief function argument with no prior is preferable.

The simple repeated measurements theory of Section 2 extends easily to the case of a common non-Gaussian error distribution, and again the three methods reach identical inferences. But more general linear models are unable to incorporate non-Gaussian error distributions and still produce a canonical answer shared across the three approaches. Nor does the normal error theory have an unambiguous extension to situations where error variances and covariances are undetermined a priori, and so must be assessed from data. Thus the ecumenical core of examples where a canonical answer exists is really very small, and a choice among models cannot be avoided.

Consider, for example, the original problem of Bernoulli and Bayes, to estimate a population frequency p given a sample of n. Pearson (1921) called this the fundamental problem of practical statistics, and advocated a Bayesian approach. Fisher (1956) concluded that the fiducial argument could not be used, implying that no posterior probability assessment was possible in the absence of a prior distribution.

Dempster (1966, 1968) gave what seemed then and still seems a natural belief function formulation. But evidently, different modelling strategies can be argued out and rational people can disagree.

3. Discussion

Walley (1987) bases his criticism of belief function methodology on a rejection of the product-intersection rule. In effect, he argues that Fisher's principles of likelihood, or much the same, sufficiency should take precedence. In my opinion, a critic of the product-intersection rule must deal carefully with the impressive credentials of the rule. For example, Walley objects to the way certain belief function models for sampling inference violate the likelihood principle. Notice, however, that the definition of a likelihood itself involves repeated application of the rule, so Walley must be advocating a much more complicated system with a variety of rules to be invoked in different circumstances. It was to bypass the severe difficulties put forth by Fisher's critics (Dempster, 1963, 1964) that I explicitly decided 20 years ago that within the belief function framework the likelihood principle per se is not compelling.

For me, as for Fisher, models must be evaluated on an example-by-example basis, drawing on knowledge bases appropriate to the realities of actual practice. In particular, I take independence very seriously as a subjectivist judgement to be made case by case. Components of evidence, whether Boolean or probabilistic or both, are independent if and only if resolution of the uncertainty described by any subset of the components leaves unaltered the assessments of uncertainty embodied in the remaining subset.

Although not yet prominent in statistical practice, the theory of belief functions, so named by Shafer (1976), is a hardy plant. Apart from Shafer (1982), there have been few papers recently in mainline statistical literature. But a search of the literature of Artificial Intelligence (AI) over the past five years would turn up many references (e.g., Pearl, 1988). One factor that favors the reemergence of belief functions in AI is the close connection with Boolean calculus. Another factor is the concern in AI for 'reasoning' about specific circumstances. A complex system in AI is often represented by a network whose nodes correspond to attributes and whose connections among nodes represent locally determined Boolean or probabilistic knowledge among small sets of attributes. Consideration of such networks from a belief function perspective has led to a major advance in belief function methodology which is just beginning to find its way into print (Shenoy and Shafer (1986), Kong (1986), Shafer, Shenoy, and Mellouli (1987), and Dempster and Kong (1988)). The new theory lays bare a simple computational process for tree-structured networks which promises to facilitate experimentation with many examples. So far most of the theory is developed for categorical attributes, and generalizes Bayesian theory being developed almost in parallel (Pearl (1986), and Lauritzen and Spiegelhalter (1988)). I plan to report elsewhere on an application to normal measurement models, based on the recognition that the widely used Kalman filter provides a direct illustration of the basic algorithm for propagation and fusion of belief functions in a network.

Acknowledgment

Work facilitated by support from ARO Contract DAAL03-86-K-0042.

References

Barnard, G.A. (1987). "R. A. Fisher—A True Bayesian?" *International Statistical Review*, Vol. 55, 183-189.

Bernoulli, J. (1713). *Ars Conjectandi.*

Dempster, A.P. (1963). "Further Examples of Inconsistencies in the Fiducial Argument," *Annals of Mathematical Statistics*, Vol. 34, 884-891.

Dempster, A.P. (1964). "On the Difficulties Inherent in Fisher's Fiducial Argument," *Journal of the American Statistical Association*, Vol. 60, 420-436.

Dempster, A.P. (1966). "New Methods for Reasoning Towards Posterior Distributions Based on Sample Data," *Annals of Mathematical Statistics*, Vol. 37, 355-374.

Dempster, A.P. (1967). "Upper and Lower Probabilities Induced by a Multivalued Mapping," *Annals of Mathematical Statistics*, Vol. 38, 325-339.

Dempster, A.P. (1968). "A Generalization of Bayesian Inference" (with discussion), *Journal of the Royal Statistical Society, Series B*, Vol. 30, 205-247.

Dempster, A.P. and A. Kong (1988). "Uncertain Evidence and Artificial Analysis," *Journal of Statistical Planning and Inference*, forthcoming.

Fisher, R.A. (1922). "On the Mathematical Foundations of Theoretical Statistics," *Philosophical Transactions of the Royal Society of London, Series A*, Vol. 222, 309-368.

Fisher, R.A. (1956). *Statistical Methods and Scientific Inference.* Edinburgh: Oliver and Boyd (2nd ed.,1959, 3rd ed., 1973).

Fisher, R.A. (1958a). "The Nature of Probability," *Centennial Review*, Vol. 2, 261-274 (Reprinted in J.H. Bennett (ed.) *The Collected Papers of R.A. Fisher*, Vol. 5, 384-397, The University of Adelaide 1974).

Fisher, R.A. (1958b). "Mathematical Probability in the Natural Sciences," *Technometrics*, Vol. 1, 21-29 (Reprinted in J.H. Bennett (ed.), *The Collected Papers of R.A. Fisher*, Vol. 5, 398-406, The University of Adelaide 1974).

Fisher, R.A. (1962). "Confidence Limits for a Cross-Product Ratio," *Australian Journal of Statistics*, Vol. 4, 41.

Gauss, C.F. (1809). *Theoria Motus Corporum Celestium.*

Kong, A. (1986). "Multivariate Belief Functions and Graphical Models," Ph.D. Thesis, Department of Statistics, Harvard University.

Lane, D.A. (1980). "Fisher, Jeffries, and the Nature of Probability," in S.E. Fienberg and D.V. Hinkley (eds.), *R.A. Fisher: An Appreciation.* New York: Springer-Verlag, 148-160.

Lauritzen, S.L. and D.J. Spiegelhalter (1988). "Local Computations with Probabilities on Graphical Structures and Their Application to Expert Systems" (with discussion), *Journal of the Royal Statistical Society, Series B*, Vol. 50, 157-224.

Pearl, J. (1986). "Fusion, Propagation and Structuring in Belief Networks," *Artificial Intelligence*, Vol. 9, 241-288.

Pearl, J. (1988). *Probabilistic Reasoning in Intelligent Systems.* Morgan Kaufmann.

Pearson, K. (1921). "The Fundamental Problem of Practical Statistics," *Biometrika*, Vol. 13, 1-16.

Porter, Theodore M. (1986). *The Rise of Statistical Thinking 1820-1900.* Princeton University Press.

Savage, L.J. (1954). *Foundations of Statistics.* New York: Wiley.

Shafer, G. (1976). *A Mathematical Theory of Evidence.* Princeton, NJ: Princeton University Press.

Shafer, G. (1982). "Belief Functions and Parametric Models" (with discussion), *Journal of the Royal Statistical Society, Series B*, Vol. 44, 322-352.

Shafer, G., P.P. Shenoy, and K. Mellouli (1987). "Propagating Belief Functions in Qualitative Markov Trees," *International Journal of Approximate Reasoning*, Vol. 1, 349-400.

Shenoy, P.P. and G. Shafer (1986). "Propagating Belief Functions with Local Computations," *IEEE Expert*, Vol. 1, 43-52.

Stigler, S.M. (1986). *The History of Statistics. The Measurement of Uncertainty Before 1900.* Harvard University Press.

Todhunter, I. (1865). *History of the Mathematical Theory of Probability.* MacMillan (Reprinted: Chelsea 1949, 1965).

Wald, A. (1950). *Statistical Decision Functions.* New York: Wiley.

Walley, P. (1987). "Belief Function Representations of Statistical Evidence," *Annals of Statistics*, Vol. 15, 1439-1465.

Bayesian and Likelihood Methods in Statistics and Econometrics
S. Geisser, J.S. Hodges, S.J. Press and A. Zellner (Editors)

A THEORY OF BAYESIAN DATA ANALYSIS

Bruce M. Hill
University of Michigan

Abstract

This article addresses some of the basic philosophical and practical issues that arise in attempting to integrate the Bayesian theory with data analysis. It argues that both conventional data analytic approaches—which appear largely to abandon probability—and conventional pre-data Bayesian theory—which largely takes models as given—can benefit from one another. At the center of this discussion is the distinction between pre-data considerations, such as arise in the design of experiments, and post-data considerations, such as arise in actual decision-making in the light of the data.

1. Introduction

Bayesian data analysis is concerned with the type of data manipulations, transformations, and just plain playing with the data, that any serious scientist engages in during the statistical (or other) analysis of his data. It is largely a post-data procedure, rather than a pre-data procedure, since even when it is desirable to think through such matters quite carefully prior to obtaining the data, in many real world experiments time and other constraints would provide limits on such activities. Compare Hacking (1967), or the discussion in Hodges (1987, p. 291) concerning how much is enough. Bayesian data analysis goes beyond the mere data manipulations, however, and attempts to integrate the theory of subjective probability with such data analysis. In this respect it differs from other data-analytic approaches, which appear, more or less, to abandon probability. In this article I shall attempt further to elucidate the theory of Bayesian data analysis begun in Hill (1985-86, 1987a, 1987b, 1988b). See also Hill (1970a, 1975a) for earlier thoughts on the subject with regard to tests of significance, and Smith (1986). The Bayesian theory of tests of significance that originated with H. Jeffreys (1961), and was developed by Jimmie Savage in Savage (1962), and in the beautiful article "Bayesian statistical inference for psychological research," by Edwards, Lindman and Savage (1963), was the starting point for my own attempts to integrate the Bayesian theory with data analysis. Such an integration could be viewed as a synthesis of the empiricism-pragmatism of John Locke, David Hume, Charles Peirce, and William James, with the rationalistic tradition of Plato, Descartes, Leibniz, Kant, and others.

The purpose of this article is to address some of the basic philosophical and practical issues that arise in attempting to integrate the Bayesian theory with data analysis. Failure to address these issues may have, in the past, led to serious deficiencies in both of these approaches. It will be argued that both conventional data analysis and conventional pre-data Bayesian theory can benefit from one another.

2. Inadequacy of Pre-Data Theories

Perhaps the greatest single source of confusion in modern statistics is due to the failure to distinguish pre-data considerations, such as arise in the design of experiments, from post-data considerations, such as arise in actual decision-making in the light of the data. Sequential analysis provides an excellent example. See Anscombe (1963, p. 381) for a very forceful and convincing analysis of such confusion, and discussion of the waste of time and effort spent on sequential analysis, which he calls a "hoax."[1]

Consider, for example, a sequential stopping rule, N, that depends only upon the observations, as almost all such rules studied do. Suppose that the data of the experiment consists of the fact that one stopped at time $N = n$, and that the actual observations were $X_1 = x_1, \ldots, X_n = x_n$. If a parametric model is employed, say with parameter θ, then we have

$$\Pr\{\text{data} \mid \theta\} = \Pr\{N = n \mid X_1 = x_1, \ldots, X_n = x_n,\ \theta\} \\ \times \Pr\{X_1 = x_1, \ldots, X_n = x_n \mid \theta\}.$$

Since the stopping rule depends only upon the observations, and since we did in fact stop at time n, and not before, the first factor on the right-hand side must be identically unity, and thus does not depend upon θ. For example, if the stopping rule were to stop at the first time that the sample mean, $\overline{X}$, exceeded a specified constant, c, then given the actual observations $X_1 = x_1, \ldots, X_n = x_n$, and θ, it would be absolutely certain that we must stop precisely at time n, irrespective of

[1] It is not the procedure of sequentially observing the data that is being condemned, but rather interpretation of the data according to a body of non-Bayesian statistical technique known as sequential analysis.

the value of θ. The second factor on the right-hand side is simply the likelihood function for θ, based on a fixed sample size n. This means that on a post-data basis, i.e., given the data, the information obtained from a sequential experiment that actually stopped at time n, is logically equivalent to the information contained in a *fixed* sample size experiment with n observations, together with a logically certain event. Somehow or other, sequential analysts purport to extract information out of this logically certain event, over and above the information contained in the fixed sample size experiment. This appears to have some connection with arguments for perpetual motion, and is perhaps one of the reasons why Anscombe calls the subject a hoax. Savage (1961, 3.23), Savage (1962, p. 18-20), and Edwards, Lindman, and Savage (1963, Section 8) provide further discussion of such matters. Also see Barnard (1947), who apparently first understood the true nature of 'sequential analysis,' and Berger and Wolpert (1988, 4.2) and Berger (1985, 7.7) for more recent discussions. (To the best of my knowledge, no sequential analyst has yet attempted to answer these 25 year old *logical* objections to their subject. In medical trials the subject has, unfortunately, been used extensively. In practice, often real-world constraints, such as time, patience, and funding, change the true stopping rule, so that it becomes completely unknown. In this case the theory underlying sequential analysis is not even germane to the analysis of the data.)

That the stopping rule is irrelevant for inference and decision-making follows trivially from the likelihood principle, or from the restricted likelihood principle of Hill (1987a, 1988a). However, the analysis that I have given here has some new aspects. It does not depend upon the likelihood principle, but rather upon the willingness of the reader to acknowledge that a logically certain event cannot provide information, *in any meaningful sense,* with respect to empirical questions. See also Hill (1989) for discussion of the concept of an 'analytic' argument in philosophy. All probability based methods of inference, including those violating the restricted likelihood principle, focus attention upon the probability of the data, given the parameter. My argument is that no matter how sequential analysts purport to utilize such distributions, it is necessary for them to pretend to obtain information from a logically certain event, over and above that stemming from the corresponding fixed sample size experiment. Note that on a pre-data basis, the event in question is only *conditionally* logically certain. Indeed, one might not stop at all. However, *given the actual observations*, it follows necessarily that one must stop at precisely time n, so that the event in question is logically certain. This highlights the critical nature of the distinction between pre-data and post-data considerations.

I have used this example to illustrate the difference between a pre-data approach, such as sequential analysis, and a post-data approach, such as the likelihood or Bayesian approach. It shows how a pre-data consideration, such as choice of a stopping rule, can become misleading and/or irrelevant on a post-data basis. Of course, the choice of a stopping rule can be important in the design of a study, and relates to the question 'how much is enough.' It may also be remarked, however, that sequential design, which in principle should have been a serious subject, does not seem to have contributed as much as it might have to real-world problems either, perhaps because of the same underlying confusion. Anscombe (1963, p. 382) says: "In fact, sequential experiments are a most stimulating and provoking topic for the statistical theorist to meditate on. Too little attention of the right sort has been paid to them."

Although sequential analysis is a particularly blatant example, there are many other situations where the pre-data considerations are quite different from the post-data considerations. Some examples arising in econometrics are discussed in Hill (1985-86, p. 218). More generally, it can be argued that the conventional classical theory of statistical inference is almost entirely a pre-data theory. For example, the confidence approach of Neyman and others is based upon a pre-data evaluation of the probability of coverage. Does it have any relevance on a post-data basis? If so, how does it allow for situations where, on a post-data basis, it is obvious that no reasonable person could have much confidence in the quoted confidence coefficient? Such a lack of confidence can arise either for logical reasons, such as in the Fieller-Creasy problem, or simply because of common-sense.

The fiducial argument of Sir Ronald Fisher, which appears to have historically preceded the confidence argument, was an important first step towards a genuine post-data approach. Here Fisher formulated the idea that after seeing the data one might wish to retain certain probability statements as still valid. Initially he believed that fiducial probability involved a new concept of probability, but he later acknowledged, in a footnote, Fisher (1959, p. 51), that "Probability statements derived by arguments of the fiducial type have often been called statements of 'fiducial probability'. This usage is a convenient one so long as it is recognized that the concept of probability involved is entirely identical with the classical probability of the early writers, such as Bayes. It is only the mode of derivation which was unknown to them." See Zabell (1988) for a fascinating account of the comedy of errors, tragic for statistics in the twentieth century, that took place between the discovery of the innovative fiducial argument by Fisher, and his eventual understanding of its connection with the Bayesian approach, as in the quotation. In Hill (1988b) the role of the fiducial argument in Bayesian nonparametric inference is discussed.

The Bayesian approach provides a framework in which the meaning and validity of an intuitively brilliant argument, such as the fiducial argument, can be critically examined. Some pre-data probability evaluations will remain valid, in the sense that they are also Bayesian post-data evaluations, and some will not. As will be argued in Section 3, there is no objective way to state, on a pre-data basis, which will be retained and which will be dropped. This, in fact, is precisely where data analysis enters the picture.

In the pre-data design situation one may usefully employ a statistical model to get a rough idea of the type of experiment or quantity of data needed to provide a serious answer to a real-world question. In the post-data situation it is necessary to check, in some way or other, the approximate validity of the model, using appropriate diagnostic procedures, if necessary to abandon the original model, and perhaps to replace it with a new one. Such a post-data model might then be used to obtain inferential and decision procedures, given the data. The pre-data considerations, such as initial models and/or prior distributions, may or may not be deemed relevant after exploring the data.

It was argued in Hill (1985-86, Section 2) that conventional pre-data theories of statistical inference, such as the Neyman-Pearson approach, break down completely when considered in the context of real-world data analysis. For in order that confidence coefficients, p-values, etc., have any meaning at all, these would have to be evaluated *conditional* upon all the diagnostics actually used, including their order, and even upon the thoughts that cross one's mind during the analysis of the data. Plainly such conditional probabilities are both unknown and unknowable. Hence even if the conventional theories were not rejected for the many other reasons that Bayesians have put forth, such as incoherency, inadmissibility, the failure to incorporate realistic prior knowledge, etc., they would have to be rejected as being totally inapplicable in the real-world, except in those rare cases where someone is rash enough to give total certainty to the pre-data model that he has selected.

The Bayesian approach also faces some serious challenges in the context of post-data analysis of the data. As argued in Hill (1985-86, p. 223), the saving grace for the Bayesian approach is that the likelihood function, even when it has been formulated through the process of data analysis, remains precisely the same as if it had been specified *a priori*. The 'prior' distribution is, of course, no longer a prior distribution, since the parameters may not have been even thought of prior to the data. However, in this situation the Bayesian can do a post-data robustness and sensitivity analysis, as in Hill (1980b). See also the related ideas in Berger (1984, 1987). In other words, having perhaps formulated a new model, with new parameters, one can examine the sensitivity of the conclusions to variations in the 'prior' distribution for the parameters of the new model. It may be the case that for decision-making purposes the conclusions are quite clear. If not, it means that reasonable people with different 'priors' would come to quite different conclusions, given the available data, and this is important to know. Compare the discussion in Hill (1985-86, p. 241), and comments by the four discussants.

3. Extreme Data

Suppose that a vector of observations Y, in an n-dimensional Euclidean space, is thought of as being the sum of a parameter vector θ, and a vector of errors ϵ. The use of capital Y indicates here the pre-data status of the observations, i.e., Y has not yet been observed, although it may have already been determined. The data of the experiment is $\{Y = y\}$, so y consists of the observed value of Y. Suppose that a Bayesian regards θ as marginally independent of ϵ. (Here I do not have in mind the conventional assumption of conditional independence, given some other parameters, such as scale parameters, but rather the definition of independence, i.e., the joint distribution factors appropriately.) Let $f(\theta)$ and $g(\epsilon)$ be the marginal prior densities for θ and ϵ, respectively. This means, for example, that if there were an unknown scale parameter, say σ, in connection with the distribution of ϵ, then it would already have been integrated out to obtain $g(\epsilon)$ as the marginal density, as discussed in Hill (1969a).

Clearly the posterior density for θ, given the data $\{Y = y\}$, is

$$f''(\theta) \propto f(\theta) \times g(y - \theta).$$

Similarly, the posterior density for ϵ is

$$g''(\epsilon) \propto g(\epsilon) \times f(y - \epsilon).$$

This model for the realized errors was put forth in Hill (1969a). Hill (1969b) obtained the limiting posterior distributions for extreme data, and was eventually published in Hill (1974a). Note that so far we have only given the ordinary Bayesian posterior distributions in terms of f and g, which are pre-data specifications of the marginal prior distributions for the parameter and error vectors, respectively. The first post-data consideration that arises is that as soon as y is realized (but not necessarily observed), there is complete symmetry with regard to the logical status of θ and ϵ. Thus on a post-data basis, ϵ consists of the *actual realized* errors in the experiment. These errors are unknown, but simply form a vector of n numbers, just like θ, at least in the case where the parameters have a physical meaning.

In conventional non-Bayesian statistical theory it is sometimes argued that after a coin has been flipped, but with the result unknown, that probability is no longer relevant. The result is said to be either heads or tails, and that is supposed to be all there is to say. From a Bayesian point of view, of course, probability remains relevant, since it is used to describe the uncertainty of an individual, and in this example (without further information) the state of uncertainty remains the same. Compare de Finetti (1974, Ch. 2). In the same way, even though the 'errors' have now been realized, for a subjective Bayesian the pre-data density, g, would be just as relevant after y has been realized as before, at least in the absence of further information or thought. Of course such information is often available, in the form of covariates, but we do not consider this case here. It may be noted that there are actually three conceptual stages involved. The first stage is the pre-data stage, before the errors and/or parameters are realized, and Y is determined. The second stage is after both are realized and $Y = y$ is determined, but before y is observed, as in the coin flipping example, after the coin is flipped but before one knows the outcome. The third stage is the post-data stage, after y is observed and the data analysis has taken place.

A simple concrete example may be helpful. One walks in a northern city in December, and although it appears quite balmy, say about 60 degrees Fahrenheit, a bank thermometer reads 25 degrees. Let us suppose that it is known from experience that this particular thermometer is usually accurate to within a few degrees. The data is $y = 25$, and we ignore other information, such as the dress of people on the street. Which does one believe? Although the bank thermometer is usually accurate, it may have gone haywire. Also, perhaps one is having a fever or some other form of delusion. The problem is to separate out the component of y due to the true temperature, θ, from the component due to error, ϵ. There is complete symmetry between the status of θ and ϵ, given the data, $\{Y = y\}$, and it is only the character of the distributions determined by f and g that allows one to differentiate the 'errors' from the parameters.

In Hill (1969b, 1974a, Section 4), a basic theorem is proved relating to this situation, in the n-dimensional case. The question that is addressed concerns what happens to the posterior distributions for θ and for ϵ when y is extreme in some sense. My theorem relates to the case in which the Euclidean norm of y is large. The theorem states that only three possible limiting situations can occur, *no matter what the densities f and g may be.* The first possibility is that when $\|y\|$ goes to ∞, the posterior density of $\theta - y = -\epsilon$ converges to the *prior* density for $-\epsilon$, i.e., $g(-\epsilon)$. The second possibility is that as $\|y\|$ goes to ∞ the posterior density of $\epsilon - y$ converges to the *prior* density for $-\theta$, i.e., to $f(-\theta)$. By the symmetry between parameters and realized errors alluded to earlier, the second possibility follows automatically from the first. The third possibility is that no limit exists.

A little thought shows that the first case corresponds to both classical non-Bayesian inference, and also to the Bayesian situation when stable or precise estimation in the sense of L.J. Savage occurs, as discussed in Savage (1962, Section 4). In the Bayesian case, the likelihood function for θ is then extremely sharp relative to the prior density for θ, so that one more or less believes the data, and regards y as being approximately equal to θ, with an error whose magnitude is determined by g. This occurs, for example, with normal data and a Cauchy or t prior distribution for θ. A key point to understand, however, is that there is nothing in the logic of the situation that requires this to occur. Indeed, just the opposite can occur, as in the second possibility, in which case one instead regards the *error*, ϵ, as about equal to y. This second possibility actually corresponds to the intuitive interpretation of ϵ as being an 'outlier.' In the first possibility it is $\theta - y$ that has a limiting distribution, while in the second possibility it is $\epsilon - y$ that has a limiting distribution. Both have reasonable subjectivistic interpretations.

Remarks 1 through 8 of Hill (1974a, p. 570-573) attempt mathematically and philosophically to characterize these two possibilities in terms of the relative sharpness of f and g. Thus basically what underlies possibility 1 is the view that errors are in some appropriate sense believed to be 'small' relative to the parameters. This may or may not be the case. In the bank thermometer example, I think most of us will ordinarily tend to rely upon our prior judgment. In other examples, where great caution is taken with respect to potential sources of bias, such as in the celebrated Michelson-Morley experiment of physics, we will tend to believe the data. See Hill (1988a) for a discussion of the Michelson-Morley experiment in this context. Of course the Michelson-Morley experiment was exceptional, and in most experiments the possibility of a serious bias must be taken more seriously. Such bias can then be viewed as giving rise to a relatively diffuse distribution for the errors, as perhaps occurs in the bank thermometer example.

The main point that I am making is that it is only careful consideration of one's marginal prior densities, f and g, with respect to their relative sharpness, that can allow one to make a reasoned decision as to which of the two possible limiting distributions one wishes to accept. (It should be emphasized, as in Remark 5, that a third alternative is that no limiting posterior distribution for θ exists, as occurs when both f and g are of the normal form. I personally regard this case, although possible, and sometimes appropriate, as ordinarily of lesser interest. The assumption that a limiting posterior distribution exists can be viewed as representing a form of stability in one's outlook.) As mentioned earlier classical non-Bayesian inference implicitly concerns only possibility 1, and also some Bayesians have emphasized this case in order to achieve bounded risk.

In the post-data situation, the situation becomes even more complicated. After playing with the data, one may find that either f or g or both have to be regarded as mixtures. For example, it may become apparent that g should include an extremely fat-tailed component, representing an even more diffuse 'outlier' than previously anticipated, even if this was not explicitly thought of prior to obtaining the data. In the bank thermometer example, this component might be used to represent some special form of breakdown in the mechanism. Such a phenomenon will be discussed further in Section 4 to illustrate the process of Bayesian data analysis. Remark 9 of Hill (1974a, p. 572) shows that even when both f and g are represented as mixtures, the basic theorem still holds, but one must compare the fattest-tailed component of f with the fattest-tailed component of g. The one of these that is sharpest wins, in the sense that if the fattest component of g is sharp relative to the fattest component of f, then it is $\theta - y$ that has a limiting distribution. Otherwise, it is $\epsilon - y$ that has a limiting distribution. Although the 'what if' method, or "device of imaginary results" of Good (1965, p. 19), can be of value, it seems clear that it would not ordinarily be possible, before examining the data, to know which limiting distribution will eventually be thought to be appropriate. This is partly due to the labor involved in trying to assess all the components of f and g a priori, and partly due to the fact that the specific methods of data analysis used may trigger off a chain of thought that leads to some previously unsuspected understanding of the data. Eventually, one must make some decision, and this implicitly involves a post-data assessment of the various components of f and g.

For some related articles, see Dawid (1973), which deals with the limiting posterior expectation when $n = 1$, and Umbach (1978).

4. Methodology for Model Selection

Let us now attempt to characterize Bayesian data analysis as distinct from more conventional forms of data analysis. This will be done for the important problem of model selection. It will be argued that all the ingenious techniques to analyse and display data that have been developed by scientists and others for centuries, automatically become part of Bayesian data analysis. One may perform such searches and displays in any way whatsoever, giving free rein to scientific creativity. The part that is uniquely Bayesian only occurs after, as a result of such techniques, one has reformulated old models, or formulated new models for the data. Such models may have been thought of a priori, or may have arisen partly or entirely through the process of data analysis. Sometimes, in relatively simple examples, such data-instigated models can actually be convincingly obtained through use of Bayes's theorem. However, in realistically complicated examples, the computations would be prohibitive. Furthermore, as I argued earlier, the process of analysing the realized data may trigger off a chain of thought that was not foreseen beforehand. From this point of view the use of data-analytic techniques might be viewed as a computationally efficient way to arrive at new models. Whether or not such models could also have been achieved through use of Bayes's theorem seems to be a moot point.

After the conventional forms of data analysis have been completed, suppose one arrives at one or more models that command some degree of credibility. If one wishes to make a statement with operational content, other than a purely descriptive one about the particular data set that has been analysed, then one must deal in some fashion or other with uncertainty about either models or parameters or both. The work of de Finetti, Savage, and others, makes it abundantly clear that there is no way to do so without bringing in subjective probability explicitly; and of course once this is done, one presumably wishes to avoid so far as possible, Dutch-books and other forms of incoherence and irrationality. There is often no need to make more than a small number of post-data probability assessments, i.e., those that are sufficient for the purpose at hand. This purpose might be to choose amongst a few available decisions, or to specify, approximately, a predictive distribution for future observations, or to make inferences about conventional parameters. De Finetti's fundamental theorem of probability (1974, p. 111) shows that any coherent collection of probability specifications can always be extended coherently, but in practice it is rare for there to be a need to make more than a few such specifications.

It is the attempt to integrate the data-analytic search procedures of conventional data analysis with the subjectivistic theory of probability and decision-making that is unique to Bayesian data analysis. Conventional data analysis appears simply to ignore the problem of coherent judgment, and apart from technological advances in the display of the data, returns to the pre-probabilistic frame of mind with regard to the fundamental questions of induction and inference. Such data analysis was, of course, very welcome in the statistical climate of the past quarter century, which had degenerated into a sterile and largely meaningless form of mathematical exercise that was stifling to scientists in many fields. See the article by Salsburg (1985) for an example in the medical area. But it is difficult to see how real progress in the problems of inference and decision-making can be made without some operationally meaningful concept of probability, and at present the subjective Bayesian concept appears to be the only viable one.

How then can the Bayesian approach achieve the integration about which I am speaking? The primary jewel in this crown is the principle of stable estimation of Jimmie Savage, as in Savage (1961, 1962), and its extension to the comparison of hypotheses, as in Edwards, Lindman and Savage (1963). At present these constitute the chief way in which consensus can be achieved other than by fiat. We begin by discussing the role of the likelihood principle in Bayesian data analysis.

Suppose that as a result of data analysis one finds a new model for the data, or modifies a previous one. Suppose this new model is denoted by M, and that it contains a parameter vector θ. It is assumed that, conditional upon M and θ, the probability for the realized data is completely specified. This will indeed be taken as our definition of a model. In other words, in order for M to represent a model in our sense, it must be such that conditional upon M and its parameters, the probability for the data is completely specified. The likelihood function for θ, conditional on the model M, is then $\Pr(\text{data} \mid M, \theta)$. Note that this likelihood function is precisely the same, whether M has been arrived at through a data-search procedure, or whether M had been specified a priori. This follows immediately from our definition of a model, since given M and θ the probability of the data is uniquely specified by M. The 'data' for this evaluation of the likelihood function can be taken as any subset and/or transformation of the original data, no matter how arrived at.

If only one model appears viable after data analysis, then a question that arises is whether it is worthwhile to use this model for various purposes, such as prediction, inference, or decision-making. This is a fairly subtle question that implicitly involves considerations as to whether the departures from this model are so large as to make its use unwise, even though no alternative model has as yet been formulated. However, one can usefully consider inference and decision-making, conditional upon the truth of M, even when one considers it unlikely that M is true. Similarly, if after data analysis two or more models emerge as being thought worthy of consideration, then the above analysis can be extended to yield posterior odds for each, conditional on the truth of at least one of these models. Suppose, for example, that M_1 and M_2 are two such models, and that the associated parameters are θ_1 and θ_2. These models may be nested or unnested. Let $L_i(\theta_i) = \Pr(\text{data} \mid M_i, \theta_i)$ be the associated likelihood functions, for $i = 1, 2$. Finally, let $\pi_i(\theta_i) = P(\theta_i \mid M_i)$ be candidate 'prior' densities for θ_i, say relative to Lebesgue measure, for $i = 1, 2$. As argued above, such likelihood functions do not in any way depend upon the fact that the models may have been developed through data analysis. Note that unlike the case in which there is only a single model under consideration, one must here include all constants of proportionality in the definition of the likelihood function.

The Bayes factor is then Pr(data | M_1)/Pr(data | M_2), where Pr(data | M_i) = $\int L_i(\theta_i) \times \pi_i(\theta_i) d\theta_i$, for $i = 1, 2$. The posterior odds for M_1 versus M_2 are obtained by multiplying the Bayes factor by the 'prior odds,' i.e., by $\Pr(M_1)/\Pr(M_2)$. A great deal is known about the behavior of the Bayes factor in cases where the models have been specified in advance. Edwards, Lindman, and Savage (1963) give a number of approximations and bounds for this factor. Thus once the problem is reduced to this form, we have available a variety of ways to deal with the purely mathematical or computational aspects of the problem. What I would like to discuss here, however, is the question as to the validity of such an analysis, when one or both models have been derived through data analysis. This in fact is the key question that faces one in applying the Bayesian theory in practice.

There are two separate issues. The first concerns the validity of the likelihood principle in this context. In Hill (1987a, 1988a) I have argued that the formulation of the likelihood principle by Birnbaum was incorrect in an essential way, namely in trying to give an abstract objective definition of 'the evidence,' which as the example in these articles shows, cannot be done without greatly delimiting the concept of evidence. Furthermore, even with respect to the limited concept of evidence formulated in my restricted likelihood principle, the evidence is always only relative to a specific model or models. However, the real power of the (reformulated) likelihood principle is best seen in connection with the data-instigated models that I have discussed above. For, as I have argued above, once a model has been formulated, whether pre- or post-data, *the likelihood function for the parameters of that model, conditional upon the truth of that model, does not in any way depend upon the circumstances under which that model was discovered.* This simple logical fact constitutes, I believe, the only truly 'objective' feature of statistical practice. The second issue concerns how, within the subjectivistic theory, one can live with this fact. Clearly, one must somehow discount some of the adhoc models discovered through data analysis. The subjective Bayesian approach can only do this through the choice of the $\Pr(M_i)$, and the $\pi_i(\theta_i)$. It is the logical status, and practical aspects of such evaluations that must now be discussed.

With regard to the $\Pr(M_i)$, I believe that when one or both of the models have been formulated through data analysis, then it is ordinarily appropriate to assess or reassess these probabilities *after* the process of data analysis that gave rise to them. For example, if only M_1 had been thought of prior to the data analysis, this would suggest that M_2 must have been given negligible a priori probability. However, I think this is largely irrelevant. The results of the data analysis have suggested that one was in error in neglecting M_2, and it would *now* be appropriate to give it a non-negligible probability, prior to evaluating the post-data odds. In other words, one should interpret the $\Pr(M_i)$ as the probabilities that one would give to the two models, conditional upon the truth of at least one, *after* the data analysis that gave rise to M_2, but *prior* to the use of the Bayes factor to update the corresponding odds to become the overall post-data posterior odds.

This violates the classical version of Bayes's theorem, but I think it is the sensible way to proceed. What has been suggested may be viewed in the following light. Most of the time in life we do not use Bayes's theorem in updating our opinions. Even if it were thought wise in principle to do so, as for example in the theories of de Finetti and Savage, it would ordinarily be computationally prohibitive. In certain special situations, for example in the narrow but relatively precise problems of science, we attempt so far as possible to follow the Bayesian approach. This is done largely because it is the only approach that even attempts to deal seriously with the problem of rational updating of belief, and rational discourse. But it does not follow that it is sensible *always* to use Bayes's theorem in the updating of opinions. For this reason, I have separated the data-analytic process of discovering or reformulating models, from the formal Bayesian analysis after such models have been formulated. Having done so, I am now in a position to take advantage of the rational and persuasive aspects of the Bayesian approach in updating my odds for the two hypotheses, given the 'data' that is now being explicitly considered. Although the overall data includes experiences with, and results of, the data analysis, it is useful to separate this into two parts, only one of which is dealt with through Bayes's theorem. This may appear to be only an attempt to get the best of both worlds, the empiricist world of the data-analytic school, and the rationalist world of the Bayesian school, but it is difficult to see what alternative there is to such a procedure, other than total subjectivism and adhockery. It should be observed in this context that the Bayesian approach has always had an arbitrary element in it as to the time point at which one proceeds to make a formal Bayesian analysis. I am suggesting that often the appropriate time point is following the process of data analysis.

If one adopts this point of view, then some of the more troublesome aspects of the Bayesian approach can be largely bypassed. The Bayesian no longer needs to justify how the pre-data value for $\Pr(M_2)$, which must have been negligible, has suddenly become enormously larger. This could happen purely through Bayes's theorem, as discussed in Hill (1970a), but it need not. It could also occur through the informal updating involved in the data analysis. For scientific purposes, as opposed to explicit decision-making purposes, it might now be appropriate to evaluate the 'prior' probability for M_1, given that at least one of the two models or hypotheses is true, as 1/2. It should be understood here that when we speak of a model as being true, ordinarily we mean approximately true. The earth is neither planar nor spherical, but the spherical approximation is generally more useful. In the same way presumably neither of the M_i is literally true, but one or the other may provide a more satisfactory approximation, and it is important to know which.

Similarly, the prior distribution for θ_i, given M_i, can only be truly a priori, if M_i has been specified in advance. However, as I see it this need not necessarily cause any serious problems. For example, as soon as M_i has been stated, it may be abundantly clear that there are things one can say about the value of θ_i based simply on the meaning of M_i and θ_i. For example, it might be clear that there are reasonable grounds for regarding θ_i as diffuse relative to the likelihood function for θ_i, thus setting the stage for stable estimation. Although there are obviously some subtle aspects to such a procedure of speaking about opinions concerning θ_i, given M_i, acting as though one had not already observed and analysed the data, it appears to be the only thing that can usefully be done. Furthermore, the difficulties are primarily psychological, and arise from the fact that having seen the data one must nevertheless attempt to erase it from one's mind. The degree of success that can be achieved will depend upon the circumstances. In the thermometer example, which will be returned to in the next section, I think this can be done quite adequately. It should be noted that the basic difficulty arises not only through data analysis, but to a lesser extent in *any* Bayesian application. The reader of an article making a Bayesian analysis of data will not ordinarily have considered his a priori distributions, although he may have some definite opinions. *Thus in any case the force of a Bayesian analysis of data must depend upon an agreement among scientists that specific prior distributions and likelihood functions are pertinent to the problem, and can be considered on their own merits, even after the data has been observed.* See also Leamer (1978, Ch. 9) and my review of his book in Hill (1980c) for further discussion.

When, for example, it can be agreed that a particular model, M, is relevant for inference about θ, and also that it is reasonable to take the prior distribution for θ, given M, as being diffuse relative to the likelihood function specified by M, then Savage's principle of stable estimation applies, given M, and leads to useful approximations to the post-data distribution for θ. This appears to be the primary method by which consensus can be obtained as to empirical matters, whether for decision-making, inference, or prediction. (It would perhaps be better in this context not to speak of the prior and posterior distributions, but rather of weighting the realized likelihood function by some function $w(\theta)$, as in Barnard, Jenkins and Winston (1962).) From my point of view, the output of a Bayesian analysis of data should include the likelihood function (or, in high dimensions, characteristics thereof), together with a formal Bayesian analysis of the data using one or more prior distributions for the parameters. Through a sensitivity analysis obtained by varying the prior distribution, one can attempt to see what aspects, if any, of the conclusions are robust to the specific form of prior knowledge assumed. See Hill (1980b). The justification would rest in a consensus that on the one hand the Bayesian method for revaluation of probabilities is rational, or at any rate the best we now have, and on the other hand that the specific prior distributions and likelihood functions being employed are plausible and worthy of consideration. Obviously there is no possibility of *proving* that particular distributions are valid for everyone, so the force of the argument must stem from some agreement that the distributions being employed are reasonable for the problem at hand. Conventional classical inference, as interpreted from a Bayesian viewpoint, demands that the prior distribution be taken as diffuse or improper, even when it is ridiculous to do so. This is consensus by fiat, and is a high price to pay for such consensus.

If we agree that none of our models is likely to be true, then the question of real importance is whether, given the data, one thinks the departures from the best model or models one has are so large as to make them not worthy of use. This question cannot be answered without explicit consideration of the purposes for which the model is to be used, and the utility of using the model for such purposes, as opposed to not doing anything. See Dickey and Kadane (1980). One might believe that a model one has is not true, but that it will still get you to the moon. On the other hand, one might regard it as highly improbable that use of the model will be worthwhile. If we don't think that any of the models we now have are adequate to achieve our purposes, then one presumably starts to think hard about formulating new models. It is here, of course, that data analysis and various diagnostic techniques can be of the greatest value. It is my thesis that except in extremely simple situations this process is not usefully viewed as merely an application of Bayes's theorem. But when and if such new models are found, I believe it is entirely appropriate and useful to resume the conventional Bayesian mode of analysis, making use of the knowledge gained through data analysis just as though it were the usual type of background or a priori knowledge. The justification for such a procedure would lie in yet another aspect of the Bayesian paradigm, namely the attempt to maximize *post-data* expected utility. In my opinion, this aspect is the more fundamental and overrides even use of Bayes's theorem. Of course, in many situations use of Bayes's theorem follows from such maximization. See Hacking (1967) for a discussion of some related issues.

Some will of course dislike the subjectivity involved in all such considerations. However, I know of no way to avoid it, other than to sweep it under the carpet, SUTC, as Jack Good says. The distinguished philosopher and psychologist, William James (1896, p. 97), puts it quite well:

> Objective evidence and certitude are doubtless very fine ideals to play with, but where on this moonlit and dream-visited planet are they to be found? I am, therefore, myself a complete empiricist so far as my theory of human knowledge goes. I live, to be sure, by the practical faith that we must go on experiencing and thinking over our experience, for only thus can our opinions grow more true; but to hold any one of them—I absolutely do not care which—as if it never could be reinterpretable or corrigible, I believe to be a tremendously mistaken attitude, and I think that the whole history of philosophy will bear me out.

Despite such subjectivity, I believe that Bayesian analyses can have every bit as much impact in obtaining a post-data consensus of opinion as if the model had been specified a priori. Indeed, an M that has been found and confirmed in some sense on the basis of the data, is in many ways a much sounder basis for inference than a speculative a priori M that has not been so founded. When only one such model has been formulated, then all our inferences must be conditional on the truth of that model. One can, of course, add an M_1, etc, if the data support such additions. In this case all inference is conditional upon the truth of one from amongst this finite set of models. One might describe scientific progress as consisting of the refinement of an undifferentiated model, such as the complement of some initial M into specific alternatives such as M_1, M_2, etc. Such alternatives are typically found through the process of creative data analysis and hard thought. The Bayesian approach can then, at any point in time, be used for decisions, inference and predictions, using the models currently taken seriously. In this way the subjective Bayesian approach, as integrated with data analysis, can provide a relatively objective and *reasoned* form of argument with regard to model selection. By contrast, the conventional form of data analysis either does not deal at all with the question of model selection, or else relies upon total subjectivism, since it cannot hope to show that there might be a consensus as to the evaluation of probabilities when it does not even have an operationally meaningful concept of probability to work with.

5. Examples of Bayesian Data Analysis

Let us return to the bank-thermometer example to illustrate the procedure of Bayesian data analysis. Suppose that five hours later, during which time it appears to have cooled down noticeably, one returns and finds the thermometer still reading 25. At this point, whether one had consciously thought about it a priori or not, the thought suggests itself that perhaps the thermometer is simply frozen stuck at 25. Suppose in fact that one had not consciously thought of this hypothesis beforehand. One can nonetheless reason as follows. Let M denote the original model of

Section 3, and let M_1 denote the model that states that the thermometer is frozen. Upon reflection, one decides that it would have been reasonable to have attached a non-negligible prior probability, say around .03, to the hypothesis that the bank thermometer would be frozen at some unspecified value. Before seeing the number 25, of course one would not have much information as to what the number would be, but taking a range of say 100 degrees Fahrenheit, one might take the 'a priori' probability of the thermometer being frozen at 25 to have been about .0003. From a post-data point of view, the question of interest is whether or not the thermometer is frozen stuck. If so, then it can only be at 25, and it is not the .0003, but simply the .03 that is the relevant 'prior' probability or weight for the hypothesis under consideration. The datum $y = 25$ would then be used to revise this 'a priori' probability in accord with Bayes's theorem. This illustrates how careful one must be in dealing with post-data hypotheses or models in the Bayesian framework. Note that in this example, even though the hypothesis M_1 was only thought of after seeing the data, the probability of .03, which would be based on experience and judgment, seems just as compelling as though the evaluation had been made beforehand. It is my opinion that the psychological effect of seeing the data can vary greatly in problems of post-data Bayesian evaluations of probability, and that it will have little effect in problems such as this, where once the model has been formulated, one can easily refer the question to related experience.

Now I think that most people will have a clear post-data preference for M_1 in this case, even without doing the formal Bayesian analysis. Of course, if one returned still again late in the evening when it was even more noticeably cooler, and discovered that the thermometer was still at 25, one would then become nearly certain of the truth of M_1. This illustrates the wisdom, sometimes, of the classical statistician's recommendation to take more data. On the other hand, the weakness of this point of view, and the strength of the Bayesian approach, is seen in circumstances where one must act without the possibility of taking more data. The classical statistician who eschews the use of prior knowledge obviously has no basis for even being suspicious of the thermometer, since the value 25 is certainly a possible temperature. Yet even without the confirmatory data of later observations, which would make M_1 almost certain, a Bayesian might see M_1 as highly probable, and be prepared to act accordingly.

The formal Bayesian analysis of the problem sheds some light on how such a conclusion can be arrived at, and how it can be justified to others. Plainly this depends upon the precise specification of f and g in model M, the original model. In accord with my definition of a model, M implies a specific choice of the distribution of errors, represented by g. Because of the symmetry between errors and parameters in the present example, it would be well here to include the specification of f as part of the model as well. Once these have been specified one would simply calculate the posterior odds in favor of M versus M_1 as in the Jeffreys-Savage theory of hypothesis testing. In the case at hand the model M largely discredits itself because of the fact that one's initial opinions about θ and ϵ are such as to make the observed value 25 highly improbable. In other words, for temperatures around 60 or so, which are regarded as highly probable a priori, it would take an improbably low ϵ to yield the result 25. Since M may be interpreted as the hypothesis that the thermometer is functioning normally, in which case the distribution of ϵ is reasonably well known from past experience, it follows that an effective evaluation of the posterior odds can be made. A sensitivity analysis would reveal whether the conclusions are in fact reasonably robust to the precise choice of f and g, and to the probabilities selected for M and M_1.

The model M_1 is a degenerate type of model that explains the data perfectly. The reason that a Bayesian is not necessarily led to adopt such a 'perfect' model is because he may discount such a model due to its low a priori probability. It is one of the great advantages of the Bayesian approach that such discounting can occur, which can prevent one from simply using maximum-likelihood estimates in their most adhoc form. See Hill (1975b) for a striking example. The conventional non-Bayesian theory does not appear to have an adequate way to deal with such things, since it foregoes the use of subjective judgments and a priori probabilities. For example, in the problem of polynomial regression it remains an unsolved problem, within the conventional framework, as to why one does not fit an nth degree polynomial to $n+1$ data points, thus obtaining a perfect fit. The Bayesian perspective offers a simple answer. Such a fit would require the errors to be identically 0, i.e ϵ to be the 0 vector. This may be viewed as highly improbable a priori. For example, as suggested in Hill (1969a) one might use a spherically symmetrical 'volcanic' prior distribution with density g for ϵ. This is a distribution having a crater centered at the origin, minimum inside the crater at the origin, and with mode along a ridge at some specified positive distance from the origin. For such a prior distribution the most probable value of ϵ is far removed from the origin, and this tends to prevent one from choosing a model for which ϵ is 0. Thus this constitutes another example of how the Bayesian approach can provide a discounting for 'perfect fit' models.

In the thermometer example it might be argued, of course, that the conclusion is self-evident, and the problem hardly worth the effort to make a careful Bayesian post-data analysis. However, with only minor changes, this example would apply equally well to nuclear disasters such as occurred at Three-Mile-Island and Chernobyl, or to the space-shuttle crash. All of these disasters are examples of where there is a conflict between the data and a priori judgements, and where some thought could have averted the disasters. Often engineers cite remarkably low a priori estimates of probabilities for such accidents. These do not appear to be based on experience or sensible forms of data analysis. Confusion as to the meaning of probability versus conditional probability, and pre-data versus post-data considerations, presumably also plays a role. Although careful data analysis requires substantial expertise, the failure on the part of administrators and others to comprehend even the most basic facts about data analysis and decision-making appears to be partly responsible for many easily preventable foulups.

One of the criticisms that can be made of the conventional Bayesian approach is that it focuses too much attention on the a priori aspects, and not enough on such strategies as 'take more data.' There is a sense in which this is a highly appropriate criticism. Plainly it is foolish to devote an overly large time to the evaluation of a prior distribution. This could never be done perfectly in any case, and there is an important practical question as to when to cease such activity, and simply explore the data, which will often suggest entirely new avenues and hypotheses. Unfortunately there are no hard and fast answers here. In the above examples it is clear that not enough a priori thought had been given so that a quick response could have been made. Decision-makers are sometimes lulled into wishful thinking that certain probabilities are very tiny, when in fact simple Bayesian calculations would reveal otherwise. Similarly, it might pay to consider utilities more carefully than is customary.

Here are some concrete examples of Bayesian data analysis. Hill (1963) employed the three-parameter log normal distribution to model incubation periods for small pox. Although there was some previous theory suggesting the appropriateness of the log normal model, this obviously could not be taken for granted, and had to be checked for the data at hand. I employed stable estimation, and plotted both the marginal likelihood function for the thresh-hold parameter γ, and what is now called the profile likelihood function. These turned out to be nearly proportional. A table of observed and expected values was also presented, showing some discrepancies, but relatively minor in view of the large sample size (310). The chi-square goodness of fit statistic was obviously going to be misleading because of the large sample size, and was not presented. (There would have been no harm in giving it, other than that it is customarily misinterpreted and used to reject possibly useful models.) I believe that the stable estimation argument is entirely valid in this case. I would describe the thought process as follows. Based upon both other data on incubation periods, and on the observable characteristics of the data set, I made the decision that the log normal model would serve a useful purpose in describing the data set, in making inference about the parameters, and for predictive purposes. This aspect of my inferential procedure was not based on Bayes's theorem per se. Rather it was a result of a fairly complex procedure of data analysis. I did not believe that the log normal model was literally true, but rather was implicitly saying that I regarded the departures as being sufficiently small that this model could be usefully employed. Once having made the decision to employ the model, I believed that the inferential procedure was fairly straight-forward, and used the stable estimation argument. It goes without saying that others would be free to substitute their own 'prior' distribution or weighting function.

Another example of Bayesian data analysis occurs in the variance components problem, as in Hill (1967, 1970a). Here I broadened the conventional one-way model to allow for correlated residuals. This was an attempt to explain the familiar fact that the conventional unbiased estimator for the between variance component is often negative. In this example I evaluated the posterior odds in the Jeffreys-Savage sense, in favor of the original model versus the broadened model. Thus this example is one in which a new model, the one with correlated errors, is formulated on the basis of the data, and is then compared with the original model for the data. This is a fairly subtle comparison. See Chaloner (1987) for a review of some aspects of the components of variance problem.

More complex examples of Bayesian data analysis occur in the Mosteller and Wallace (1964) analysis of the Federalist papers, in the analysis of inference about the tails of distributions in Hill (1975a), in work on Zipf's Law in Hill (1970b, 1974b, 1980a), in Hill and Woodroofe (1975) and in Chen (1980), and in the Bayesian survival analysis of Chen et al. (1983). An important example in which the data analysis forms an integral part of the inferential theory occurs with mixtures of distributions, as in Hill (1987b). In all these examples the data are quite complex, and model specification procedures require substantial creative efforts.

There has not been room here to do more than suggest the nature of Bayesian data analysis. However, I believe that a primary stumbling block for both conventional data analysis and for conventional Bayesian statistics in the past has been the failure of each to address the concerns of the other, and to take advantage of the achievements of the other. I see no way to avoid either the data analysis, or the Bayesian approach, if one is to make any form of progress in either the practice or the theory of statistics and decision-making. I have tried to suggest above how each can benefit from the other.

6. Concluding Remarks

As the Bayesian approach becomes dominant, which now appears to be only a matter of a short time, it is important that we learn from the mistakes of the past. Bayesian thinking emerged from the writings of some of the greatest thinkers in history, including B. Pascal, D. Bernoulli, T. Bayes, P. Laplace, C. Gauss, H. Poincáre, E. Borel, F. Ramsey, B. de Finetti, H. Jeffreys, and L.J. Savage. In many cases, it was a confused or perhaps even perverted misunderstanding of the writings of these distinguished people, that led to the type of thinking that we Bayesians have been arguing against these many years. At a certain point in history it might have been the state of the Bayesian art to use maximum-likelihood estimates, and in the absence of adequate computational facilities, to derive an asymptotic distribution for the maximum-likelihood estimator. Nowadays we can do much better. In low dimensional cases we simply plot the *realized* likelihood function, weight it, and integrate numerically, if need be. In high dimensional situations we learn concepts and techniques to deal with the display and understanding of the information, such as in Hill (1975a).

Certainly Fisher's fiducial argument, and to some extent even the Neyman-Pearson theory, can be seen as approximations to Bayesian procedures. Compare the discussion in Hill (1988b, Section 4). George Barnard has written often and well on this subject. See Barnard (1985) for references. The present day Bayesian approach, such as has been solidified in some of the textbooks that have been written recently, will perhaps in a few years also be seen as only a crude approximation to more realistic Bayesian procedures. Some of the important problems that we have still to come to grips with include the questions of time coherency, randomization, and other alternatives to the conventional Bayesian approach. See, for example, Diaconis and Zabell (1986), Goldstein (1983), Lane and Sudderth (1985), and Zellner (1988). In my opinion, the real mistake of the past was to take some crude approximation to the Bayesian approach, which may have arisen historically due to a variety of real-world constraints and limitations, to be the final answer. We can do much better.

Acknowledgment

This work was supported by the U.S. Air Force under grant AFOSR-87-0192. The U.S. government is authorized to reproduce and distribute reprints for Governmental purposes notwithstanding any copyright notation thereon.

My thanks to Frank Anscombe for a number of helpful comments.

References

Anscombe, F. (1963). "Sequential Medical Trials," *Journal of the American Statistical Association*, Vol. 58, 365-383.

Barnard, G. (1947). Review of "Sequential Analysis" by A. Wald, *Journal of the American Statistical Association*, Vol. 42, 658-664.

Barnard, G. (1979). "Pivotal Inference and the Bayesian Controversy" (with discussion) in *Bayesian Statistics*, by J.M. Bernardo, M.H. Degroot, D.V. Lindley, and A.F.M. Smith, eds., Amsterdam: North-Holland.

Barnard, G. (1985). Discussion of "In Defense of the Likelihood Principle: Axiomatics and Coherency," by J. Berger, in *Bayesian Statistics 2*, J.M. Bernardo, M.H. Degroot, D.V. Lindley, and A.F.M. Smith, eds., Amsterdam: North-Holland.

Barnard, G., G. Jenkins, and C. Winsten (1962). "Likelihood Inference and Time Series" (with discussion), *Journal of the Royal Statistical Society A*, Vol. 125, 321-372.

Berger, J. (1984). "The Robust Bayesian Viewpoint" (with discussion), in *Robustness of Bayesian Analysis*, J. Kadane, ed., Amsterdam: North-Holland, 63-124.

Berger, J. (1985). *Statistical Decision Theory and Bayesian Analysis, Second Edition.* New York: Springer-Verlag.

Berger, J. (1987). "Robust Bayesian Analysis: Sensitivity to the Prior," Purdue University Technical Report No. 87-10.

Berger, J. and R. Wolpert (1988a). *The Likelihood Principle, Second Edition.* Lecture Notes-Monograph Series, Volume 6, Institute of Mathematical Statistics.

Berkson, J. (1938), "Some Difficulties of Interpretation Encountered in the Application of the Chi-Square Test," *Journal of the American Statistical Association*, Vol. 33, 526-542.

Chaloner, K. (1987). "A Bayesian Approach to the Estimation of Variance Components for the Unbalanced One-Way Random Model," *Technometrics*, Vol. 29, 323-337.

Chen, W-C. (1980). "On the Weak Form of Zipf's Law," *Journal of Applied Probability*, Vol. 17, 611-622.

Chen, W-C., B.M. Hill, J. Greenhouse, and J. Fayos (1985). "Bayesian Analysis of Survival Curves for Cancer Patients Following Treatment" (with discussion), in *Bayesian Statistics 2*, J.M. Bernardo, M.H. Degroot, D.V. Lindley, and A.F.M. Smith, Amsterdam: North-Holland.

Dawid, A.P. (1973). "Posterior Expectations for Large Observations," *Biometrika*, 664-666.

De Finetti, B. (1974). *Theory of Probability, Volume 1.* New York: John Wiley & Sons.

Diaconis, P. and S. Zabell (1986). "Some Alternatives to Bayes's Rule," *Proceedings of the Second Conference on Political Economy, University of California, Irvine*, B. Grofman and G. Owen, eds., Greenwich, CT: JAI Press, 25-38.

Dickey, J. and J. Kadane (1980). "Bayesian Decision Theory and the Simplification of Models," in *Evaluation of Econometric Models*, J. Kmenta and J. Ramsey, eds., Academic Press, 245-268.

Edwards, W., H. Lindman, and L.J. Savage (1963). "Bayesian Statistical Inference for Psychological Research" (with discussion), *Psychological Review*, Vol. 70, 193-242. Reprinted in *Robustness of Bayesian Analysis*, J. Kadane, ed., Amsterdam: North-Holland: 1-62.

Fisher, R.A. (1959). *Statistical Methods and Scientific Inference*,. Second Edition, New York: Hafner Publishing Co.

Freedman, D. (1986). "A Case Study in Nonresponse: Plaintiff vs California State Board of Equalization," *Journal of Business and Economic Statistics*, Vol. 4, 125-126.

Goldstein, M. (1983). "The Prevision of a Prevision," *Journal of the American Statistical Association*, Vol. 78, 817-819.

Good, I.J. (1965). *The Estimation of Probabilities*, MIT Research Monograph No. 30.

Hacking, I. (1967). "Slightly More Realistic Personal Probability," *Philosophy of Science*, Vol. 34, 311-325.

Hill, B.M. (1963). "The Three-Parameter Log Normal Distribution and Bayesian Analysis of a Point-Source Epidemic," *Journal of the American Statistical Association*, Vol. 58, 72-84.

Hill, B.M. (1967). "Correlated Errors in the Random Model," *Journal of the American Statistical Association*, Vol. 62, 1387-1400.

Hill, B.M. (1969a). "Foundations for the Theory of Least Squares," *Journal of the Royal Statistical Society, Series B*, Vol. 31, 89-97.

Hill, B.M. (1969b). "Inference About Many Parameters in the Theory of Least Squares," unpublished manuscript, written at Lancaster University, UK.

Hill, B.M. (1970a). "Contrasts Between Bayesian and Classical Inference in the Analysis of Variance and Testing of Models," *Bayesian Statistics*, R. Meyer and R. Collier, eds., Itasca, IL: F.E. Peacock.

Hill, B.M. (1970b). "Zipf's Law and Prior Distributions for the Composition of a Population," *Journal of the American Statistical Association*, Vol. 65, 1220-1232.

Hill, B.M. (1974a)."On Coherence, Inadmissibility, and Inference About Many Parameters in the Theory of Least Squares," in *Studies in Bayesian Econometrics and Statistics in Honor of L.J. Savage*, S. Fienberg and A. Zellner, eds., Amsterdam: North-Holland, 555-584.

Hill, B.M. (1974b). "The Rank Frequency Form of Zipf's Law," *Journal of the American Statistical Association*, Vol. 69, 1017-1026.

Hill, B.M. (1975a). "A Simple General Approach to Inference About the Tail of a Distribution," *Annals of Statistics*, Vol. 3, 1163-1174.

Hill, B.M. (1975b). "Aberrant Behavior of the Likelihood Function in Discrete Cases," *Journal of the American Statistical Association*, Vol. 70, 717-719.

Hill, B.M. (1979). "Posterior Moments of the Number of Species in a Finite Population, and the Posterior Probability of Finding a New Species," *Journal of the American Statistical Association*, Vol. 74, 668-673.

Hill, B.M. (1980a). "Invariance and Robustness of the Posterior Distribution of Characteristics of a Finite Population, with Reference to Contingency Tables and the Sampling of Species," in *Bayesian Analysis in Econometrics and Statistics: Essays in Honor of Harold Jeffreys*, A. Zellner, ed., Amsterdam: North-Holland, 383-395.

Hill, B.M. (1980b). "Robust Analysis of the Random Model and Weighted Least Squares Regression," in *Evaluation of Econometric Models*, J. Kmenta and J. Ramsey, eds., Academic Press, 197-217.

Hill, B.M. (1980c). Review of *Specification Searches*, by E. Leamer, *Journal of the American Statistical Association*, Vol. 75, 252-253.

Hill, B.M. (1984). Discussion of "The Robust Bayesian Viewpoint," by J. Berger (with discussion), in *Robustness of Bayesian Analysis*, J. Kadane, ed., Amsterdam: North-Holland, 63-124.

Hill, B.M. (1985-86). "Some Subjective Bayesian Considerations in the Selection of Models" (with discussion), *Econometric Reviews*, Vol. 4, No. 2, 191-288.

Hill, B.M. (1986). "Comment: A Bayesian Approach to the Nonresponse Problem, Using Covariates à la Freedman," *Journal of Business and Economic Statistics*, Vol. 4, 125-126.

Hill, B.M. (1987a). "The Validity of the Likelihood Principle," *The American Statistician*, Vol. 41, 95-100.

Hill, B.M. (1987b). "Parametric Models for A_n: Splitting Processes and Mixtures," Unpublished technical report, The University of Michigan.

Hill, B.M. (1988a). "On the Validity of the Likelihood Principle," in *Statistical Decision Theory and Related Topics IV*, Vol. 1, S.S. Gupta and J.O. Berger, eds., New York: Springer-Verlag, 119-132.

Hill, B.M. (1988b). "De Finetti's Theorem, Induction, and $A(n)$, or Bayesian Nonparametric Predictive Inference" (with discussion), in *Bayesian Statistics 3*, J.M. Bernardo, M.H. DeGroot, D.V. Lindley, and A.F.M. Smith, eds., Oxford University Press, 211-241.

Hill, B.M. (1989). Comment on "An Alternative Interpretation of Freedman's Nonresponse Case Study" (with discussion), by R.K. Steinhorst and C. Randall Byers, *Journal of Business and Economic Statistics*, Vol. 7, 27-33.

Hill, B.M. and M. Woodroofe (1975). "Stronger Forms of Zipf's Law," *Journal of the American Statistical Association*, Vol. 70, 212-219.

Hodges, J. (1987). "Uncertainty, Policy Analysis and Statistics" (with discussion), *Statistical Science*, Vol. 2, 259-291.

James, W. (1896). "The Will to Believe," in *Essays in Pragmatism*, The Hafner Library of Classics, Number 7, ed. by Alburey Castell, 1966.

Jeffreys, H. (1961). *Theory of Probability,* Third Edition, Oxford University Press.

Lane, D. and W. Sudderth (1985). "Coherent Predictions are Strategic," *Annals of Statistics*, Vol. 13, 1244-1248.

Leamer, E. (1978). *Specification Searches.* New York: John Wiley.

Mosteller, F. and D.L. Wallace (1964). *Inference and Disputed Authorship: The Federalist.* Reading, MA: Addison-Wesley Publishing Co. Inc.

Salsburg, D.S. (1985). "The Religion of Statistics as Practiced in Medical Journals," *The American Statistician*, Vol. 39, 220-223.

Savage, L.J. (1961). "The Subjective Basis of Statistical Practice," Unpublished manuscript, Department of Mathematics, The University of Michigan.

Savage, L.J. (1962). *The Foundations of Statistical Inference, A Discussion.* London: Methuen & Co. Ltd.

Smith, A.F.M. (1986). "Some Bayesian Thoughts on Modelling and Model Choice," *The Statistician*, Vol. 35, 97-102.

Umbach, D. (1978). "On the Approximate Behavior of the Posterior Distribution for an Extreme Multivariate Observation," *Journal of Multivariate Analysis*, Vol. 8, 518-531.

Zabell, S. (1988). "R.A. Fisher on the History of Inverse Probability," Technical Report, Departments of Mathematics and Statistics, Northwestern University. To appear in *Statistical Science*, (1989).

Zellner, A. (1988). "Optimal Information-Processing and Bayes' Theorem," *The American Statistician*.

Bayesian and Likelihood Methods in Statistics and Econometrics
S. Geisser, J.S. Hodges, S.J. Press and A. Zellner (Editors)

CAN/MAY BAYESIANS DO PURE TESTS OF SIGNIFICANCE?

James S. Hodges
The RAND Corporation

Abstract

Barnard has often argued that the likelihood can't be used to test hypotheses without alternatives; significance tests must be. This paper reviews the writings of Barnard and Bayesians on the subject; refutes Barnard's argument that Bayesians cannot test against all alternatives, but argues that it is unnecessary; then refutes Bayesian arguments against exploratory use of significance tests by constructing an appealing alternative distribution and prior probabilities on the null and alternative that yield (to arbitrary accuracy) the P–value as the posterior probability of the null. The message is that Bayesians can and should use P–values in an exploratory role.

1. Introduction

1.1 The Role of Significance Tests

One recurring theme in George Barnard's writing[1] is a distinction between situations in which several hypotheses are to be compared in the light of some data, and situations in which the plausibility of a single hypothesis is to be evaluated. For the former case, Barnard argued (1949, 1962, 1972a, 1975, 1985; Barnard, Jenkins, and Winsten 1962) that likelihood ratios or some transform of them should be used to compare the hypotheses. For the latter case, no such comparison is available, and Barnard (1962, 1972a, 1975, 1985) argued that this creates a need for a method that does not rely solely on the likelihood. This need is met by significance tests.

In Barnard's view (1962, 1972a, 1972b, 1975, 1980a, 1985), the quintessential role of significance tests is to assess "whether agreement between this hypothesis and the data is so bad that we must begin to exercise our imaginations to try to think of an alternative that fits the facts better" (1972a, p. 129). "If the data do not fit that [hypothesis], then it is worthwhile going ahead. If [the hypothesis] is consistent with the data let us not waste our time" (1962, p. 85). This implies that the test

[1]References to Barnard's papers will be given by the date of the paper only.

is conducted "*before* seriously considering alternatives" (1980a, p. 305), and thus with alternatives that are, at best, vaguely specified (1975, p. 260). Moreover, the need to conduct the test before considering alternatives apparently implies that any attempt to use a Bayesian approach to this problem would require the specification of *all* possible alternative hypotheses, which Barnard considers an impossible task (e.g. 1985, p. 5; see also Anscombe 1963. Many Bayesians consider this a misperception.). In this respect, among others, Barnard views Bayesian methods as deficient.

The usual multiplicity of differing views can be found among those generally identified as Bayesians. A great deal of Bayesian writing has treated individual significance tests in isolation, as devices for assessing nebulous things like the "evidence" relating to a hypothesis. For example, long articles by Pratt (1965) and Berger and Delampady (1987) make only passing mentions of an exploratory role for significance tests, and Berger and Sellke (1987) makes none at all. The emphasis in these papers is on the relationship between P-values and Bayesian hypothesis tests of a single value of a parameter against a simple or composite alternative stated in terms of that parameter. But the quotes above clearly point to the intrinsically exploratory role of significance tests (see also Anscombe 1963, Box 1980, Andrews 1985), and so this large body of work is largely irrelevant to the issue at hand.

Lindley (1980, 1983) argued "that it does not make any sense to test a hypothesis without alternatives in mind" (1983, p. 435). This approach avoids Barnard's criticism by dismissing Barnard's problem. But any birdwatcher knows that this solution is not acceptable: sometimes the bird you see is plainly not of any species you know, and your methods must be capable of telling you so. Jeffreys is often associated with the position just attributed to Lindley (see, e.g. Hill 1986, p. 226), although his actual position is that it does not make any sense to *reject* a hypothesis in a significance test without one to replace it (Jeffreys, 1961, pp. 383-398).

> Is it of the slightest use to reject a hypothesis until we have some idea of what to put in its place? If there is no clearly stated alternative, and the null hypothesis is rejected, we are simply left without any rule at all, whereas the null hypothesis, though not satisfactory, may at any rate show some sort of correspondence with the facts. [p. 390] ... There has not been a single date in the history of the law of gravitation when a modern significance test would not have rejected all laws and left us with no law. Nevertheless the law did lead to improvement for centuries ... [p. 391]

Nonetheless, Jeffreys (1961) recognized the need to assess deficiencies in a hypothesis even if no alternative were available:

> The fundamental idea [of classical significance tests], and one that I should naturally accept, is that a law should not be accepted on data that themselves show large departures from its predictions. [p. 384] ... If the null hypothesis is not altogether satisfactory we can still point to the apparent discrepancies as possibly needing further attention, and attention to their

> amount gives an indication of the general magnitude of the errors likely to arise if it is used; and that is the best we can do (p. 391)

that is, without an alternative hypothesis in favor of which the null might be rejected.

Lindley (1980) and several others (e.g. Zellner 1980, or Berger and Wolpert 1984, section 4.4) have countered that in any significance test, some alternative is lurking in the background, and the test can only be improved by making the alternative explicit. (Sections 3.2 and 3.3 of this paper develop this point, though to a different end.) Barnard (1965, 1980b) offered counterexamples to this, arising in physics; Dempster (1971), Dawid (1980), Geisser (1980, 1989) and others expressed similar views. Box (1980) offered a particularly forceful rebuttal:

> [S]uppose I have an office that looks onto, say, Oxford Street in London, normally thronged with people. One day I look out of the window at 11 o'clock in the morning and notice that there are only two people in the whole street. My initial reaction surely is that on the null (*status quo*) model this is an unusual event possibly worthy of further investigation. Alternative models that might explain the phenomenon come later. ... [T]he basis of the initial reaction, which requires no alternatives, is surely that I have (or could have) looked out of the window on many previous occasions and rarely have I (would I have) seen as few as two people in the street. The motivation is economy of effort and is employed by all of us hundreds of times in our daily lives—when the null model is plausible I will not worry, but when data make it implausible perhaps I should be concerned. (p. 429)

The notion here is logically prior to and more primitive than the specifics of Bayesian or classical tests. The idea is: if I see something odd—if Oxford Street is empty or if people are packed into it so tightly that they cannot move—*then* I am moved to seek an explanation for this odd observation.

Hill (1986) allowed that the only purpose of a test without alternatives is "to help one decide whether it is better to continue to use the only model one has, ... which may or may not be adequate for the purposes one has in mind, or whether it is better to search for new and improved models or hypotheses" (p. 229). (See also Hill's discussion of Barnard 1980a.) Like Barnard, he found this to be "a very different activity from that involved in comparing two sharply formulated hypotheses, [that] should not be confused with the latter" (p. 229), although he did not accept significance tests as a consequence. Hill (1974) offered a Bayesian competitor to significance tests for linear models with additive errors, although his approach could be applied to other cases; his method specifies a large but not exhaustive class of alternative hypotheses, and thus is apparently vulnerable to Barnard's and Anscombe's criticisms.

Smith (1986) distinguished two forms of model criticism:

> *global criticism*, which basically asks the question "should we abandon the current framework completely, despite the fact that we have nothing at all to propose in its place?"; [and]

local criticism, which asks "should the model be modified or extended in this or that particular direction?" (p. 97)

The latter is plainly susceptible to Bayesian methods; the former "does not appear to make sense in a Bayesian context without some form of assumption about alternative models" (p. 98). In Smith's scheme, these forms of model criticism are applied once a formal framework for a given problem situation has been determined. In the earlier stage in the scientific learning cycle in which the formal framework is constructed, "much of this activity is very definitely 'informal' and not necessarily within the purview of Bayesian thinking, which requires a more or less structured framework" (p. 97). Smith specifically excluded "conventionally sanctioned goodness-of-fit criteria (e.g. deviance)" from the collection of techniques that are legitimate at this stage. One can infer that an acceptable approach to scientific learning would include poking through the data with informal exploratory techniques, such as plots, until some particular formal model is postulated, after which only elaborations and simplifications of this model would be considered, and then only in a Bayesian fashion. Presumably, if doubts begin to arise about the adequacy of the overall framework, the proper response is not a test of significance but a return to the informal exploratory phase. Smith (1986) does not indicate the means by which these doubts might begin to arise.

Box (1980) pointed out that diagnostic plots—a mainstay of Smith's exploratory phase—are animated by the same logic as a significance test. These "plots are designed to make manifest certain 'features' of the data that would rarely be extreme, if the model were true," so that using them to motivate searches for alternative models is just like using a significance test. But this did not bother Box; indeed, he argued (Box 1980) that criticism of models cannot be done within the Bayesian framework ("the difficulty with any attempt to use the Bayesian half of the model alone is that it is eternally conditional. We can move the conditionality around but we cannot lose it," p. 427), so that Bayesian methods must yield to sampling methods for these purposes. (Dempster 1971 made a similar argument.) Box advocated using the predictive distribution of the observed data, conditional only on the model, that is, after integrating out the parameters. In various places in the article he suggested using tail areas calculated under the predictive distribution, or examining its value at the observed data, much as Jeffreys suggested. Geisser (1989) offered several other ways to use predictive distributions for the special case of individual observations that may be discordant with a tentatively entertained model.

Others have taken a pragmatic position: if one's object is to catch certain kinds of deviations from a model, then postulate an alternative that will result in a Bayesian hypothesis test that catches those deviations and don't worry about whether the alternative is a plausible model of anything. Thus West (1986) devised a Bayesian method for monitoring the adequacy of a dynamic linear model. It uses a single alternative at each time t, constructed by increasing the variance of the model entertained at time t. This alternative allows detection of outliers and abrupt shifts in the level of the time series, but it is not motivated by any belief in the truth of the alternative as a description of the world.

Leonard (1975), Zellner (1975), and Chaloner and Brant (1988) went a step farther by in effect choosing capacious alternatives that not only allow the detection of a large number of model failures, but suggest correctives as well. Zellner (1975) treated the unknown actual residuals of a linear regression (*i.e.* actual dependent values minus the unknown correct regression) as the objects of interest. These residuals are linear functions of the data and the unknown parameters, so that once the data have been observed, their joint posterior distribution—conditional on the current regression model—can be obtained easily from the posterior of the regression coefficients. The posterior distribution of the realized residuals can be examined for evidence of model failures, and also to search for alternative models. (Hill 1969 examined other uses of the distribution of the realized residuals.)

Leonard (1975) and Chaloner and Brant (1988) treated realized residuals as unknown parameters in similar ways. Chaloner and Brant said that their approach—which is essentially Zellner's, but with a different emphasis—"is a general exploratory method of investigating whether there is a problem when no alternative model is immediately apparent" (p. 2). Zellner used very similar words to describe his method, although he noted that once his procedure had been used for exploration, one could "compute posterior odds ratios relating to variants of a standard model ... that are helpful in model choice" (p. 130; also Dempster 1971, p. 61).

In sum, significance tests fill an exploratory role, and Bayesian researchers have generally agreed that there is a role to be filled. Some have accepted ordinary significance tests or variants on them, while others have attempted Bayesian surrogates in the apparent absence of any possibility of finding a palatable direct analog. However, little systematic scrutiny has been given to Barnard's assertion that no analog is possible, and no-one has been able to explain significance tests in satisfactory Bayesian terms.

1.2 Answering Barnard Directly

Though their other merits are considerable, none of these approaches addresses Barnard's criticism head on, leaving the impression that Barnard has indeed found a problem that Bayesian methods cannot handle. In section 2.1, I show that for a large class of problems, it is possible to specify all of the alternatives to a given model; in fact it is quite simple. However, it is also quite unnecessary, ever, and the simpler construction—the Bayesian analog to significance tests—is given in Section 2.2. In section 3.1, this Bayesian construction is shown to have the same structure as a significance test a la Barnard. Many Bayesian writers (e.g. the contributions of Dickey, Good, Hill, Leonard, and Zellner to the discussion of Barnard 1980a) have attempted to interpret P–values from significance tests as approximations to Bayesian quantities. For pure significance tests, though—as Barnard (1980a, p. 317) pointed out—such approximations require a specification of an alternative, which the authors listed above did not give. In sections 3.2, I show that classical significance tests are equivalent to any of a very large class of Bayesian significance tests in the sense that they produce the same ranking of possible data outcomes according to their discrepancy from the model. In section 3.3, the P–value, viewed as a function of the data, is shown to be, to arbitrary accuracy, the posterior probability of the null model, for a specific alternative and a specific prior probability of the null. Section 4 discusses these results and their implications for Bayesian arguments against the use of significance tests as exploratory tools.

The main point here is intended mostly for Bayesian readers. Bayesian analogs to "alternative-free" tests are straightforward to define, and it turns out that in a large class of cases, classical significance tests are essentially the same as these Bayesian tests. Given the exploratory—and thus rough-and-ready—role that significance tests are intended to fill, it follows that little can be gained by a finicky concern over logical niceties and an attendant refusal to make exploratory use of the abundance of non-Bayesian computer software, especially when competing Bayesian software is non-existent or much more difficult to use. To anticipate the results further, West's (1986) method, cited above, is an archetype of the Bayesian analog to significance tests. Like a significance test, West's test tells its user that something is wrong, but it doesn't say what; the user must figure that out some other way. We have some Bayesian methods that *do* suggest what is wrong (also cited above), so perhaps the message to be drawn from all this is that it is time we stopped worrying about significance tests and spent our energies developing more methods like those of Leonard, Zellner, and Chaloner and Brant.

2. *Can* Bayesians Do Pure Tests of Significance?

2.1. A Bayesian Can Specify All Possible Alternatives for Some Problems

Let $X = (x_1, x_2, \ldots, x_n)$ be an n–vector of observables, each of which can take a finite number of values. For ease of exposition, let each x_j be a binary observable, taking the values 0 and 1; extension to the more general case is straightforward. (Berger and Wolpert 1984 allow countable sample spaces, but restrict themselves to independent and identically distributed random variables.) Any probability distribution for X, with arbitrary dependencies, can be specified as a 2^n–vector p as follows. Label the coordinates of p with the 2^n patterns of 0's and 1's, so that $p = (p_{00\ldots0}, p_{10\ldots0}, \cdots, p_{11\ldots0}, p_{11\ldots1})$. The coordinates of p are in the closed interval from zero and one, they sum to one, and each coordinate gives the probability of observing the value of X that is its subscript. Call the collection of such p's S; it is the space of all probability distributions for X. Any alternative to a given hypothesis regarding the distribution of X can be represented as a subset of S, and so all alternatives are captured in it. (Diaconis 1977, Meeden 1986, Draper 1988, and others have used this construction for different purposes.) Thus, existing Bayesian hypothesis testing ideas can (not to say should) be used in this manner to construct an explicit test of a model against all alternatives.

For example, consider testing whether, conditional on θ, the x_j are independent and identically distributed Bernoulli random variables with $Pr(x_j = 1 \mid \theta) = \theta$. In the manner of Bayesian hypothesis tests, let H_o be this Bernoulli model with a beta distribution as the prior for θ:

$$f(\theta|a,b) = B(a,b)^{-1}\theta^{a-1}(1-\theta)^{b-1} \quad \text{for} \quad \theta \in (0,1) \tag{1}$$

where $B(a,b)$ is the beta function. (In practice, of course, the prior would depend on the situation.) For the alternative, let H_1 be S, the space of all models, with the flat prior distribution $f(p) = \Gamma(2^n)$ on the simplex S. (In practice one would almost certainly use another prior, as will be seen below.) Let the prior probabilities for H_o and H_1 be π and $1-\pi$, respectively.

By a straightforward application of familiar Bayesian theory,

$$Pr(H_o|X) = Q\ Pr(X|H_o)Pr(H_o) = Q\pi B(a,b)^{-1}B(a+Y, b+n-Y) \tag{2}$$

and

$$Pr(H_1|X) = Q\ Pr(X|H_1)Pr(H_1) = Q(1-\pi)2^{-n} \tag{3}$$

where Q is a proportionality constant, $Y = \Sigma x_j$, and

$$\int_S p_X f(p)dp = \Gamma(2^n)/\Gamma(2^n+1) = 2^{-n} \tag{4}$$

(This last is easy to derive directly, without computing the integral.) Thus, for $a = b = 1$, the posterior odds ratio $Pr(H_o|X)/Pr(H_1|X)$ is

$$\frac{\pi}{1-\pi}\ \frac{2^n\Gamma(Y+1)\Gamma(n-Y+1)}{\Gamma(n+2)} \tag{5}$$

which, as it turns out, has a number of unpalatable properties. The most important—which holds for all choices of a and b—is that it depends on the data only through Y, so that according to this test one's posterior beliefs about the tenability of the binomial model depend only on the sufficient statistic under that model.

Apart from exemplifying the point that Bayesians can indeed test against all alternatives, this example illustrates once again that the outcome of a Bayesian hypothesis test must depend on which prior is used conditional on the null and the alternative hypotheses. Parenthetically, it also illustrates that the naive flat prior on S is a piece of information with less than obvious implications, which should be used with great caution.

Contrary to Barnard's assertion, then, a Bayesian can sometimes specify all alternatives and, in the usual manner, test a precisely specified hypothesis against them. But a reading of Barnard's papers on this subject suggests that this concern might not be satisfied by S and the constructed hypothesis test, and that he might raise several objections at this point. The first might be that this construction fails as a prospective way of encompassing all possibilities, because it will not capture unanticipated outcomes like the flipped coin that is lost in the flipping (1949, p. 136). But Barnard's (1947) discussion of a Fisherian significance test regarding seeds clearly permits the *post hoc* construction of the sample space, so this objection will not hold.

A more interesting objection is that the combination of S and a prior distribution on it cannot properly represent the nature of uncertainty about hypotheses as yet unconceived or inconceivable (1985, p. 5). But the construction permits positive chunks of prior probability to be assigned to subspaces corresponding to specific parametric models; and is it unreasonable that point hypotheses elsewhere are given prior (and thus posterior) probability zero? Posterior probability in S can still aggregate in the neighborhood of such hypotheses, and besides, the failure to assign positive probability to a specific model *a posteriori* does not seem to be a disadvantage; it may even be preferable.

For a third possible objection, Barnard's papers show a strong disinclination to place prior probabilities on hypotheses as a matter of course, so that he might object to the whole idea of placing a prior distribution on S, judging it to be incapable of sufficiently precise specification and thus arbitrary and unworthy of the name "probability" (1985, p. 5). But this would hold the Bayesian approach to a higher standard than that applied to significance tests by Anscombe (1963), which Barnard (1972a) endorsed. This point is discussed further in section 3.1.

2.2 But It Is Not Necessary to Specify All Possible Alternatives

It is possible to specify all alternatives for some problems, but the derivation in the last section shows that there is no need to do so, for this type of problem or for any other. Recall that

$$Pr(H_1|X) \propto Pr(X|H_1)Pr(H_1) = (1-\pi)2^{-n}, \tag{6}$$

so that the posterior probability of H_1 depends only on the marginal probability of the data given H_1, that is, on the *a priori* predictive probability of the data given H_1. Thus, the only new thing that a Bayesian needs to make an analog to the pure significance test is an alternative predictive distribution for the data—not a parametric alternative or even a class of parametric alternatives. (This fact was alluded to by Pratt 1965 in his rejoinder to the discussion.) In this sense, West's (1986) "model monitoring" method is the archetype of the Bayesian significance test.

3. The Similarity of Bayesian and Conventional Significance Tests

3.1 Both Tests Rank the Possible Data Outcomes by Their "Discrepancy"

A conventional significance test requires two ingredients: a hypothesized model and a function that assigns a "discrepancy" to the possible data outcomes (1962, 1980a; Cox 1977; Anscombe 1963 uses the term "test criterion"). The discrepancy function sorts all possible data outcomes from the least to the most discrepant; when the actual data are observed, the P–value is computed as the probability, under the hypothesized model, of observing an outcome as discrepant as or more discrepant than that which was actually observed. A conventional significance test, then, has two products: a ranking of the possible outcomes and a method of attaching a probability to the sorted outcomes. The choice of this discrepancy function is quite arbitrary; it depends on the model failures one considers likely or interesting (Anscombe 1963). To Jeffreys (1961), this was a disadvantage, because for any set of data, some function is likely to look odd. But as will become clear shortly, Bayesian methods are no more immune to this criticism than classical tests.

A Bayesian significance test, as defined in section 2, requires four ingredients: the same hypothesized model as in the conventional significance test, plus a prior distribution on the parameters of the hypothesized model, a predictive distribution that specifies the alternative hypothesis and a prior probability of the null hypothesis. (Actually, three ingredients suffice: predictive distributions for the null and alternative, and the prior probability of the null.) The prior distribution on the parameters and the alternative predictive distribution produce, as in the conventional case, a criterion for ranking the possible data outcomes from least to most discrepant—where in this case increasing discrepancy corresponds to decreasing posterior probability of the hypothesized model—but altogether these ingredients yield a different method for tacking probabilities onto the ranked outcomes.

This structural similarity removes the force of the third objection to the construction of section 2.1, that priors on S (and thus alternative predictive distributions) are arbitrary. The arbitrariness of the alternative predictive distribution corresponds precisely to the arbitrariness of the discrepancy function of a conventional significance test. If the latter is tolerable for the limited legitimate purpose of significance testing, the former must be tolerable as well. This structural similarity also removes the force of Jeffreys' criticism (noted above) of classical tests based on the arbitrariness of the discrepancy function. It is possible to argue that Jeffreys' criticism still holds as long as the alternative models are chosen before looking at the data, but this requirement is incompatible with the exploratory function; if any kind of test is to be useful in exploration, it must be used to consider things that were not envisioned before the data were analyzed.

The foregoing shows that both Bayesian and conventional significance tests produce rankings of the possible data outcomes and formulae for assigning a measure of discrepancy calibrated as a probability. This prompts two questions: (i) can sorting algorithms produced by significance tests be duplicated by Bayesian tests, and (ii) if so, are the P–values posterior probabilities for any interesting constructed alternative and prior for the null model? The answers will be shown in sections 3.2 and 3.3 to be (i) yes, always, and (ii) to an arbitrary degree of accuracy, for significance tests in which the P–value is distributed as a uniform random variable under the null hypothesis. These results extend those in Cox (1962, p. 84-85) and DeGroot (1973). Many Bayesian authors show similar relations between classical significance tests of a null hypothesis and Bayesian hypothesis tests of the same null against a specific alternative. For example, Zellner (1984) gives an example of a Bayesian test of a hypothesis about a regression parameter in which the P–value is identical to the posterior probability that the hypothesis is true. The correspondence between Bayesian and classical methods in this literature is different from the one to be developed now.

3.2 The Significance Test's Ranking Can Always Be Reproduced Exactly

Let X be a random variable (scalar, vector, matrix, or whatever) and let x be a realization of X. Let $T = T(x)$ be a test statistic or test criterion that can be calculated from x, and suppose without loss of generality that it is to be used in a test of significance in which larger values will be treated as less favorable to the null model than smaller values. Let $P(x)$ be the P–value computed for the observation x from the null model; P is actually a function of T alone, but in the sequel the argument of P will be given as T or x as convenience and clarity dictate. For now, this need not be an exact P–value (although in section 3.3, it will), but it must be a function of the data and not of any unknown parameters. Let $F(x)$ be the marginal distribution function of these data x under the hypothesized model, and for convenience let it have a density $f(x)$ with respect to Lebesgue measure. (This does not affect the generality of the result.) This distribution is marginal with respect to the prior for any unknown parameters of the null model; no restrictions are required on that prior except that $F(x)$ must exist.

Let $\phi(T)$ be any positive monotone increasing function that can take T as an argument, and define an alternative density for X by $p_a(x) = K_\phi f(x)\phi(T(x))$, where K_ϕ is the constant of proportionality that makes p_a a probability distribution. This alternative places more mass in those regions where the test statistic is large, which mimics the intuition behind selecting the test statistic. (DeGroot 1973 used a very similar construction for a somewhat different purpose; Cox 1962 used a special case.) The requirement that K_ϕ be non-zero places a further constraint on $\phi(T)$, which I will henceforth assume to be satisfied. This is not generally an onerous restriction; for statistics T having moment generating functions under the null model, for example, the exponential function will be permissible. If π is the prior probability of the null model, then the posterior probability given x is

$$\left[1 + \frac{1-\pi}{\pi}\frac{p_a(x)}{f(x)}\right]^{-1} = \left[1 + \frac{1-\pi}{\pi} K_\phi \phi(T)\right]^{-1}, \tag{7}$$

so that the ranking produced by this alternative is equivalent to the ranking given by $\phi(T)$, and thus, by the monotonicity of ϕ, equivalent to the ranking given by T itself.

Thus the ranking given by the significance test can be reproduced by a Bayesian significance test, indeed, by any Bayesian significance test in a large class.

3.3. Usually the Significance Test's P–Value Is, To Arbitrary Accuracy, a Posterior Probability

For cases in which an exact continuous distribution can be given for the test statistic T, the P–value $P(T)$ has a uniform distribution under the null, conditional on any unknown parameters. Thus, for these cases, $P(T)$ has a uniform distribution with respect to $F(x)$, the marginal distribution of the observation X under the null. For the remainder of this section, I will restrict consideration to test statistics and P–values having this property. This includes all of the familiar tests associated with the analysis of variance, for example. It excludes approximate P–values and P–values for discrete distributions; results similar to those below can be produced for at least some of these cases, but I will not do so here.

For significance tests in this restricted class, construct an alternative distribution by specifying $\phi(T) = P(T)^{\delta-1} - 1$ in the notation of section 3.2, where $0 < \delta < 1$. Then $K_\phi^{-1} = E(P(T)^{\delta-1}) - 1$, where the expectation is with respect to the marginal distribution $F(x)$, that is, where $P(T)$ has a uniform distribution. Set the prior probability of the null hypothesis, π_ϕ, so that it satisfies $(1 - \pi_\phi)K_\phi/\pi_\phi = 1$; this gives $\pi_\phi = K_\phi/(1+K_\phi)$. Because we are only considering cases for which $P(T)$ has a uniform distribution, $K_\phi^{-1} = (1-\delta)/\delta$ and $\pi_\phi = \delta$. For this alternative distribution and this prior probability on the null hypothesis, the posterior probability of the null is $P(T)^{1-\delta}$. That is, the P–value can be made arbitrarily close to the posterior probability of the null model, whatever x is, for a specific alternative and prior probability of the null that do not depend on the data. In the limit as δ goes to zero, however, the constructed alternative becomes improper and π_ϕ goes to zero.

4. So What? (*May* Bayesians Do Pure Tests of Significance?)

As section 1.1 made clear, some people associated with the Bayesian viewpoint have no difficulty with the use of significance tests for exploratory purposes. Others continue to object because significance tests appear to be in violation of the likelihood principle (Lindley 1980, Berger and Wolpert 1984), or because they are apparently otherwise incapable of a satisfactory Bayesian explanation. It appears that the basis of these continuing objections has been weakened considerably.

Pratt (1965) argued that significance tests suppress uncertainty by hiding the uncertainty about the implicit alternative hypothesis. But section 3.2 showed that for a very broad class of possible alternative hypotheses, the ranking produced by the resulting Bayesian significance test is identical to that of the Barnardian significance test. To insist that this class is not large enough is to insist on a logical point beyond any practical implication; certainly this class of alternatives captures the vague notion that moves a statistician to consider a particular test statistic. Continuing objection can only be based on the assignment of the specific posterior probability, described in section 3.3.

The first point here, shown in section 3.3, is that for a large class of commonly used significance tests, the P–values *are*, to arbitrary accuracy, posterior probabilities for a specific constructed alternative hypothesis and prior probability of the null. So the only objection that seems to remain is that the P–value is only *a* posterior probability, not *my* posterior probability in some given instance. But what would one's posterior probability be in this situation? To quote Dempster (1971, p. 60):

> In principle ... the concept of a prior probability of a null hypothesis clashes with the very concept of a null hypothesis as a tentative stopping place on

> the way to devising a model in accord with the known facts. One knows that no null hypothesis could ever be true, but hopes to find one that fits and illumines the facts. Even to conceive of a numerical degree of certainty attached to such an entity is awkward, letting alone the practical question of specifying the number. In practice, when the data are such that judgments of significance are borderline, the dependence of the posterior probability of the null hypothesis on its prior probability will be crucial, and the user will be led to agonize over a largely fictitious aspect of his prior knowledge at the expense of constructively examining his available data.

This is, ultimately, the point: why waste any energy over the remaining difference? Berger and Delampady (1987), among others, argue that one can always construct an alternative in cases apparently without one, and that one should do so and then conduct the Bayesian test. But the P–value is precisely the result of such a construction; so why not use it for exploration — especially when we have so many readily available tools for computing P–values, and almost no tools for the competing Bayesian quantities? It is a counsel of inefficiency to insist that we should not use what is available and should use what is not available, particularly when the latter offers no practical advantage and the most ethereal of logical advantages.

4.1 A Footnote

Many Bayesians (e.g., Pratt 1965) have found it objectionable that significance tests will, for a large enough sample, find any deficiency in a model, no matter how small. It is odd that this should be considered a disadvantage; what are we to make of a tool for finding problems that fails to detect them under advantageous circumstances? To quote again from Dempster (1971, p. 62):

> Some may hold that the beauty of science lies in a conjunction of truth and simplicity, and certainly there are way stations along the path of science which give that impression, but I would hold ... that the real world, even the real world of some quite restricted scientific phenomenon, is endlessly complex. ... The real life process of constructing a model may be roughly conceived as operating between a floor and a ceiling where increasing altitude means increasing *complexity* of the model. The floor pushing complexity upward consists in the need for the model adequately to fit the known facts of the phenomenon under study. The ceiling, however, is a much less well understood logical construct which I should like to label *confusion*. Confusion arises because too many dimensions of an overly complex model are insufficiently determined by the available facts, so that predictions and insights from the model are muffled. I believe, in other words, that there are inherent limitations on the process of estimation which imply that restricted fixed data simply cannot determine a broadly acceptable posterior distribution on the many parameters of a highly complex model. Thus, the processes of model searching and estimation interact through the determination of a comfortable altitude of complexity at which the available supply of factual data can be reasonably absorbed and interpreted.

As a body of data gets larger, should we not expect, indeed, desire that it will reveal deficiencies in any currently held model? For various reasons we may choose to ignore the revealed deficiencies, but that does not mean that we are better off ignorant of them.

Acknowledgements

While writing this paper I received the thoughtful comments of Seymour Geisser, Bruce Hill, John Pratt, and Arnold Zellner. They do not necessarily share and are not responsible for the judgments and interpretations expressed in this paper.

References

Andrews, D.F. (1985). Discussion of Barnard (1985).

Anscombe, F.J. (1963). "Tests of Goodness of Fit" (with discussion), *Journal of the Royal Statistical Society, Series B*, Vol. 25, 81-94.

Barnard, G.A. (1947). "The Meaning of a Significance Level," *Biometrika*, Vol. 34, 179-182.

Barnard, G.A. (1949). "Statistical Inference" (with discussion), *Journal of the Royal Statistical Society, Series B*, Vol. 11, 115-149.

Barnard, G.A. (1962). Prepared contribution and discussion in *Foundations of Statistical Inference*, L.J. Savage, ed., London: Methuen.

Barnard, G.A. (1965). Comment on Pratt (1965).

Barnard, G.A. (1972a). Review of I. Hacking, "The Logic of Statistical Inference," *British Journal for the Philosophy of Science*, Vol. 23, 123-190.

Barnard, G.A. (1972b). "The Unity of Statistics" (with discussion), *Journal of the Royal Statistical Society, Series A*, Vol. 135, 1-14.

Barnard, G.A. (1975). Comment on J.D. Kalbfleisch, "Sufficiency and Conditionality," *Biometrika*, Vol. 62, 251-268.

Barnard, G.A. (1980a). "Pivotal Inference and the Bayesian Controversy" (with discussion), in *Bayesian Statistics*, J.M. Bernardo, M.H. DeGroot, D.V. Lindley, A.F.M. Smith, eds., Valencia, Spain: University Press.

Barnard, G.A. (1980b). Comment on Box (1980).

Barnard, G.A. (1985). "A Coherent View of Statistical Inference" (with discussion), Technical Report, University of Waterloo, Department of Statistics and Actuarial Science.

Barnard, G.A., G.M. Jenkins, and C.B. Winsten (1962). "Likelihood Inference and Time Series" (with discussion), *Journal of the Royal Statistical Society, Series A*, Vol. 125, 321-372.

Berger, J.O. and M. Delampady (1987). "Testing Precise Hypotheses" (with discussion), *Statistical Science*, Vol. 2, 317-352.

Berger, J.O. and T. Sellke (1987). "Testing a Point Null Hypothesis: The Irreconcilability of P-Values and Evidence," *Journal of the American Statistical Association*, Vol. 82, 112-122.

Berger, J.O. and R.L. Wolpert (1984). "The Likelihood Principle," Institute of Mathematical Statistics, Hayward, CA.

Box, G.E.P. (1980). "Sampling and Bayes Inference in Scientific Modelling and Robustness" (with discussion), *Journal of the Royal Statistical Society, Series A*, Vol. 143, 383-430.

Chaloner, C. and R. Brant (1988). "A Bayesian Approach to Outlier Detection and Residual Analysis," *Biometrika*, Vol. 75, No. 4, 651-659.

Cox, D.R. (1962). *Discussion in Foundations of Statistical Inference*, L.J. Savage, ed., London: Methuen.

Cox, D.R. (1977). "The Role of Significance Tests" (with discussion), *Scandinavian Journal of Statistics*, Vol. 4, 49-70.

Dawid, A.P. (1980). Comment on Box (1980).

Dempster, A.P. (1971). "Model Searching and Estimation in the Logic of Inference" (with discussion) in *Foundations of Statistical Inference*, V.P. Godambe and D.A. Sprott, eds., Toronto: Holt, Rinehart, and Winston.

DeGroot, M.H. (1973). "Doing What Comes Naturally: Interpreting a Tail Area as a Posterior Probability or as a Likelihood Ratio," *Journal of the American Statistical Association*, Vol. 68, 966-969.

Diaconis, P. (1977). "Finite Forms of de Finetti's Theorem on Exchangeability," *Synthese*, Vol. 36, 271-281.

Draper, D.C. (1988). "Rejoinder," *Statistical Science*, Vol. 3, No. 2, May, 266-271.

Geisser, S. (1980). Comment on Box (1980).

Geisser, S. (1989). "On Predictive Tests of Discordancy," in *Bayesian and Likelihood Methods in Statistics and Econometrics: Essays in Honor of George Barnard*, S. Geisser, J.S. Hodges, S.J. Press, A. Zellner, eds., Amsterdam: North-Holland.

Hill, B.M. (1969). "Foundations for the Theory of Least Squares," *Journal of the Royal Statistical Society, Series B*, Vol. 31, No. 1, 89-97.

Hill, B.M. (1974). "On Coherence, Inadmissibility, and Inference About Many Parameters in the Theory of Least Squares," in *Studies in Bayesian Econometrics and Statistics: In Honor of Leonard J. Savage*, S.E. Fienberg, A. Zellner, eds., New York: North-Holland/American Elsevier.

Hill, B.M. (1986). "Some Subjective Bayesian Considerations in the Selection of Models" (with discussion), *Econometric Reviews*, Vol. 4, 191-246.

Jeffreys, H. (1961). *Theory of Probability*, third edition. London: Oxford University Press.

Leonard, T. (1975). "Bayesian Estimation Methods for Two-Way Contingency Tables," *Journal of the Royal Statistical Society, Series B*, Vol. 37, 23-37.

Lindley, D.V. (1980). Comment on Box (1980).

Lindley, D.V. (1983). "The Role of Randomization in Inference," *Philosophy of Science Association 1982*, Vol. 2, 431-446 (Philosophy of Science Association).

Meeden, G. (1986). "Sufficiency and Partitions of the Class of All Possible Discrete Distributions," *The American Statistician*, Vol. 40, 42-44.

Pratt, J.W. (1965). "Bayesian Interpretation of Standard Inference Statements" (with discussion), *Journal of the Royal Statistical Society, Series B*, Vol. 27, 169-203.

Smith, A.F.M. (1986). "Some Bayesian Thoughts on Modelling and Model Choice," *The Statistician*, Vol. 35, 97-102.

West, M. (1986). "Bayesian Model Monitoring," *Journal of the Royal Statistical Society, Series B*, Vol. 48, 70-78.

Zellner, A. (1975). "Bayesian Analysis of Regression Error Terms," *Journal of the American Statistical Association*, Vol. 70, 138-144.

Zellner, A. (1980). Comment on Barnard (1980a).

Zellner, A. (1984). "Posterior Odds Ratios for Regression Hypotheses: General Considerations and Some Specific Results," in *Basic Issues in Econometrics*, Chicago: University of Chicago Press.

Bayesian and Likelihood Methods in Statistics and Econometrics
S. Geisser, J.S. Hodges, S.J. Press and A. Zellner (Editors)

CONGLOMERABILITY, COHERENCE AND COUNTABLE ADDIVITY

David A. Lane
University of Minnesota

Abstract

This paper offers answers to two foundational questions relating to the Law of Total Probability:

- Is the Law always available? That is, for a given partition of the sample space and unconditional probability distribution, is there necessarily a set of conditional probability distributions that yield the unconditional distribution when the Law is applied?
- When the Law is available, what are the consequences of violating it?

1. Introduction

In this paper, I will offer answers to two foundational questions relating to the Law of Total Probability. Statisticians and probabilists of all foundational persuasions use the Law as a fundamental tool for probability assessment. That is, to determine the probability of a proposition A, one often finds it convenient to introduce a set $\mathcal{H}$ of mutually exclusive, exhaustive propositions; then to evaluate a probability distribution μ over $\mathcal{H}$ and, for each h in $\mathcal{H}$, $P(A \mid h)$; and finally to invoke the Law to assess the probability for A by the formula

$$P(A) = \int P(A \mid h) d\mu(h). \tag{1}$$

The questions relating to the Law that I will consider are the following:

- Is the Law always available? That is, for a given $\mathcal{H}$ and unconditional probability P, is there necessarily a set of conditional probabilities $\{P(\cdot \mid h)\}$, such that P can be obtained according to equation (1)?
- When the Law is available, must it bc respected? And if so, why? That is, what are the consequences of violating it?

2. Is the Law Always Available?

2a. The Phenomenon of Nonconglomerability

Most of the recent statistical literature concerning the availability of the Law of Total Probability has centered around the notion of *nonconglomerability.* The inferential importance of nonconglomerability derives from the fact that when nonconglomerability obtains, the Law fails.

To define nonconglomerability, suppose that Ω is a nonempty set, $\mathcal{F}$ and $\mathcal{G}$ are algebras of subsets of Ω with $\mathcal{F}$ contained in $\mathcal{G}$, and $\mathcal{H}$ is a $\mathcal{G}$-measurable partition of Ω. In addition, P is a probability measure on $(\Omega, \mathcal{G})$ with $\mathcal{H}$-marginal μ, and for each h in $\mathcal{H}$, $P(\cdot \mid h)$ is a probability measure on $(\Omega, \mathcal{F})$ with support in h. Finally, suppose that for all A in $\mathcal{F}$ and h in $\mathcal{H}$, $P(A \cap h) = P(A \mid h) \cdot P(h)$. Then:

(a) P is *$\mathcal{H}$-disintegrable* with *disintegration* $\{P(\cdot \mid h)\}$ if equation (1), the Law of Total Probability, holds for all A in $\mathcal{F}$.

(b) P is *conglomerable* with respect to $\{P(\cdot \mid h)\}$ if for all A in $\mathcal{F}$.

$$\inf_{h \in \mathcal{H}} P(A \mid h) \leq P(A) \leq \sup_{h \in \mathcal{H}} P(A \mid h). \tag{2}$$

If (2) fails for some A in $\mathcal{F}$, then P is *nonconglomerable.*

Clearly, if P is nonconglomerable, it cannot be disintegrable and so the Law fails.

2b. Nonconglomerability and Noncountable Additivity I

For most statisticians and probabilists, nonconglomerability seems pathological. Others, following de Finetti, disagree, as will be explained shortly. In most of the recent discussions of this question, the examples of nonconglomerability that have been put forward involve noncountably additive probability distributions. As a result, some statisticians have taken the position that violations of the Law of Total Probability are just an artifact of the introduction of noncountably additive distributions into inferential discourse. For example, consider Dickey's comments to Hill (1980):

> I simply do not understand the practical need for merely finite additivity.... It seems to me that countable additivity, conglomerability, and proper integrable distributions enable us to treat real problems realistically, without worrying that the mathematics itself will deal us an unpleasant surprise.

It would of course be difficult to make this argument rigorously: what exactly does Dickey mean by "practical", "real" and "realistically"? But even putting that question to one side, his position is untenable. In fact, nonconglomerability cannot be avoided by restricting to countably additive distributions. Quite the opposite:

noncountably additive distributions may be required to *rescue* the Law of Total Probability in situations that start with "realistic" countably additive assessments.

Consider the following example, due to Dubins. Let P be the usual fair coin-toss measure on $\Omega = \{0,1\}^\infty$. Next, let $\mathcal{H}$ consist of the atoms of the tail sigma-field on Ω (that is, ω and ω' are in the same partition set if, eventually, they have all the same coordinates). Then there are finitely additive $\mathcal{H}$-disintegrations of P, but no countably additive ones (Dubins (1977), Dubins and Heath (1983)). That is, starting with a natural, much-used countably additive P and a natural, much-studied partition $\mathcal{H}$, the Law of Total Probability is not available without introducing noncountably additive distributions.

2c. Essential Nonconglomerability

If we admit noncountably additive distributions as conditional probabilities, there can be no bar to considering them as unconditional probabilities as well. That done, it is easy to find examples of a noncountably additive probability P and partition $\mathcal{H}$ such that P is *essentially nonconglomerable* with respect to H: that is, there exists *no* set of conditional probabilities $\{P(\cdot \mid h)\}$ that makes P conglomerable.

Here is an example: Suppose Ω is the nonnegative integers, P_1 is a countably additive distribution on Ω supported by the even integers and giving each even integer positive probability, and P_2 is a diffuse distribution supported by the odd integers (that is, $P_2(\{i\}) = 0$ for every integer i). Let $P = (P_1 + P_2)/2$, and for each i define the partition element h_i to be $\{2i, 2i+1\}$. To satisfy the Law of Total Probability for the event $\{2i\}$, $P(\{2i\} \mid h_i)$ must equal 1 for each i; but then the Law fails when A contains a subset of the odd integers with positive P_2 - probability.

It is possible, though harder, to find examples of essential nonconglomerability involving countably additive P. For examples and some characterizations, see Blackwell and Dubins (1975) and Maitra and Ramakrishnan (1988).

2d. Epistemic Nonconglomerability

Several authors have presented examples of nonconglomerable assessments that are not essential, but which the authors have argued represent a valid encoding of a reasonable state of belief. Such assessments are examples of *epistemic nonconglomerability.* I will briefly sketch two of these.

The first example is due to de Finetti (1974) and Hill (1980). Suppose a point is selected at random on Ω, the surface of the unit sphere in R^3: an assessor describes his uncertainty about the location of the point by the uniform distribution on Ω. Next, the assessor is asked to describe his uncertainty about the longitude of the selected point, relative to a specified pair of poles and a reference great circle; he announces a uniform distribution on $(-\pi, \pi)$ for this quantity. Now the assessor considers where he would believe the point to be if he were *told* the exact longitude on which the point was located. Hill and de Finetti argue that the assessor might well decide to select a uniform distribution for the latitude of the point, given the longitude. In fact, Hill asks the assessor to consider what his response would be if he were only told that the point lay on a specified great circle, without designated poles; and if the uniform distribution were assessed on that great circle, why should the designation of poles change that assessment? All these assessed distributions are of course countably additive (on the relevant Borel sets), but they violate the Law of Total Probability, since the uniform distribution on Ω is the product of Lebesgue measure on longitude with a measure on latitude that has density $\cos(x)/2$ with respect to Lebesgue measure on $(-\pi/2, \pi/2)$.

The second example is due to Scozzofava (1981,1984). It concerns the first digit problem: take a large heterogeneous collection of statistical data and calculate the observed frequency of the nine possible first digits; these frequencies are often found to be close of $f(k) = \log_{10}(1 + 1/k)(1 \leq k \leq 9)$, rather than approximately uniform. Scozzafava describes a finitely additive nonconglomerable distribution on the set Ω consisting of integers and rational decimal fractions such that P(first digit $= k) = \log_{10}(1 + 1/k)$, while if it is *known* that the numbers come from a bounded interval in Ω of the form $(10^n, 10^{n+1})$, the distribution on the first digit is uniform. He claims that this distribution satisfactorily expresses uncertainty about the first digit in the "bounded" and "unbounded" cases.

2e. Nonconglomerability and Noncountable Additivity II

Suppose the partition $\mathcal{H}$ is countable. Then, if P is countably additive, P must be conglomerable (regardless of whether or not the conditional assessments $\{P(\cdot \mid h)\}$ are countably additive). De Finetti (1972) conjectured, and Schervish, Seidenfeld and Kadane (1984) and later Hill and Lane (1985) proved, a converse to this fact: if P is not countably additive, then there exists a countable partition $\mathcal{H}$ with respect to which P is not conglomerable. In this sense, noncountable additivity and nonconglomerability are inseparable companions.

3. Should the Law Be Respected?

3a. Coherence

Whether or not the Law of Total Probability should be respected depends on one's views about the nature and purpose of inference. Most authors who have concerned themselves recently with the validity of the Law, including Hill (1980), Scozzafava (1984), Kadane et al. (1986), Seidenfeld (1985) and Regazzini (1987), have adopted the foundational framework developed by de Finetti (1972, 1974). According to this framework, the role of inference is to help the inferrer to decide how to act in the face of uncertainty. The inferrer must quantify his uncertainty; that is, for each uncertain proposition bearing upon the consequences of his possible acts, the inferrer must measure his degree of belief in the proposition, where belief is equated to his propensity to act as though the proposition were true. To establish the proper terms in which to carry out this measurement, de Finetti introduced simple localized acts whose consequences are expressed in economic terms, and used these acts to define the degree of belief of propositions of interest. With this device, inconsistencies in reasoning about uncertainty can be concretized in terms of their adverse economic consequences (that is, as dollar losses). Specifically, de Finetti defined the probability of a proposition whose truth can be determined in the future in terms of the price at which the assessor is neutral between buying and selling a ticket worth \$1 if the proposition turns out to be true, \$0 otherwise, and he defined conditional probability as a price for a ticket as described, except that the purchase price is refunded if the conditioning proposition turns out to be false. These definitions can be extended in a kind of economic metaphor even to propositions that are not ultimately observable.

With these definitions, a set of assessed probabilities represents consistent reasoning in the face of uncertainty if no transaction for tickets at the prices that correspond to the probabilities can result in a sure loss to the assessor, where a transaction involves buying or selling a finite number of tickets. Such sets of assessments are called *coherent*. The restriction to a finite number of sales or purchases is essential to make the metaphor operational. With the stroke of a pen, you can make an infinite number of probability assessments, but there is no physically realizable way to exchange more than a finite number of dollars.

Thus, within this foundational framework, the Law of Total Probability should be respected in any situation in which an appropriate economic metaphor for the inferential problem under consideration results in economic loss if the Law is violated. Otherwise, the Law may be violated with impunity.

De Finetti considered the situation in which an assessor announces his probabilities for a set of propositions Ω, as well as his values for the probabilities of these propositions conditional on each proposition in ζ, a subset of Ω. He then showed that this set of assessments is coherent if all the unconditional probabilities are consistent with a single finitely additive distribution on Ω and if the multiplicative law for conditional probabilities holds. Thus, the Law of Total Probability must hold for all finite partitions. However, the Law need not apply when the partition is infinite, because an infinite number of sales and purchases would be required to take advantage of the seeming inconsistency in a nonconglomerable set of assessments. This position is also adopted by the other authors referenced in this section.

In the next two sections, I will argue to uphold the Law, even when the partition $\mathcal{H}$ is infinite. Goldstein (1983) comes to a similar conclusion about the validity of the Law and even relies on quite similar mathematics in support of his position, but his inferential interpretations of the objects under discussion is quite different from mine. In particular, he requires an assessor to evaluate a prevision for his own future previsions, the interpretation of which is problematic in terms of either my or de Finetti's economic metaphors.

3b. Prospective Probabilities and Updating Opinion

De Finetti's argument that the Law need not be respected with respect to infinite partitions is of course unimpeachable mathematically. The question is whether the economic metaphor on which the argument is based corresponds to any inferential problem of interest. The difficulty concerns the meaning of conditional probability. Do the rules for economic transactions in de Finetti's metaphor adequately translate the roles that conditional probability plays in inference?

We are now all familiar with the truism that all probabilities are conditional—on everything that the assessor believes to be true (or, more precisely, that he chooses to act as though he believes to be true). The conditional probabilities that figure in de Finetti's definition and result have a different character: they are evaluated *as though* the evaluator believes that the conditioning proposition is true, even though he *explicitly asserts* (through his "unconditional" probability assessment) that this is not the case.

What does it mean to evaluate such quantities? Bearing in mind that probability assessment is never easy, and so it makes no sense to evaluate probabilities that you do not know how to use in inferential problems, we can rephrase the question: what do you intend to *do* with these conditional measurements of belief? I know only two kinds of situations in which it is inferentially advantageous to evaluate a probability conditionally on a proposition whose truth the evaluator is unwilling to assume. The first, discussed in this section, has to do with updating opinion through time, and the second, discussed in the next section, concerns "vantage points" from which it is possible to assess probabilities for propositions that are not directly accessible to the evaluator's knowledge and experience.

To set the inferential scene in which updating takes place, suppose the assessor believes that tomorrow he will observe a quantity X whose value is informative about another quantity, Y. In addition, suppose that the assessor believes that *nothing else* he will learn in the interim should affect his opinions about Y, currently encoded in a probability distribution P. In such a situation, the assessor may choose to evaluate distributions $P(\cdot \mid X = x)$ for each possible value of the quantity X. The purpose of these assessments is to discipline and direct his response to the actual value of X when he learns it tomorrow: $P(\cdot \mid X = x)$ represents an opinion about Y that he currently *contracts* to adopt tomorrow, should he observe x, and should his belief that he will learn nothing else relevant about Y turn out to be true.

Why should he agree now to specify his future opinions? Because he knows how easy it is for the shock of the new to overwhelm previously accrued information bearing on Y, as encoded in P, and he believes that the distancing from that shock that he achieves through the prospective assessment, considering each candidate new observation in turn, can help him achieve a successful reconciliation between old and new information. Evaluations of this sort can also help the assessor to decide whether it is worth observing X at all, in terms of its effects on his opinions about Y; but in this "design" context also, $P(\cdot \mid X = x)$ must be interpreted as a contract on a future opinion. In addition, these distributions, along with P, can also be assessed *retrospectively*, if the assessor wants to guard against overreacting to an observation he has just made. Carrying out such a retrospective analysis is a reasonable response when something other than X *is* learned in the interim, with X and that "something else" playing the role that X plays in the discussion here. (Bruce Hill's paper in this volume gives a more leisurely discussion of this dynamic process of constructing and evaluating probabilities, as a continuing dialogue between an assessor's opinions, the data and the requirements of coherence).

If we regard $P(A \mid X = x)$ as a contract to adopt a particular opinion about the proposition A (referring to the value of Y) if and when X turns out to equal x, what should be the economic interpretation for this quantity? Today, before X is observed, the ticket priced at $P(A \mid X = x)$ can be reserved with no cash payment (since the opinion it represents is not in force today); if it is reserved, and $X = x$ tomorrow, the opinion is in effect and so the price must be turned over to the seller, in which case if A turns out be true, the seller must pay the buyer \$1 and otherwise nothing; while if the ticket is reserved and $X \neq x$ tomorrow, the reservation is cancelled and no money changes hands. That is: tickets are actually purchased only when the opinions they represent are in effect; an assessor willing to commit himself to an opinion in advance will accept reservations for the future purchase of sale of tickets at his prespecified price. I claim that this formulation captures in economic terms exactly what the assessor means when he evaluates $P(A \mid X = x)$ prior to observing X.

With this formulation, the Law of Total Probability must obtain. That is, suppose the assessor evaluates P on $\mathcal{F}$ (a set of propositions about Y), and, for each possible value x for X, $P(\cdot \mid X = x)$ on $\mathcal{F}$. In the economic metaphor, the bettor is allowed to buy a finite number of tickets from the P-book; and, for each x, to contract for a finite number of tickets from the $P(\cdot \mid X = x)$-book—these contracts become operative when and if x is observed. Because only one x can be observed, such a system of contracts, purchases and sales involves only a finite number of cash transactions. Unless the assessments satisfy $P(A) = \int P(A \mid X = x) d\mu(x)$ for some finitely additive distribution μ on the values of X and all propositions A in $\mathcal{F}$, there exists a system of contracts, purchases and sales as described above that will guarantee the assessor a *loss of at least* \$$c$ (for any prespecified amount c), no matter which x is observed or which of the propositions in $\mathcal{F}$ are true. Thus, in this sequential learning context, assessments that violate the Law of Total Probability are incoherent. Proofs of this assertion are presented in Lane and Sudderth (1984, 1985).

This result has implications for the examples of epistemic nonconglomerability discussed above. The nonconglomerable distributions discussed in both examples will result in sure loss if the assessor will gain information sequentially. That is, in the first example suppose that the assessor will first learn the longitude of the selected point tomorrow, and then the latitude (and hence the exact location) of the point the day after tomorrow, so Y corresponds to the location of the point on the sphere, and X to its longitude. Then assessing Y as uniform *and* Y given $X = x$ as uniform on latitude, as described by de Finetti and Hill, is incoherent. Similarly, in the second example, if the assessor first learns the order of magnitude (X) of a "randomly selected" piece of statistical data and then its first digit (Y), he would be incoherent if he assessed Y unconditionally to have the $\log_{10}(1 + 1/k)$ distribution, but conditionally on order of magnitude to have a uniform distribution on $\{1, \ldots, 9\}$.

In both cases, if the evaluator wishes to respond to X coherently, he must think harder about how X and Y are actually being generated. In the first problem, if he believes that the point is in fact selected according to the uniform distribution on the sphere, then he should revise his conditional assessments; while if he believes the selector first chooses longitude uniformly and then latitude uniformly, he should revise his unconditional assessment. In the second problem, a more careful modelling of how the large heterogeneous collection of statistical data in question was generated—which would probably force some attention on the problem of selecting units for particular classes of measurement—should persuade the assessor that for his particular problem one of the two distributions he first used was inappropriate, either conditionally or unconditionally.

3c. Putative Probabilities and Statistical Models

Suppose $\mathcal{F}$ is a set of observable propositions whose probabilities must be evaluated for some inferential problem, but are difficult to evaluate directly. Suppose also that Θ is a set of propositions that the assessor believes to be mutually inconsistent and exhaustive, and for each θ in Θ and A in $\mathcal{F}$, $P(A \mid \theta)$ is directly accessible to the assessor's knowledge and experience and so easy to assess. Suppose also that *which* θ is true is unobservable, so the interpretation of conditional probability as a contract discussed above could not be made operational (when would the money change hands?). Rather, $P(A \mid \theta)$ has a merely *putative* meaning: *if* the evaluator lived in a world in which θ is true and is otherwise like his own, what would be his degree of belief in A? That is, while he is confused about the propositions in $\mathcal{F}$ in the world he actually inhabits, the assessor can imagine how he would believe from the *vantage points* offered by the elements of Θ.

Now suppose that the assessor also evaluates his probability distribution P on the elements of $\mathcal{F}$ directly (difficult though this task may be). What relation need these evaluations bear to those conditional on the elements of Θ? Suppose a bettor is allowed to purchase or sell a finite number of P-priced tickets. Since the assessor is confident about his ability to evaluate probabilities for the elements of $\mathcal{F}$ from every θ-vantage point, and he believes that one of these vantage points gives a view of the world he actually inhabits, he may evaluate the worth of the bettor's transaction with him from each θ-vantage point in turn. That is, if L represents his loss from the transaction, he can evaluate with confidence $E(L \mid \theta)$ for each θ in Θ. If *each* of these is greater than \$$c$, then, whichever θ, coupled with the rest of his general background information, actually describes the world in which the assessor lives, from the vantage point of that world the assessor faces a loss of at least \$$c$.

This anticipated loss represents an inconsistency between his *direct* assessment P and his assessments from the vantage points in Θ. It is proved in Lane and Sudderth (1984) that this anticipated loss can be avoided if and only if there is a finitely additive probability distribution μ on Θ such that $P(A) = \int P(A \mid \theta) d\mu(\theta)$ for all A in $\mathcal{F}$; that is, if the (difficult) unconditional assessment P is linked to the (accessible) conditional assessments $\{P(\cdot \mid \theta)\}$ by the Law of Total Probability.

This result responds to a question raised by Piccinato in his published comment to Hill (1980). He asked

> How to deal with statistical models if we must get rid of conglomerability? Of course the ground for accepting or refusing conglomerability (or complete additivity) is a logical one, and must be independent from the mentioned question.... For 'statistical model' I mean as usual a set of

> probability distributions on the space of possible outcomes, possibly indexed by a parameter whose actual value is unknown.... Can we live with nonconglomerability without giving up statistical models?

Of course, some statistical models have attributes like the set Θ discussed above: the elements in the parameter space can be interpreted as propositions about the world, one of which the observer believes to be true, and it is possible to confidently assess putative probabilities for observables from the vantage point of each element of the model. When these conditions are satisfied, the logical ground that Piccanato seeks is the relevance of the evaluations of the anticipated loss from the vantage point of each parameter value as defined above; to avoid that loss, conglomerability must hold for the probabilities of observables with respect to their putative probabilities evaluated given parameter values.

3d. Some Objections and Answers

In the previous two subsections, I have argued that in the situations in which conditional probabilities play a role in inference, as prospective probabilities and putative probabilities, their evaluation should respect the Law of Total Probability with respect to finitely additive distributions μ. This point of view is criticized by Kadane, Schervish and Seidenfeld (1986), who take the position that

> ...the attempt to modulate nonconglomerability is misguided. If nonglomerability is necessary for 'consistency', then nothing less than countably additive distributions suffice with denomerable partitions, and even 'proper' priors may suffer nonconglomerability in non-denumerable partitions... .

I disagree with their position for two reasons. First, I can think of no inferential problem in which it is reasonable to evaluate probabilities conditional on all possible events, and I believe that consistency tests should be applied only to the probabilities that play a role in particular inferential problems. Secondly, I do not understand the relevance of the "called-off bet" economic formulation of conditional probability, with purchase price handed over before the opinion represented by the conditional probability is relevant. Rather, I believe that the actual role that conditional probabilities play in inference should be the basis of their economic interpretation, and I have attempted to isolate two such roles and provide their respective (and different) interpretations in the two previous sections.

Several authors have objected to the criterion of anticipated loss as a condition for coherence developed in the previous section. I share this objection when it is applied to situations in which a statistical model has been adopted "for convenience", with no presumption that probabilities given θ really describe the evaluator's putative beliefs in a world he believes might be his own and in which he can comfortably measure his uncertainty. In such cases, I cannot interpret probabilities given θ at all, much less decide how consistent they are with probabilities that actually do measure some of the evaluator's beliefs. But if putative probabilities are available as described in the previous section, then evaluating expectations from the same vantage points seems reasonable and meaningful; and if I face a \$c loss in all possible worlds, than in particular I do *in this one* as well.

References

Blackwell, D. and L. Dubins (1975). "On Existence and Non- Existence of Proper, Regular, Conditional Distributions," *Annals of Probability* Vol. 3, 741-752.

de Finetti, B. (1972). *Probability, Induction and Statistics.* New York: Wiley.

de Finetti, B. (1974). *Theory of Probability, Volume 1.* New York: Wiley. (English translation, first published in Italian, 1970).

Dubins, L. (1977). "Measurable, Tail Disintegrations of the Haar Integral Are Purely Finitely Additive," *Proceedings of the American Mathematical Society*, Vol. 62, 34-36.

Dubins, L. and D. Heath (1983). "With Respect to Tail Sigma- Fields, Standard Measures Possess Measurable Disintegrations," *Proceedings of the American Mathematical Society*, Vol. 88, 416-418.

Goldstein, M. (1983). "The Prevision of a Prevision," *Journal of the American Statistical Association*, Vol. 78, 817-819.

Hill, B. (1980). " On Some Statistical Paradoxes and Nonconglomerability" (with discussion), in *Bayesian Statistics: Proceedings of the First International Meeting Held in Valencia (Spain)*, J.M. Bernardo, M.H. DeGroot, D.V. Lindley, and A.F.M. Smith, eds., Valencia, Spain: University Press, 39-69.

Hill, B. and D. Lane (1985). "Conglomerability and Countable Additivity," *Sankhya*, Vol. 47, 366-379.

Kadane, J., M. Schervish and T. Seidenfield (1986). "Statistical Implications of Finitely Additive Probability," in *Bayesian Inference and Decision Techniques, Essays in Honor of Bruno de Finetti*, P. Goel and A. Zellner, eds., Amsterdam: Elsevier, 59-78.

Lane, D. and W. Sudderth (1984). "Coherent Predictive Inference," *Sankhya*, Vol. 46, 166-185.

Lane, D. and W. Sudderth (1985). "Coherent Predictions Are Strategic," *Annals of Statistics*, Vol. 13, 1244-1248.

Maitra, A. and S. Ramakrishnan (1988). "Factorization of Measures and Normal Conditional Distributions," *Proceedings of the AMS*, in press.

Regazzini, E. (1987). "De Finetti's Coherence and Statistical Inference," *Annals of Statistics*, Vol. 15, 845-864.

Schervish, M., T. Seidenfeld and J. Kadane (1984). "The Extent of Nonconglomerability of Finitely Additive Probabilities," *Zeitschrift fur Wahrscheinlichkeitstheorie*, Vol. 66, 205-226.

Scozzafava, R. (1981). "Nonconglomerability and the First Digit Problem," *Statistica*, Vol. 41, 561-565.

Scozzafava, R. (1984). "A Survey of Some Common Misunderstandings Concerning the Role and Meaning of Finitely Additive Probabilities in Statistical Inference," *Statistica*, Vol. 44, 21-45.

Seidenfeld, T. (1985). "Coherence, 'Improper Priors', and Finite Additivity," *Econometric Reviews*, Vol. 4, 81-91.

IV. SELECTED APPLICATIONS

Bayesian and Likelihood Methods in Statistics and Econometrics
S. Geisser, J.S. Hodges, S.J. Press and A. Zellner (Editors)

ON BAYES SMOOTHING IN A TIME VARYING REGRESSION MODEL

Siddhartha Chib, S. Rao Jammalamadaka, and Ram C. Tiwari
University of Missouri, University of California, and University of North Carolina

Abstract

This paper considers the Bayes smoothing problem for a time varying regression model with unknown noise variance parameters. Using the prediction error decomposition, an approach is developed that provides the exact pdf of the regression coefficients and the noise variance parameters. In addition, the paper provides a discussion, and an example, of Monte Carlo numerical integration procedures that can be used to generate estimates of the moments of the smoothed coefficients.

1. Introduction

In this paper we are concerned with the smoothing problem in a bivariate time varying regression (TVR) model when the variance parameters of the error terms are unknown. The model we consider is the following

$$y_t = x_t\beta_t + \epsilon_t, \qquad \beta_t = g_t\beta_{t-1} + \eta_t \qquad (t = 1, \ldots, n) \tag{1}$$

where ϵ_t are iid $N(0, \sigma^2)$, the η_t are iid $N(0, \lambda\sigma^2)$, $(\lambda, \sigma^2 > 0)$, and ϵ_t and η_s are uncorrelated for all t and s. We assume that $\{g_t\}$ is a fixed, nonstochastic sequence of scalars. We further assume that the two variance parameters, σ^2 and λ, are unknown.

An important issue in such models has been the choice of β_0, the initial value at start up time 0. We assume that the parameter β_0 is unknown with its uncertainty expressed by a normal distribution $N(\hat{\beta}_{0|0}, \sigma^2 R_{0|0})$, where the hyperparameters, $\hat{\beta}_{0|0}$ and $R_{0|0}$, are known.

The smoothing problem for models similar to (1) above have been analyzed by several authors including Anderson and Moore (1979), Chow (1983), and Engle and Watson (1988), using classical methods, and by Broemeling (1985) and Broemeling, Diaz and Yusoff (1985) using Bayesian methods.

In this paper, we provide a Bayes solution to the smoothing problem. Essentially the problem of smoothing is concerned with making inferences about the $n \times 1$ regression parameter, $\beta = (\beta_1, \ldots, \beta_n)$ given an available data record $y^n = (y_1, \ldots, y_n)$. In order to solve this problem when the noise variance parameters $\theta = (\sigma^2, \lambda) : 2 \times 1$ are unknown we develop a two step Bayesian procedure that involves deriving the posterior pdf of β given θ and y^n and the marginal posterior pdf of θ given y^n. Then the posterior pdf of β given y^n is derived by integrating out θ using Monte Carlo (MC) integration techniques (cf. Geweke, 1987, Zellner, van Dijk and Bauwens, 1988). Note that the marginal posterior pdf of θ is given by

$$\pi(\theta|y^n) = \pi(\theta) \cdot \ell_{obs}(\theta|y^n) / \int \pi(\theta)\ell_{obs}(\theta|y^n)d\theta, \tag{2}$$

where $\pi(\theta)$ is the prior of θ at $t = 0$, and $\ell_{obs}(\theta|y^n)$, is the observed likelihood function which is computed using the prediction error decomposition (PED) of Schweppe (1965) and Harvey (1981). The PED relies on one pass of the Kalman filter (KF) recursion of β_t given θ.

As will become evident below, a critical aspect of our procedure, as in any Bayesian method, is the formulation of a suitable prior of σ^2 and λ. Although, it would be useful to have a prior that yields closed form results no such prior of λ is available. We recommend adopting a noninformative prior since in practice variance parameters are usually the most difficult to specify.

The plan of the paper is as follows. In Section 2 we briefly review the prediction error decomposition, providing results for the filtering problem when θ is known. In Section 3 we derive the main results including the posterior pdf of θ, and the unconditional posterior of β given y^n. Section 4 contains a numerical example that illustrates the methods developed in the paper, while Section 5 provides some concluding comments.

2. The Prediction Error Decomposition

Unlike the previous Bayesian treatments of this model, our approach makes considerable use of the prediction error decomposition (PED) and a particular vector-matrix representation of (1). The PED is described in this section, and the vector matrix representation in Section 3.

First, a few words about the notations used. The mean and variance of β_t given information up to time s are denoted by $\hat{\beta}_{t|s}$ and $\hat{\sigma}^2_{t|s}$ respectively, i.e.,

$$\hat{\beta}_{t|s} = E(\beta_t|y^s); \quad \hat{\sigma}^2_{t|s} = E((\beta_t - \hat{\beta}_{t|s})^2|y^s) \qquad s = 1, \ldots, n. \tag{3}$$

Note that when parameter $\lambda(\theta)$ is also involved in the conditioning in (3), $\hat{\beta}_{t|s}$ and $\hat{\sigma}^2_{t|s}$ are denoted by $\hat{\beta}_{t|s,\lambda}(\hat{\beta}_{t|s,\theta})$ and $\hat{\sigma}^2_{t|s,\lambda}(\hat{\sigma}^2_{t|s,\theta})$ respectively.

Likewise the mean and variance of y_t given information up to time s is given by $\hat{y}_{t|s}$ and $\hat{f}_{t|s}$, respectively, i.e.,

$$\hat{y}_{t|s} = E(y_t|y^s); \quad \hat{f}_{t|s} = E((y_t - \hat{y}_{t|s})^2|y^s), \quad s = 1, \dots, n. \tag{4}$$

In this context, smoothing is the evaluation of $\hat{\beta}_{t|n}, t = 1, 2, \dots, n$, and filtering is the evaluation of $\hat{\beta}_{t|t}, t = 1, 2, \dots, n$.

Under the assumptions of the model, it is now possible to derive the conditional distributions of β_t and y_t, given information up to $t-1$ and θ. In particular, if we suppose that the posterior distribution of β_{t-1} given y^{t-1} and θ is

$$\beta_{t-1}|y^{t-1}, \theta \sim N(\hat{\beta}_{t-1,t-1,\lambda}, \quad \sigma^2 R_{t-1|t-1,\lambda}),$$

then we can derive, using (1), that

$$\beta_t|y^{t-1}, \theta \sim N(\hat{\beta}_{t|t-1,\lambda}, \quad \sigma^2 R_{t|t-1,\lambda}) \tag{5}$$

and

$$y_t|y^{t-1}, \theta \sim N(\hat{y}_{t|t-1,\lambda}, \quad \sigma^2 \tilde{f}_{t|t-1,\lambda}), \tag{6}$$

where

$$\hat{\beta}_{t|t-1,\lambda} = g_t \hat{\beta}_{t-1|t-1,\lambda} \quad ; \quad R_{t|t-1,\lambda} = g_t^2 R_{t-1|t-1,\lambda} + \lambda,$$

and

$$\hat{y}_{t|t-1,\lambda} = x_t \hat{\beta}_{t|t-1,\lambda} \quad ; \quad \tilde{f}_{t|t-1,\lambda} = x_t^2 R_{t|t-1,\lambda} + 1.$$

Using (5), (6) and the Bayes theorem,, the conditional distribution of β_t given y^t and θ is

$$\beta_t|y^t, \theta \sim N(\hat{\beta}_{t|t,\lambda}, \quad \hat{\sigma}^2_{t|t,\theta} = \sigma^2 R_{t|t,\lambda}), \tag{7}$$

where

$$\hat{\beta}_{t|t,\lambda} = \hat{\beta}_{t|t-1,\lambda} + R_{t|t-1,\lambda} x_t \tilde{f}^{-1}_{t|t-1,\lambda}(y_t - x_t \hat{\beta}_{t|t-1,\lambda})$$

$$R_{t|t,\lambda} = [R_{t|t-1,\lambda} - R^2_{t|t-1,\lambda} x_t^2 (\tilde{f}_{t|t-1,\lambda})^{-1}] \tag{8}$$

are the Kalman filtering updating equations for the mean and variance of β_t. Notice that in (8), σ^2 is not involved in the estimate of β_t.

With these results, the likelihood function of θ can be expressed in terms of the conditional distribution of $y_t|y^{t-1}, \theta, \quad t = 1, \dots, n$, using the multiplicative theorem of total probability as:

$$\begin{aligned}\ell(\theta|y^n) &\propto f(y_1,\ldots,y_n|\theta)\\&\propto f(y_1|\theta)f(y_2|y^1,\theta)\ldots f(y_n|y^{n-1},\theta)\\&\propto \left(\frac{1}{\sigma^2}\right)^{n/2}\prod_{t=1}^{n}[\tilde{f}_{t|t-1,\lambda}]^{-\frac{1}{2}}e^{-\frac{1}{2\sigma^2}\sum_{t=1}^{n}(y_t-\hat{y}_{t|t-1,\lambda})^2/\tilde{f}_{t|t-1,\lambda}}\end{aligned} \tag{9}$$

where $\tilde{f}_{t|t-1,\lambda}$ and $\hat{y}_{t|t-1,\lambda}$ are produced by the KF, for each value of λ. It is clear from (9) that if we let the prediction error be denoted by $\tilde{y}_{t|t-1,\lambda}$ where

$$\tilde{y}_{t|t-1,\lambda} = y_t - \hat{y}_{t|t-1,\lambda}, \tag{10}$$

then the likelihood function is produced in terms of the distribution of the prediction errors. It is for this reason that (9) is referred to as the prediction error decomposition.

REMARK 1: In classical analysis, estimation of the parameter, θ, is accomplished by maximizing the likelihood function. The usual procedure is to fix β_0, and then adopt a Newton-Raphson algorithm to successively update the parameter values until the log-likelihood is maximized. Unless the likelihood function is well behaved, such a procedure can sometimes locate the wrong maxima, especially when the sample size is small.

3. The Smoothing Problem

3.1 Posterior of θ

In order to make inferences about the vector of regression parameters $\beta = (\beta_1,\ldots,\beta_n) : n\times 1$ when the variance parameter θ is unknown, we need to derive the marginal posterior of θ and the conditional posterior of β given y^n and θ. Then the unconditional posterior pdf of β is obtained by integrating out θ from the latter distribution, i.e.,

$$\pi(\beta|y^n) = \int_\theta \pi(\beta|y^n,\theta)\pi(\theta|y^n)d\theta \tag{11}$$

where $\pi(\theta|y^n)$ is the posterior pdf of θ specified in (2).

It is the purpose of this subsection to show that the posterior pdf of θ in (11) can be derived using the PED just described. It is, of course, necessary to specify a useful prior for the variance parameter θ that would allow the integration in (11) to be performed conveniently. It turns out that an informative Inverted Gamma pdf of σ^2 with known hyperparameters combines nicely with the likelihood function. However, as far as the second variance parameter, λ, is concerned no useful prior seems available. The best practical strategy appears to be to use some locally uniform prior pdf of λ. We prefer the latter Bayesian method to the alternative of specifying λ and ignoring the prior uncertainty about its value.

In the light of the previous comments, we assume that σ^2 and λ are *a priori* independent, with the prior of σ^2 an informative, Inverted Gamma($\frac{v^*}{2}, \frac{\delta^*}{2}$) density with shape parameter $\frac{v^*}{2}$ and scale parameter $\frac{\delta^*}{2}$, and that of λ specified in general form as $\pi(\lambda)$, where $\pi(\lambda)$ is possibly some locally uniform pdf. Thus, the joint prior density of (σ^2, λ) is given by

$$\pi(\sigma^2,\lambda) \propto (\sigma^2)^{\frac{v^*}{2}+1}e^{-\frac{\delta^*}{2\sigma^2}} \cdot \pi(\lambda), \quad \sigma^2,\lambda > 0, \tag{12}$$

v^* and $\delta^* > 0$ known. If we combine this prior with the observed likelihood function in (9) and apply (2), then using (10) the posterior pdf of (σ^2, λ) is

$$\pi(\sigma^2,\lambda|y^n) \propto \pi(\sigma^2,\lambda)\ell_{obs}(\sigma^2,\lambda|y^n)$$

$$\propto \pi(\lambda)\left[\prod_{t=1}^{n}(\tilde{f}_{t|t-1,\lambda})^{-\frac{1}{2}}\right](1/\sigma^2)^{\frac{v^*+n}{2}+1}e^{-\frac{1}{2\sigma^2}(\delta^*+\sum_{t=1}^{n}(\tilde{y}_{t|t-1,\lambda})^2/\tilde{f}_{t|t-1,\lambda})} \tag{13}$$

where the normalizing constant is $1/\int \pi(\theta)\ell_{obs}(\theta|y^n)d\theta$. Observe that the posterior pdf in (13) can be broken up into two pieces as

$$\pi(\sigma^2,\lambda|y^n) = \pi(\sigma^2|y^n,\lambda)\cdot\pi(\lambda|y^n),$$

where the conditional posterior pdf of σ^2 given λ and y^n is an updated Inverted Gamma
($\frac{v^*+n}{2}, \frac{\delta_\lambda^{**}}{2}$) density with $\delta_\lambda^{**} = \delta^* + \sum_{t=1}^{n}\tilde{y}_{t|t-1,\lambda}^2/\tilde{f}_{t|t-1,\lambda}$.

The second piece, the marginal posterior of λ given y^n, is obtained by integrating out σ^2 from (13). Thus we have the following.

THEOREM 1: *Let the prior of (σ^2, λ) be as in (12). Then the conditional posterior distribution of σ^2 given y^n and λ is*

$$\sigma^2|y^n,\lambda \sim \textit{Inverted Gamma}\left(\frac{v^*+n}{2},\frac{\delta_\lambda^{**}}{2}\right), \tag{14}$$

and the marginal posterior pdf of λ given y^n is

$$\pi(\lambda|y^n) \propto \pi(\lambda)\left[\prod_{t=1}^{n}\tilde{f}_{t|t-1,\lambda}^{-1/2}\right]\left(\delta^*+\sum_{t=1}^{n}(\tilde{y}_{t|t-1,\lambda})^2/\tilde{f}_{t|t-1,\lambda}\right)^{(-v^*+n)/2}. \tag{15}$$

REMARK 2: If we wish to obtain the posterior pdf's when the prior of σ^2 is diffuse and is given by Jeffreys pdf: $\pi(\sigma^2) \propto 1/\sigma^2$, all that has to be done is to set $v^* = 0$ and $\delta^* = 0$ in (12)-(15).

REMARK 3: Unfortunately the posterior of λ in (15), for any prior $\pi(\lambda)$, does not belong to a known parametric family of densities. Nonetheless, (15) can be analyzed numerically, and the normalizing constant determined using Monte Carlo integration procedures. The inverse of the normalizing constant, as stated before, is the integral of the RHS of (15) over λ. In any given application, it will be necessary to approximate $\pi(\lambda|y^n)$ by a parametric density. Approximating pdf's that are symmetric are probably not good choices in small samples since in the example discussed in the next section, the pdf of $\lambda|y^n$ is shown to be skewed.

3.2 The Conditional Smoothing Problem

We now derive the posterior distribution of β given y^n and θ using a vector-matrix representation of (1) that allows us to appeal to known results from Bayes analysis of linear models. The vector of smoothed estimates of β given θ is then the posterior mean of this distribution.

If we let $\varepsilon = (\varepsilon_1, \varepsilon_2, \ldots, \varepsilon_n)' : n \times 1$, $\eta = (\eta_1, \eta_2, \ldots, \eta_n)' : n \times 1$, and $X = diag(x_1, \ldots, x_n) : n \times n$, equation (1) can be expressed in vector-matrix form as

$$y^n = X\beta + \varepsilon, \quad \varepsilon \sim N_n(0, \sigma^2 I_n) \tag{16}$$

with the evolution equation of the regression parameters summarized as

$$H\beta = e\beta_0 + \eta, \qquad \eta \sim N_n(0, \lambda\sigma^2 I_n), \tag{17}$$

where the matrix H is given in the simple form:

$$H = \begin{bmatrix} 1 & 0 & & 0 \\ -g_2 & 1 & & 0 \\ 0 & -g_3 & 1 & 0 \\ \cdot & & & \cdot \\ \cdots & & & \cdots \\ 0 & 0 & -g_n & 1 \end{bmatrix} : n \times n,$$

and $e = (g_1, 0, \ldots, 0)$ is an $n \times 1$ vector. Under this representation of the TVR model, it is relatively straightforward to derive the posterior pdf of β given y^n and θ. Equation (17), along with the assumption that the distribution of β_0 is $\beta_0|\theta \sim N_k(\hat{\beta}_{0|0}, \sigma^2 R_{0|0})$ imply that the prior of β given θ is

$$\beta|\theta \sim N_n(\beta^* = H^{-1}e\hat{\beta}_{0|0}, A_\theta^{*-1} = \sigma^2 \tilde{A}_\lambda^{*-1})$$

where $\tilde{A}_\lambda^{*-1} = H^{-1}eR_{0|0}e'H^{-1'} + \lambda H^{-1}H^{-1'}$. Consequently using standard calculations from Bayes analysis of linear models, we have that the distribution of β given y^n and θ is multivariate normal with mean vector $\hat{\beta}_\lambda : n \times 1$, and variance matrix $\sum_\theta : n \times n$, i.e.,

$$\beta|y^n, \theta \sim N_n(\hat{\beta}_\lambda, \Sigma_\theta) \tag{18}$$

where the posterior mean is the matrix weighted average

$$\hat{\beta}_\lambda = (\tilde{A}_\lambda^* + X'X)^{-1}(\tilde{A}_\lambda^*\beta^* + X'y^n), \tag{19}$$

and the posterior variance, which does not depend on y^n, is

$$\Sigma_\theta = \sigma^2(\tilde{A}_\lambda^* + X'X)^{-1} \equiv \sigma^2\Sigma_\lambda. \tag{20}$$

The smoothed estimate of β, given θ, is given by (19). This estimator of β is, optimal under the squared error loss function. Note that the smoothed value of β, i.e., $\hat{\beta}_\lambda$ does not depend on σ^2.

The marginal posterior density of β given λ and y^n can be obtained using (14), and is given by

$$\pi(\beta|y^n, \lambda) = \int_{\sigma^2>0} \pi(\beta|y^n, \sigma^2, \lambda)\pi(\sigma^2|y^n, \lambda)d\sigma^2,$$

where from (18), $\pi(\beta|y^n, \sigma^2, \lambda)$ is the pdf of a $N_n(\hat{\beta}_\lambda, \sum_\theta)$ distribution. Hence

$$\begin{aligned}\pi(\beta|y^n, \lambda) &\propto \int_{\sigma^2>0} (1/\sigma^2)^{\frac{v*+2n}{2}+1} e^{-\frac{1}{2\sigma^2}[(\beta-\hat{\beta}_\lambda)'\Sigma_\lambda^{-1}(\beta-\hat{\beta})+\delta_\lambda^{**}]} d\sigma^2 \\ &\propto [\delta_\lambda^{**} + (\beta - \hat{\beta}_\lambda)'\Sigma_\lambda^{-1}(\beta - \hat{\beta}_\lambda)]^{-(v^*+2n)/2} \end{aligned} \tag{21}$$

which is the kernel of a multivariate t density with mean $\hat{\beta}_\lambda$, dispersion $\tilde{\delta}_\lambda^{**} \sum_\lambda$, and $v^* + n$ degrees of freedom, where $\tilde{\delta}_\lambda^{**} \equiv \delta_\lambda^{**}/(v * +n)$.

3.3 The Unconditional Smoothing Problem

At this point, the marginal posterior of β can be derived using the results in Sections (3.1) and (3.2). Since the marginal posterior of λ, and the conditional posterior of β given λ, is available, the unconditional posterior of β is obtained as

$$\pi(\beta|y^n) = \int_\lambda \pi(\beta|y^n, \lambda)\pi(\lambda|y^n)d\lambda$$

where $\pi(\lambda|y^n)$ is given in (15) and $\pi(\beta|y^n, \lambda)$ in (21). Although the above integral cannot be performed analytically, we can state the following result.

THEOREM 2: *Let $\pi(\lambda|y^n)$ be as in (15). Then in the TVR model, the marginal posterior distribution of β is*

$$\pi(\beta|y^n) \propto \int \left[1 + \frac{1}{\tilde{\delta}_\lambda^{**}(v^* + n)}(\beta - \hat{\beta}_\lambda)'\Sigma_\lambda^{-1}(\beta - \hat{\beta}_\lambda)\right]^{-(v^*+2n)/2} \cdot \pi(\lambda|y^n)d\lambda. \quad (22)$$

REMARK 4: If one is interested in the pdf of some subvector of β, for example, the most current value β_n, the posterior is obtained as follows. If we denote the appropriate subcomponents of $\hat{\beta}_\lambda$ and $\sum_\lambda$ by $\hat{\beta}_{n,\lambda}$ and $\sum_{n,\lambda}$ respectively, then from (21) and Theorem 2, the pdf of β_n is

$$\pi(\beta_n|y^n) \propto$$
$$\int_\lambda \left[1 + \frac{1}{\tilde{\delta}_\lambda^{**}(v^* + n)}(\beta_n - \hat{\beta}_{n,\lambda})'\Sigma_{n,\lambda}^{-1}(\beta_n - \hat{\beta}_{n,\lambda})\right]^{-(1+v^*+n)/2} \cdot \pi(\lambda|y^n)d\lambda. \quad (23)$$

An important issue at this stage is how the pdf in (22), and also (23), should be analyzed. The most effective method is to use Monte Carlo (MC) integration procedures that have been recently applied by Zellner and Rossi (1984), Zellner, Van Dijk and Bauwens (1988), and Chib, Tiwari and Jammalamadaka (1988), among others. The general method is developed in Klock and van Dijk (1978), and Geweke (1987). Application of that methodology to the specific problem at hand can proceed as follows. Suppose we are interested in obtaining the unconditional moments of the elements of β, for example the mean and variance. We can use well known formulas to compute marginal moments from conditional moments. For example, to compute the unconditional mean of β, we have

$$\hat{\beta} = E^{\pi(\beta|y^n)}[\beta] = \int_\beta \beta\pi(\beta|y^n)d\beta$$
$$= \int_\lambda \left[\int_\beta \beta\pi(\beta|y^n, \lambda)d\beta\right] \pi(\lambda|y^n)d\lambda = \int_\lambda \hat{\beta}_\lambda \pi(\lambda|y^n)d\lambda. \quad (24)$$

This can be computed numerically as follows. Let $h(\lambda)$ be an *importance density* with the property that it closely approximates $\pi(\lambda|y^n)$. Further $h(\lambda)$ should be a pdf *from which it is relatively easy to simulate.* Let $\lambda_1, \ldots, \lambda_N$ be N random draws of λ from $h(\lambda)$ where N is a suitably large number. Then the MC estimate of (24) is

$$\hat{\beta} \simeq \frac{1}{N}\sum_{i=1}^{N} \hat{\beta}_{\lambda_i} \frac{\pi(\lambda_i|y^n)}{h(\lambda_i)}, \quad (25)$$

where $\hat{\beta}_{\lambda_i}$ is $\hat{\beta}_\lambda$ evaluated at $\lambda = \lambda_i$, and $\pi(\lambda_i|y^n)$ is the posterior of λ evaluated at $\lambda = \lambda_i$, i.e.,

$$\pi(\lambda_i|y^n) =$$

$$\frac{\pi(\lambda_i)(\Pi_{t=1}^n \tilde{f}_{t|t-1,\lambda_i}^{-1/2})(\delta^* + \sum_{t=1}^n \tilde{y}_{t|t-1,\lambda_i}^2/\tilde{f}_{t|t-1,\lambda_i})^{-(v^*+n)/2}}{\frac{1}{N}\sum_{i=1}^N \pi(\lambda_i)(\Pi_{t=1}^n \tilde{f}_{t|t-1,\lambda_i}^{-1/2})(\delta^* + \sum_{t=1}^n \tilde{y}_{t|t-1,\lambda_i}^2/\tilde{f}_{t|t-1,\lambda_i})^{-(v^*+n)/2}/h(\lambda_i)}. \quad (26)$$

Note that the denominator of (26) is the MC estimate of the normalizing constant of $\pi(\lambda|y^n)$.

Now the marginal covariance matrix of β is defined as the sum of the expectation of the conditional variance and the variance of the conditional expectation, i.e.,

$$\sum = V^{\pi(\beta|y^n)}[\beta] = \int_\lambda \frac{v^* + n}{(v^* + n - 2)} \tilde{\delta}_\lambda^{**} \Sigma_\lambda \pi(\lambda|y^n) d\lambda$$

$$+ \int_\lambda (\hat{\beta}_\lambda - \hat{\beta})(\hat{\beta}_\lambda - \hat{\beta})' \quad \pi(\lambda|y^n) d\lambda. \quad (27)$$

where $\hat{\beta}$ is defined in (24). Each of these terms may be approximated by sums as in (25).

When we consider the method described above, it is clear that we have to find a suitable importance density $h(\lambda)$. If the underlying pdf of λ were symmetric, it may be possible to let the importance density be a student t density with parameters suitably chosen. This strategy will not work when the pdf of λ is nonsymmetric, as in the example in the next section. Although we have not been able to test this assertion in a wide array of examples, we are of the opinion that the pdf of λ is nonsymmetric, unless the sample size is large. Thus, symmetric importance functions are unlikely to be adequate for this problem.

4. An Example

In this section we report some numerical results from a Monte Carlo experiment. The idea is to illustrate how the methods described in the paper work in practice.

The data for the experiment is generated according to the model

$$y_t = x_t\beta_t + \varepsilon_t, \qquad \varepsilon_t \sim N(0,1),$$

$$\beta_t = .5\beta_{t-1} + \eta_t, \qquad \eta_t \sim N(0, 0.5), \qquad t = 1, 2, \ldots, 15 \quad (28)$$

with $\beta_0 \sim N(.9, 1)$, and x_t distributed as iid Uniform(0, 1) random variables. In terms of the model of the paper, we have set $\hat{\beta}_{0|0} = .9$; $\sigma^2 = 1$; $\lambda = 0.5$; and $R_{0|0} = 1$. Fifteen observations are generated from model (28).

Using equation (13), and a prior of (σ^2, λ) that is proportional to $1/\sigma^2$, the joint posterior pdf of σ^2, λ and λ is reproduced in Figure 1. It is clear that the joint pdf has the appropriate shape with most probability concentrated in the parameter space that generated the data.

Figure 1: Joint Posterior Density Function

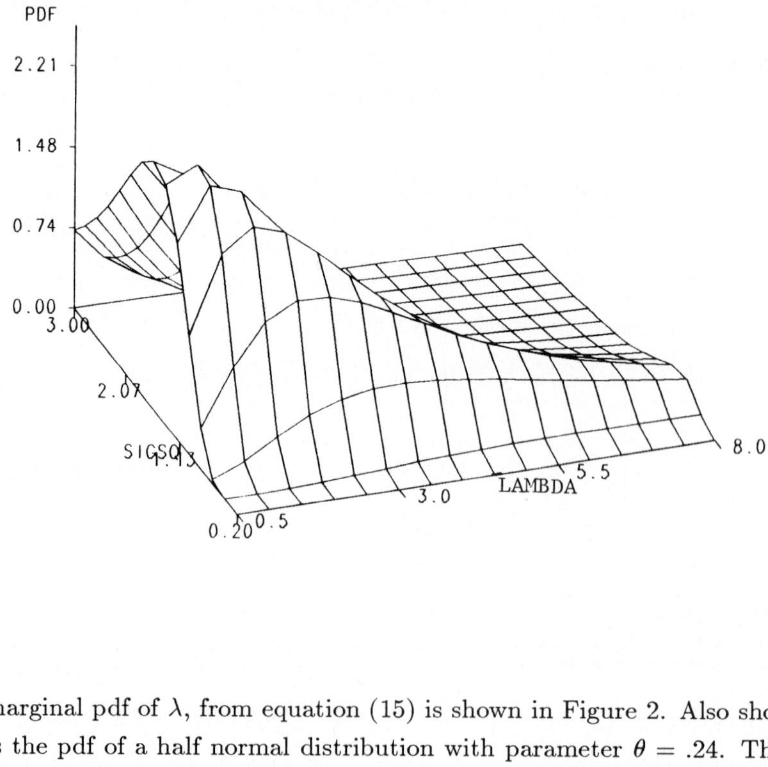

The marginal pdf of λ, from equation (15) is shown in Figure 2. Also shown in Figure 2 is the pdf of a half normal distribution with parameter $\theta = .24$. This pdf has the form

$$h(\lambda) = \frac{2\theta}{\pi} e^{-\theta^2 \lambda^2/\pi}, \qquad 0 \leq \lambda < \infty, \tag{29}$$

and appears to be an appropriate importance density.

Note that the pdf of λ is skewed with fairly heavy tails, and a pronounced peak in the region around $\lambda = 0.5$.

Using the half normal importance density specified in (29) we have computed the conditional and unconditional smoothed estimates of β as described in (25) and (26).

Figure 2: Posterior PDF of Lambda and Half Normal Approximation

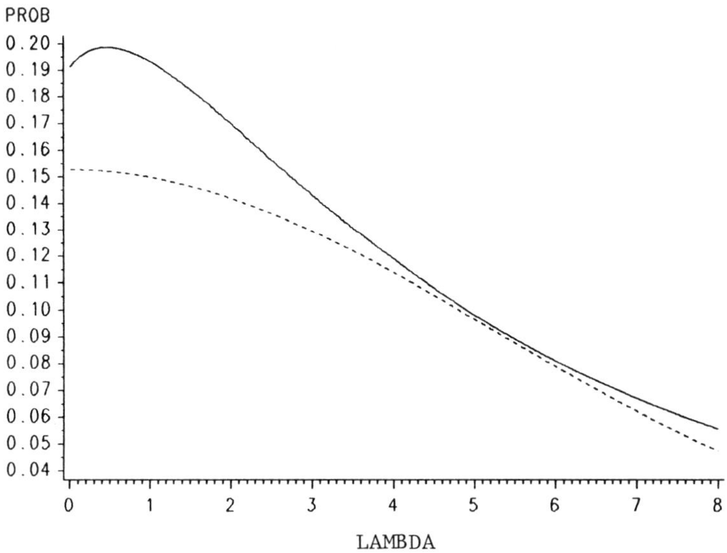

Legend: $\pi(\lambda \mid y^n)$: ———————— ; $h(\lambda)$: – – – – –.

[The normalizing constant of $\pi(\lambda \mid y^n)$ is found by Monte Carlo integration. The importance function is the uniform density on $(0, 6)$ which was used to produce 500 random draws of λ.]

At the suggestion of a referee, the estimates are calculated from repeated samples, not from one Monte Carlo run. By keeping β_t and x_t fixed at the values in the previous examples, 30 samples of size 15 are generated from the model in (28). *The estimates reported in Table 1 are averages of the estimates from the* 30 *samples.* Also reported is the conditional variance of β (i.e., diag $\sigma^2 \sum_\lambda$) with $\sigma^2 = 1$ and $\lambda = 0.5$, and the unconditional variance of β using (27). The results are summarized in Table 1 which has seven columns. The first consists of the true values of the components of β, and second contains the filtered estimates using $\lambda = 0.5$. The third and fourth provide the conditional smoothed estimates of β (i.e., $\hat{\beta}_\lambda$), and the conditional variance of β. The last two columns contain the unconditional estimates of β and its variance.

In interpreting the evidence in Table 1, we should first note the satisfactory performance of the unconditional estimate. On graphing in Figure 3, the absolute

Table 1

obs	β	$\hat{\beta}_{t\|t,\lambda=0.5}$	$R_{t\|t,\lambda=0.5}$	$\hat{\beta}_{\lambda=0.5}$	$diag\sum_\theta$	$\hat{\beta}$	$diag\sum$
1	1.14560	0.72006	0.43161	0.63597	0.42027	0.80254	0.54337
2	-0.49108	0.26409	0.55852	0.12317	0.51788	-0.05372	1.20456
3	-0.71234	-0.15107	0.44494	-0.19072	0.42641	-0.46455	0.72449
4	-0.88160	-0.18723	0.47735	-0.18449	0.47137	-0.39093	0.85193
5	-0.57651	-0.11087	0.58570	-0.08649	0.57908	-0.20567	1.49513
6	0.58402	-0.07145	0.62570	-0.00162	0.61419	-0.00878	1.76890
7	0.37801	-0.04192	0.65443	0.11079	0.60574	0.23466	1.81598
8	1.47084	0.27726	0.50172	0.28875	0.46337	0.61048	0.84762
9	0.45402	0.14029	0.39500	0.16727	0.38702	0.18634	0.56644
10	0.22205	0.16462	0.52929	0.15194	0.52534	0.32790	1.22710
11	1.15013	0.08290	0.63229	0.05201	0.60979	0.12943	1.85377
12	-0.19034	0.06482	0.59747	-0.02284	0.56058	-0.00588	1.36721
13	-0.59954	-0.14579	0.48406	-0.15813	0.47502	-0.34160	0.83063
14	0.26274	-0.08539	0.61649	-0.10455	0.56152	-0.20503	1.58431
15	-0.24382	-0.08335	0.40660	-0.08335	0.40660	-0.10558	0.59966

Note: The estimates reported in the table are averages calculated from 30 samples of size 15. The number of replications from the half normal pdf in (29) is $N = 500$. The joint prior of λ and σ^2 is taken to be proportional to $1/\sigma^2$. Thus in the calculations, $\upsilon* = 0$ and $\delta^* = 0$.

deviations of the estimates from the true parameter, we can conclude that the unconditional estimate is, in general, closer to the true value than both the filtered and conditional estimates.

Figure 3: Absolute Parameter Deviations

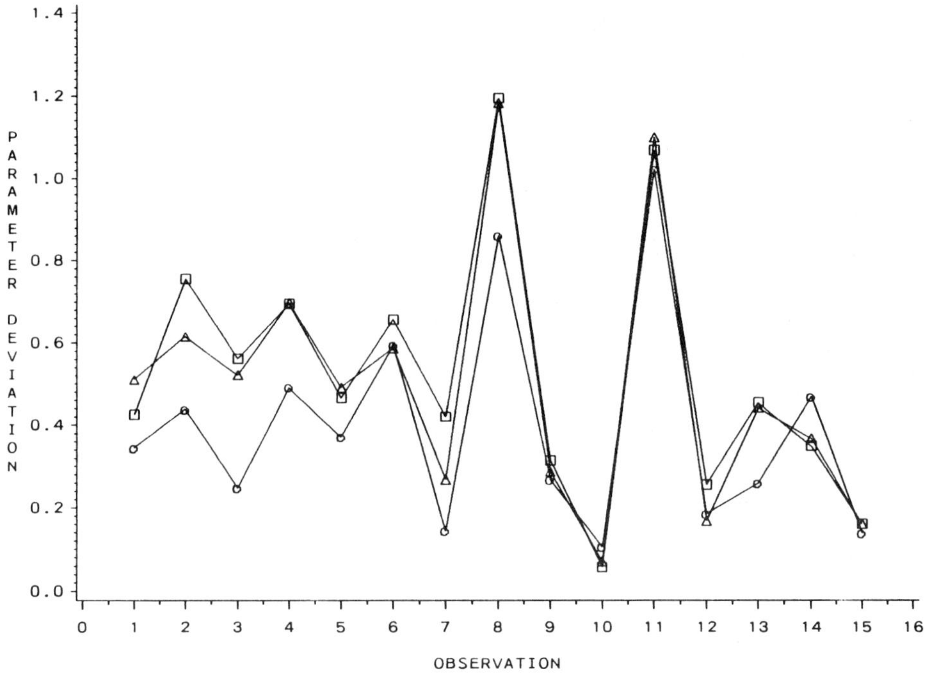

Legend : $0 = |\hat{\beta} - \beta|$; $\Delta = |\hat{\beta}_{\lambda=0.5} - \beta|$; $\square = |\hat{\beta}_{t|t,\lambda=0.5} - \beta_t|$

We also point out that the filtered and conditional estimates in the table above are computed using the value of λ that generated the data. When λ is unknown, as in most applications, these estimates can still be computed but with a value of λ that is likely to be incorrect. In such circumstances, it is more reasonable to bear the extra computational burden, and compute the unconditional estimate, than to use the conditional estimate based on a misspecified value of λ. To emphasize this point, we have provided in Table 2, the sampling mean squared error (MSE) of the conditional estimate using two values of λ, i.e., $\lambda = 0.5$ and $\lambda = 5$, and the MSE of the unconditional estimate. The MSE of the optimal estimate, $\beta_{\lambda=0.5}$, is, of course, smaller than that of $\hat{\beta}$. However, the MSE of the conditional estimate using an incorrect value of λ is larger than that of $\hat{\beta}$. Although the results below are based on a limited sampling investigation, we confirmed that similar results are obtained in other examples. To conserve space, those results are not reported.

Table 2: Sampling Mean Squared Error of the Estimates
Computed from 30 samples of size 15

obs	$\hat{\beta}_{\lambda=0.5}$	$\hat{\beta}_{\lambda=5}$	$\hat{\beta}$
1	0.42341	0.70180	0.53698
2	0.44656	0.81684	0.45471
3	0.41962	1.02542	0.63801
4	0.58891	1.06727	0.71121
5	0.28103	1.04921	0.46559
6	0.37140	1.52418	0.66745
7	0.10165	0.51738	0.19686
8	1.50176	1.14964	1.17638
9	0.25482	0.79170	0.46704
10	0.13425	1.57139	0.60643
11	1.25848	1.55346	1.34032
12	0.12108	1.60701	0.72236
13	0.31260	1.01436	0.58318
14	0.19325	0.85835	0.41076
15	0.23863	1.14518	0.70925

5. Conclusion

This paper considers the Bayes smoothing problem in the TVR model and presents several exact results that can be put to use in applications. It is shown that Monte-Carlo numerical procedures can be implemented relatively easily to compute the results in the paper. We are confident that the multiple covariate case can be handled along the lines described in the paper with only a little extra difficulty. In future work, we plan to describe the analysis to cover the multiple covariate case.

Finally, we point out that the results in this paper can be directly used to find the exact distribution of the coefficients in the filtering problem. After noticing from (7) that the posterior distribution of $\beta_t|t, \lambda$ is univariate-t, λ can be integrated out using the results in Section 3. We suppress the details.

Acknowledgments

We are grateful to the anonymous referees for many useful comments on an earlier draft of the paper. We would also like to thank Benjaporn Thangkasemvanthana for providing assistance with the numerical computations reported in the paper.

Partial support for Professor Chib's work was provided by a University of Missouri Summer Research Fellowship.

References

Anderson, B.D., and J.B. Moore (1979). *Optimal Filtering*, New Jersey: Prentice Hall, Inc.

Broemeling, L. (1985). *Bayesian Analysis of Linear Models*, New York: Marcel Dekker.

Broemeling, L., J. Diaz and A.M. Yusoff (1985). "Some Bayesian Solutions for Problems of Adaptive Estimation in Linear Dynamic Systems," *Communications in Statistics: Theory and Methods,* Vol. 14, No. 2, 401-418.

Chib, S., R.C. Tiwari, and S.R. Jammalamadaka (1988). "Bayes Prediction in Regressions with Elliptical Errors," *Journal of Econometrics*, Vol. 38, 349-360.

Chow, G.C. (1983). *Econometrics*, New York: McGraw Hill.

Engle, R.F. and M.W. Watson (1988). "The Kalman Filter: Applications to Forecasting and Rational-Expectations Models," in *Advances in Econometrics*, T.F. Bewley, ed., New York: Fifth World Congress, Cambridge University Press.

Geweke, J. (1987). "Bayesian Inference in Econometric Models Using Monte Carlo Integration", mimeo.

Harvey, A.C. (1981). *Time Series Models*, Oxford: Phillip Allan.

Klock, T. and H. van Dijk (1978). "Bayesian Estimates of Equation System Parameters: An Application of Integration by Monte-Carlo", *Econometrica*, Vol. 46, 1-19.

Schweppe, F.C. (1965). "Evaluation of Likelihood Functions for Gaussian Signals", *I.E.E.E. Transactions on Information Theory*, Vol. 11, 61-70.

Zellner, A., L. Bauwens, and H.K. Van Dijk (1988). "Bayesian Specification Analysis and Estimation of Simultaneous Equation Models Using Monte Carlo Methods," *Journal of Econometrics*, Vol. 38, 39-72.

Zellner, A., and P.E. Rossi (1984). "Bayesian Analysis of Dichotomous Quantal Response Models," *Journal of Econometrics*, 25, 365-393.

Bayesian and Likelihood Methods in Statistics and Econometrics
S. Geisser, J.S. Hodges, S.J. Press and A. Zellner (Editors)
© Elsevier Science Publishers B.V. (North-Holland), 1990

AN ANALYSIS OF QUESTIONNAIRE DATA ON WORK DESIGN AND JOB SATISFACTION: A CASE STUDY IN THE USE OF SIMPLE GRAPHICAL DISPLAYS

L. Denby, R. Gnanadesikan, J.R. Kettenring, and J.W. Suzansky
AT&T Bell Laboratories, Bellcore, Bellcore, Work Systems Associates, Inc.

Abstract

A 44-item questionnaire was used to collect data about telephone operators' perceptions of their jobs before and after changes were made to their work environment. The data were analyzed via simple graphical displays that provide an informal "internal comparison" of the effects of the changes. These techniques avoid some of the pitfalls of more traditional and complex multivariate procedures, such as factor analysis, for such data.

1. Introduction

The objective of this paper is to describe the analysis of a set of questionnaire data concerning the perception of telephone operators of their jobs. The analysis relies heavily on simple graphical tools which provide a graphical *internal comparison* (Gnanadesikan 1977, Section 6.3.1) of the effects of experimental changes that were made to the work environment. These informal methods, it is argued, are more appropriate for these data than more traditional and more complex tools, such as factor analysis and multivariate analysis of variance, that are often applied in such situations.

The spirit in which these graphical methods have been applied is consonant with Barnard's views on applied statistics. For example, Barnard (1987) is critical of those "who so often seem to think that the scientific merit of their work is somehow enhanced, rather than the opposite, by their use of complex procedures when simple ones would do." Also, Barnard's interest in the application of mathematics and statistics to the social sciences (see, e.g., Barnard, 1968a,b) provides a contact point for the particular application considered here and for the general philosophy and emphases.

The study which gave rise to the operator data was done in the old Bell System prior to the divestiture of AT&T. It follows a general approach that was used to tackle problems of work design and job satisfaction. The approach relies heavily on questionnaire surveys for determining the perceptions of the employees themselves about their jobs. The resultant data are analyzed for identifying such things as important dimensions of the job and for diagnosing which aspects of the job need to be altered or improved for enhancing job satisfaction and hence performance (hopefully). The operator data set arose in this way and was based upon a questionnaire, the *Job Dimensions Checklist* (JDC), developed by Suzansky (1974).

As a diagnostic and evaluative tool, the JDC questionnaire is administered to a group of employees whose jobs are nominally the same (e.g., operators). Then some specific changes are made in the job or job environment either in the light of the findings from the questionnaire or for other extraneous reasons, such as changes in administrative methods. After a period of time, the questionnaire is administered again to the group and the resulting data are studied for assessing the effects of the changes that had been made to the job or environment. Another interest in the data is to use them for identifying "basic" dimensions, if there are any such, of job quality and satisfaction.

In this paper, two graphical techniques, difference histograms and probability plots, that were used while studying the operator data are emphasized. A difference histogram is simply a plot of the difference between two histograms, each normalized by its own sample size. A probability plot is a graph of the ith largest observation in a sample of size n against the $100(i-1/2)/n$ percentage point of a reference distribution (Wilk and Gnanadesikan, 1968). In both plots, the shape of the configuration is judged relative to a line. If the difference histogram is concentrated about the zero line, the two distributions are similar. If the probability plot is reasonably straight, with zero intercept and unit slope, then the empirical distribution of the data and the reference distribution are similar. There are many advantages of these techniques. They stick close to the data, require few assumptions, allow checks on assumptions, reveal unanticipated peculiarities, and aid digestion of complex data.

The remainder of the paper is divided into three sections: statement of the experimental design features of the data, presentation of the analysis of the data from both a factor analysis and a graphical analysis of variance perspective, and closing discussion.

2. Experimental Design

2.1 Description of Questionnaire

Table 1 shows the first 15 of the 44 questions that constitute the JDC questionnaire. On each question the subject is asked to respond on a five-point scale anchored with labels at three of the five points, and the response is in two parts—firstly as it pertains to "your job" and secondly as it pertains to a perceived "best job." The difference between the scores for "best job" and "your job" may be considered as an indicator of job satisfaction.

Table 1. The First 15 Questions From the JDC Questionnaire

1. How much planning do you do in your job?
2. How hard do you usually find your work?
3. How complete a product results from your job?
4. How often do you find out how good a job you are doing?
5. From day to day, how varied is your job?
6. What level of skill is required to perform your job?
7. How do you usually perform your job?
8. How clear is the relation of your job's products to your company's products and goals?
9. How would you describe your usual discussions with your co-workers?
10. How well does your company keep you informed of opportunities for advancement?
11. How often do you find you can use something learned on your job in other situations outside the job?
12. How often do you meet with your co-workers socially outside of work?
13. What kind of information do you get on the quality and quantity of your job performance?
14. How varied is your pace of doing the work in your job?
15. How often do you talk with your clients or customers?

2.2 Data

A carefully designed experiment, involving both treatment and control groups, was used to collect data from the operators using the JDC questionnaire. There were 14 offices involved, two each from seven telephone companies. Seven of the offices (one from each company) were assigned to be a control group and the remaining were used as a treatment group. The pre-trial administration of the questionnaire was in the fall of 1975 while the post-trial administration was carried out in the spring of 1976.

Data were available (see Table 2) from 1141 operators in all from the 14 offices, with 562 of these being from the pre-trial administration and 579 from the post-trial one. Of the 562 pre-trial set, 303 belonged to the treatment group and 259 to the control group, while the corresponding breakdown for the 579 post-trial set was 298 and 281 respectively.

Table 2. Number of Offices and Respondents

	Group		
Period	Treatment	Control	Total
# offices	7	7	14
# respondents (pre-trial)	303	259	562
# respondents (post-trial)	298	281	579
# matches (pre and post)	137	97	234

For practical reasons, offices (i.e., groups of operators) were assigned to the control and treatment groups, with all the operators in an office rather than individual operators being assigned to one of these. There are questions of carefully matching the offices in such a procedure to avoid possible biases in the results. The design of the study had indeed paid attention to these considerations and, moreover, the degree to which it succeeded in avoiding pitfalls can be studied via the data themselves (discussed later).

Another issue in the data is the degree to which the pre- and post-trial sets of operators are "comparable" given the elapsed time between the two administrations of the questionnaire. There were, in fact, a substantial number of perfect matches, i.e., the same operator filled out the questionnaire in the pre- and post-trial situations. Specifically, 97 operators (37%) in the pre-trial control group were also in the post-trial and 137 operators (45%) in the pre-trial treatment group. (This matching feature will be dealt with in the latter part of Section 3.3.)

2.3 Treatment

The information from the initial administration of the questionnaire indicated some dissatisfaction among the respondents about the feedback that they were receiving concerning their performance. The "treatment," therefore, consisted of the supervisors providing feedback not only on things done poorly but also on things done well by the operators and, furthermore, giving such feedback more frequently, if possible on a weekly basis. *Thus, a major purpose of the study was to determine whether this "treatment" had any effects that were reflected in the pre- versus post-trial responses of the operators.*

3. Analyses and Results

3.1 Initial Considerations

At the outset, there were actually two related objectives in mind: identifying underlying dimensions in the data and then assessing the treatment effects in terms of these dimensions.

The motivation for dimensionality reduction comes not only from the awkwardness of dealing with 44 dimensions but also from a long tradition of successful applications of factor analysis in the present context of work design and job satisfaction (see, e.g., Lawler and Hall 1970; Roberts, Walter, and Miles 1971; Suzansky 1974; Waters, Roach, and Batlis 1974; and Weaver 1978). In so far as stable and interpretable factors are present in the data, they can provide convenient points of focus for explaining the treatment effects and tying the findings to previous studies.

The initial analysis of the data was, in fact, done in this manner and highlights are presented in Section 3.2. For a variety of reasons, the results were disappointing. In retrospect, the usual conditions needed for a successful factor analysis were not met, and the derived dimensions only obscured the later analyses based upon them.

These experiences with the factor analytic approach to dimensionality reduction led to consideration of alternative techniques that were more closely aligned with the categorical nature of the data. Highlights of this new analysis are presented in Section 3.3. In terms of the two original objectives, it dodges the objective of dimensionality reduction and focuses instead on direct assessment of the treatment effects through a simultaneous analysis of all of the 44 variables or questions.

3.2 Factor Analysis

To address the desire for dimensionality reduction, a standard principal factor analysis (Harman 1967; Johnson and Wichern 1982) was performed on the 44×44 sample correlation matrix derived from the 562 pre-trial responses to the 44 "your job" questions. The reduction that is being sought can also be viewed in terms of the number of parameters involved. There are 946 correlation parameters in the underlying population correlation matrix, and the factor analysis is a way of imposing simplifying structure so as to bring this number down.

The convention of using factors associated with eigenvalues larger than unity led to 18 factors that accounted for 65% of the variance. Further simplification was achieved by applying a varimax rotation to the factor loadings. Then, *subjectively*, the 18 factors were used to develop six subgroups of them which were labelled as six basic dimensions—*closure, control, difficulty, feedback, skill*, and *variety*—and 37 of the 44 questions were assigned to one and only one of these six dimensions for further analyses. For instance, the closure dimension consisted of items about the completeness of the output of the job as it relates to the output of the company as a whole. The comparison of the pre- and post-trial responses was then made by assessing, via standard two-sample t-tests, the differences in the means of the responses on questions assigned to each dimension. A separate t-test was made for each dimension.

The factor analysis, as well as the method for comparing pre- and post-trial results, was not very satisfactory. Difficulties arose in attempting to interpret the results. One problem is the substantial fraction (over 35%) of variability unexplained even after including 18 factors. A second is the fact that identification of the basic dimensions and assignment of the initial items to one of these dimensions were far from unequivocal partly because the reduction of dimensionality was not a strong one and also because there was no clear indication that every item was exclusively associated with a specific dimension. These problems are a consequence of the weak correlation structure in the data: no pairwise correlation was as large as 0.4 in magnitude.

Since factor analysis was really used as a vehicle for reducing the data to six clusters of 37 questions, the same type of objective might be achieved more directly by application of cluster analysis methods. Indeed, this approach was attempted too, but the groupings of questions that resulted were neither very strong nor very stable. Again, the reason is clear: the lack of high correlations—the measures upon which the clustering was based—meant that there were no strong affinities among the responses to the questions.

The difficulty of interpretation is highlighted even more in the t-statistics comparisons of pre- and post-trial results in terms of the six dimensions. Figure 1 shows a plot of the percent change, from pre- to post-trial, in the mean scores of the items (without consideration of their standard deviations) in each of the six dimensions. The percent changes for the treatment and control groups are shown separately, and when the t-statistic for the change (calculated as a standard two-sample statistic for the 1141 operators) in a dimension is statistically significant at a level of significance of 5% or smaller the point is circled. Additionally, if the change for the treatment group is significantly different from that for the control group, the dimension label is enclosed in a box. As seen in Figure 1, the change in the feedback dimension for the treatment group is positive and statistically significant as one might expect. However, the same is true on the closure dimension for the treatment group. Moreover for five of the six dimensions the changes in the treatment and control groups are significantly different from each other. While hypotheses, such as the desire to compete on the part of the control group, may be put forth for explaining some of these differences, they are neither simple nor convincing. In the light of the previous discussion about the real reduction of dimensionality achieved by the factor analysis, one has to wonder whether the method of analysis has induced a smearing out of interesting effects across several of the dimensions.

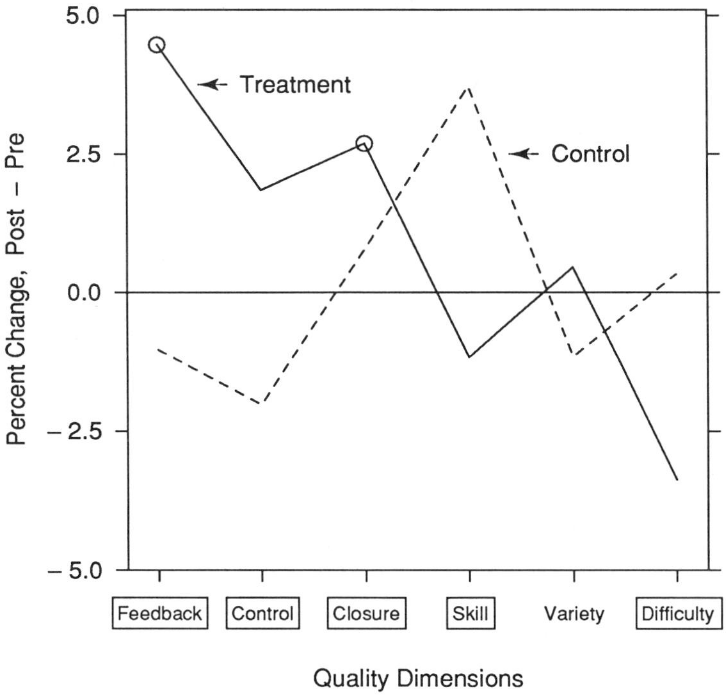

Figure 1: Plot of the percent change, from pre- to post-trial, in the mean scores of the items in each of the six dimensions obtained from the factor analysis.

3.3 Graphical Analysis of the Data

A variety of simple plots were made which revealed unanticipated peculiarities with the data. For example, simple histograms revealed that the responses to a question tended to have "saw-toothed" distributions, with much heavier density for the three categories labelled on the questionnaire. They also showed that the distributions were often very skewed with many respondents choosing either 1 or 5 in scoring the questions. These peculiarities raise additional doubts about the appropriateness of product-moment correlation coefficients, and factor analyses based on them, as tools for this study. Similar comments apply to multivariate analysis of variance, which may appear at first to be a suitable method for these data.

A plot which is especially effective for focusing on the effects of the trial is a display of the differences between the pre- and post-trial histograms (more precisely the relative frequencies) for the treatment and control groups. Figures 2a and 2b display such difference histograms for questions #4 and #13. Because the difference histograms for "your job" in the treatment group show marked departures from the zero line, there appear to be effects due to the trial for these questions. Specifically, for question #4 (see Table 1 for wording), a higher percentage of operators in the treatment group answered in the first category, "very rarely," before the trial, and a higher percentage chose the middle response of "about once a week" after the trial. For question #13, a higher percentage in the treatment group answered in the first category, "only learn of things done wrong or too slow," and the third category, "sometimes on all things done, sometimes only on errors," before the trial, and a higher percentage fell into the fifth category, "describes all things done, right or wrong," after the trial. Of the 44 questions, these two showed the most striking differences in the treatment group's answers before and after the trial. The other plots in Figures 2a and 2b are typical of all the questions and suggest that the trial affected neither the "best job" responses of the treatment group nor the control group responses for either "your job" or "best job" in a substantial way.

Unfortunately, the inspection of such plots for all questions, while useful, tends to be a laborious, and possibly hit-and-miss, means for identifying effects. What is needed is a method for consolidating the information from all 44 questions on a simple plot so that the relative magnitudes of the differences can be assessed statistically. A natural way to do this is with a probability plot. It requires few assumptions about the data and allows direct assessment not only of the treatment effects but also of the assumptions involved.

Consider, for instance, the pre- and post-trial responses to a particular question about "your job". These can be summarized in the following form:

Response Levels

	1	2	3	4	5	totals
pre-trial	n_{11}	n_{12}	n_{13}	n_{14}	n_{15}	$n_{1.}$
post-trial	n_{21}	n_{22}	n_{23}	n_{24}	n_{25}	$n_{2.}$
totals	$n_{.1}$	$n_{.2}$	$n_{.3}$	$n_{.4}$	$n_{.5}$	$n_{..}$

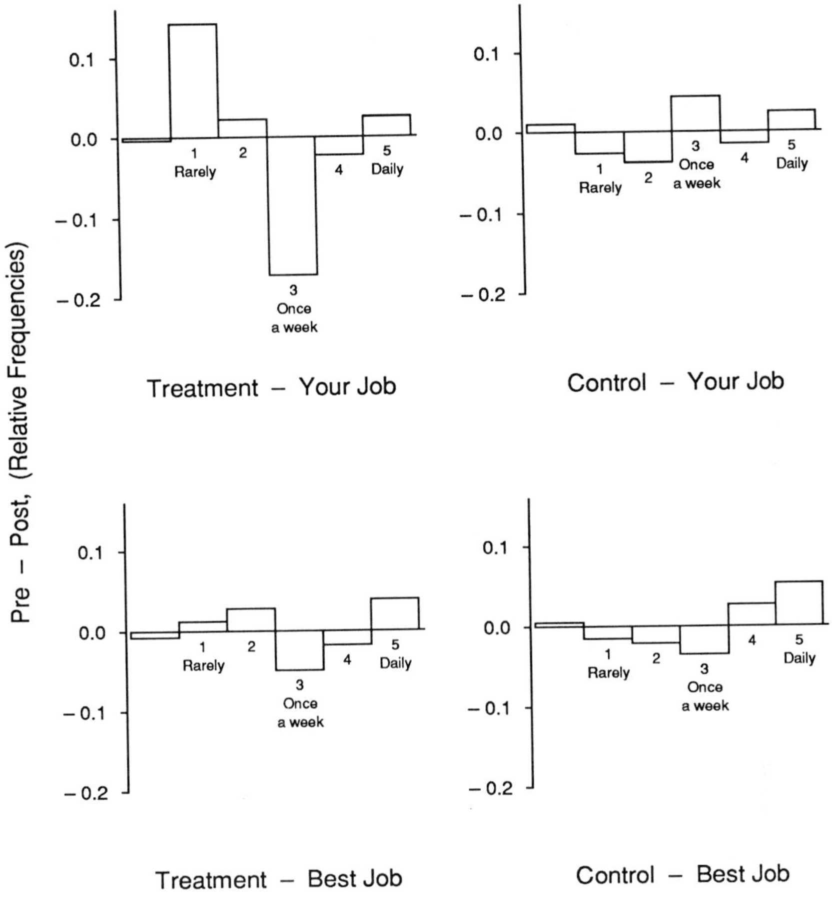

Figure 2a: Difference histograms for question #4 based upon answers to the "your job" and "best job" questions by the treatment and control groups.

To measure the similarity or homogeneity between the pre- and post-trial distributions, one can compute

$$\chi^2_{(4)} = \frac{\sum_{j=1}^{5} n^2_{1j}/n_{.j} - n^2_{1.}/n_{..}}{(n_{1.}/n_{..})(1 - n_{1.}/n_{..})}.$$

If the two distributions are the same (i.e., there are no treatment effects) and assuming that the pre- and post-samples can be regarded as statistically independent of each other (i.e., ignoring the existence of matches), then $\chi^2_{(4)}$ should be distributed

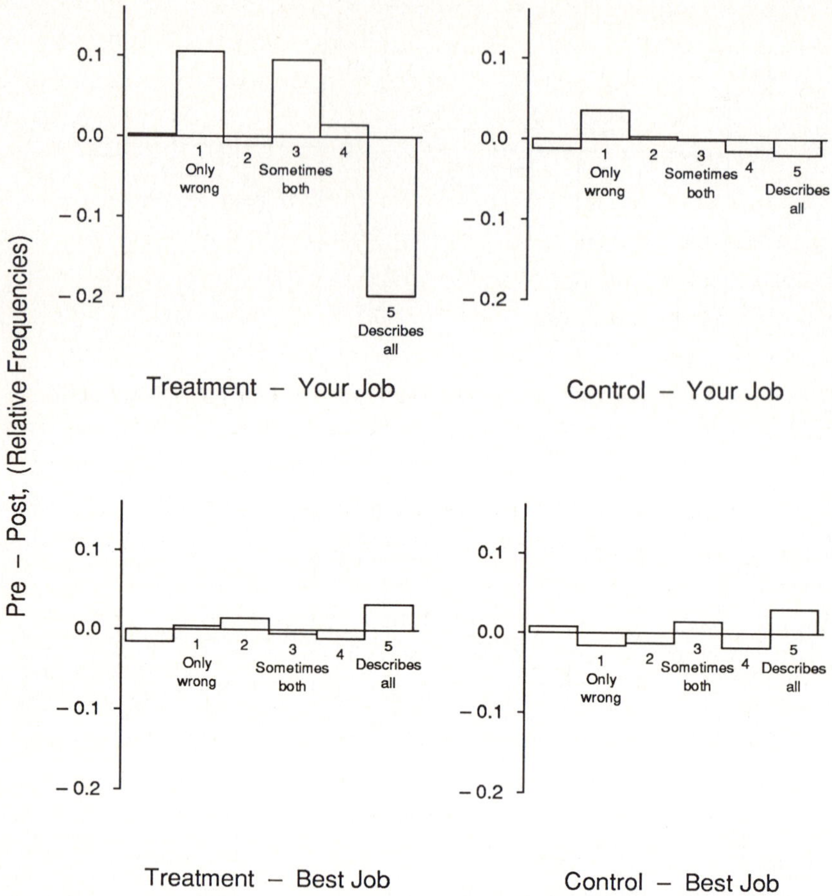

Figure 2b: Same display as Figure 2a for question #13.

approximately as a chi-square random variable with four degrees of freedom. One can thus carry out a formal test of significance of the null hypothesis that there are no treatment effects by referring the computed value of $\chi^2_{(4)}$ for each question to a chi-square distribution with four degrees of freedom. Given that there are 44 χ^2-values corresponding to the 44 questions, and that there are several possible sets of such 44 values depending on different ways of partitioning the data (see discussion below), there is the thorny problem of choosing the significance level for the multiple comparisons involved. The choice is complicated by the presence of varying degrees of association and dependence both among the responses to the questions and between the samples.

From a data analytic viewpoint, what one wants for identifying treatment effects is a means of comparing the relative magnitudes of the 44 χ^2-values so as to identify those that are clearly large relative to others that vary over an expected (in some sense) range. It is not statistical significance that is of interest in many problems (especially ones involving large data sets wherein any deviation at all may turn out to be statistically significant) but rather it is in *internal comparisons* among a collection of comparable quantities. For such comparisons, one can adopt the "null" view that the 44 values should behave like a random sample from a chi-square distribution with four degrees of freedom, and then check the extent to which this view is reasonable by making a Q-Q probability plot (Wilk and Gnanadesikan 1968) in which the ith largest χ^2-value is plotted against the corresponding quantile of the chi-square distribution with four degrees of freedom. If there are no treatment effects and no severe violations of the other assumptions, then the plot should be reasonably straight with a zero intercept and unit slope. If there are treatment effects associated with some of the questions, then these will result in larger χ^2-values which will tend to stand out on the plot in contrast to the linear pattern of the others. Systematic departures from assumed conditions, such as independence between samples and across questions, tend to induce more subtle changes in the plot which will permeate the entire configuration.

There are many Q-Q plots of this form which are of interest, but two of the most vital ones are the comparisons of pre- vs. post-trial responses for the treatment group and for the control group. Figures 3 and 4 show the Q-Q plots for these cases. In Figure 3, although the configuration is basically linear, with a near zero intercept and slope slightly larger than one, two points stand out as being distinctly different from the others. If the visual assessment needs the crutch of a more formal evaluation, one can superimpose the error bars shown on the plot to demonstrate that the two largest χ^2-values are "too large." (The bars extend to plus and minus two standard errors of the order statistics.)

The deviant points correspond to #4 "How often do you find out how good a job you are doing?" and #13 "What kind of information do you get on the quality and quantity of your job performance?". Both of these are directly related to the "treatment" which attempted to improve the comprehensiveness and increase the frequency of feedback to the operators, as previously discussed. In the factor analysis, these are two of the six questions which were assigned to the feedback dimension. Recall that the difference histograms, shown in Figures 2a and 2b, for these two questions showed the most striking effects of the trial.

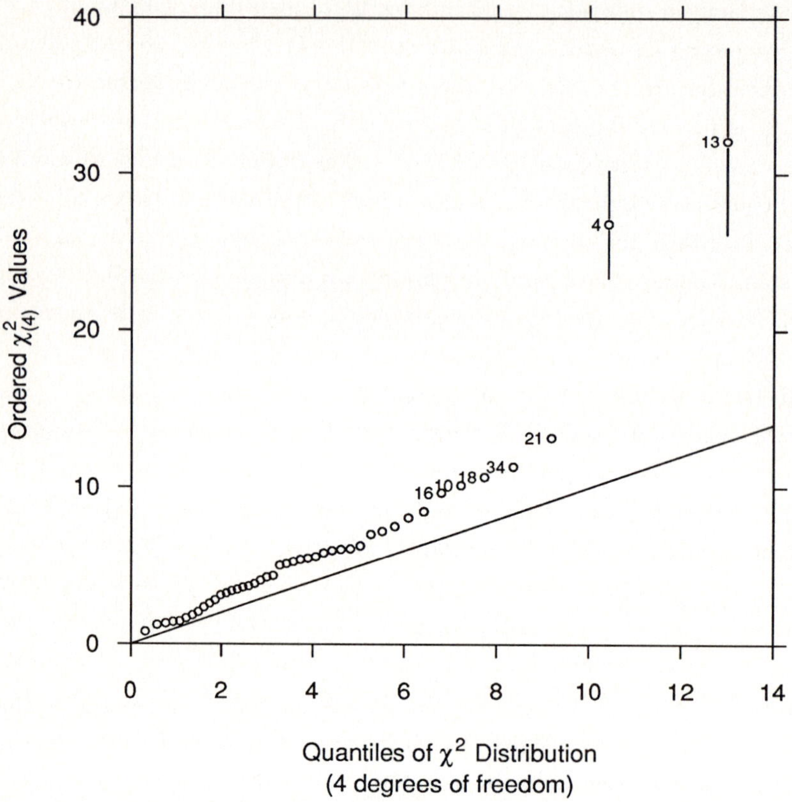

Figure 3: Q-Q probability plot of the 44 values of $\chi^2_{(4)}$ comparing the pre- and post-trial responses of the treatment group to the "your job" questions.

The corresponding Q-Q plot for the control group is displayed in Figure 4. The plot is remarkably well approximated by a line emanating from the origin with unit slope. The main statistical message from this plot is that the null statistical assumptions underlying it are appropriate ones.

Several other plots of this type can be made which provide additional insights. To check on the comparability of the treatment and control groups, one can compare the "your job" or the "best job" responses prior to the trial and the "best job" responses after the trial for the two groups. These plots confirmed that the two groups are reasonably comparable.

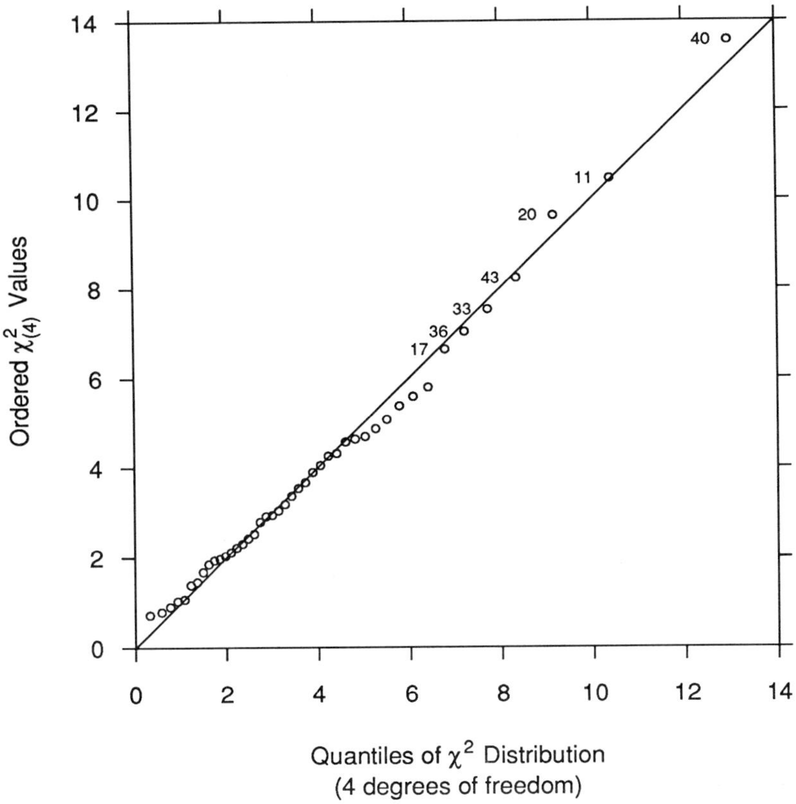

Figure 4: Q-Q probability plot of the 44 values of $\chi^2_{(4)}$ comparing the pre- and post-trial responses of the control group to the "your job" questions.

One aspect of the data which has not been explicitly accounted for in the pre- vs. post-trial comparisons, illustrated in Figures 3 and 4, is the substantial number of matches of operators between these two situations. In principle, a more sensitive analysis should be attainable by calculating appropriate χ^2-statistics for the matched and unmatched parts of the data separately, adding these together, and making probability plots based on these 44 sums.

It is convenient to think of the matched data in the form of a two-way table with entries n_{ij}, the number responding in category i before the trial and category j after it; row totals $n_{i.}$, column totals $n_{.j}$; and overall total $n_{..}$. To illustrate with the "your job" responses, one can then compare the pre- and post-trial distributions using the statistic (Stuart 1955; Kendall and Stuart 1961, p. 589)

$$\chi^2_{[4]} = \mathbf{n}'\mathbf{V}^-\mathbf{n}$$

where $\mathbf{n} = (n_{1.} - n_{.1}, \ldots, n_{5.} - n_{.5})'$; $\mathbf{V} = ((v_{ij}))$ with $v_{ii} = n_{i.} + n_{.i} - 2n_{ii}$ and $v_{ij} = -(n_{ij} + n_{ji})$, $i \neq j$; and $\mathbf{V}^-$ is a generalized inverse of $\mathbf{V}$. A generalized inverse of $\mathbf{V}$ is used because its rank is only four. The statistic can be calculated by omitting a row from $\mathbf{n}$ and the corresponding row and column from $\mathbf{V}$ and using a regular inverse on the reduced version of $\mathbf{V}$. Like $\chi^2_{(4)}$, $\chi^2_{[4]}$ should be distributed approximately as a chi-square random variable with four degrees of freedom if there are no treatment effects. The sum of $\chi^2_{(4)}$, calculated from the unmatched group alone, and $\chi^2_{[4]}$ should be distributed as a chi-square with eight degrees of freedom.

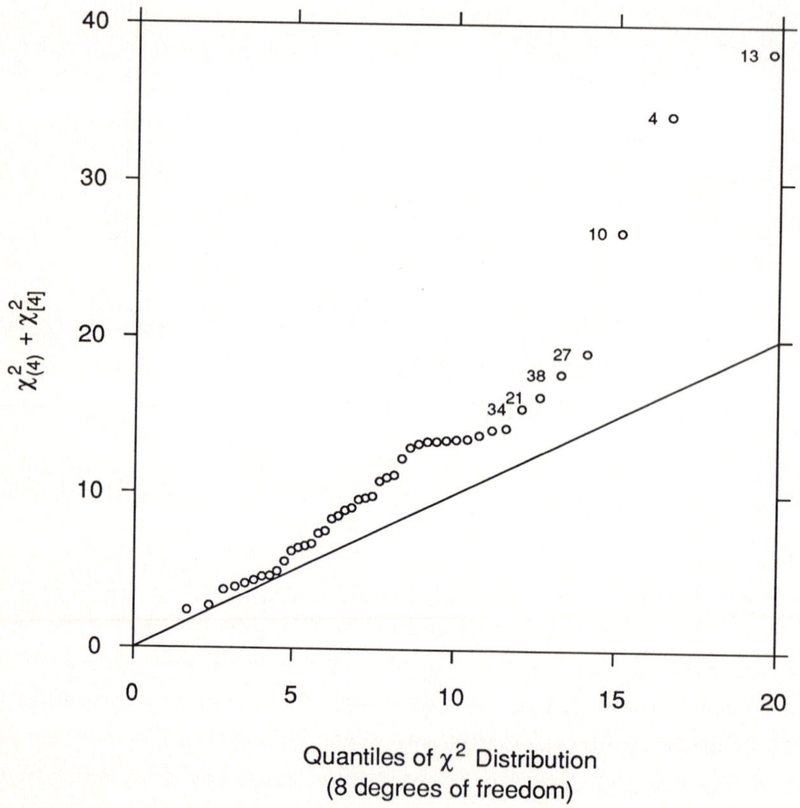

Figure 5: Q-Q probability plot of the 44 values of the sum of $\chi^2_{(4)}$ for the unmatched data and $\chi^2_{[4]}$ for the matched data comparing the pre- and post-trial responses of the treatment group to the "your job" questions.

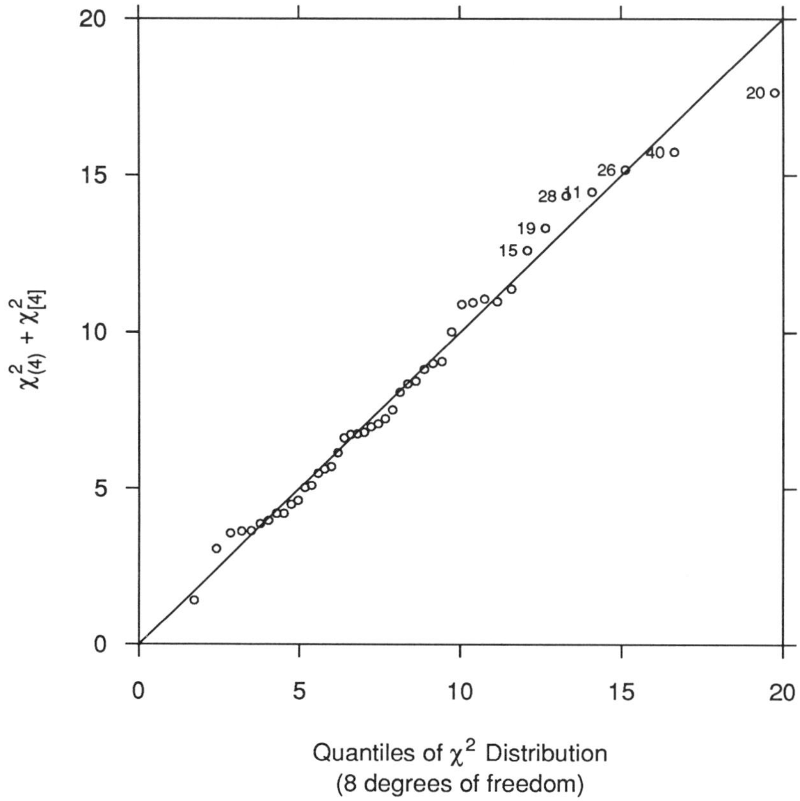

Figure 6: Same display as Figure 5 for the control group.

Figures 5 and 6 show the resulting Q-Q plots corresponding to Figures 3 and 4. The points which stand out for the treatment group are #4 and #13, as in Figure 3, plus #10 "How well does your company keep you informed of opportunities for advancement?". The appearance of #10 may be a side effect of the feedback treatment. The main part of the configuration in Figure 5 is bumpier than in Figure 3. A line through it would have slope slightly larger than one. The picture for the control group in Figure 6 is a bit more ragged than in Figure 4, but it is still well approximated by a line with unit slope coming out of the origin.

The raggedness in these two figures may be due to the presence of many cell counts which are very small or even zero—a situation which limits the usefulness of the approach of adding the matched and unmatched χ^2-values for this data set. Moreover, several cases arose where the rank of $\mathbf{V}$ was actually three instead of four because an entire row and corresponding column were zero. These questions were simply dropped for this analysis.

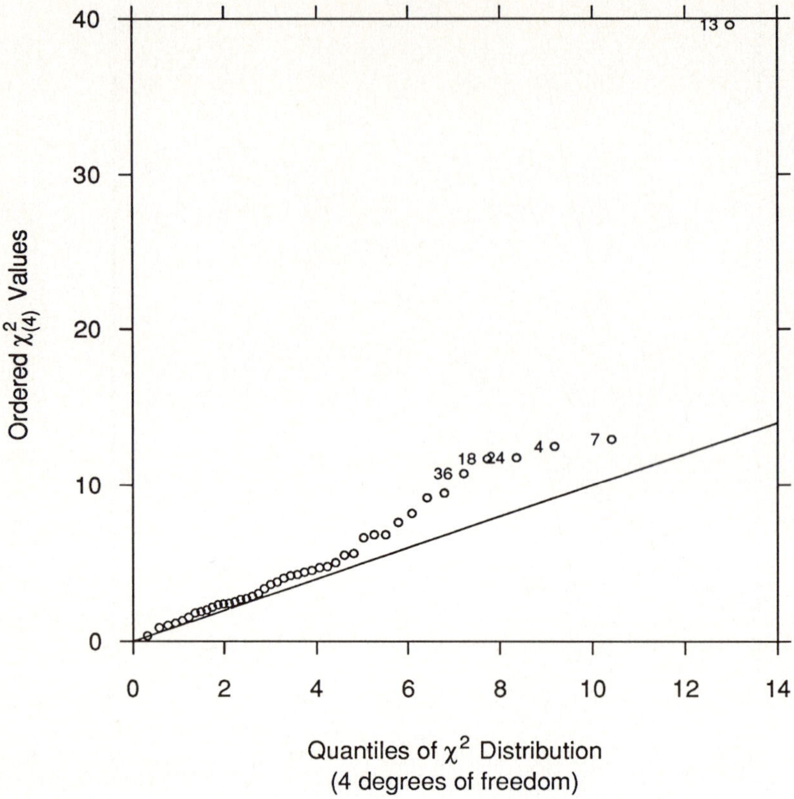

Figure 7: Q-Q probability plot of the 44 values of $\chi^2_{(4)}$ comparing the pre- and post-trial absolute difference between the "your job" and "best job" responses of the treatment group.

In brief, Figures 3-6, plus others not shown, collectively point to these main findings: the treatment and control groups were reasonably well matched initially and the trial itself had a pronounced effect, statistically speaking, on the operators' perception of the frequency and comprehensiveness of feedback which they received.

As a direct measure of job satisfaction, the absolute difference between the "your job" and "best job" responses can be computed. These values will range from zero, for no difference or a high degree of satisfaction, to four, for a low degree of satisfaction. Their distributions can be compared using χ^2-statistics and Q-Q plots as before. The plots for the pre- vs. post-trial responses are shown in Figure 7 for the treatment group and Figure 8 for the control group. The control group plot does not reveal any unusually large values, but question #13 stands out in a most striking way in the configuration for the treatment group. The reason behind the whopping value for

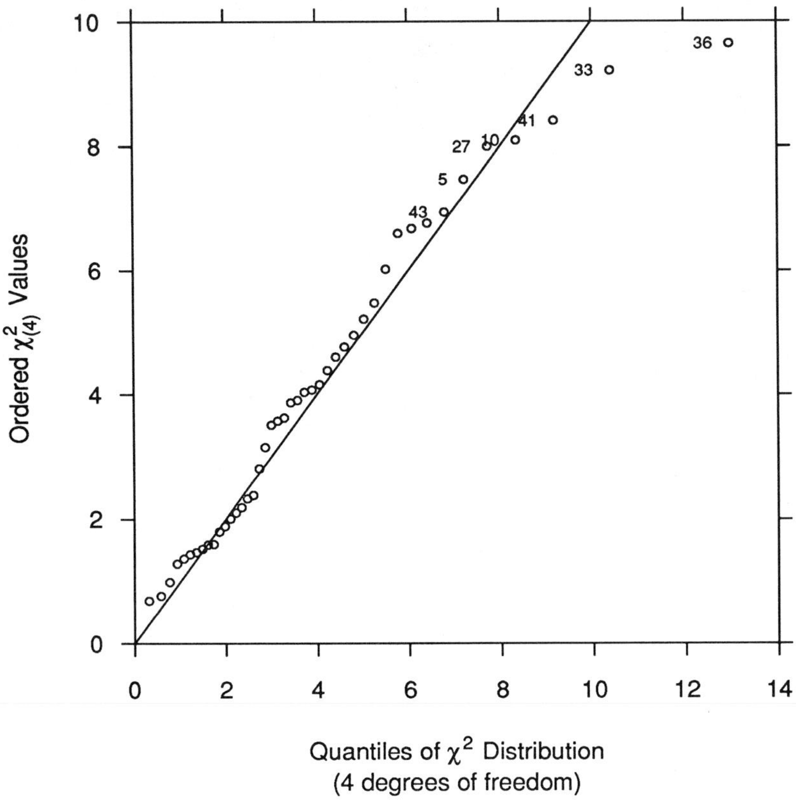

Figure 8: Same display as Figure 7 for the control group.

#13 in Figure 7 is that the number of operators with the same responses for "your job" and "best job" jumped sharply after the trial to include over half of those in the post-trial treatment group. A similar but much less dramatic change occurred also for #4. Thus, in addition to substantially improving the frequency (#4) and comprehensiveness (#13) of feedback, as judged by the "your job" responses, the attempt to improve the comprehensiveness of the feedback had particular payoff from the point of view of job satisfaction.

4. Conclusion

The analysis of the questionnaire data about operators' perceptions of their current and ideal jobs has provided clear insights on the effects of the intervention on their work environment. Specifically, an important finding was that the attempt to increase the frequency and comprehensiveness of the feedback provided to operators concerning their performance was a success as measured by the change in responses of the operators to the questionnaire. Also, for job satisfaction, the increased comprehensiveness of the feedback turned out to be of prime importance.

The major thrust of the work has been its reliance on relatively simple graphical techniques as alternatives to more complex methods such as factor analysis. The graphical procedures were useful both in uncovering unanticipated data anomalies (e.g., the tendency for the univariate and bivariate distributions of responses to be concentrated in particular regions) and in assessing the effects of the trial. These methods yielded more direct and simpler interpretations than the factor analysis whose propensity for smearing out the effects of the trial into several of the recovered dimensions or factors had, in part, created interpretational difficulties.

The utility of the graphically-oriented statistical methodology in this case study suggests that it could be applied advantageously to similar data situations involving lengthy questionnaires.

Acknowledgment

The authors wish to thank the referees for their comments.

References

Barnard, G.A. (1968a). "Computers, Statistics, and Politics," in *The Future of Statistics*, D.G. Watts, ed., New York: Academic Press, 39-43 (esp. p. 40, para. 3, last sentence).

Barnard, G.A. (1968b). "Measurement in the Social Sciences," in *The Future of Statistics*, D.G. Watts, ed., New York: Academic Press, 291-307 (esp. pp. 293, 300-301).

Barnard, G.A. (1987). Personal Communication.

Gnanadesikan, R. (1977). *Methods for Statistical Data Analysis of Multivariate Observations*. New York: Wiley.

Harman, H.H. (1967). *Modern Factor Analysis* (2nd ed.). Chicago: University of Chicago Press.

Johnson, R.A. and D.W. Wichern (1982). *Applied Multivariate Statistical Analysis*. Englewood Cliffs, NJ: Prentice-Hall.

Kendall, M.G. and A. Stuart (1961). *The Advanced Theory of Statistics*, Vol. 2. New York: Hafner.

Lawler, E.E., III and D.T. Hall (1970). "Relationship of Job Characteristics to Job Involvement, Satisfaction, and Intrinsic Motivation," *Journal of Applied Psychology*, Vol. 54, 305-312.

Roberts, K.H., G.A. Walter, and R.E. Miles (1971). "A Factor Analytic Study of Job Satisfaction Items Designed to Measure MASLOW Need Categories," *Personnel Psychology*, Vol. 24, 205-220.

Stuart, A. (1955). "A Test for Homogeneity of the Marginal Distributions in a Two-Way Classification," *Biometrika*, Vol. 42, 412-416.

Suzansky, J.W. (1974). "The Effects of Individual Characteristics as Moderating Variables of the Relation Between Job Design Quality and Job Satisfaction," Ph.D. thesis, Stevens Institute of Technology, Hoboken, NJ.

Waters, L.K., D. Roach, and N. Batlis (1974). "Organizational Climate Dimensions and Job-Related Attitudes," *Personnel Psychology*, Vol. 27, 465-476.

Weaver, C.N. (1978). "Job Satisfaction as a Component of Happiness Among Males and Females," *Personnel Psychology*, Vol. 31, 831-840.

Wilk, M.B. and R. Gnanadesikan (1968). "Probability Plotting Methods for the Analysis of Data," *Biometrika*, Vol. 55, 1-17.

Bayesian and Likelihood Methods in Statistics and Econometrics
S. Geisser, J.S. Hodges, S.J. Press and A. Zellner (Editors)

LEGAL LIKELIHOODS AND A PRIORI ASSESSMENTS: WHAT GOES WHERE?

Stephen E. Fienberg
Carnegie Mellon University

Abstract

Many authors have suggested that the Bayesian approach should serve as a normative theory for legal decisionmaking in courtroom settings. Here we review some key aspects of the specification of prior distributions and likelihoods in this legal context, and we discuss them in the context of two well-known "paradoxes," the paradox of the gatecrasher and the case of the blue bus. One of the key issues in each example is specification of the prior and of the likelihood, and in distinguishing what information goes where. We conclude with a discussion of the critical forensic evidence in a recent criminal case and alternative ways to use that evidence in specifying likelihoods and priors.

1. Introduction

Because the resolution of legal disputes deals with the evaluation of evidence and the making of decisions, it should not surprise us that many authors have argued for the use of statistical methods in decisionmaking in courtroom settings and that many witnesses (especially expert witnesses) in actual legal cases have quantified the evidence they presented in probabilistic terms (e.g., see Fienberg, 1989, and Weinstein, 1988). At the level of legal decisionmaking the literature, both legal and statistical, is dominated by the advocates of the Bayesian or subjective approach (e.g., see Good, 1950; Kaplan, 1968; Finkelstein and Fairley, 1970, 1971; Fairley, 1973; Lindley, 1977 a,b; Evett, 1977, 1978, 1984; Kaye, 1982, 1988; Fienberg and Kadane, 1983; and Fienberg and Schervish, 1986) but advocates of other approaches still exist (e.g., see Cohen, 1977 and Shafer, 1986). In the present paper, we adopt the Bayesian perspective and the Bayesian interpretation of standards of proof set forth in Fienberg and Kadane (1983) and utilized in Fienberg and Schervish (1986).

In Section 2, we briefly summarize the Bayesian apparatus as it pertains to the evaluation of legal evidence. Then, in Section 3, we turn to two "classic" paradoxes often referred to in the legal literature, the paradox of the gatecrasher and the case of the blue bus. Both of these paradoxes raise provocative issues regarding the specification of priors and likelihoods associated with the "guilt" or "liability" of the defendant. In many ways these examples point out the ambiguities of language, similar to those described by Bayarri, DeGroot, and Kadane (1988), which make the separation of what is prior and what is likelihood difficult at best.

In Section 4, we consider the quantitative forensic evidence introduced in a recent criminal case and explore how such evidence might have been used to create a suitable likelihood function. The same difficulties encountered in Section 3 emerge here, and these are compounded by additional problems of specification that relate to multiple criminal charges and the selection of evidence. We explore several alternative likelihood specifications.

George Barnard has not, to my knowledge, written about the specification and interpretation of likelihoods in a legal context. Nonetheless, given his longstanding interests in issues involving the use of likelihood functions, it is my hope that he will find the problems described in this paper intriguing.

2. Standards of Legal Proof

From the Bayesian perspective the legal decisionmaker, be it a judge or a member of a jury, begins with an *a priori* assessment of the legal matter at issue and uses a subjective evaluation of the evidence presented to update that prior to produce an a posteriori assessment using Bayes' Theorem. Two different standards of proof apply in Anglo-American law for criminal and civil cases. We describe the Bayesian interpretation of these here based on the approach set forth in Fienberg and Kadane (1983). This background helps set the context for the discussion of quantitative evaluation of evidence that follows in Sections 3 and 4.

2.1 The Criminal Standard

A juror or judge in a criminal case is typically asked to determine whether the defendant is guilty "beyond a reasonable doubt," beginning from the presumption that the defendant is "innocent until proven guilty." How does this standard get interpreted from the Bayesian perspective?

There is general agreement among legal commentators that the meaning of the standard can vary with the crime as well as with other aspects of the trial. Thus, each juror (or judge) begins in a given case with a prior probability of "guilt of the defendant," $P_j(G)$, and a personal threshold probability, b_j, such that

$$0 < P_j(G) < b_j < 1. \tag{1}$$

The quantity $P_j(G)$ is formulated independently for each trial, but may depend on general information about the nature of charges, as well as external information such as the percentage of defendants found to be guilty when charged with similar crimes. Although precision might suggest the use of the notation $P_j(G \mid D_j)$, where D_j is the jth juror's information, we opt here for the simpler notation where the conditioning is implicit in the subscript. There are also some legal prescriptions about what information a juror is allowed to condition upon, but in practice jurors will often ignore instructions of this sort.

If the probability that juror j will find the defendant guilty after seeing the evidence, E, presented at the trial is written as $P_j(G \mid E)$, then, at a minimum, requiring that the defendant be convicted only when guilty beyond a reasonable doubt can be represented by asking for a guilty vote from the jth juror (or verdict in the case of a judge), if and only if

$$P_j(G \mid E) \geq b_j. \tag{2}$$

Legal commentary suggests that b_j should be well above $\frac{1}{2}$, perhaps quite close to 1 (for further discussion see Fienberg and Schervish, 1986). Thus, in principle, a trial in which no evidence is presented would lead to the unanimous acquittal of the defendant. A reviewer has suggested that the choice of the threshold probability need not be fixed, i.e., it could vary as evidence is presented. We find it somewhat difficult to reconcile this proposal with our understanding of legal prescripts but we think the idea bears greater scrutiny. Another reader noted that he thought that b_j in principle should be the same for all jurors, even though in practice it varies from person to person. The law uses vague legal phrases that make it difficult to get unanimity, even in principle, on this issue.

2.2 The Civil Standard

The standards of proof in civil cases are different from those in criminal cases, reflecting the different relationships between the two parties involved. In a civil case, a juror or judge is required to "find" for the plaintiff if such a finding is supported by "a preponderance of the evidence," a "balance of probabilities," or is "more probable than not."

As in criminal cases, the civil standard varies somewhat from trial to trial and possibly from juror to juror. Thus juror j in a civil case begins with a prior probability of "finding for the plaintiff," $P_j(L)$ where L connotes liability, and a personal threshold probability, d_j, such that

$$0 < P_j(L) < d_j < 1. \tag{3}$$

After seeing the evidence E presented at trial the juror updates using Bayes' Theorem to produce $P_j(L \mid E)$ and votes in favor of a finding for the plaintiff if and only if

$$P_j(L \mid E) \geq d_j. \tag{4}$$

The civil standard is not as demanding as the criminal one and thus the two threshold probabilities must satisfy

$$b_j > d_j. \tag{5}$$

In the sense that a civil case involves the settlement of a legal dispute between two parties of equal standing, some legal commentators (e.g., see Appendix A in Fienberg, 1989) argue that d_j should be "close" to $\frac{1}{2}$ (i.e., slightly greater than $\frac{1}{2}$) and that jurors should start with the common prior probability $P_j(L) = \frac{1}{2}$. But there is an asymmetry between plaintiff and defendant in a civil case and, since in the absence of evidence the juror should find for the defendant, Fienberg and Kadane (1983) argue that the legal theories really presuppose $P_j(L) \leq \frac{1}{2}$, rather than $P_j(L) = \frac{1}{2}$. Moreover they and others argue there is no formal restriction suggesting how close d_j should be to $\frac{1}{2}$. Indeed the notion of "burden of proof" that rests on the plaintiff suggests some gap between $P_j(L)$ and d_j that needs to be overcome by the evidence.

3. Two Classsical Legal Paradoxes

In their attempts to argue against the use of probability in the course of legal decisionmaking various authors have attempted to present highly stylized descriptions of legal disputes in which the use of probability is seemingly problematic. We reconsider two of the best-known examples of this genre described by Tribe (1971) and by Cohen (1977), both of which are civil in nature. In each case, we argue that there is no problem with the use of probability; however, we also note that, in both examples, there is more than one way to interpret the information in a probabilistic fashion. The real problem in these examples comes in deciding what information really belongs in the prior, i.e., $P_j(L)$, and what information forms the evidence and thus belongs in the likelihood. The discussion here is drawn from Fienberg (1986).

3.1 The Paradox of the Gatecrasher

Cohen poses the "paradox of the gatecrasher" as follows:

> Consider, for example, a case in which it is common ground that 499 people paid for admission to a rodeo, and that 1,000 are counted on the seats, of whom A is one. Suppose no tickets were issued and there can be no testimony as to whether A paid for admission or climbed over the fence. So by any plausible criterion of mathematical probability there is a .501 probability, on the admitted facts, that he did not pay. The mathematicist theory would apparently imply that in such circumstances the rodeo organizers are entitled to judgement against A for the admission-money, since the balance of probability (and also the difference between prior and posterior probabilities) would lie in their favour. But it seems manifestly unjust that A should lose his case when there is an agreed mathematical probability of as high as .499 that he in fact paid for admission.
>
> Indeed, if the organizers were really entitled to judgement against A, they would presumably be equally entitled to judgement against each person in the same situation as A. So they might conceivably be entitled to recover 1,000 admission-moneys, when it was admitted that 499 had actually been paid. The absurd injustice of this suffices to show that there is something wrong somewhere. But where?

Although Cohen is not explicit, he presents this example in the context of a civil dispute and thus the focus on $\frac{1}{2}$ as a threshold probability. In order to extract the misattributions of the role of Bayes' Theorem and the nature of probability from Cohen's case description, we ask: What is the evidence? This question is equivalent to asking: What is the likelihood ratio?

$$\frac{P(E \mid L)}{P(E \mid \bar{L})}?$$

Once we determine the likelihood ratio we can attempt to answer the question: What do we need to assume about $P_j(L)$ in order to reach Cohen's conclusion that "there is a .501 probability, on the admitted facts, that he did not pay," and thus that the rodeo organizers are entitled to a finding against A?

One way to view the evidence, E, is that it consists of the background information (that 1000 people attended the rodeo and 499 have paid to enter) and the fact that the defendant A attended the rodeo. We label these two components E_1 and E_2. The background information, E_1, consists of mere numbers that have no probability content for the issue at hand but are important nonetheless in understanding and framing the problem. Because E_1 is constant from this perspective, if we are to ask about $P(E_1 \mid \bar{L})$ and $P(E_1 \mid L)$, the conditioning on L and $\bar{L}$ does not seem to matter, i.e.,

$$P(E_1 \mid \bar{L}) = P(E_1 \mid L) = P(E_1). \tag{6}$$

Similarly, we would argue that

$$P(E_1 \mid \bar{L} \cap E_2) = P(E_1 \mid L \cap E_2) = P(E_1). \tag{7}$$

The evidence E_2, however, is crucial and explains why the gatecrasher case might not be viewed as just a case of "naked statistical evidence" (see Kaye, 1979). Rather, all of the useful evidence is encapsuled in the information that A was present at the rodeo. (In an actual case this information would come from eyewitnesses.) If there are N people who represent a reference population from which the rodeo attendees were drawn at random, then for non-gatecrashers we have the conditional probability:

$$P(E_2 \mid \bar{L}) = \frac{499}{N - 501}.$$

For gatecrashers, the conditional probability of E_2 is:

$$P(E_2 \mid L) = 1,$$

i.e., given that A is guilty, he attended the rodeo with probability 1. In this interpretation, we think of each non-gatecrasher in the reference population as being "equally likely to have been selected to have attended the rodeo," and we take $P(E_2 \mid L)$ to describe the population of this selection process. (If A has a special identification or label that distinguishes him, then the equally likely argument may be invalid.)

We can now put together the likelihood ratio:

$$\begin{aligned}\frac{P(E \mid L)}{P(E \mid \bar{L})} &= \frac{P(E_1 \cap E_2 \mid L)}{P(E_1 \cap E_2 \mid \bar{L})}\\ &= \frac{P(E_2 \mid L)P(E_1 \mid L \cap E_2)}{P(E_2 \mid \bar{L})P(E_1 \mid \bar{L} \cap E_2)}\\ &= \frac{1}{499/(N-501)}. \qquad (8)\end{aligned}$$

If the posterior odds, odds $(L \mid E)$, are to be 501/499, as Cohen and others suggest, then we can conclude that the prior odds, odds (L) must be $501/(N - 501)$. By working backwards from the odds version of Bayes' Theorem we find the following:

$$\text{odds}\,(L \mid E) = \frac{P(E \mid L)}{P(E \mid \bar{L})} \times \text{odds}\,(L). \qquad (9)$$

These prior odds imply that $P(L)$ equals the base rate of gatecrashers in the reference population. If N is large, then this prior probability is extremely small, and is not in accord with the commonly stated view that the law assumes we begin with prior odds of 1.

On the other hand, if we were to assume $P(L) = \frac{1}{2}$ as suggested by legal commentators, i.e. if we assume prior odds of 1, then the posterior odds of guilt become

$(N - 501)/499$. For a reference population of $N = 100{,}000$, the posterior odds are staggeringly large, approximately 199 to 1. This might suggest that we should not automatically assume a prior probability of $\frac{1}{2}$ in a civil case, irrespective of what we accept as given in that case.

One can argue that we have misspecified the problem, but where has the misspecification occurred? One possible answer is in the choice of the reference population. After all, we know that 1,000 people attended the rodeo and, in a sense, these people form the reference population. Suppose we alter slightly the problem description and note that the defendant was selected at random from among those in attendance at the rodeo. Thus we should take $N = 1{,}000$. But then the likelihood ratio is:

$$\frac{N-501}{499} = \frac{1,000-501}{499} = 1.$$

Moreover, there is available no "relevant evidence" under the standard of Rule 401 of the Federal Rules of Evidence, i.e. the evidence yields no change from the prior $P(L)$ to the posterior $P(L \mid E)$. Thus, there is no way to make a finding for the plaintiff if we accept the framework of Section 2. Martin (1986) makes an argument similar to this one, noting that the "evidence" is only base-rate information and thus there is no observable evidence in the formal legal sense.

We might also pick N to ensure a prior odds ratio of 1. If $N = 1{,}002$, then the prior odds are 1, the likelihood ratio is $501/499$, and the posterior odds are $501/499$. This likelihood, which Brilmayer (1986) seems to think is so intuitive and which Cohen implicitly assumes, is not very sensible absent an argument for the choice of $N = 1{,}002$.

Given the formulation of the gatecrasher paradox presented here one can see that:

(i) the "common ground" in Cohen's hypothetical case is not specific enough to ensure an unambiguous interpretation and likelihood specification,

(ii) there are several plausible subjective criteria that utilize Bayes' Theorem but that do not lead to a posterior odds of $501/499$, and

(iii) "the balance of probability" (and the difference between prior and posterior probabilities) does not necessarily lie in favor of the rodeo organizers.

The evidence in the gatecrasher problem is probabilistic only if we make explicit some random selection mechanism whereby defendants are chosen from some population. The probabilistic nature of such selection mechanisms is often not part of the evidence at all, or at least its link to the evidence is not made explicit to the legal decision maker. A classic criminal case where the selection mechanism used to identify the defendants, an interracial couple, was not made explicit at trial is

People v. Collins (68 Cal. 2d 319, 438 P. 2d33, 66 Cal. Rptr. 497 (1968 en banc)). This case is discussed by Finkelstein and Fairley (1970) and now is used in many introductory statistical textbooks to illustrate a very different statistical issue, the inappropriateness of multiple probabilities of dependent events.

3.2 The Case of the Blue Bus

Tribe (1971) posed a problem in the "blue bus case":

> ...Plaintiff is negligently run down by a blue bus. The question is whether the bus belonged to the defendant. Plaintiff is prepared to prove that defendant operates four-fifths of all the blue buses in town. What effect, if any, should such proof be given?

Again we must ask: what is the evidence, E? The evidence appears to consist of (1) the fact the plaintiff was hit by a blue bus, and (2) the high proportion of blue buses in town that are operated by the defendant. We refer to these components as E_1 and E_2. Unfortunately, the only way to use the evidence E is via a model for which the occurrence of E is influenced by whether or not the bus that struck the plaintiff actually belongs to the defendant, i.e., the hypothesis L.

There is no clear probabilistic mechanism that suggests itself here for the defendant as there was with random selection from the attendees for the gatecrasher problem. Instead, we need to posit a random mechanism for blue buses passing the location at which the plaintiff was run down. Now, if all blue buses in town are equally likely to pass this location during some time interval, we can write the likelihood ratio as:

$$\frac{P(E \mid L)}{P(E \mid \bar{L})} = \frac{4}{1}.$$

In writing the likelihood ratio in this form we assume that a randomly selected blue bus, from the population of the blue buses in town, struck the defendant. The equally likely assumption here is a key one and in a real trial we would expect some form of empirical evidence on its reasonableness.

Finally, to get the posterior odds against the defendant, we need to quantify the prior odds. If we take them as 1:1, the common argument made about this example—that the "naked evidence" leads to a posterior probability of .8 that defendant is liable—turns out to correspond to the Bayesian solution.

Others, however, have argued that, if the information about the blue buses in town is thought of as base rate information, commonly known in advance of the trial, then this information contributes to the prior odds and not at all to the likelihood. This argument might even lead to a prior probability of $P_j(L) = .8$, which some view to be impermissibly high. Whatever the prior probability, however, in this scenario there is no "relevant" evidence and thus no finding for the plaintiff is possible. For the base rate information to be relevant it needs to have probabilistic value and this can happen only with an explicit probabilistic mechanism. Had we been told about the existence of buses of other colors in town by Tribe in his case description, this second argument might be especially compelling. (Suppose, for example, that there are 400 buses in town—300 red and 100 blue—and the defendant was identified solely on the grounds that he owned more blue buses than anyone else in town (even though he owns only 20% of all of the buses).) The issue here relates to that considered in the actual criminal case discussed in the next section.

Martin (1986), in commenting on the blue bus case, argues that the information can be interpreted as observable evidence presented at trial—namely that the plaintiff was struck by a blue bus. She argues that, if there are 100 blue buses and 300 red ones in town and if the defendant owns 80 and 280 of them respectively, then

$$\frac{P(E \mid L)}{P(E \mid \bar{L})} = \frac{80/360}{20/40} = \frac{4}{9}.$$

If one now takes prior odds of 1:1, the likelihood reduces the probability of a finding for the plaintiff! Martin notes that if there were only 100 red buses with 20 owned by the defendant, the posterior probability is .8 and we are back in a position similar to the first argument.

4. The Likelihood of Rape: A Case Study

One might argue that the difficulties in specifying the likelihood function in the two hypothetical cases presented in the previous section arise primarily because they are highly stylized and presented out of the context of the actual trial setting. While this may well be true in part, we would argue that many of the same difficulties arise in actual legal cases. Indeed, Tribe's case of the blue bus is modeled after the well-known case of *Smith v. Rapid Transit* (317 Mass. 469, N.E. 2d 754 (1945)) that is widely cited in texts on evidence.

In this section we describe in detail the quantitative evidence introduced in a recent criminal trial in Pittsburgh and we use it to illustrate many of the same specification issues that were raised in Section 3. The case involves the use of forensic identification evidence linked to blood and other genetic markers. Berry and Geisser (1986) describe the increasing use of blood testing for hereditary factors in paternity cases to infer that a particular man is the father, and the role played by Bayes' Theorem in the presentation of such evidence. While many of the problems they describe are of some relevance in this case (e.g., the assumption of independence among markers, the choice of a reference population, errors in testing, additional genetic information, etc.) the use of genetic marker information in the present context is even more complex due to the fragmentary nature of the evidence involved and the multiple crimes the defendant is alleged to have committed.

The Case

In a 1987 criminal trial in Common Pleas Court in Pittsburgh, PA, Joseph Jamieson was found guilty of committing seven rapes that took place in the Shadyside district of the city over a period from April 18, 1985 to January 30, 1986. These rapes were linked together by the police primarily because of the locations, because the assailant wore a make-shift mask or ski mask, and because of a "common pattern" in each rape. The pattern was not "discovered" by the police, however, until the fourth or fifth case and later analysis of blood types and enzymes also linked the cases. Only one of the victims actually saw the assailant and might have been able to identify him.

At trial, the key witness for the prosecution was a worker in the Allegheny County Crime Laboratory who analyzed body secretion evidence taken from the victims as well as from the bedsheets and clothes in some of their departments. She compared this evidence from the scenes of the crimes with information about the blood of the defendant, testifying about the "match" in quantitative terms. The defense countered this testimony with that of a statistician who attempted to explain, among other things, how the probabilistic information presented by the forensic expert should be interpreted. This interpretation was Bayesian in nature although the terms Bayesian, personal probability, and prior probability were not explicitly used. While there were several other aspects of the evidence and eyewitness testimony from the victims, following the trial verdict several members of the jury told reporters that they were particularly impressed by the evidence presented by the forensic expert. For the remainder of this section we focus on the quantitative information presented by the forensic expert and explore how it might be assessed in the Bayesian context described in Section 2. Because this is a criminal case, the standard of proof required is that associated with the phrase "beyond a reasonable doubt."

The Evidence

The basic evidence presented by the forensic expert is summarized in Table 1. For each case we list the results of forensic tests for the ABO blood system and the PGM enzyme markers, as well as information on blood and enzyme markers for the victim. (In the Appendix, we provide some details on these two genetic systems to enable the reader to follow the calculations.) This information was then "combined"

TABLE 1

Evidence from Forensic Analysis of Body Fluids
from Defendant, Victims, and Evidence at Scenes of Crimes

Source	ABO	Secretor	PGM	Attibutable to Assailant
Victim A	O	No	2^+1^+	B (secretor)
Swab	B and H	Yes	–*	?
Swab	B and H	Yes	–	
Semen	B and H	Yes	–	
Bathrobe	B and H	Yes	2–1	
Victim B	B	Yes	1^+1^-	B or O (secretor) or Nonsecretor
Swab	B and H	Yes	$2^+1^+1^-$	2^+ or 2^+1^+ or 2^+1^-
Swab	B and H	Yes	–	
Victim C	A	No	2^+1^+	B (secretor)
Bedsheet	B and H	Yes	2–1	?
Victim D	AB	Yes	2^+1^+	?
Swab	A, B, and H	Yes	$2^+1^+1^-$	2^+1^- or 1^+1^- or 1^-
Victim E	O	Yes	1^+1^-	B (secretor)
Swab	B and H	Yes	2^+1^-	2^+ or 2^+1^-
Pantyhose	H	Yes	–	
Victim F	A	Yes	1^+	AB or B (secretor)
Swab	A, B, and H	Yes	$2^+1^+1^-$	2^+1^-
Victim G	O	Yes	1^+	B (secretor)
Swab	B and H	Yes	1^+	?
Defendant	B	Yes	2^+1^-	

* There was no activity on the enzyme test for this marker.

to produce information on blood and marker information attributable to the assailant listed in the final column of Table 1. The basic principle used in combining the information from the victim and the evidence at the scene of the crime is that only blood marker information not possibly attributed to the victim can be attributable to the assailant. For example, let us consider Victim D, who is type AB, a secretor, and of PGM type 2^+1^+. A swab from the rape kit, used to gather evidence when she reported the rape, yielded evidence of antigens A, B, and H. These could all have come from the victim because she is a secretor, and thus we can infer nothing about the ABO blood type of the assailant from this information. The swab also showed activity for PGM 2^+, 1^+, and 1^-. The 2^+ and 1^+ could be attributable to the victim but not the 1^-. Thus, if the assailant had subtype 2^+1^-, 1^+1^-, or 1^-, we would have been able to detect the evidence on the swab. We therefore infer that the assailant of Victim D is of unknown ABO type and is of PGM subtype 2^+1^-, 1^+1^- or 1^-.

The estimated frequencies of occurrence of the ABO and PGM types in the population of Allegheny County are listed in Table 2. The forensic laboratory worker who testified for the prosecution in this case used these proportions along with the ABO and PGM information attributable to the defendant, as listed in the final column of Table 1, to produce the estimated proportions in Table 3. For example, the assailant of Victim F is presumed to be a type AB secretor or type B secretor of PGM type 2^+1^-. The proportion of the population of type AB is .05 and of type B is .10, while .8 of the population are secretors. The proportion of PGM type 2^+1^- is .04. Thus the proportion of the population of Allegheny County with characteristics of the assailant of Victim F is

$$(.05 + .10) \times .10 \times .04 = .0048.$$

The final part of the testimony of the forensic expert as it relates to the evidence summarized in Tables 1 and 3 consisted of the following answers to questions posed by the prosecutor:

[Q] Now, if we were to conclude—if this Court were to conclude that a single person was the perpetrator by virtue of modus operandi in all of these cases, in each one of these seven cases—first of all, is it possible, given your blood working on this case, that one person could have deposited all seven stains in this case?

[A] Yes.

[Q] Okay. And if one person deposited all seminal stains in this case, are you able to put forth a probable genetic breakdown of their blood type and enzyme factors?

[A] Yes, I would. If one person would have done—committed all these sexual assaults—it would be my opinion that it would be a person that would be a type B secretor, PGM 2^+1^-.

[Q] And, again, what was the blood type and the enzyme factors involved in the blood you examined from Joseph Jamieson?

[A] Joseph Jamieson is a Type B secretor. His enzyme is PGM 2^+1^-.

TABLE 2

Population Proportions Corresponding to
Genetic Marker Groups and Types

Genetic Marker	Type	Proportions*
ABO	A	.39
	B	.10
	O	.45
	AB	.05
Secretor	Yes	.80
	No	.20
PGM	1	.59
	2–1	.36
	2	.04
PGM (subtypes)	1^+	.37
	1^+1^-	.20
	1^-	.02
	2^+1^+	.20
	2^+1^-	.04
	2^-1^+	.08
	2^-1^-	.02
	2^+	.03
	2^+2^-	.02
	2^-	.01

*Proportions within categories should add to 1, but are occasionally less due to rounding error.

Sources: (1) ABO, secretor and PGM data are based on sample genetic information from Grunbaum et al. (1980) and population information for Allegheny County.
(2) PGM subtype data are based on sample genetic information from Nelson (1984) and population information for Allegheny County.

[Q] What percentage of the male population, again, could have deposited that kind of seminal stain with that blood breakdown?

[A] 0.32 percent.

[Q] Three in a thousand?

TABLE 3

Population Proportions Corresponding to Genetic Marker Information Attributable to Assailants in Seven Rape Cases

Victim	Genetic Markers Attributable to Defendant	Proportion of Population with Attributable Marker
A	B (secretor)	.08
B	B or O (secretor) or Nonsecretor, PGM 2^+, 2^+1^+, or 2^+1^-	.17
C	B (secretor)	.08
D	PGM 2^+1^-, 1^+1^- or 1^-	.26
E	B (secretor), PGM 2^+ or 2^+1^-	.022
F	AB or B (secretor), PGM 2^+1^-	.0048
G	B (secretor)	.08
Composite	B (secretor); PGM 2^+1^-	.0032

[A] Yes.

[Q] If one person did all of these crimes, would they be within that percentage of the population as described of .32 percent?

[A] Yes.

[Q] Is Joseph Jamieson within that .32 percent of the population, that same .32 percent of the population?

[A] His genetic markers indicate he is .32 percent of the population.

Interpreting the Evidence

We note that the foregoing exchange between the prosecutor and the forensic expert contains no explanation of how the jurors are to interpret the proportion .0032. Moreover, while the proportions in the final column of Table 1 relate to the separate cases, the final probability of .0032 is based on *the assumption that all seven rapes were committed by the same person.*

Two pieces of additional evidence that were presented at trial were:

(a) The defendant, Joseph Jamieson, was originally arrested on charges unrelated to the seven rapes and then his blood was tested. When the police determined that his blood type matched the composite of the assailants in the seven rapes they charged him with these crimes and attempted to gather additional evidence;

(b) When the series of rapes occurred police believed that it included nine cases of rape. Only after the forensic laboratory reported the results of the blood tests for the victims and the evidence did the police exclude two of the nine cases.

In addition, a potentially important piece of evidence was not allowed:

(c) Following the arrest of the suspect additional rapes with similar descriptions occurred, but the defendant was prevented from presenting this information at trial for technical legal reasons.

The naive Bayesian interpretation of the evidence presented in this case would be as follows. Let G be the event that the defendant committed all seven rapes, and let E be the "composite" blood evidence presented by forensic expert. Then

$$\frac{P(E \mid G)}{P(E \mid \bar{G})} = \frac{1}{.0032} = 312.5 \tag{10}$$

and, even with prior odds as low as 1 to 16, the posterior odds that the defendant is guilty are 19.5 to 1, i.e., $P(G \mid E) > .95$. Now, if we add the information that the assailant was a male, age 25 to 54, etc., the value of $P(G \mid E)$ is even higher.

If, however, the blood evidence was really used to identify the defendant, then he is not distinguishable from all other males in Allegheny County in the broad 25–54 age group reported by the victims. Then we know a priori that there are about 1150 males, 25–54 years of age, with blood of "B (secretor), PGM 2^+1^-" status, and the defendant is indistinguishable from the other 1149. Now the nature of the evidence is changed. We label this version of the evidence as E^*. Then

$$\frac{P(E^* \mid G)}{P(E^* \mid \bar{G})} = 1, \tag{11}$$

since the blood identification should be thought of as "known" at the start of trial when $P(G)$ is assessed. Moreover we might even want to think of using $P(G) = 1/1150$. (Note the parallel here with the blue bus case from Section 3.) This interpretation was included as part of the testimony of the statistical expert presented by the defense.

There were seven separate crimes at issue in this trial and the preceding discussion has treated them as if they were one. Suppose we leave aside (temporarily) the issue of the original identification of the defendant. For each crime we have somewhat different blood identification evidence. Let G_i be the event that the defendant is guilty of crime i, $i = 1, 2, \ldots, 7$, and let E_i be the blood identification evidence in the ith case, $i = 1, 2, \ldots, 7$. The focus of the discussion at trial is on

$$G' = \cap_{i=1}^{7} G_i \tag{12}$$

based on evidence

$$E' = \cap_{i=1}^{7} E_i. \tag{13}$$

What we know is that

$$P(E_i \mid G_i) = 1 \qquad i = 1, 2, \ldots, 7, \tag{14}$$

and we can use the proportions listed in Table 3 to correspond to $P(E_i \mid \bar{G}_i)$. Moreover, for the joint event G' we have

$$P(E' \mid G') = 1. \tag{15}$$

We could now conceive of proceeding in one of three different ways, by assessing

(i)

$$\frac{P(E' \mid G_i)}{P(E' \mid \bar{G}_i)} \text{ for } i = 1, 2, \ldots, 7, \tag{16}$$

(ii)

$$\frac{P(E' \mid G')}{P(E' \mid \bar{G}')}, \tag{17}$$

or

(iii)

$$\frac{P(E' \mid \overline{\cap_{i=1}^{7} \bar{G}_i})}{P(E' \mid \cap_{i=1}^{7} \bar{G}_i)}. \tag{18}$$

Each of these likelihood ratios has pieces which we can evaluate numerically with ease, and pieces for which the evaluation is not at all clear. For (i), we know

$$\frac{P(E_i \mid G_i)}{P(E_i \mid \bar{G}_i)} = \frac{1}{P(E_i \mid \bar{G}_i)}, \tag{19}$$

but we are unclear how to assess

$$\frac{P(\cap_{j \neq i} E_j \mid E_i \cap G_i)}{P(\cap_{j \neq i} E_j \mid E_i \cap \bar{G}_i)}. \tag{20}$$

For (ii), we have $P(E' \mid G') = 1$ but the calculation of the denominator is complex. Finally, for (iii), if we assume that the a priori assessments of guilt in the seven cases are separate, and thus that

$$P(\cap_{i=1}^{7} \bar{G}_i) = \Pi_{i=1}^{7} P(\bar{G}_i) \tag{21}$$

Then we can evaluate the denominator of (iii) in the form

$$P(E' \mid \cap_{i=1}^{7} \bar{G}_i) = \Pi_{i=1}^{7} P(E_i \mid \bar{G}_i) \tag{22}$$

but have difficulty with the numerator which can be rewritten as

$$P(E' \mid \cup_{i=1}^{7} G_i). \tag{23}$$

Which of the three likelihood approaches corresponds to the legally correct structure? Despite the single trial and the attempt of the prosecution to merge the seven crimes, each of the seven crimes corresponded to a separate legal charge and the jury was required to enter separate verdicts on each. Thus the seven likelihoods in approach (i) seem to be the appropriate ones, with seven independently assessed priors, $P(G_i), i = 1, 2, \ldots, 7$. These priors might reasonably be expected to be the same. Nonetheless, the conflation of the charges, the common scenarios for the crimes, the evidence and its evaluation suggest features of both approaches (ii) and (iii). After all, because of the common modus operandi, the state may be entitled to use evidence from the other six rapes in each of the seven cases. Moreover, if we choose to interpret the a priori identification of the defendant by means of his blood markers, then we are back to the alternative interpretation for the initial naive presentation of the likelihood and there is no further information contained in the likelihoods for any of the three approaches.

An issue ignored throughout the trial, and in our discussion up to this point, is the fragmentary nature of the blood identification evidence. In none of the seven cases did the information attributable to the assailant actually match that of the defendant. Intuitively, we expect to find information on "exact matches" in all seven cases far more compelling than the fragmentary matches displayed in Table 1. To come to grips with this aspect of the data, we need to determine how it might affect the assessment of quantities such as those in expressions (19), (20), and (23).

None of the likelihoods in expressions (16) to (18) take account of the possible selection of the seven crimes in question from a larger set of similar crimes, but where the information in the other crimes were excluded from presentation at trial either because the blood evidence was not in full agreement with that of the defendant or because the crimes occurred after the arrest of the defendant. Surely this should affect the credibility of the evidence, but how? Furthermore, the choice of the genetic markers to be used in the forensic tests raises further issues regarding statistical specification and interpretation as does possible errors in the tests themselves, and the use of sample data from North Carolina and California to estimate the probabilities in Table 2. Berry and Geisser (1986) discuss these latter issues in the context of paternity cases and their discussion is relevant here as well.

There is an interesting postscript to this case. Subsequent to the trial the Federal Bureau of Investigation and some local jurisdictions in the United States have begun to use a relatively new DNA typing technology for blood, hair, and semen samples that is known as "DNA fingerprinting" (e.g., see Jeffreys, Wilson, and Thein, 1985). It is widely believed that this approach "can be used to identify any individual if enough DNA can be sequenced" (Koshland, 1988) and that "DNA typing produces unique patterns that are individual specific" (Hicks, 1989). As this method works its way into the courts we are likely to see the disappearance of genetic evidence of the sort used in the Jamieson case. The *Wall Street Journal* recently (March 17, 1989) reported that DNA fingerprinting has been permitted in 79 criminal cases (usually by the prosecution) in 24 different states. But there are major questions regarding the reliability of such DNA testing and the empirical support for the method's apparent effectiveness in identifying individuals (Berry, 1989; Cohen, 1988; Koshland, 1988). Thus the types of issues raised in our discussion of the present case will almost certainly be relevant to the evaluation of the new forensic evidence.

5. Summary

The trial description and the quantitative evidence presented in Section 4 suggest that problems of likelihood specification and a priori assessment that occur in real world settings are far more complex than those that arise in the highly stylized examples of Section 3. The richness of detail available in real world settings does little to clarify the naive statistical questions

(a) What is the likelihood function?

(b) What is the a priori assessment of the probability of guilt?

These questions are similar to those raised by Bayarri, DeGroot, and Kadane (1988) from a more technical perspective.

Until we are better able to sort out issues such as the ones described in this paper, the quantitative assessment of legal evidence and the use of the Bayesian or any other probabilistic framework for legal decisionmaking will remain highly controversial.

Acknowledgments

This research was supported in part by the National Science Foundation under Grant No. SES-8701606 to Carnegie Mellon University. A version of the material was presented at the *Thirty-third NBER-NSF Seminar on Bayesian Inference in*

Econometrics, October 24–25, 1986, at the University of California, Riverside. I thank David Kaye and an anonymous referee for helpful comments and suggestions.

References

Bayarri, M.J., M.H. DeGroot, and J.B. Kadane (1988). " What is the Likelihood Function?" (with discussion), in S.S. Gupta and J.O. Berger, eds., *Proceedings of the Fourth Purdue Symposium on Decision Theory and Related Topics*, New York: Springer-Verlag, 3–27.

Berry, D.A. (1989). "Inferences in Criminal and Paternity Cases Using Highly Polymorphic DNA Sequences, unpublished manuscript.

Berry, D.A. and S. Geisser (1986). "Inference in Cases of Disputed Paternity," in M.H. DeGroot, S.E. Fienberg, and J.B. Kadane, eds., *Statistics and the Law*, New York: Wiley, 353–382.

Brilmayer, L. (1986). "Second Order Evidence and Bayesian Logic," *Boston University Law Review*, Vol. 66, 673–692.

Cohen, J.E. (1988). "DNA Fingerprinting for Forensic Identification: Problems in the Analysis and Interpretation of Data," unpublished manuscript.

Cohen, L.J. (1977). *The Probable and the Provable.* Oxford: Clarendon Press.

Elandt-Johnson, R.C. (1971). *Probability Models and Statistical Methods in Genetics.* New York: Wiley.

Evett, I.W. (1977). "The Interpretation of Refractive Index Measurements," *Forensic Science*, Vol. 9, 209–217.

Evett, I.W. (1978). "The Interpretation of Refractive Index Measurements, II," *Forensic Science International*, Vol. 12, 37–47.

Evett, I.W. (1984). "A Qualitative Theory of Interpreting Transfer Evidence in Criminal Cases," *Applied Statistics*, Vol. 33, 25–32.

Fairley, Wm.B. (1973). "Probabilistic Analysis of Identification Evidence," *Journal of Legal Studies*, Vol. II, 493–513.

Fienberg, S.E. (1986). "Comment: Gatecrashers, Blue Buses, and the Bayesian Representation of Legal Evidence," *Boston University Law Review*, Vol. 66, 693–699.

Fienberg, S.E., ed. (1989). *The Evolving Role of Statistical Assessments as Evidence in the Courts*, Report of the Panel on Statistical Assessments as Evidence, National Research Council, New York: Springer-Verlag.

Fienberg, S.E. and J.B. Kadane (1983). "The Presentation of Bayesian Statistical Analyses in Legal Proceedings," *The Statistician*, Vol. 32, 88–98.

Fienberg, S.E. and M.J. Schervish (1986). "The Relevance of Bayesian Inference for the Representation of Statistical Evidence and for Legal Decisionmaking," *Boston University Law Review*, Vol. 66, 771–798.

Finkelstein, M. and W. Fairley (1970). "A Bayesian Approach to Identification Evidence," *Harvard Law Review*, Vol. 83, 235–314.

Finkelstein, M. and W. Fairley (1971). "A Comment on 'Trial by Mathematics'," *Harvard Law Review*, Vol. 84, 1801–1809.

Good, I.J. (1950). *Probability and the Weighing of Evidence.* London: Griffin, 66–67.

Grunbaum, B.W., S. Selvin, B.A. Myhre, and N. Pace (1980). "Distribution of Gene Frequencies and Discrimination Probabilities for 22 Blood Genetic Systems in Four Racial Groups," *Journal of Forensic Sciences*, Vol. 25, 428–444.

Hicks, J.W. (1989). "DNA Typing: A Unique Weapon Against Crime," *The Scientist*, January 23, 1982, 12–13.

Jeffreys, A.J., V. Wilson, and S.L. Thein (1985). "Individual-Specific 'Fingerprints' of Human DNA," *Nature*, Vol. 314, 67–73.

Kaplan, J. (1968). "Decision Theory and the Factfinding Process," *Stanford Law Review*, Vol. 20, 1065–1083.

Kaye, D. (1979). "The Paradox of the Gatecrasher and Other Stories," *Arizona State Law Journal*, 101–109.

Kaye, D. (1982). "Statistical Evidence of Discrimination" (with discussion), *Journal of the American Statistical Association*, Vol. 77, 773–792.

Kaye, D. (1988). "What is Bayesianism? A Guide for the Perplexed," *Jurimetrics Journal*, Vol. 28, 161–177.

Koshland, D.E., Jr. (1988). "A Tale of Two Techniques," *Science*, Vol. 242, 993.

Lindley, D.V. (1977a). "A Problem in Forensic Science," *Biometrika*, Vol. 64, 207–213.

Lindley, D.V. (1977b). "Probability and the Law," *The Statistician*, Vol. 26, 203–212.

Martin, A.W. (1986). "Comment on Brilmayer," *Boston University Law Review*, Vol. 66, 709–714.

Nelson, M.S. (1984). "A Computer Assisted Population Frequency Study of 14 Polymorphic Blood Grouping Systems in North Carolina," *J. Forensic Sciences*, Vol. 29, 762–773.

Race, R.R. and R. Sanger (1968)., *Blood Groups in Man*, (5th edition). Oxford: Blackwell.

Shafer, G. (1986). "The Construction of Probability Arguments," *Boston University Law Review*, Vol. 66, 799–816.

Tribe, L. (1971). "Trial by Mathematics: Precision and Ritual in the Legal Process," *Harvard Law Review*, Vol. 84, 1329–1393.

Weinstein, J. (1988). "Litigation and Statistics," *Statistical Science*, Vol. 3, 286–297.

Appendix: Detection of Information from ABO Blood Group and PGM Marker Systems

The ABO blood group and its detection plays a special role in the case study of Section 4. Here we present a brief summary of the information necessary to follow the calculations and discussion there. This material is drawn primarily from Elandt-Johnson (1971) and Race and Sanger (1968).

The ABO blood system has three "alleles," A, B, and 0 with six "genotypes" (AA, AO, BB, BO, AB, OO). The two alleles A and B are each dominant to O but codominant to each other. The usual method used to determine the blood group of an individual is based on the seriological reactions of the red blood cells with special reagents, anti-A and anti-B. Anti-A reacts with genotypes AA, AO, and AB, while anti-B reacts with BB, BO, and AB. If blood cells react to only anti-A the blood is said to be of type A (i.e. genotype AA or AO); if it reacts to only anti-B it is said to be of type B (i.e., genotype BB or BO); if it reacts to both it is said to be of type AB (i.e., genotype AB); and if it reacts to neither it is said to be of type O (i.e. genotype OO). The determination of *ABO* blood group type for the defendant and the seven victims comes from this kind of test procedure done directly on blood samples.

Evidence of *ABO* blood group antigens is not confined to red blood cells, but shows up in other body fluids such as saliva, semen, and vaginal secretions. The occurrence of such secretions is controlled by another "secretion gene," *Se*, with two alleles Se and se, and three genotypes: SeSe, Sese, sese. The SeSe and Sese types are known as "secretors" and evidence of A and B substances shows up their other bodily fluids.

Closely related to the ABO system is another gene, H, which has two alleles, H and h, and three genotypes: HH, Hh, hh. The H allele is dominant and the frequency of h is extremely low. Thus almost all individuals are either HH or Hh; it is of virtually no value to test blood samples for this gene. The genes from the H blood group are thought to interact with those from the ABO group in secretors in order to produce the antigen A or the antigen B in the saliva and bodily fluids of secretors. The antigen H shows up in the fluids of such individuals as well. In secretors of type O, however, the antigen H is all that one can detect in bodily fluids.

To check whether an individual is a secretor his/her saliva is tested for the presence of either antigen A, B, or H. The presence of ABO or H substances on evidence found at the scene of the crime can occur only via the bodily fluids of a secretor.

The phosphoglucomutase system (PGM) has three distinguishable phenotypes, 1-1, 2-1, 2-2. These can be broken down into the following subtypes:

Phenotype	*Subtype*
1–1	1^+
	1^+1^-
	1^-
2–1	2^+1^+
	2^+1^-
	2^-1^+
	2^-1^-
2–2	2^+
	2^+2^-
	2^-

Both phenotype and subtypes can be readily identified from analysis of blood, and evidence of this genetic marker is also found in other bodily fluids. The occurrence of components of the PGM system in other bodily fluids is *not* affected by the secretor status of the individual.

The detection of the PGM phenotypes and subtypes in the forensic laboratory is carried out by means of an enzyme test. Even though bodily fluids deposited on material contain information on PGM markers the test may be sensitive enough to detect the marker.

Bayesian and Likelihood Methods in Statistics and Econometrics
S. Geisser, J.S. Hodges, S.J. Press and A. Zellner (Editors)

SEEMINGLY UNRELATED REGRESSION WITH LINEAR HIERARCHICAL STRUCTURE

Yoel Haitovsky
Hebrew University of Jerusalem

Abstract

This paper generalizes Zellner's Bayesian analysis of the seemingly unrelated regression (SUR) model. We present the SUR model as a three stage hierarchical model, in which the regression model is the first stage, the vectors of regression parameters for the SU regressions vary around a common vector in the second stage, and the unknown common vector forms the third stage. We examine several cases: with and without normality assumptions on the first two stages (with the third stage left uninformative) and covariance known and unknown. An illustrative numerical example is presented.

1. Introduction

Consider a system of m linear seemingly unrelated regression equations (SUR)

$$\underset{\sim}{y}_i = \underset{\sim}{X}_i \underset{\sim}{\beta}_i + \underset{\sim}{\epsilon}_i, \qquad i = 1, \ldots, m \tag{1}$$

where $\underset{\sim}{y}_i$ is a $T \times 1$ vector of T observations on the variables to be explained, $\underset{\sim}{X}_i$ is a $T \times k$ matrix, with full column rank, of T observations on k explanatory exogenous variables, $\underset{\sim}{\beta}_i$ is a $k \times 1$ vector of k regression coefficients associated with them, and $\underset{\sim}{\epsilon}_i$ is a $T \times 1$ vector of T unobservable disturbances with

$$\left.\begin{aligned} E(\underset{\sim}{\epsilon}_i) &= \underset{\sim}{0} \\ E(\underset{\sim}{\epsilon}_i \underset{\sim}{\epsilon}_j') &= \gamma_{ij} \underset{\sim}{I}_T \end{aligned}\right\} \tag{2}$$

Such a system of equations has become very popular in econometric studies and has been used, among other things, to analyze systems of consumer demand equations, to combine cross section and time series data, to include inter-regional correlations in the explanations of an economic activity in different geographical locations, etc.

If $\underset{\sim}{X}_i = \underset{\sim}{X}_j$ for all $i, j = 1, \ldots, m$, the multivariate regression model is obtained as a special case.

All m regression equations can be combined into one system of equations which can be compactly represented as

$$\underset{\sim}{y} = \underset{\sim}{X}\underset{\sim}{\beta} + \underset{\sim}{\epsilon} \tag{1a}$$

where $\underset{\sim}{y} = (\underset{\sim}{y}_1', \underset{\sim}{y}_2', \ldots, \underset{\sim}{y}_m')'$, the stacked $mT \times 1$ vector of T observations on all the m equation's dependent variables, $\underset{\sim}{X} = \text{Diag}\{\underset{\sim}{X}_1, \underset{\sim}{X}_2, \ldots, \underset{\sim}{X}_m\}$ an $mT \times mk$ block diagonal matrix of observations on the explanatory variables, $\underset{\sim}{\beta} = (\underset{\sim}{\beta}_1', \underset{\sim}{\beta}_2', \ldots, \underset{\sim}{\beta}_m')'$ the $mk \times 1$ stacked vector of all m equation's regression coefficients and $\underset{\sim}{\epsilon} = (\underset{\sim}{\epsilon}_1', \underset{\sim}{\epsilon}_2', \ldots, \underset{\sim}{\epsilon}_m')'$, the $mT \times 1$ vector of all the system's disturbances, with

$$E(\underset{\sim}{\epsilon}) = \underset{\sim}{0} \quad \text{and} \quad E(\underset{\sim}{\epsilon}\underset{\sim}{\epsilon}') = \underset{\sim}{\Gamma} \otimes \underset{\sim}{I}_T, \tag{2a}$$

where $\underset{\sim}{\Gamma}$ is an $m \times m$ positive definite symmetric (PDS) matrix with typical elements γ_{ij}, $i, j = 1, \ldots, m$, and $\otimes$ denotes the usual Kronecker product of matrices. The (generalized) least squares solution was obtained by Zellner (1962) and is given by

$$\hat{\underset{\sim}{\beta}}_{LS} = \{\underset{\sim}{X}'(\underset{\sim}{\Gamma}^{-1} \otimes \underset{\sim}{I}_T)\underset{\sim}{X}\}^{-1}\underset{\sim}{X}'(\underset{\sim}{\Gamma}^{-1} \otimes \underset{\sim}{I}_T)\underset{\sim}{y}. \tag{3}$$

When normality of the disturbances is assumed, $\hat{\underset{\sim}{\beta}}_{LS}$ will also be the maximum likelihood estimate for likelihood functions of the form $\underset{\sim}{\epsilon}'(\underset{\sim}{\Gamma}^{-1} \otimes \underset{\sim}{I}_T)\underset{\sim}{\epsilon}$. It is also the Gauss-Markov estimator of $\underset{\sim}{\beta}$ given $\underset{\sim}{\Gamma}$. See Zellner (1962). In practice, however, $\underset{\sim}{\Gamma}$ is not known and thus is replaced by its consistent estimator $\underset{\sim}{G}$, with a typical element g_{ij}, based on the restricted residuals obtained by estimating each equation separately by ordinary least squares, namely,

$$g_{ij} = \frac{1}{T}\underset{\sim}{y}_i'[\underset{\sim}{I}_T - \underset{\sim}{X}_i(\underset{\sim}{X}_i'\underset{\sim}{X}_i)^{-1}\underset{\sim}{X}_i'][\underset{\sim}{I}_T - \underset{\sim}{X}_j(\underset{\sim}{X}_j'\underset{\sim}{X}_j)^{-1}\underset{\sim}{X}_j']\underset{\sim}{y}_j. \tag{4}$$

Divisors other than T were proposed to obtain unbiased estimates [Zellner and Huang (1962)] or minimum mean squared error estimates in the class of sum of squared restricted residuals [Srivastava and Dwivedi (1979)].

Finally, the variance of $\hat{\underset{\sim}{\beta}}_{LS}$ is obtained by

$$V(\hat{\underset{\sim}{\beta}}_{LS}) = \{\underset{\sim}{X}'(\underset{\sim}{\Gamma}^{-1} \otimes \underset{\sim}{I}_T)\underset{\sim}{X}\}^{-1} \tag{5}$$

which can be estimated by replacing $\underset{\sim}{\Gamma}$ by $\underset{\sim}{G}$, as given by (4).

Bayesian analyses of the SUR model was first applied by Zellner (1971), who employed a diffuse prior about the parameters of the form

$$p(\underset{\sim}{\beta}, \underset{\sim}{\Gamma}^{-1}) = p(\underset{\sim}{\beta}) \cdot p(\underset{\sim}{\Gamma}^{-1}) \; \alpha \; |\underset{\sim}{\Gamma}^{-1}|^{-\frac{m+1}{2}}$$

Adding to it a multivariate normal distribution assumption on the model disturbances, and since essentially SUR is a generalized regression model, the conditional mean and covariance matrix give precisely the generalized least squares estimators for the SUR coefficients and its associated variance matrix that Zellner obtains in studying the SUR model from the sampling theoretic point of view.

In the present paper we generalize Zellner's Bayesian approach. In Section II we present SUR as a three stage hierarchical model with and without the normality assumption on the first two stages. The third stage is left uninformative. In Section III we regard the covariances as unknown and derive their estimates. In Section IV we present an illustrative numerical example.

2. Hierarchical Structure

It will be convenient now to regard the m-equation model as representing individuals or firms whose behavior is explained by the set of explanatory variables $\underset{\sim}{X}_i$. Suppose now that the researcher regards the group of individuals or firms represented by the m equations as being homogeneous, in the sense that the reaction of each individual to each of the explanatory variables differs from all other individuals in the group by a random factor. It can be expressed as a random vector $\underset{\sim}{\nu}_i (i = 1, \ldots, m)$ of deviations from a common vector $\underset{\sim}{\xi}$. The researcher might also have prior information about $\underset{\sim}{\xi}$, which he can use in the estimation of the $\underset{\sim}{\beta}_i$'s, or otherwise the last prior may be uninformative. The latter can be treated as a special case of the former, as was shown by Lindley and Smith (1972), and indeed we shall restrict our attention here to this special case. The researcher's basic model and his information about its parameters can be expressed as a two stage linear hierarchical model with a third uninformative stage.

Assuming normality, this model is written as:

$$\begin{aligned}
&\underset{\sim}{y} \mid \underset{\sim}{\beta},\ \underset{\sim}{\Gamma} \sim N(\underset{\sim}{X}\underset{\sim}{\beta}, \underset{\sim}{\Gamma} \otimes \underset{\sim}{I}_T);\\
&\underset{\sim}{\beta} \mid \underset{\sim}{\xi},\ \underset{\sim}{\Omega} \sim N(\underset{\sim}{l}_m \otimes \underset{\sim}{\xi}, \underset{\sim}{\Omega});\\
&\underset{\sim}{\xi} \text{ is uniform on } R^k.
\end{aligned}$$

Then the joint (improper) pdf is proportional to

$$|\underset{\sim}{\Gamma} \otimes \underset{\sim}{I}_T|^{-\frac{1}{2}} |\underset{\sim}{\Omega}|^{-\frac{1}{2}} \exp\left[-\frac{1}{2}(\underset{\sim}{y} - \underset{\sim}{X}\underset{\sim}{\beta})'(\underset{\sim}{\Gamma}^{-1} \otimes \underset{\sim}{I}_T)(\underset{\sim}{y} - \underset{\sim}{X}\underset{\sim}{\beta}) \right.$$
$$\left. -\frac{1}{2}(\underset{\sim}{\beta} - \underset{\sim}{l}_m \otimes \underset{\sim}{\xi})'\underset{\sim}{\Omega}^{-1}(\underset{\sim}{\beta} - \underset{\sim}{l}_m \otimes \underset{\sim}{\xi}) \right]$$

where $\underset{\sim}{l}_m$ is a $m \times 1$ vector of ones.

The integration with respect to $\underset{\sim}{\xi}$ is performed by first decomposing the second term in the exponent of eqn. (6):

$$\begin{aligned}&(\underset{\sim}{\beta} - \underset{\sim}{l}_m \otimes \underset{\sim}{\xi})'\underset{\sim}{\Omega}^{-1}(\underset{\sim}{\beta} - \underset{\sim}{l}_m \otimes \underset{\sim}{\xi})\\ &= (\underset{\sim}{\beta} - \underset{\sim}{l}_m \otimes \underset{\sim}{\bar{\xi}})'\underset{\sim}{\Omega}^{-1}(\underset{\sim}{\beta} - \underset{\sim}{l}_m \otimes \underset{\sim}{\bar{\xi}}) + [\underset{\sim}{l}_m \otimes (\underset{\sim}{\xi} - \underset{\sim}{\bar{\xi}})\underset{\sim}{\Omega}^{-1}[\underset{\sim}{l} \otimes (\underset{\sim}{\xi} - \underset{\sim}{\bar{\xi}})]\end{aligned}$$

where

$$\bar{\xi} = [(\underset{\sim}{l}'_m \otimes \underset{\sim}{I}_k)\underset{\sim}{\Omega}^{-1}(\underset{\sim}{l}_m \otimes \underset{\sim}{I}_k)]^{-1}(\underset{\sim}{l}'_m \otimes \underset{\sim}{I})k)\underset{\sim}{\Omega}^{-1}\underset{\sim}{\beta} \tag{6}$$

and then by straightforward integration, the second term in the exponent in (6) becomes

$$\begin{aligned}(\underset{\sim}{\beta} - \underset{\sim}{l}_m \otimes \underset{\sim}{\xi})'\underset{\sim}{\Omega}^{-1}(\underset{\sim}{\beta} - \underset{\sim}{l}_m \otimes \underset{\sim}{\xi}) &= \underset{\sim}{\beta}'\underset{\sim}{\Omega}^{-1}\underset{\sim}{\beta} + (\underset{\sim}{l}_m \otimes \underset{\sim}{\bar{\xi}})'\underset{\sim}{\Omega}^{-1}(\underset{\sim}{l}_m \otimes \underset{\sim}{\bar{\xi}})\\ &= \underset{\sim}{\beta}'\underset{\sim}{Q}\underset{\sim}{\beta}\end{aligned}$$

where

$$\underset{\sim}{Q} = \{\underset{\sim}{I}_{km} - \underset{\sim}{\Omega}^{-1}(\underset{\sim}{l}_m \otimes \underset{\sim}{I}_k)[(\underset{\sim}{l}_m \otimes \underset{\sim}{I}_k)'\underset{\sim}{\Omega}^{-1}(\underset{\sim}{l}_m \otimes \underset{\sim}{I}_k)]^{-1}(\underset{\sim}{l}_m \otimes \underset{\sim}{I}_k)'\}\underset{\sim}{\Omega}^{-1}. \tag{7}$$

and the joint (improper) pdf of $\underset{\sim}{y}$ and $\underset{\sim}{\beta}$ is

$$\alpha \exp\left[-\frac{1}{2}\{\underset{\sim}{y}'(\underset{\sim}{\Gamma}^{-1} \otimes \underset{\sim}{I}_T)\underset{\sim}{y} - 2\underset{\sim}{\beta}'\underset{\sim}{X}'(\underset{\sim}{\Gamma}^{-1} \otimes \underset{\sim}{I}_T)\underset{\sim}{y} + \underset{\sim}{\beta}'(\underset{\sim}{X}'(\underset{\sim}{\Gamma}^{-1} \otimes \underset{\sim}{I}_T)\underset{\sim}{X} + \underset{\sim}{Q})\underset{\sim}{\beta}\}\right].$$

The posterior distribution of $\underset{\sim}{\beta}$ given $\underset{\sim}{y}$ is

$$N\left[(\underset{\sim}{X}'(\underset{\sim}{\Gamma}^{-1} \otimes \underset{\sim}{I}_T)\underset{\sim}{X} + \underset{\sim}{Q})^{-1}\underset{\sim}{X}'(\underset{\sim}{\Gamma}^{-1} \otimes \underset{\sim}{I}_T)\underset{\sim}{y}, (\underset{\sim}{X}'(\underset{\sim}{\Gamma}^{-1} \otimes \underset{\sim}{I}_T)\underset{\sim}{X} + \underset{\sim}{Q})^{-1}\right]. \tag{8}$$

The posterior mean and variance can serve as Bayes point estimates:

$$\underset{\sim}{\beta}^* = [\underset{\sim}{X}'(\underset{\sim}{\Gamma}^{-1} \otimes \underset{\sim}{I}_T)\underset{\sim}{X} + \underset{\sim}{Q}]^{-1}\underset{\sim}{X}'(\underset{\sim}{\Gamma}^{-1} \otimes \underset{\sim}{I}_T)\underset{\sim}{y} \tag{9}$$

with[1]

$$\text{cov}(\underset{\sim}{\beta}^*) = [\underset{\sim}{X}'(\underset{\sim}{\Gamma}^{-1} \otimes \underset{\sim}{I}_T)\underset{\sim}{X} + \underset{\sim}{Q}]^{-1} \tag{10}$$

If normality is not assumed eqns. (9) and (10) can be interpreted as linear empirical Bayes estimators.

A convenient way to obtain the linear empirical Bayes estimators is to set the model in a least-squares framework. For this matter, if the prior on the last stage vector of parameters is uninformative, one starts by estimating the last vector. This estimator is derived by letting $\underset{\sim}{\beta}^*$, the linear empirical Bayes estimator of $\underset{\sim}{\beta}$, play the role of the dependent variable and $\underset{\sim}{l}_m \otimes \underset{\sim}{l}_k$ being the design matrix in the least squares estimate, namely,

$$\underset{\sim}{\xi}^* = \{(\underset{\sim}{l}'_m \otimes \underset{\sim}{I}_k)\underset{\sim}{\Omega}^{-1}(\underset{\sim}{l}_m \otimes \underset{\sim}{I}_k)\}^{-1}(\underset{\sim}{l}'_m \otimes \underset{\sim}{I}_k)\underset{\sim}{\Omega}^{-1}\underset{\sim}{\beta}^* \tag{11}$$

where starred symbols denote linear empirical Bayes estimators. See Haitovsky (1986). Given $\underset{\sim}{\xi}^*$, equation (11) can now be combined with (1a) to yield a least squares representation

$$\begin{bmatrix} \underset{\sim}{y} \\ \underset{\sim}{l}_m \otimes \underset{\sim}{\xi}^* \end{bmatrix} = \begin{bmatrix} \underset{\sim}{X} \\ \underset{\sim}{l}_{km} \end{bmatrix} \underset{\sim}{\beta} + \begin{bmatrix} \underset{\sim}{\epsilon} \\ \underset{\sim}{\nu} \end{bmatrix}$$

with

$$E\begin{bmatrix} \underset{\sim}{\epsilon} \\ \underset{\sim}{\nu} \end{bmatrix} = \underset{\sim}{0} \quad \text{and} \quad \text{cov}\begin{bmatrix} \underset{\sim}{\epsilon} \\ \underset{\sim}{\nu} \end{bmatrix} = \begin{bmatrix} \underset{\sim}{\Gamma} \otimes \underset{\sim}{I}_T & \underset{\sim}{0} \\ \underset{\sim}{0} & \underset{\sim}{\Omega} \end{bmatrix}$$

where $\underset{\sim}{\nu} = (\underset{\sim}{\nu}'_1, \ldots, \underset{\sim}{\nu}'_m)'$ and hence

[1]Note that the multivariate regression model can be deduced from the SUR model by letting $\underset{\sim}{X} = \underset{\sim}{I}_m \otimes \overline{\underset{\sim}{X}}$, where $\underset{\sim}{X}_i = \underset{\sim}{X}_j = \overline{\underset{\sim}{X}}$ for all $i, j = 1, \ldots, m$, hence obtaining

$$\underset{\sim}{\beta}^* = [\underset{\sim}{I}_m \otimes \overline{\underset{\sim}{X}}'\overline{\underset{\sim}{X}} + (\underset{\sim}{\Gamma} \otimes \underset{\sim}{I}_k)\underset{\sim}{Q}]^{-1}(\underset{\sim}{I}_m \otimes \overline{\underset{\sim}{X}}')\underset{\sim}{y}$$

with $\text{cov}(\underset{\sim}{\beta}^*) = (\underset{\sim}{\Gamma}^{-1} \otimes \overline{\underset{\sim}{X}}'\overline{\underset{\sim}{X}} + \underset{\sim}{Q})^{-1}$.

$$\underset{\sim}{\beta}^* = \left\{ \begin{bmatrix} \underset{\sim}{X} \\ \underset{\sim}{I}_{km} \end{bmatrix}' \begin{bmatrix} \underset{\sim}{\Gamma} \otimes \underset{\sim}{I}_T & \underset{\sim}{0} \\ \underset{\sim}{0} & \underset{\sim}{\Omega} \end{bmatrix}^{-1} \begin{bmatrix} \underset{\sim}{X} \\ \underset{\sim}{I}_{km} \end{bmatrix} \right\}^{-1}$$

$$\begin{bmatrix} \underset{\sim}{X} \\ \underset{\sim}{I}_{km} \end{bmatrix}' \begin{bmatrix} \underset{\sim}{\Gamma} \otimes \underset{\sim}{I}_T & \underset{\sim}{0} \\ \underset{\sim}{0} & \underset{\sim}{\Omega} \end{bmatrix}^{-1} \begin{bmatrix} \underset{\sim}{y} \\ \underset{\sim}{l}_m \otimes \underset{\sim}{\xi}^* \end{bmatrix}$$

$$= [\underset{\sim}{X}'(\underset{\sim}{\Gamma}^{-1} \otimes \underset{\sim}{I}_T)\underset{\sim}{X} + \underset{\sim}{\Omega}^{-1}]^{-1}[\underset{\sim}{X}'(\underset{\sim}{\Gamma}^{-1} \otimes \underset{\sim}{I}_T)\underset{\sim}{y} + \underset{\sim}{\Omega}^{-1}(\underset{\sim}{l}_m \otimes \underset{\sim}{\xi}^*)].$$

Finally, substituting (11) in the last equation after some algebra, eqns. (9) and (10) are obtained.

As will be seen later, the estimation of the $(km+1)km/2$ different terms of Ω requires a lot of prior information. Thus, a simplification is warranted. We suggest as a compromise between the unrestricted $\underset{\sim}{\Omega}$ matrix and the too restricted prior $\underset{\sim}{\Omega}_{ij} = \underset{\sim}{\Sigma}$ when $i = j$ and $\underset{\sim}{0}$ otherwise used elsewhere [see Smith (1973)], to impose more structure on $\underset{\sim}{\Omega}$, as in Haitovsky (1987) in the form

$$E(\underset{\sim}{\nu}_i \underset{\sim}{\nu}_j') = \underset{\sim}{\Omega}_{ij} = \nu_{ij}\underset{\sim}{\Sigma}, \qquad i,j = 1,\ldots,m,$$

where $\underset{\sim}{\Omega}_{ij}$ is the $k \times k$ submatrix of $\underset{\sim}{\Omega}$, and the $k \times k$ $\underset{\sim}{\Sigma}$ matrix and $m \times m$ $\underset{\sim}{V} \equiv (\nu_{ij})$ matrix are PDS. In short,

$$\underset{\sim}{\Omega} = \underset{\sim}{V} \otimes \underset{\sim}{\Sigma}. \tag{12}$$

As a result $\underset{\sim}{Q}$ reduces to

$$\underset{\sim}{Q} = \underset{\sim}{V}^{-1}\underset{\sim}{W} \otimes \underset{\sim}{\Sigma}^{-1}$$

where $\underset{\sim}{W} = \underset{\sim}{I}_m - (\underset{\sim}{l}_m'\underset{\sim}{V}^{-1}\underset{\sim}{l}_m)^{-1}\underset{\sim}{J}_m\underset{\sim}{V}^{-1}$ and $\underset{\sim}{J}_m = \underset{\sim}{l}_m\underset{\sim}{l}_m'$, an $m \times m$ matrix of ones.

Likewise, (11) simplifies to

$$\underset{\sim}{\xi}^* = (\underset{\sim}{l}_m'\underset{\sim}{V}^{-1}\underset{\sim}{l}_m)^{-1}(\underset{\sim}{l}_m'\underset{\sim}{V}^{-1} \otimes \underset{\sim}{I}_k)\underset{\sim}{\beta}^* \tag{13}$$

with

$$\mathrm{cov}(\underset{\sim}{\xi}^*) = \{(\underset{\sim}{l}_m \otimes \underset{\sim}{I}_k)\underset{\sim}{X}'(\underset{\sim}{\Gamma}^{-1} \otimes \underset{\sim}{I}_T)\underset{\sim}{X} + \underset{\sim}{V} \otimes \underset{\sim}{\Sigma}]^{-1}(\underset{\sim}{l}_m \otimes \underset{\sim}{I}_k)\}^{-1}. \tag{14}$$

Again, if conditional normality is assumed as before, the posterior distribution of $\underset{\sim}{\xi}$ given $\underset{\sim}{\beta}$ is normal with (13) and (14) as its mean and variance.

3. Covariances Unknown

Turning now to the realistic situation where the covariances are unknown, one may choose to derive Type II maximum-likelihood estimators. It is achieved by maximizing the joint probability function which, assuming normality and $\underset{\sim}{\Omega} = \underset{\sim}{V} \otimes \underset{\sim}{\Sigma}$ is proportional to,

$$
\begin{aligned}
&|\underset{\sim}{\Gamma} \otimes \underset{\sim}{I}_T|^{-\frac{1}{2}} |\underset{\sim}{V} \otimes \underset{\sim}{\Sigma}|^{-\frac{1}{2}} \operatorname{etr}\Big\{ -\frac{1}{2}(\underset{\sim}{y} - \underset{\sim}{X}\underset{\sim}{\beta})'(\underset{\sim}{\Gamma}^{-1} \otimes \underset{\sim}{I}_T)(\underset{\sim}{y} - \underset{\sim}{X}\underset{\sim}{\beta}) \\
&\qquad -\frac{1}{2}(\underset{\sim}{\beta} - \underset{\sim}{l}_m \otimes \underset{\sim}{\xi})'(\underset{\sim}{V}^{-1} \otimes \underset{\sim}{\Sigma}^{-1})(\underset{\sim}{\beta} - \underset{\sim}{l}_m \otimes \underset{\sim}{\xi})\Big\} \\
&= |\underset{\sim}{\Gamma}|^{-\frac{T}{2}} |V|^{-\frac{k}{2}} |\underset{\sim}{\Sigma}|^{-\frac{m}{2}} \operatorname{etr}\{-\frac{1}{2}\underset{\sim}{\Gamma}^{-1}\underset{\sim}{E}'\underset{\sim}{E} - \frac{1}{2}\underset{\sim}{\Sigma}^{-1}\underset{\sim}{\Xi}\underset{\sim}{V}^{-1}\underset{\sim}{\Xi}'\},
\end{aligned}
\tag{15}
$$

where etr denotes the exponent of the matrix trace, and $\underset{\sim}{E}$ and $\underset{\sim}{\Xi}$ are obtained by "matricizing" the $\underset{\sim}{y} - \underset{\sim}{X}\underset{\sim}{\beta}$ and $\underset{\sim}{\beta} - \underset{\sim}{l}_m \otimes \underset{\sim}{\xi}$ vectors, respectively. That is,

$$\operatorname{vec}\{\underset{\sim}{E}\} = \underset{\sim}{y} - \underset{\sim}{X}\underset{\sim}{\beta} \quad \text{and} \quad \operatorname{vec}\{\underset{\sim}{\Xi}\} = \underset{\sim}{\beta} - \underset{\sim}{l}_m \otimes \underset{\sim}{\xi}.$$

In the derivation of the two matrices in the exponent in (15) use was made of the well known relation

$$\operatorname{vec}'\{\underset{\sim}{Z}\}(\underset{\sim}{A} \otimes \underset{\sim}{C}) \operatorname{vec}\{\underset{\sim}{Z}\} = \operatorname{tr}\{\underset{\sim}{A}\underset{\sim}{Z}'\underset{\sim}{C}\underset{\sim}{Z}\} = \operatorname{tr}\{\underset{\sim}{Z}\underset{\sim}{A}\underset{\sim}{Z}'\underset{\sim}{C}\}$$

where $\underset{\sim}{A}$, $\underset{\sim}{C}$ and $\underset{\sim}{Z}$ are properly dimensioned matrices. Maximizing (15) with respect to $\underset{\sim}{\Gamma}^{-1}$, $\underset{\sim}{V}^{-1}$, $\underset{\sim}{\Sigma}^{-1}$, $\underset{\sim}{\beta}$ and $\underset{\sim}{\xi}$ by partially differentiating its logarithmic function with respect to the maximizing quantities, the following estimators are obtained:

$$
\begin{aligned}
\hat{\underset{\sim}{\Gamma}} &= \frac{1}{T}\underset{\sim}{E}^{*'}\underset{\sim}{E}^{*} \\
\hat{\underset{\sim}{V}} &= \frac{1}{k}\hat{\underset{\sim}{\Xi}}'\hat{\underset{\sim}{\Sigma}}^{-1}\hat{\underset{\sim}{\Xi}} \\
\hat{\underset{\sim}{\Sigma}} &= \frac{1}{m}\hat{\underset{\sim}{\Xi}}\hat{\underset{\sim}{V}}^{-1}\hat{\underset{\sim}{\Xi}}'
\end{aligned}
\tag{16}
$$

where $\text{vec}\{\underset{\sim}{E}^*\} = \underset{\sim}{y} - \underset{\sim}{X}\hat{\underset{\sim}{\beta}}^*$, and $\hat{\underset{\sim}{\beta}}^*$ and $\hat{\underset{\sim}{\xi}}^*$ are as in (9) and (13), but with the covariance matrices replaced by their estimates.[1]

Note that $\hat{\underset{\sim}{\Gamma}}$ and $\underset{\sim}{G}$ differ only by the way the regression residuals are estimated. Whereas, the latter is based on the restricted residuals obtained by estimating each equation separately by least squares, $\hat{\underset{\sim}{\Gamma}}$ is based on the unrestricted residuals.

Note further that the matrix of order $k \times m$

$$\hat{\underset{\sim}{\Xi}} = \hat{\underset{\sim}{B}}^* - \underset{\sim}{l}'_m \otimes \underset{\sim}{\xi} = \hat{\underset{\sim}{B}}^*[\underset{\sim}{I}_m - (\underset{\sim}{l}'_m\hat{\underset{\sim}{V}}^{-1}\underset{\sim}{l}_m)^{-1}\hat{\underset{\sim}{V}}\underset{\sim}{J}_m]$$

where $\text{vec}\{\hat{\underset{\sim}{B}}^*\} = \hat{\underset{\sim}{\beta}}^*$, is of rank $\leq \min\{k, m-1\}$, since its rows lie in the $m-1$ dimensional space,[2] resulting in either $\hat{\underset{\sim}{\Sigma}}$ or $\hat{\underset{\sim}{V}}$ being a singular matrix.

A Bayesian approach will overcome the singularity. It requires specification of prior distributions for the covariance matrices. It will be done through the conjugate family, which is the Wishart distribution. Assuming $\underset{\sim}{\Gamma}^{-1}$, $\underset{\sim}{V}^{-1}$ and $\underset{\sim}{\Sigma}^{-1}$ are independent, Wishart distributed with γ, ν, and σ degrees of freedom and $\underset{\sim}{G}$, $\underset{\sim}{R}$, and $\underset{\sim}{S}$ matrices, respectively, the following densities (up to a proportionality factor) are obtained:

$$|\underset{\sim}{\Gamma}|^{-\frac{1}{2}(\gamma-m-1)}\text{etr}\{-\frac{1}{2}\underset{\sim}{\Gamma}^{-1}\underset{\sim}{G}\}$$
$$|\underset{\sim}{V}|^{-\frac{1}{2}(\nu-m-1)}\text{etr}\{-\frac{1}{2}\underset{\sim}{V}^{-1}\underset{\sim}{R}\}$$

[1]In deriving (16), use was made of the following relations:

$$|\underset{\sim}{A} \otimes \underset{\sim}{C}| = |\underset{\sim}{A}|^s|\underset{\sim}{C}|^t \text{ for } \underset{\sim}{A} : t \times t \text{ and } \underset{\sim}{C} : s \times s,$$

$$\frac{\partial \ln|\underset{\sim}{G}|}{\partial \underset{\sim}{G}} = \frac{\partial \ln|\underset{\sim}{G}|}{\partial|\underset{\sim}{G}|}\frac{\partial|\underset{\sim}{G}|}{\partial \underset{\sim}{G}} = 2\underset{\sim}{G}^{-1} - \text{Diag}\{\underset{\sim}{G}^{-1}\}$$

and

$$\frac{\partial \text{ tr}\{\underset{\sim}{H}\underset{\sim}{G}\}}{\partial \underset{\sim}{G}} = 2\underset{\sim}{H}' - \text{Diag}\{\underset{\sim}{H}\},$$

where $\underset{\sim}{G}$ is a symmetric matrix (and hence $\underset{\sim}{H}$ is square) and $\text{Diag}\{\cdot\}$ is the matrix diagonalization operator.

[2]Since $\underset{\sim}{W} = \underset{\sim}{I}_m - (\underset{\sim}{l}'_m\underset{\sim}{V}^{-1}\underset{\sim}{l}_m)^{-1}\underset{\sim}{V}^{-1}\underset{\sim}{J}_m$ is an idempotent (nonsymmetric) matrix with rank and trace equal to $m-1$.

and

$$|\underset{\sim}{\Sigma}|^{-\frac{1}{2}(\sigma-k-1)}\text{etr}\{-\frac{1}{2}\underset{\sim}{\Sigma}^{-1}\underset{\sim}{S}\}.$$

The joint distribution of all the quantities is obtained by multiplying the joint probability function (15) by the three prior densities above. The integration with respect to $\underset{\sim}{\xi}$ is performed as in Section II. Notice that now

$$\begin{aligned}&(\underset{\sim}{\beta}-\underset{\sim}{l}_m\otimes\underset{\sim}{\xi})'(\underset{\sim}{V}\otimes\underset{\sim}{\Sigma})^{-1}(\underset{\sim}{\beta}-\underset{\sim}{l}_m\otimes\underset{\sim}{\xi})\\&=(\underset{\sim}{\beta}-\underset{\sim}{l}_m\otimes\overline{\underset{\sim}{\xi}})'(\underset{\sim}{V}\otimes\underset{\sim}{\Sigma})^{-1}(\underset{\sim}{\beta}-\underset{\sim}{l}_m\otimes\overline{\underset{\sim}{\xi}})+(\underset{\sim}{l}'_m\underset{\sim}{V}^{-1}\underset{\sim}{l}_m)(\underset{\sim}{\xi}-\overline{\underset{\sim}{\xi}})'\underset{\sim}{\Sigma}^{-1}(\underset{\sim}{\xi}-\overline{\underset{\sim}{\xi}}),\end{aligned}$$

where now

$$\overline{\underset{\sim}{\xi}}=(\underset{\sim}{l}'_m\underset{\sim}{V}^{-1}\underset{\sim}{l}_m)^{-1}(\underset{\sim}{l}'_m\underset{\sim}{V}^{-1}\otimes\underset{\sim}{I}_k)\underset{\sim}{\beta}\qquad\text{[The counterpart of (13)]},$$

and after integration with respect to $\underset{\sim}{\xi}$ the joint posterior density for $\underset{\sim}{\beta}$, $\underset{\sim}{\Gamma}$, $\underset{\sim}{V}$ and $\underset{\sim}{\Sigma}$ is proportional to:

$$\begin{aligned}&|\underset{\sim}{\Gamma}|^{-\frac{1}{2}(T+\gamma-m-1)}|\underset{\sim}{V}|^{-\frac{1}{2}(k+\nu-m-1)}|\underset{\sim}{\Sigma}|^{-\frac{1}{2}(m+\sigma-k-2)}(\underset{\sim}{l}'_m\underset{\sim}{V}^{-1}\underset{\sim}{l}_m)^{-\frac{1}{2}k}\\&\cdot\text{etr}\{-\frac{1}{2}[\underset{\sim}{\Gamma}^{-1}\underset{\sim}{G}+\underset{\sim}{V}^{-1}\underset{\sim}{R}+\underset{\sim}{\Sigma}^{-1}\underset{\sim}{S}+\underset{\sim}{\Gamma}^{-1}\underset{\sim}{E}'\underset{\sim}{E}+\underset{\sim}{\Sigma}^{-1}\underset{\sim}{\Xi}\underset{\sim}{V}^{-1}\underset{\sim}{\Xi}']\}.\end{aligned}\tag{17}$$

The integration with respect to $\underset{\sim}{\xi}$ resulted in the loss of one degree of freedom in the estimation of $\underset{\sim}{\Sigma}$.

Following Lindley and Smith (1972) we derive the modal Bayesian estimates by differentiating the log joint posterior density partially with respect to $\underset{\sim}{\beta}$, $\underset{\sim}{\Gamma}^{-1}$, $\underset{\sim}{V}^{-1}$ and $\underset{\sim}{\Sigma}^{-1}$ obtaining $\hat{\underset{\sim}{\beta}}^*$ as before (equation (9)) with $\underset{\sim}{\Gamma}$, $\underset{\sim}{\Sigma}$, and $\underset{\sim}{V}$ replaced by their estimates:[4]

---

[4]The first two expressions were obtained using the differentiation formulae in footnote 2, and the last one from the differentiation:

$$(k+\nu-m-1)\frac{\partial\ \ln|\underset{\sim}{V}^{-1}|}{\partial\underset{\sim}{V}^{-1}}-\frac{k\partial\ \ln\underset{\sim}{l}'_m\underset{\sim}{V}^{-1}\underset{\sim}{l}_m}{\partial\underset{\sim}{V}^{-1}}-\frac{\partial\ \text{tr}\{\underset{\sim}{V}^{-1}(\underset{\sim}{R}+\underset{\sim}{\Xi}'\underset{\sim}{\Sigma}^{-1}\underset{\sim}{\Xi})\}}{\partial\underset{\sim}{V}^{-1}}=\underset{\sim}{0}$$

where

$$\begin{aligned}\frac{\partial\ \ln\underset{\sim}{l}'_m\underset{\sim}{V}^{-1}\underset{\sim}{l}_m}{\partial\underset{\sim}{V}^{-1}}&=\frac{\partial\ \ln\underset{\sim}{l}'_m\underset{\sim}{V}^{-1}\underset{\sim}{l}_m}{\partial\underset{\sim}{l}^T_m\underset{\sim}{V}^{-1}\underset{\sim}{l}_m}\cdot\frac{\partial\underset{\sim}{l}'_m\underset{\sim}{V}^{-1}\underset{\sim}{l}_m}{\partial\underset{\sim}{V}^{-1}}\\&=2(\underset{\sim}{l}'_m\underset{\sim}{V}^{-1}\underset{\sim}{l}_m)^{-1}\underset{\sim}{l}_m\underset{\sim}{l}'_m-(\underset{\sim}{l}'_m\underset{\sim}{V}^{-1}\underset{\sim}{l}_m)^{-1}\underset{\sim}{I}_m,\end{aligned}$$

because of the symmetry of $\underset{\sim}{V}$, obtaining $\hat{\underset{\sim}{V}}=(k+\nu-m-1)^{-1}k(\underset{\sim}{l}'_m\hat{\underset{\sim}{V}}^{-1}\underset{\sim}{l}_m)^{-1}\underset{\sim}{J}_m+\underset{\sim}{R}+\hat{\underset{\sim}{\Xi}}'\hat{\underset{\sim}{\Sigma}}^{-1}\hat{\underset{\sim}{\Xi}}]$ and after some algebra the final expression for $\underset{\sim}{V}$ is obtained.

$$\begin{cases} \hat{\Gamma} = (T+\gamma-m-1)^{-1}[G+E^{*'}E^{*}] \\ \hat{\Sigma} = (m+\sigma-k-2)^{-1}[S+\hat{\Xi}\hat{V}^{-1}\hat{\Xi}'] \\ \hat{V} = (k+\nu-m-1)^{-1}\{k[(\nu-m-1)l_m'(R+\hat{\Xi}'\hat{\Sigma}^{-1}\hat{\Xi}^{-1})l_m]^{-1}J_m \\ \qquad +R+\hat{\Xi}'\hat{\Sigma}^{-1}\hat{\Xi}\}. \end{cases} \tag{18}$$

The correction of the number of degrees of freedom in the estimation of Γ is familiar from LS estimation theory and those for $\hat{\Sigma}$ and $\hat{V}$ come natural in view of the theorem in the appendix in Haitovsky (1987).

The Bayesian modal estimates can be calculated using an iterative procedure: initial values for $\hat{\beta}^*$, $\hat{\beta}_0^*$, can be computed using least squares regression estimates, from which the initial values for $\hat{\Gamma}$ and for $\hat{\xi}^*$ (equation (13)) can be computed, using for the latter a simple average of the m subvectors of $\hat{\beta}_0^*$ (i.e., letting $V = I_m$ in equation (13)). Next, $\hat{\Xi}_0$ is computed, from which $\hat{\Sigma}_0$ is obtained after suppressing $\hat{V}$ to I_m. Having computed $\hat{\Xi}_0$ and $\hat{\Sigma}_0$, $\hat{V}_0$ can be calculated, completing the first round, from which iterations can start.

4. An Illustrative Example

Our method will be applied now to panel data, i.e., a time series of cross section data. We consider here the investment equation developed by Grunfeld (1958). In this model the ith firm's current gross investment at time t, y_{ti}, is dependent on the value of the firm's outstanding shares at the beginning of the year $x_{1i,t-1}$, and on its beginning-of-year capital stock, $x_{2i,t-1}$. That is, the micro investment function for a firm is

$$E(y_{ti}) = \beta_{0i} + \beta_{1i}x_{1i,t-1} + \beta_{2i}x_{2i,t-1} \quad (i = 1, \ldots, m;\ t = 1, \ldots, T).$$

We use here annual data for the years 1935-54 on eleven corporations[2] in the United States. Thus in our example $m = 11$, $T = 20$. The β_{0i}'s are removed by centering the observations about the respective firm means, so that $\beta_i' = (\beta_{1i}, \beta_{2i})$ and $\xi' = (\xi_1, \xi_2)$.

The purpose of the illustrative example is not to explore the suitability of Grunfeld's investment model, but *given* the model, to illustrate the feasibility of the application of the method described in the paper to real life data and to compare our estimates ($\hat{\beta}_i^*$ and $\hat{\xi}_j^*$) with Swamy's (1970) random coefficient estimates ($\hat{\beta}_i$ and $\hat{\xi}_j$) and with Smith's (1973) more restrictive Bayes estimates (β_i^* and ξ_j^*). We shall refer

[5]The corporations are (1) General Motors, (2) U.S. Steel, (3) General Electric, (4) Chrysler, (5) Atlantic Refineries, (6) IBM, (7) Union Oil, (8) Westinghouse, (9) Goodyear, (10) Diamond Match, (11) American Steel Foundries.

to our estimates as the "Linear Hierarchical" estimates, to distinguish them from Smith's estimates.

Swamy's basic concern is with deriving appropriate weighting schemes to be applied to the ordinary least squares regression coefficients in obtaining $\underset{\sim}{\xi}$. Smith's method, on the other hand, differs from ours in two ways; in his: (a) $\underset{\sim}{\Gamma} = \text{Diag}\{\gamma_{11}, \ldots, \gamma_{mm}\}$ and (b) $\underset{\sim}{V} = \underset{\sim}{I}_m$. Thus, Smith's method can be considered a special case of ours.

In order to obtain maximum comparability with Smith's estimates, we follow Smith and choose $\gamma = 0$, $\underset{\sim}{G} = \underset{\sim}{0}, \sigma = 1$, $\underset{\sim}{R} = 0.001\underset{\sim}{I}_m$, $\underset{\sim}{S} = 0.001\underset{\sim}{I}_2$. ν, which does not exist in Smith's method due to his choice of $\underset{\sim}{V} = \underset{\sim}{I}_m$, is set to 25.

Table 1 lists the least squares and the Bayesian and linear hierarchical modal estimates for each of the eleven firms together with the respective estimates of the within-firm variances σ_i^2. We use the estimated interfirm variances for each $\beta_i (i = 1, 2)$, relative to that computed from the least squares, to measure the pulling-toward-the-mean of Smith's and our estimates.

As can be seen from the Table, Smith's pulling-toward-the-mean of his β_{1i}s seems to be excessive. All his β_is values converge to a common value. Our method results in a more sensible pulling for both β_1 and β_2, except for the sign change of $\beta_{1,10}$, which should be judged very insignificant by any measure. The aggregate parameters ξ_1 and ξ_2 are given for all three methods in Table 2. Ours seem to be closer to Swamy's values than to Smith's.

Table 2
Estimates of the Aggregate Parameters

	ξ_1	ξ_2
Swamy	0.084	0.196
Bayes Mode	0.064	0.208
Linear Hierarchical Mode	0.077	0.167

Finally, the pattern of convergence of the regression coefficients is traced throughout the iterations and depicted in Figures 1 and 2. The numbers 1 to 9, and the letters B and C refer to firms numbered 1 to 11 respectively, while A refers to $\hat{\xi}$. The iteration numbers are listed along the abscissa. Iteration 0 refers to the initial values, i.e., the least squares estimates. As can be seen from these figures, the estimates converge very smoothly to their solutions. Furthermore, for all practical purposes,

Table 1

Estimates of Investment Function Parameters and Variances of Least Squares $(\hat{\beta}_i, \hat{\sigma}^2)$ Smith's Bayes (β_i^*, σ^{*2}), and Linear Hierarchical Model $(\hat{\beta}_i^*, \hat{\sigma}^{*2})$

Firm	$\hat{\beta}_1$	β_1^*	$\hat{\beta}_1^*$	$\hat{\beta}_2$	β_2^*	$\hat{\beta}_2^*$	$\hat{\sigma}^2$	σ^{*2}	$\hat{\sigma}^{*2}$
1	0.119	0.064	0.1145	0.371	0.393	0.3963	7537.2	9113.0	7955.9
2	0.175	0.064	0.1338	0.390	0.308	0.2656	8320.7	9223.6	8783.0
3	0.027	0.064	0.0308	0.152	0.147	0.1204	695.6	884.4	734.3
4	0.078	0.064	0.0674	0.316	0.318	0.2996	157.8	154.2	166.5
5	0.162	0.064	0.1199	0.003	0.037	0.0159	73.5	82.2	77.6
6	0.131	0.064	0.1263	0.085	0.290	0.0721	58.5	70.8	61.7
7	0.088	0.064	0.0839	0.123	0.125	0.1345	79.3	76.0	83.7
8	0.053	0.064	0.0517	0.093	0.073	0.0422	93.6	91.4	98.5
9	0.075	0.064	0.0778	0.082	0.091	0.0637	74.1	70.9	78.3
10	0.005	0.064	-0.0092	0.437	0.389	0.3268	1.1	1.3	1.1
11	0.066	0.064	0.0558	0.084	0.112	0.1041	8.8	8.4	9.3
pulling toward the mean relative to least squares*	1	0	0.693	1	0.757	0.730			

*Computed by $\sum_i(\beta_i^* - \overline{\beta}^*)^2 / \sum_i(\hat{\beta}_i - \overline{\hat{\beta}})^2$ for β_i^* and $\hat{\beta}_i^*$.

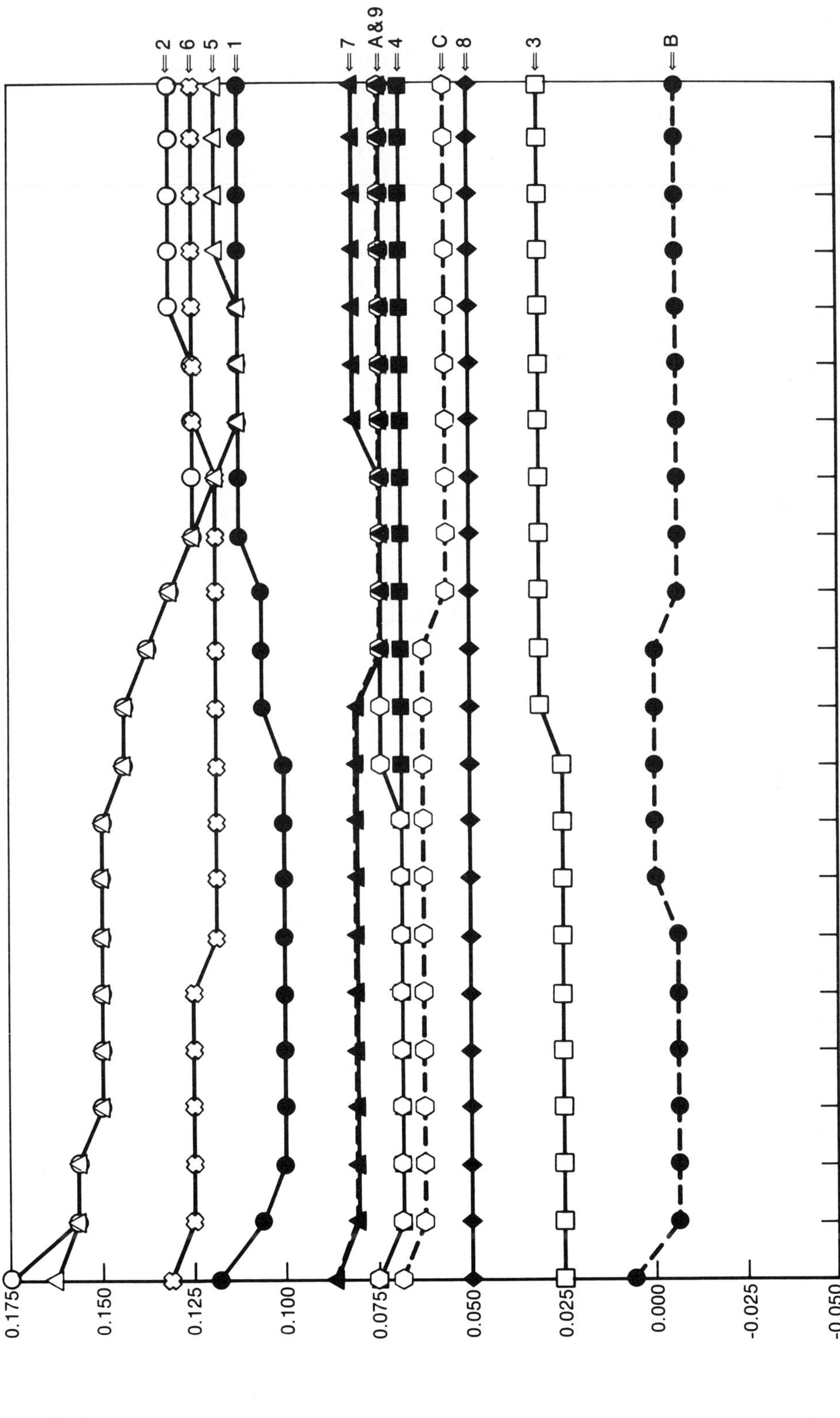

⇐2
⇐6
⇐5
⇐1
⇐7
⇐A&9
⇐4
⇐C
⇐8
⇐3
⇐B
0.175
0.150
0.125
0.100
0.075
0.050
0.025
0.000
-0.025
-0.050
0
1
2
3
4
5
6
7
8
9
10
11
12
13
14
15
16
17
18
19
20
21

Figure 1.

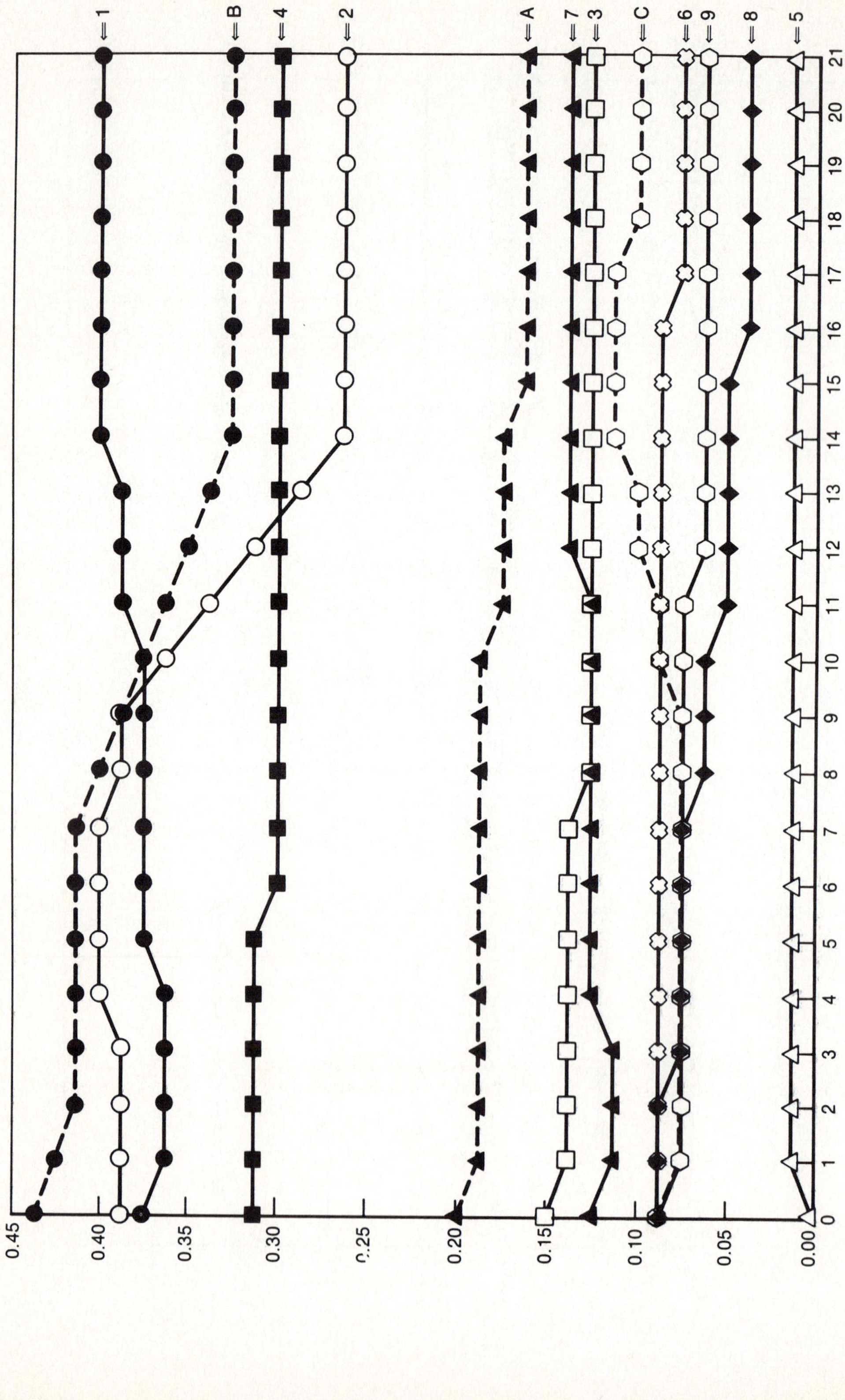
⇐1
⇐B
⇐4
⇐2
⇐A
⇐7
⇐3
⇐C
⇐6
⇐9
⇐8
⇐5
0.45
0.40
0.35
0.30
0.25
0.20
0.15
0.10
0.05
0.00
0
1
2
3
4
5
6
7
8
9
10
11
12
13
14
15
16
17
18
19
20
21

Figure 2.

convergence is achieved in seventeen steps. It continues to iterate because of our strict convergence criteria. For such a complex problem, the speed of convergence is very satisfactory.

Acknowledgments

This research was supported by a NSF grant to the University of Southern California, where the research was conducted. I wish to thank Professor Seymour Geisser, the editor who handled the paper, for his useful comments, and the anonymous referee for spotting an error in an earlier draft and whose comment resulted in an improved exposition. The computer programming was skillfully performed by Dr. Wolfgang Polasek.

References

Grunfeld, Y. (1952). "The Determinants of Corporate Investment," unpublished Ph.D. dissertation, University of Chicago.

Haitovsky, Y. (1986). "The Linear Hierarchical Model and its Applications in Econometric Analysis," in *Bayesian Inference and Decision Techniques with Applications: Essays in Honor of Bruno de Finetti*, P.K. Goel and A. Zellner, eds., Amsterdam: North-Holland, 119-138.

Haitovsky, Y. (1987). "On Multivariate Ridge Regression," *Biometrika*, Vol. 74, 563-570.

Lindley, D.V. and A.F.M. Smith (1972). "Bayes Estimates for the Linear Model" (with discussion), *Journal of the Royal Statistical Society, Series B*, Vol. 34, 1-41.

Smith, A.F.M. (1973). "A General Bayesian Linear Model," *Journal of the Royal Statistical Society, Series B*, Vol. 35, 67-75.

Srivastava, V.K. and T.D. Dwivedi (1979). "Estimation of Seemingly Unrelated Regression Equations. A Brief Survey," *Journal of Econometrics*, Vol. 10, 15-32.

Swamy, P.A.V.B (1970). "Efficient Inference in a Random Coefficient Regression Model," *Econometrica*, Vol. 38, 311-323.

Zellner A. (1962). "An Efficient Method of Estimating Seemingly Unrelated Regressions and Test for Aggregation Bias," *Journal of the American Statistical Association*, Vol. 57, 348-368.

Zellner, A. (1971). *Introduction to Bayesian Inference in Econometrics*. New York: Wiley and Sons.

Zellner, A. and D.S. Huang (1962). "Further Properties of Efficient Estimators for Seemingly Unrelated Regression Equations," *International Economic Review*, Vol. 3, 300-313.

Bayesian and Likelihood Methods in Statistics and Econometrics
S. Geisser, J.S. Hodges, S.J. Press and A. Zellner (Editors)

COMPARING BAYESIAN AND NON-BAYESIAN LIMITED INFORMATION ESTIMATORS

Hiroki Tsurumi
Rutgers University

Abstract

In this paper we derive conditional and unconditional limited information Bayesian estimators for structural coefficients of a simultaneous equation. We use sampling experiments to compare these new estimators to other Bayesian and non-Bayesian limited information estimators. The sampling experiments show that while no estimator dominates all the others in all the cases considered, the conditional and unconditional Bayesian estimators and Zellner's extended minimum expected loss estimator tend to outperform the other estimators as the degree of simultaneity increases.

1. Introduction

Earlier the author derived analytical expressions for the marginal posterior probability density functions (pdf's) for structural coefficients of a simultaneous equation [Tsurumi (1985)]. In this paper we derive conditional and unconditional limited information Bayesian estimators, and compare their performance to those of other Bayesian and non-Bayesian limited information estimators in sampling experiments. The sampling experiments are designed to control for some key parameters that are often present in economic data: multicollinearity, the correlation between stochastic regressors and the error term, and reduced form coefficient of determination. We show the relationship between these parameters and the noncentrality parameter of the exact distributions of K-*class* estimators.

The sampling results show that while no single estimator dominates all the other estimators at all times, the conditional and unconditional Bayesian estimators and Zellner's extended minimum expected loss estimator tend to outperform the other estimators when the degree of simultaneity increases.

The organization of the paper is as follows. Section 2 presents the conditional and unconditional Bayesian limited information estimators, and sampling experiments are given in Section 3. Section 4 gives concluding remarks.

2. Limited Information Estimators

Let a structural equation of interest and reduced form equations be given by

$$\begin{aligned} y_1 &= Y_1\gamma_1 + X_1\beta_1 + u_1 \\ Y_1 &= X\Pi_1 + V_1 \end{aligned} \tag{1}$$

where y_1 is an $n \times 1$ vector of observations on the left-hand side endogenous variables; Y_1 is an $n \times m_1$ matrix of observations on the endogenous variables included on the right-hand side of the structural equation of interest; X_1 is an $n \times k_1$ matrix of observations on the exogenous variables included on the right-hand side of the structural equation of interest; $X = (X_1, X_2)$, where X_2 is an $n \times k_2$ matrix of observations on the exogenous variables excluded from the structural equation of interest; $k = k_1 + k_2$; u_1 is an $n \times 1$ vector of error terms. V_1 is an $n \times m_1$ matrix of reduced form error terms. γ_1 is an $m_1 \times 1$ vector of structural coefficients associated with Y_1. β_1 is a a $k_1 \times 1$ vector of structural coefficients associated with X_1, and Π_1 is a $k \times m_1$ matrix of reduced form equations.

Let us derive conditional and unconditional Bayesian estimators of the structural parameters, γ_1 and β_1 of the equation of interest. We start from the reduced form equations

$$\begin{aligned} Y_1 &= X\Pi_I + V_I \qquad (2) \\ &= (X_1, X_2)\begin{bmatrix} \Pi_{11} & \Pi_{12} \\ \Pi_{21} & \Pi_{22} \end{bmatrix} + (V_1, V_2). \end{aligned}$$

Post-multiplying (1) by the $(m_1 + 1) \times (m_1 + 1)$ matrix

$$\Lambda = \begin{bmatrix} 1 & 0 \\ -\gamma_1 & I_{m_1} \end{bmatrix}$$

we obtain

$$W = X\Pi_* + V_* \tag{3}$$

where $W = Y_I\Lambda = (y_1 - Y_1\gamma_1, Y_1)$. Equation (3) is identical to (1) with $\beta_1 = \Pi_{11} - \Pi_{12}\gamma_1$, $\beta_2 = \Pi_{21} - \Pi_{22}\gamma_1$, $u_1 = v_1 - V_1\gamma_1$, and $\beta_2 = 0$ is the identifiability condition.

If we assume that a row of V_I is distributed as $N(0, \Omega)$, where

$$\Sigma = \Lambda' \Omega \Lambda = \begin{bmatrix} \sigma_{11} & \delta' \\ \delta & \Omega_{22} \end{bmatrix}$$

and $\sigma_{11} = \text{Var}(u_{t1})$, u_{t1} being the t-th element of u_1; $\delta = \text{Cov}(u_{t1},\ Y_{t1})$, Y_{t1} being the t-th row of Y_1, and $\text{Var}(\text{Vec}(V_1')) = I \otimes \Omega_{22}$. The likelihood function is given by

$$\begin{aligned} &\ell(\gamma_1, \beta_1, \beta_2, \Pi_2, \Sigma \text{data}) \\ &\quad \propto | \Sigma |^{-n/2} \exp\left[-\frac{1}{2} \text{tr}\,(W - X\Pi_*)' \,(W - X\Pi_*)\, \Sigma^{-1}\right] \end{aligned} \tag{4}$$

and this can be decomposed as

$$\begin{aligned} &\ell\,(\gamma_1, \beta_1, \beta_2, \theta_1, \sigma_{11\cdot 2}, \Pi_1, \Sigma \mid \text{data}) \ \propto\ \sigma_{11\cdot 2}^{-n/2} \mid \Sigma_{22} \mid^{-n/2} \\ &\quad \exp\left[-\frac{1}{2\sigma_{11\cdot 2}} Q\,(y_1 - Y_1\gamma_1 - X_1\beta_1 - X_2\beta_2 - V_1\theta_1)\right] \\ &\qquad \exp\left[-\frac{1}{2} \text{tr}\ Q\left(\Omega_{22}^{-1/2}\,(Y_1 - X\Pi_1)'\right)\right] \end{aligned} \tag{5}$$

where $Q(a)$ is a quadratic form in a: $Q(a) = a'a$, and $\sigma_{11\cdot 2} = \sigma_{11} - \delta' \Omega_{22}^{-1} \delta$, and $\theta_1 = \Omega_{22}^{-1}\delta$. If we use the identifiability condition $\beta_2 = \Pi_{21} - \Pi_{22}\gamma_1 = 0$, and set Π_1 and Ω_{22} at $\tilde{\Pi}_1$ and $\tilde{\Omega}_{22} = \left(Y_1 - X\tilde{\Pi}_1\right)' \left(Y_1 - X\tilde{\Pi}_1\right)/n$, respectively, then

$$\begin{aligned} &\ell\left(\gamma_1, \beta_1, \theta_1, \sigma_{11\cdot 2} \mid \beta_2 = 0, \Pi_1 = \tilde{\Pi}_1, \Omega_{22} = \tilde{\Omega}_{22}, \text{data}\right) \\ &\quad \propto\ \sigma_{11\cdot 2}^{-n/2} \exp\left[-\frac{1}{2} Q\left(y_1 - Y_1\gamma_1 - X_1\beta_1 - \tilde{V}_1\theta_1\right)\right] \end{aligned} \tag{6}$$

where $\tilde{V}_1 = Y_1 - X\tilde{\Pi}_1$. Given the prior pdf

$$\begin{aligned} &p\left(\gamma_1, \beta_1, \theta_1, \sigma_{11\cdot 2}^{1/2}\right) \\ &\quad \propto\ \sigma_{11\cdot 2}^{-1/2} \quad -\infty < \gamma_1, \beta_1, \theta_1 < \infty, \qquad 0 < \sigma_{11\cdot 2} < \infty \end{aligned} \tag{7}$$

the posterior pdf for γ_1, β_1, and θ_1 becomes after integrating $\sigma_{11\cdot 2}$ out

$$\begin{aligned} &p\left(\gamma_1, \beta_1, \theta_1 \mid \beta_2 = 0, \Pi_1 = \tilde{\Pi}_1, \Omega_{22} = \tilde{\Omega}_{22}\right) \\ &\quad \propto\ \left[\nu s^2 + (\mu - \hat{\mu})'\, E'E\, (\mu - \hat{\mu})\right]^{-n/2} \end{aligned} \tag{8}$$

where $\nu s^2 = y_1' M_E y_1$, $\nu = n - 2m_1 - k_1$, $M_E = I - E\,(E'E)^{-1} E'$, $E = \left(Y_1, X_1, \tilde{V}_1\right)$, $\tilde{V}_1 = Y_1 - X\tilde{\Pi}_1$, $\mu = (\gamma_1', \beta_1', \theta_1')'$, and $\hat{\mu} = (E'E)^{-1} E' y_1$. If we set $\tilde{\Pi}_1 = \hat{\Pi}_1$, where

$\hat{\Pi}_1 = (X'X)^{-1} X'Y_1$, then the posterior mean of γ_1 and β_1 in (8) is given by the two-stage least squares ($2SLS$) estimator of γ_1 and β_1. Accordingly, the $2SLS$ estimator may be interpreted as a conditional Bayes estimator, given $\Pi_1 = \hat{\Pi}_1$. This is another example of the fact that, as pointed out by Zellner (1988), most non-Bayesian inference results in econometrics can be produced by Bayesian methods under special assumptions.

It is the case that the three-stage least squares ($3SLS$) estimator of γ_1 and β_1 in (1) is identical to the $2SLS$ estimator of γ_1 and β_1. The $3SLS$ estimator applied to (1) is given by

$$\hat{\theta} = \left\{Z' \left[S^{-1} \otimes X (X'X)^{-1} X'\right] Z\right\}^{-1} Z' \left[S^{-1} \otimes X (X'X)^{-1} X'\right] y \tag{9}$$

where $\theta = \left(\delta_1', \mathrm{Vec}\,(\Pi_1)'\right)'$, $\delta_1 = (\gamma_1', \beta_1')'$, and $y = \left(y_1', \mathrm{Vec}\,(Y_1')\right)'$. S is a consistent estimator of Ω. After tedious manipulation, it can be shown that

$$\begin{aligned} \hat{\delta}_1 &= \left(Z_1' D Z_1\right)^{-1} Z_1' D y_1 \\ \overline{\Pi}_1 &= (X'X)^{-1} X'Y_1 - s_{11}^{-1} (X'X)^{-1} X' (S_{12} \otimes \hat{u}_1) \end{aligned} \tag{10}$$

where $Z_1 = (Y_1, X_1)$, $D = X\left(X'X\right)^{-1} X'$, s_{11} and S_{12} are the estimators of σ_{11} and δ in Σ, respectively; $\hat{u}_1 = y_1 - Z_1\hat{\delta}_1$, and $\hat{\delta}_1$ is the $2SLS$ estimator of δ_1. An intuitive reason why the $3SLS$ estimator of γ_1 and β_1 in (1) yields the $2SLS$ estimator is that since the reduced form equations for Y_1 in (1) contain all the exogenous variables and no endogenous variables, there is no informational gain in treating the structural equation y_1 as a part of the system in jointly estimating $\gamma_1, \hat{\beta}_1$, and Π_1.

If we use $\hat{\Pi}_1$ as the conditioning value of Π_1 in (6), we obtain the $2SLS$ estimator as the posterior mean. If S_{12} in equation (10) is not zero, then $\overline{\Pi}_1$ is, at least asymptotically, more efficient than $\hat{\Pi}_1$. For this reason, we may use $\overline{\Pi}_1$ as the conditioning value of Π_1 and obtain the posterior mean of γ_1 and β_1. This conditional value is suggested on the basis of asymptotic efficiency but it is not optimal relative to the squared error loss used by Zellner (1978, 1986) to derive the minimum expected loss estimators. Let us call the estimator of γ_1 and β_1 given $\overline{\Pi}_1$ *the conditional limited information Bayesian* ($CLIB$) estimator. More precisley, the $CLIB$ estimator is given by

$$\hat{\delta}_{1,CLIB} = \begin{bmatrix} \hat{\gamma}_{1,CLIB} \\ \hat{\beta}_{1,CLIB} \end{bmatrix} = (Z_1' M_{v_1} Z_1)^{-1} Z_1' M_{v_1} y_1 \tag{11}$$

where $M_{v_1} = I - \tilde{V}_1 \left(\tilde{V}_1'\tilde{V}_1\right)^{-1} \tilde{V}_1'$, $\tilde{V}_1 = Y_1 - X\overline{\Pi}_1$, and $\overline{\Pi}_1$ is given in equation (10). The use of $\overline{\Pi}_1$ for the $CLIB$ estimator is suggested as a matter of convenience since it is easy to compute $\overline{\Pi}_1$. One may suggest, in lieu of $\overline{\Pi}_1$, a restricted reduced form (*RRF*) estimator, since it incorporates information on the inclusions and exclusions of variables in the structural equations of the system. Phillips (1983) reviews properties of *RRF* coefficient estimators using finite sample theory.

The *unconditional LIB* estimator (*ULIB*) is derived from equation (3) by first deriving the posterior pdf's for γ_{i1} and for β_{i1}, where γ_{i1} and β_{i1} are the i-th elements of γ_1 and β_1, respectively. Given the likelihood function (4) and the prior pdf of γ_1, β_1, Π_1, and Σ

$$p(\gamma_1, \beta_1, \Pi_1, \Sigma \mid \beta_2 = 0) \;\propto\; |\Sigma|^{(m_1+2)/2} \tag{12}$$

Tsurumi (1985) derives the posterior pdf for γ_{11} as

$$p(\gamma_{11} \mid \beta_2 = 0, \text{data}) \;\propto\; \frac{\left[1 + (\gamma_{11} - \hat{\gamma}_{11})^2 h_{11\cdot 2}\right]^{\nu_1/2}}{\left[1 + (\gamma_{11} - \tilde{\gamma}_{11})^2 g_{11\cdot 2}\right]^{\nu_2/2}} \sum_{i=0}^{r} p_i \binom{r}{i} \tag{13}$$

where $\hat{\gamma}_{11}$ is the 1*st* element of $\hat{\gamma}_1 = (Y_1' M Y_1)^{-1} Y_1' M y_1$, $\tilde{\gamma}_{11}$ is the 1*st* element of $\tilde{\gamma}_1 = (Y_1' M_1 Y_1)^{-1} Y_1' M_1 y_1$; $M = I - X(X'X)^{-1} X'$, and $M_1 = I - X_1 (X_1' X_1)^{-1} X_1'$, $h_{11\cdot 2} = h_{11} - H_{12} H_{22}^{-1} H_{21}$,

$$H = Y_1' M Y_1 / \nu_1 s_1^2 = \begin{matrix} \\ 1 \\ m_1 - 1 \end{matrix} \overset{\begin{matrix} 1 & \quad m_1 - 1 \end{matrix}}{\begin{pmatrix} h_{11} & H_{12} \\ H_{21} & H_{22} \end{pmatrix}}$$

$$\nu_1 s_1^2 = (y_1 - Y_1 \tilde{\gamma}_1)' (y_1 - Y_1 \tilde{\gamma}_1), \; g_{11\cdot 2} = g_{11} - G_{12} G_{22}^{-1} G_{21},$$

$$G = Y_1' M_1 Y_1 / \nu_2 s_2^2 = \begin{matrix} \\ 1 \\ m_1 - 1 \end{matrix} \overset{\begin{matrix} 1 & \quad m_1 - 1 \end{matrix}}{\begin{pmatrix} g_{11} & G_{12} \\ G_{21} & G_{22} \end{pmatrix}}$$

$\nu_2 s_2^2 = (y_1 - Y_1\tilde{\gamma}_1)'(y_1 - Y_1\tilde{\gamma}_1)$, $r = \nu_1/2$, and $\hat{p}_i$'s are given by

$$p_0 = 1$$

$$p_1 = a'a + \frac{\text{tr } D}{\nu_2 - m_1 - 2}$$

$$p_2 = (a'a)^2 + \frac{2(a'a)\text{tr } D^{-1} + 4a'D^{-1}a}{\nu_2 - m_1 - 2} + \frac{(\text{tr } D^{-1})^2 + 2\text{tr } D^{-2}}{(\nu_2 - m_1 - 2)(\nu_2 - m_1 - 4)}$$

$$p_3 = (a'a)^3 + \frac{3(a'a)^2\text{tr } D^{-1} + 12(a'D^{-1}a)(a'a)}{\nu_2 - m_1 - 2}$$
$$+ \frac{24a'D^{-2}a + 3(a'a)(\text{tr } D^{-1})^2 + 6\text{tr } D^{-2}(a'a) + 12(a'D^{-1}a)\text{tr } D^{-1}}{(\nu_2 - m_1 - 2)(\nu_2 - m_1 - 4)}$$
$$+ \frac{(\text{tr } D^{-1})^3 + 6(\text{tr } D^{-2})(\text{tr } D^{-1}) + 8\text{tr } D^{-3}}{(\nu_2 - m_1 - 2)(\nu_2 - m_1 - 4)(\nu_2 - m_1 - 6)}$$

....

$a = R(\tilde{\gamma}_2^* - \hat{\gamma}_2^*)$, $\hat{\gamma}_2^* = \hat{\gamma}_2 - H_{22}^{-1}H_{21}(\gamma_{11} - \hat{\gamma}_{11})$, $\tilde{\gamma}_2^* = \tilde{\gamma}_2 - G_{22}^{-1}G_{21}(\gamma_{11} - \tilde{\gamma}_{11}), \gamma_2 = (\gamma_{21}, \gamma_{22}, \cdots, \gamma_{m_1,1})'$. R is a $(m_1 - 1) \times (m_1 - 1)$ nonsingular matrix such that $R'R = H_{22}^* = H_{22}/[1 + (\gamma_{11} - \hat{\gamma}_{11})^2 h_{11\cdot 2}]$, $D = R'^{-1}G_{22}^* R^{-1}$, and $G_{22}^* = G_{22}/[1 + (\gamma_{11} - \tilde{\gamma}_{11})^2 g_{11\cdot 2}]$. It should be noted that $p_{i'}$s are functions of γ_{11}.

The posterior pdf of β_{11} is

$$p(\beta_{11} \mid \beta_2 = 0) \propto (\nu_2^* s_2^{*^2})^{-(\nu_2^* - m_1)/2} \sum_{i=0}^{r} q_i^* \binom{r}{i} \tag{14}$$

where $\nu_2^* = \nu_2 - 1$, $\nu_2^* s_2^{*^2} = (z_1^* - Y_1\tilde{\gamma}_1^*)' M_1^*(z_1^* - Y_1\tilde{\gamma}_1^*)$, $M_1^* = I - X_1^*(X_1^{*'}X_1^*)^{-1}X_1^{*'}$, $z_1^* = y_1 - x_1\beta_{11}$, $x_1 = (x_{11}, \cdots, x_{n1})'$, $\tilde{\gamma}_1^* = (Y_1'M_1^*Y_1)^{-1}Y_1'M_1^*z_1^*$; $q_i^{*'}$s are identical to p_i's given below (13) except that D and a are replaced by Q and b with $Q = R'^{-1}G^*R^{-1}$, $G^* = Y_1'M_1^*Y_1/\nu_2^* s_2^{*^2}$, $b = \tilde{\gamma}_1^* - \hat{\gamma}_1$.

Unconditional *LIB* estimators of γ_{11} and β_{11}, $\hat{\gamma}_{11,ULIB}$ and $\hat{\beta}_{11,ULIB}$, will be given by the posterior means if we use the quadratic loss functions, $L(\gamma_{11}, \hat{\gamma}_{11,ULIB}) = c_1(\gamma_{11} - \hat{\gamma}_{1,ULIB})^2$ and $L(\beta_{11}, \hat{\beta}_{11,ULIB}) = c_2(\beta_{11} - \hat{\beta}_{11,ULIB})^2$:

$$\hat{\gamma}_{11,ULIB} = E(\gamma_{11} \mid \cdot) = \int \gamma_{11} p(\gamma_{11} \mid \cdot)\, d\gamma_{11} \tag{15}$$

and

$$\hat{\beta}_{11,ULIB} = E(\beta_{11} \mid \cdot) = \int \beta_{11} p(\beta_{11} \mid \cdot)\, d\beta_{11} \tag{16}$$

$$= \int\int \beta_{11} p(\beta_{11}, \gamma_1 \mid \cdot)\, d\beta_{11} d\gamma_{11}$$

$$= \int\int \beta_{11} p(\beta_{11} \mid \gamma_1, \cdot)\, p(\gamma_1 \mid \cdot)\, d\beta_{11} d\gamma_1$$

$$= \int \hat{\beta}_{11} p(\gamma_1 \mid \cdot)\, d\gamma_1 = \mathbf{i}_1' \left(X_1'X_1\right)^{-1} X_1' \left(y_1 - Y_1 \hat{\gamma}_{1,ULIB}\right)$$

where $\mathbf{i}_1 = (1, 0, \cdots, 0)'$. The last equality in (16) is derived using the following result, since given γ_1, β_1 has a multiple student-t distribution:

$$E(\beta_1 \mid \cdot) = \left(X_1'X_1\right)^{-1} X_1' \left[y_1 - Y_1 \int \gamma_1 p(\gamma_1 \mid \cdot)\, d\gamma_1\right] \tag{17}$$

$$= \left(X_1'X_1\right)^{-1} X_1' \left[y_1 - Y_1 E(\gamma_1 \mid \cdot)\right]$$

If, instead of a quadratic loss function, we use the loss:

$$L\left(\theta - \hat{\theta}\right) = \begin{cases} 1 & \text{for } |\theta - \hat{\theta}| > \epsilon \\ 0 & \text{for } |\theta - \hat{\theta}| < \epsilon \end{cases} \tag{18}$$

where $\epsilon > 0$ is an arbitrarily small constant, in estimating γ_{11} and β_{11}, the posterior modal values of γ_{11} and of β_{11} are the Bayesian point estimators:

$$\hat{\gamma}_{11,PMODE} = mode\ of\ p(\gamma_{11} \mid \cdot)$$
$$\hat{\beta}_{11,PMODE} = mode\ of\ p(\beta_{11} \mid \cdot) \tag{19}$$

Zellner (1971) and Drèze (1976) derived joint posterior pdf's for a vector of coefficients of a structural equation of interest. Equation (13) and (14) in the previous section, on the other hand, are the marginal posterior pdf's for individual structural coefficients. Although Drèze's main interest is the derivation of the joint posterior pdf, he suggests as one of Bayes' estimators the posterior modal values of the joint posterior pdf of γ_1.

3. Sampling Experiments

In the previous section we derived conditional and unconditional limited information Bayesian (LIB) estimators. Asymptotically, these estimators are strongly consistent, and as far as one seeks consistency, they are not different from any other consistent limited information estimators that have been developed so far. For a finite sample case, however, one expects the performance of various limited information estimators to be different. By sampling experiments, let us compare the performance of the conditional and unconditional limited information Bayesian estimators with the following six estimators: ordinary least squares (OLS) estimator, two-stage least squares ($2SLS$) estimator, limited information maximum likelihood ($LIML$) estimator, Fuller's (1977) modified $LIML$ estimator. Zellner's diffuse prior minimum expected loss ($MELO$) estimator, and Zellner's diffuse prior extended $MELO$ estimator. More specifically, the following table summarizes the estimators that are compared by sampling experiments in this paper. The first six estimators in Table 1 can be expressed in terms of the *K-class* estimator or the *double K-class* estimator. The *double K-class* estimator is given by

$$\hat{\delta}_1(\mathbf{k}_1, k_2) = \begin{bmatrix} Y_1'Y_1 - \mathbf{k}_1\hat{V}_1'\hat{V}_1 & Y_1'X_1 \\ X_1'Y_1 & X_1'X_1 \end{bmatrix}^{-1} \begin{bmatrix} \left(Y_1 - \mathbf{k}_2\hat{V}_1\right)' y_1 \\ X_1'y_1 \end{bmatrix} \tag{20}$$

where $\hat{V}_1 = X\hat{\Pi}_1$, $\hat{\Pi}_1 = (X'X)^{-1}X'Y_1$. Equation (20) becomes the *K-class* if $\mathbf{k}_1 = \mathbf{k}_2 = \mathbf{k}$.

The relative performance of various limited information estimators has been studied in the literature from analytical as well as from sampling experimental points of view. A large number of recent finite sample research results has been conveniently presented in Anderson (1983), Mariano (1982), Phillips (1983) and Morimune (1985). In a recent study, Anderson, Kunitomo and Morimune (1986) evaluated the $2SLS$, $LIML$, $FLIM$, ZM, and Drèze's LIB estimators in addition to a few other limited information estimators on the basis of mean squared error and concentration of probability criteria computed by asymptotic expansions of underlying sampling distributions. They show that Fuller's modified $LIML$ ($FLIM$) and Drèze's LIB estimators are asymptotically third order efficient and that the $2SLS$ and related *K-class* estimators are asymptotically dominated by the $FLIM$ and Drèze's LIB estimators. The relevance of such asymptotic results for finite sample situations appears to be an open question. Anderson, Kunitomo and Morimune used a loss function different from that which generates such Bayesian estimators as ZM or ZEM, and thus their results should be evaluated according to the particular optimality criteria used in their study.

Anderson, Kunitomo and Sawa (1982) evaluate the $LIML$, $2SLS$ and $LIMLK$ (i.e. the $LIML$ estimator with a known covariance matrix) in Monte Carlo experiments, while Morimune (1985) evaluated the $LIML$, $2SLS$ and OLS estimators

Table 1: Glossary of the Limited Information Estimators

Estimator	Acronym in this paper	Reference Source
Ordinary least squares	*OLS*	
Two-stage least squares	*2SLS*	Basmann (1957),Theil (1958)
Limited information maximum likelihood	*LIML*	Anderson and Rubin (1949)
Modified *LIML*	*FLIM*	Fuller (1977)
Minimum expected loss(*MELO*)	*ZM*	Zellner (1978)
Extended *MELO*	*ZEM*	Zellner (1986)
Conditional limited information Bayesian (*LIB*)	*CLIB*	Equation (11)
Unconditional *LIB*	*ULIB*	Equations (15) and (16)
Mode of the Posterior Pdf	*PMOD*	Equation (19)

K-Class Estimator ($\mathbf{k}_1 = \mathbf{k}_2 = \mathbf{k}$)

$$OLS: \quad \mathbf{k} = 0$$
$$2SLS: \quad \mathbf{k} = 1$$
$$LIML: \quad \mathbf{k} = \hat{\lambda} = \min_{\gamma^*}\left(\frac{\gamma^{*\prime} Y_I^{\prime} M_1 Y_I \gamma^*}{\gamma^{*\prime} Y_I^{\prime} M Y_I \gamma^*}\right)$$
$$FLIM: \quad \mathbf{k} = \hat{\lambda} - \tfrac{c}{n-k}, \; c > 0 \;\; \text{(we choose } c = 4\text{)}$$
$$ZM: \quad \mathbf{k} = 1 - \tfrac{k}{n-k-m_1-2}$$
(when based on a diffuse prior)

Double K-Class Estimator

$$ZEM: \quad \mathbf{k}_1 = 1 - \tfrac{k}{n-k-m_1-2}$$
$$\mathbf{k}_2 = 1 - \tfrac{k}{(1+g)(n-k-m_1-2)}, \; g \geq 0$$
(when based on a diffuse prior)
(we choose $g = 3$)

in Monte Carlo experiments. The Monte Carlo experiments in these two studies are made by using the distributions of the normalized estimators and samples are generated using the four key parameters: $n-k$, $k_2 - m_1$, the correlation coefficent between Y_1 and u_1 in equation (1) (i.e. the measure of simultaneity), and the non-centrality parameter. Sampling experiments on the normalized estimators using the four parameters may be termed the *normalized Monte Carlo method.*

Another way to carry out sampling experiments is to set up an econometric model, generate data using the model and estimate structural parameters using the generated data. Since Summers' (1965) study, this method, which may be termed as the *direct Monte Carlo method*, has been the traditional and most prevalent method of sampling experiments in econometrics. In this paper, we shall use the *direct Monte Carlo method* to evaluate the limited information estimators that are given in Table 1. The four Bayesian estimators, the ZEM, $ULIB$, $PMOD$, and $CLIB$, have never been evaluated before in the literature, although the ZM estimator was evaluated by Park (1982) using the *direct Monte Carlo method.*

We design the sampling experiments to allow for the circumstances that are often present in econometric data analyses, namely, (i) a small sample; (ii) presence of multicollinearity; (iii) degrees of correlation between stochastic regressors and the error terms, $\delta = \mathrm{Cov}(u_{t1}, Y_{t1})$ that is given just above equation (4) and (iv) reduced form coefficient of determination $\mathbf{R}^2$. For this purpose we modify Cragg's (1967) model so that we can control the degree of correlation between stochastic regressors and the error term, δ, as well as the reduced form coefficient of determination, $\mathbf{R}^2$. The model consists of three structural equations. In the appendix, the model and sampling design are explained.

Since we are comparing limited information estimation methods, we shall evaluate one equation at a time, rather than all the equations in the system. Evaluation of one equation at a time is better suited for our purposes, since we want to control the degree of correlation between stochastic regressors and the error term of equation of interest as well as the reduced form $\mathbf{R}^2$. We evaluate various estimators of the parameters of the first equation of the model

$$y_{t1} = \gamma_{12} y_{t2} + \beta_{11} x_{t3} + \beta_{15} x_{t5} + \beta_{17} x_{t7} + u_{t1} \tag{21}$$
$$\gamma_{12} = .222,\ \beta_{11} = 6.2,\ \beta_{13} = .7,\ \beta_{15} = .96\ \beta_{17} = .06\ .$$

Equation (21) has one endogenous variable on the right hand side ($m_1 = 1$). For the case of $m_1 = 1$, Anderson, Kunitomo and Sawa (1982) and Morimune (1985) conducted sampling experiments by the *normalized Monte Carlo method*, controlling for four key parameters: $n - k$, $k_2 - m_1$, the measure of simultaneity and the noncentrality parameter. In their experiments the noncentrality parameter is treated as independent of other three parameters. Using the *direct Monte Carlo method*, we can show how the noncentrality parameter is affected by $n - k$, $k_2 - m_1$, and by the measure of simultaneity as well as by multicollinearity and the fit of the equation.

As measures of evaluation, the mean absolute deviation (MAD) and mean squared errors (MSE) for each structural coefficient are popularly used, and as the overall measures for all the coefficients in an equation or in all the equations in the system, often the rank sums of the MAD's or of the MSE's are presented. [for example, Cragg (1967) and Park (1982)]. Raj (1980) expresses mean biases and mean squared errors in relative terms by scaling these values by the values of true coefficients to account for the size differences of the coefficients. We computed MAD's and MSE's for each coefficient in the equation of interest. The results when MAD's are used are quite similar to those when MSE's are used, except that when MSE's are used the $LIML$ estimator exhibits in many cases substantially large values of MSE's, reflecting the fact that the $LIML$ estimator has no finite moments. Recently, Rao (1980) criticized the standard procedure of comparing the MSE's of alternative estimators. He suggested the use of intrinsic criteria that reflect the notion that a good estimator is one providing a high degree of concentration in the neighborhood of the true parameter value. We shall use such a measure later on. First, let us use the MAD measure, primarily because this along with the MSE is often used in sampling experiments in econometrics. We compute the relative MAD, $RMAD$, by dividing each MAD by the smallest MAD among the different estimators for a particular parameter:

$$RMAD\left(\theta_i^{(j)}\right) = MAD\left(\theta_i^{(j)}\right) / \min_k MAD\left(\hat{\theta}_i^{(k)}\right) \tag{22}$$

where $MAD\left(\theta_i^{(j)}\right) = \frac{1}{R}\sum_{i=1}^{R} |\hat{\theta}_{r,i}^{(j)} - \theta_i|$; $\hat{\theta}_{r,i}^{(j)}$ is the estimate of θ_i by the estimation method j at the rth replication. The sum of the $RMAD\left(\theta_i^{(j)}\right)$, $SRMAD^{(j)}$, is given by

$$SRAD^{(j)} = \sum_{i=1}^{m_1+k_1} RMAD\left(\theta_i^{(j)}\right)$$

The minimum value of the sum of the relative MAD's is the number of parameters we evaluate, m_1+k_1. Compared to the usual rank sum of the rankings of the MAD's, the $SRAD$'s, will convey information about how close (or far apart) the performances of different estimators.

All the experiments are based on 400 sample replications, and the $ULIB$ and $PMOD$ are obtained using numerical integration by Simpson's rule. Tables 2–5 present the sums of relative mean absolute deviations, $SRMAD$'s, for different cases of $\mathbf{R}^2 = .9$ or $.35$, $n = 100$ or 20 with or without multicollinearity and for different values of ρ_{12}^2, the squared correlation between the righthand side endogenous variable, y_{t2} and the error term, u_{t1}. Table 6 presents the top three ranked estimators for each case. The following observations may be made:

(1) No one estimator dominates all the other estimators for different values of ρ_{12}^2.

(2) For the case of $\mathbf{R}^2 = .9$, $n = 100$ or 20 without multicollinearity, the performance of all the estimators except OLS is similar when $\rho_{12}^2 \neq 0$.

Table 2: Sums of Relative Mean Absolute Deviations: No Multicollinearity; $\mathbf{R}^2 = .9$

	$\rho_{12}^2 = 0$ $SRMAD$	$\rho_{12}^2 = .3$ $SRMAD$	$\rho_{12}^2 = .5$ $SRMAD$	$\rho_{12}^2 = .7$ $SRMAD$	$\rho_{12}^2 = .9$ $SRMAD$
			$n = 100$		
OLS	5.000(1)	7.317(9)	8.573(9)	9.582(9)	10.49(9)
$2SLS$	5.071(5)	5.041(4)	5.063(4)	5.088(4)	5.115(4)
$LIML$	5.074(6)	5.039(2)	5.061(2)	5.086(2)	5.112(2)
$FLIM$	5.066(3)	5.073(6)	5.073(6)	5.100(6)	5.127(6)
ZM	5.056(2)	5.073(7)	5.115(8)	5.100(6)	5.199(8)
ZEM	5.067(4)	5.044(5)	5.069(5)	5.096(5)	5.125(5)
$CLIB$	5.074(6)	5.038(1)	5.060(1)	5.085(1)	5.112(2)
$ULIB$	5.074(6)	5.039(2)	5.061(2)	5.086(2)	5.111(1)
$PMOD$	5.122(9)	5.082(8)	5.082(7)	5.104(7)	5.139(7)
			$n = 20$		
OLS	5.001(1)	5.288(9)	5.507(9)	5.718(9)	5.929(9)
$2SLS$	5.131(5)	5.147(3)	5.180(3)	5.213(2)	5.251(2)
$LIML$	5.179(9)	5.200(7)	5.243(7)	5.285(6)	5.330(6)
$FLIM$	5.107(4)	5.144(2)	5.183(4)	5.220(3)	5.261(5)
ZM	5.022(3)	5.184(6)	5.312(8)	5.439(8)	5.569(8)
ZEM	5.017(2)	5.054(1)	5.096(1)	5.137(1)	5.183(1)
$CLIB$	5.154(7)	5.163(5)	5.193(5)	5.224(5)	5.258(3)
$ULIB$	5.179(9)	5.200(7)	5.242(6)	5.285(6)	5.330(6)
$PMOD$	5.135(6)	5.154(4)	5.178(2)	5.221(4)	5.260(4)

(1) $SRMAD$ = the sum of relative MAD's:

$$SRMAD = \sum_{i=1}^{5} MAD\left(\hat{\theta}_i^{(j)}\right) / \min_p MAD\left(\hat{\theta}_i^{(p)}\right)$$

where $MAD\left(\hat{\theta}_i^{(j)}\right)$ is the mean absolute deviation of the estimate of θ_i by the jth estimation method.

(2) The numbers in parentheses are the ranks of the $SRMAD$'s.

(3) $\rho_{12} = Corr\,(y_{t2}, u_{t1})$, the correlation coefficient between the right hand side endogenous variable and the error term.

(4) $\mathbf{R}^2$ = the coefficient of determination of the reduced form equation for the right hand side endogenous variable, y_{t2}.

(5) For each case, results are based on 400 replications.

(3) Except for the case of $\mathbf{R}^2 = .9$, $n = 100$, and without multicollinearity, ZEM is always in the top three rankings regardless of the different combinations of $\mathbf{R}^2$, n, and ρ_{12}^2, with or without multicollinearity. In many cases ZEM has the least $SRMAD$'s.

(4) When ρ_{12}^2 is higher than or equal to .5, the performance of $CLIB$, $ULIB$, and

Table 3: Sums of Relative Mean Absolute Deviations: No Multicollinearity; $\mathbf{R}^2 = .35$

	$\rho_{12}^2 = 0$ *SRMAD*	$\rho_{12}^2 = .3$ *SRMAD*	$\rho_{12}^2 = .5$ *SRMAD*	$\rho_{12}^2 = .7$ *SRMAD*	$\rho_{12}^2 = .9$ *SRMAD*
			$n = 100$		
OLS	5.000(1)	9.725(9)	11.980(9)	13.726(9)	15.365(9)
2SLS	6.136(6)	5.346(4)	5.532(5)	5.785(5)	6.229(5)
LIML	6.269(9)	5.419(7)	5.610(7)	5.875(7)	6.344(7)
FLIM	6.019(4)	5.314(2)	5.508(3)	5.758(4)	6.186(3)
ZM	5.783(2)	5.465(8)	5.842(8)	6.230(8)	6.757(8)
ZEM	5.853(4)	5.206(1)	5.418(1)	5.674(1)	6.114(1)
CLIB	6.242(7)	5.352(5)	5.514(4)	6.755(3)	6.211(4)
ULIB	6.269(8)	5.419(6)	5.609(6)	5.875(6)	6.344(6)
PMOD	6.081(5)	5.324(3)	5.506(2)	5.745(2)	6.182(2)
			$n = 20$		
OLS	5.102(1)	6.520(8)	7.427(8)	8.180(9)	8.995(9)
2SLS	6.340(6)	6.265(5)	6.405(4)	6.685(5)	7.360(5)
LIML	7.911(9)	7.588(9)	7.533(9)	7.709(8)	8.396(8)
FLIM	5.750(4)	5.894(2)	6.127(2)	6.427(3)	6.993(3)
ZM	5.156(3)	6.230(4)	6.976(7)	7.616(7)	8.287(7)
ZEM	5.000(1)	5.341(1)	5.621(1)	5.956(1)	6.613(1)
CLIB	6.702(8)	6.455(6)	6.482(5)	6.680(4)	7.348(4)
ULIB	6.569(7)	6.468(7)	6.623(6)	6.924(6)	7.700(6)
PMOD	6.093(5)	6.065(3)	6.202(3)	6.416(2)	6.962(2)

See footnotes below Table 2.

PMOD improve.

(5) When multicollinearity exists *LIML* performs the worst regardless of $\mathbf{R}^2$ or n, whereas *FLIM* performs quite well in many cases and becomes one of the top 3 ranked estimators. Since the *LIML* does not possess mean or higher moments, we expect the sensitivity of the *LIML* performance as judged by the *MAD* criterion.

Sensitivity of the *LIML* estimator to multicollinearity has been recognized. Klein and Nakamura (1962) analyze the impacts of multicollinearity on different estimators. In the case of *LIML*, they show that multicollinearity tends to yield multiple characteristic roots which in turn make it impossible to find a unique set of estimates of the normalized characteristic vector. Using the Moore-Penrose generalized inverse, we can show that perfect multicollinearity among X_1 and X_2 makes all

Table 4: Sums of Relative Mean Absolute Deviations: Multicollinearity; $\mathbf{R}^2 = .9$

	$\rho_{12}^2 = 0$ SRMAD	$\rho_{12}^2 = .3$ SRMAD	$\rho_{12}^2 = .5$ SRMAD	$\rho_{12}^2 = .7$ SRMAD	$\rho_{12}^2 = .9$ SRMAD
			$n = 100$		
OLS	5.000(1)	9.083(8)	11.193(8)	12.682(8)	14.150(8)
2*SLS*	12.291(7)	7.604(6)	7.295(5)	7.159(5)	7.257(5)
LIML	41.487(9)	68.342(9)	22.575(9)	19.495(9)	31.733(9)
FLIM	8.331(4)	6.402(2)	7.022(4)	7.556(6)	8.155(6)
ZM	6.601(2)	7.109(5)	8.440(7)	9.429(7)	10.389(7)
ZEM	6.888(3)	5.474(1)	5.777(1)	6.044(1)	6.362(3)
CLIB	14.146(8)	8.040(7)	7.323(6)	6.900(4)	6.809(4)
ULIB	10.192(6)	6.629(4)	6.378(2)	6.198(2)	6.238(2)
PMOD	10.192(5)	6.549(3)	6.451(3)	6.362(3)	6.112(1)
			$n = 20$		
OLS	5.469(2)	6.901(5)	7.836(6)	8.554(8)	9.335(8)
2*SLS*	9.302(7)	8.094(7)	8.301(7)	8.307(5)	8.496(5)
LIML	54.670(9)	32.307(9)	21.398(9)	20.936(9)	35.815(9)
FLIM	5.938(4)	6.782(3)	7.552(3)	8.145(4)	8.727(6)
ZM	5.470(3)	6.847(4)	7.751(5)	8.448(6)	9.203(7)
ZEM	5.000(1)	5.396(1)	5.616(1)	5.792(1)	5.987(1)
CLIB	10.977(8)	9.017(8)	8.806(8)	8.508(7)	8.432(4)
ULIB	6.858(5)	6.273(2)	6.573(2)	6.899(2)	7.104(2)
PMOD	7.228(6)	7.225(6)	7.641(4)	7.947(3)	8.171(3)

See footnotes below Table 2.

the characteristic roots equal to one. The minimum characteristic root is extracted from

$$| Y_I' M_1 Y_I - \lambda Y_I' M Y_I | = 0 \tag{23}$$

where $Y_I = (y_1, Y_1)$, $M_1 = I - X_1 (X_1'X_1)^{-1} X_1' = I - X_1 X_1^+$, and $M = I - X (X'X)^{-1} X' = I - XX^+$, and "+" denotes the Moore-Penrose generalized inverse. Suppose that X_1 and X_2 are generated by $X_1 = ZD_1$, and $X_2 = ZD_2$, where Z is an n $\times r$ matrix of rank r, and D_1 and D_2 are, respectively, $r \times k_1$ and $r \times k_2$ matrices of constants. Then it is easy to show that $XX^+ = X_1 X_1^+$ and thus $M = M_1$. Hence, equation (23) becomes

$$(1 - \lambda)^{m_1+1} | Y_I' M Y_I | = 0 \tag{24}$$

and we have $\lambda_i = 1$ for $i = 1, \cdots, m_1 + 1$.

Table 5: Sums of Relative Mean Absolute Deviations: Multicollinearity; $\mathbf{R}^2 = .35$

	$\rho_{12}^2 = 0$ $SRMAD$	$\rho_{12}^2 = .3$ $SRMAD$	$\rho_{12}^2 = .5$ $SRMAD$	$\rho_{12}^2 = .7$ $SRMAD$	$\rho_{12}^2 = .9$ $SRMAD$
			$n = 100$		
OLS	5.009(1)	6.868(4)	8.024(5)	8.868(6)	9.644(6)
$2SLS$	13.456(7)	9.228(7)	9.110(7)	9.190(7)	9.652(7)
$LIML$	108.13 (9)	62.284(9)	35.206(9)	31.591(9)	38.200(9)
$FLIM$	6.869(4)	6.910(5)	7.882(3)	8.635(4)	9.395(3)
ZM	5.795(2)	6.833(3)	7.916(4)	8.730(5)	9.487(5)
ZEM	6.241(3)	5.407(1)	5.656(1)	5.917(1)	6.447(1)
$CLIB$	16.806(8)	10.053(8)	9.638(8)	9.330(8)	9.721(8)
$ULIB$	8.566(5)	6.476(2)	6.991(2)	7.624(2)	8.445(2)
$PMOD$	9.484(6)	7.377(6)	8.040(6)	8.597(3)	9.398(4)
			$n = 20$		
OLS	5.382(2)	6.096(4)	6.633(4)	7.110(4)	7.760(5)
$2SLS$	8.713(7)	7.877(7)	8.007(7)	7.934(7)	8.301(7)
$LIML$	88.585(9)	36.099(9)	20.193(9)	21.552(9)	16.646(9)
$FLIM$	5.736(4)	6.187(5)	6.678(5)	7.123(5)	7.723(3)
ZM	5.388(3)	6.095(3)	6.632(3)	7.102(3)	7.749(4)
ZEM	5.000(1)	5.252(1)	5.423(1)	5.662(1)	6.267(1)
$CLIB$	10.005(8)	8.620(8)	8.603(8)	8.288(7)	8.542(8)
$ULIB$	6.419(5)	5.940(2)	6.235(2)	6.675(2)	7.399(2)
$PMOD$	6.807(6)	6.675(6)	6.930(6)	7.241(6)	7.796(6)

See footnotes below Table 2.

The fact that multiple roots tend to be generated by the existence of multicollinearity may be illustrated numerically for the cases of $\mathbf{R}^2 = .9$, $n = 100$, and $\rho_{12}^2 = .7$. There are two roots, λ_1 and λ_2, since $Y_I = (y_1, y_2)$. The means, standard deviations, maximum and minimum values of λ_1 and λ_2 for 400 replications are given by

Table 6: Top 3 Ranked Estimators for Each Case of $\mathbf{R}^2$, and n with or without Multicollinearity[(a)]

	$\rho_{12}^2 = 0$	$\rho_{12}^2 = .3$	$\rho_{12}^2 = .5$	$\rho_{12}^2 = .7$	$\rho_{12}^2 = .9$
			Without Multicollinearity		
$\mathbf{R}^2 = .9$ $n = 100$	(1)OLS (2)ZM (3)FLIM	(1)CLIB (2)LIML (3) ULIB	(1)CLIB (2)LIML (3)ULIB	(1)CLIB (2)LIML (3)ULIB	(1)ULIB (2)LIML (3)CLIB
$\mathbf{R}^2 = .9$ $n = 20$	(1)OLS (2)ZEM (3)ZM	(1)ZEM (2)FLIM (3)2SLS	(1)ZEM (2)PMOD (3)2SLS	(1)ZEM (2)2SLS (3)FLIM	(1)ZEM (2)2SLS (3)CLIB
$\mathbf{R}^2 = .35$ $n = 100$	(1)OLS (2)ZM (3)ZEM	(1)ZEM (2)FLIM (3)PMOD	(1)ZEM (2)PMOD (3)FLIM	(1)ZEM (2)PMOD (3)CLIB	(1)ZEM (2)PMOD (3)FLIM
$\mathbf{R}^2 = .35$ $n = 20$	(1)ZEM (2)OLS (3)ZEM	(1)ZEM (2)FLIM (3)PMOD	(1)ZEM (2)FLIM (3)PMOD	(1)ZEM (2)PMOD (3)FLIM	(1)ZEM (2)PMOD (3)FLIM
			With Multicollinearity		
$\mathbf{R}^2 = .9$ $n = 100$	(1)OLS (2)ZM (3)ZEM	(1)ZEM (2)FLIM (3)PMOD	(1)ZEM (2)ULIB (3)PMOD	(1)ZEM (2)ULIB (3)PMOD	(1)PMOD (2)ULIB (3)ZEM
$\mathbf{R}^2 = .9$ $n = 20$	(1)ZEM (2)OLS (3)ZM	(1)ZEM (2)ULIB (3)FLIM	(1)ZEM (2)ULIB (3)FLIM	(1)ZEM (2)ULIB (3)PMOD	(1)ZEM (2)ULIB (3)PMOD
$\mathbf{R}^2 = .35$ $n = 100$	(1)OLS (2)ZM (3)ZEM	(1)ZEM (2)ULIB (3)ZM	(1)ZEM (2)ULIB (3)FLIM	(1)ZEM (2)ULIB (3)PMOD	(1)ZEM (2)ULIB (3)FLIM
$\mathbf{R}^2 = .35$ $n = 20$	(1)ZEM (2)OLS (3)ZM	(1)ZEM (2)ULIB (3)ZM	(1)ZEM (2)ULIB (3)ZM	(1)ZEM (2)ULIB (3)ZM	(1)ZEM (2)ULIB (3)FLIM

(a) Based on results in Tables 2–5.

	Without Multicollinearity		With Multicollinearity	
	λ_1	λ_2	λ_1	λ_2
Mean	29.06	1.022	1.2328	1.0204
S.D.	5.07	.024	.1045	.0225
Min	17.26	1.00003	1.0590	1.00004
Max	47.30	1.2175	1.6822	1.1683

S.D.= standard deviation; Min = minimum value;
Max = maximum value

While the smallest characteristic root, λ_2, yields the similar values regardless of whether multicollinearity exists or not, the larger characteristic root, λ_1, yields the range of (17.22, 47.30) when multicollinearity is absent, and the range of (1.0590, 1.6822) when multicollinearity is present. Fuller's modified $LIML$ ($FLIM$), on the other hand, is not so sensitive to the existence of multicollinearity since λ_2 in the *K-class* is modified to $\mathbf{k} = \lambda_2 - 4/(n-k)$.

As shown in Anderson, Morimune and Sawa (1983), the largest characteristic root, λ_1, is proportionally related to the noncentrality parameter of the distribution of certain *K-class* estimators. Since the larger the value of the noncentrality parameter, the better is the performance of the $LIML$ estimator, the small value of λ_1 leads to the poor performance of the $LIML$ estimator. Moreover, the closeness of λ_1 to λ_2 causes a problem in testing under-identifiability of the structural equation. Koopmans and Hood's (1953) test statistic when $m_1 = 1$ is given by

$$n\left[\ln(\lambda_1) + \ln(\lambda_2)\right]$$

and if $\lambda_1 \simeq \lambda_2 \simeq 1$, then this test statistic tends to be zero, thus making it more likely to accept the null hypothesis of under-identifiability.

As shown in Anderson, Kunitomo and Sawa (1982), and Morimune (1985), among others, the distributions of *K class* estimators, for $m_1 = 1$, depend on $n-k$, $k_2 - m_1$, the correlation coefficient for simultaneity, ρ_{12}, and the noncentrality parameter which is given by

$$\eta = \Pi_{22}' X_2' M_1 X_2 \Pi_{22} / \Omega_{22} \tag{25}$$

where for $m_1 = 1$, a scalar, Ω_{22}, is the common variance of the elements of the vector V_1 in (1). The noncentrality parameter, η, is proportionately related to the sample size, $\mathbf{R}^2$, ρ_{12}^2, and multicollinearity. Table 7 gives how η is affected by n, $\mathbf{R}^2$, ρ_{12}^2, and multicollinearity. In our sampling experiments the presence of multicollinearity reduced η by the factor of over 100 to 150, depending on each case. (See the last column in Table 7.) From equation (23) it is clear that if perfect multicollinearity exists among X_1 and X_2, η becomes zero. Also it is true that the value of Π_{22} and Ω_{22} can influence the noncentrality parameter.

Table 7: Effects of Multicollinearity, $\mathbf{R}^2$, ρ_{12}^2, and Sample Sizes on the Noncentrality Parameter η

Without Multicollinearity				With Multicollinearity				
$\mathbf{R}^2$ (1)	n (2)	ρ_{12}^2 (3)	η (4)	$\mathbf{R}^2$ (5)	n (6)	ρ_{12}^2 (7)	η (8)	$\frac{(4)}{(8)}$
.9	100	0	25178.6	.9	100	0	176.4	142.8
		.3	26134.0			.3	185.4	140.9
		.5	26244.7			.5	186.2	140.9
		.7	26233.3			.7	185.7	141.3
		.9	26234.8			.9	184.5	142.2
.9	20	0	5867.4	.9	20	0	38.9	150.7
		.3	5802.8			.3	40.9	141.7
		.5	5755.7			.5	41.2	139.5
		.7	5719.7			.7	41.4	138.3
		.9	5743.9			.9	41.6	137.9
.35	100	0	1683.9	.35	100	0	13.6	124.2
		.3	1884.9			.3	16.0	118.0
		.5	1916.5			.5	16.2	118.1
		.7	1921.9			.7	16.1	119.3
		.9	1913.2			.9	15.6	122.5
.35	20	0	679.5	.35	20	0	4.9	138.8
		.3	771.8			.3	6.8	113.6
		.5	749.7			.5	7.0	107.1
		.7	705.8			.7	6.8	103.2
		.9	640.2			.9	6.3	101.1

Notes:
$\mathbf{R}^2$ = The coefficient of determination of the reduced form equation for Y_1
ρ_{12}^2 = the squared correlation between Y_{t1} and i_{t1}
$\eta = \Pi_{22}'X_2'M_1\Pi_{22}/\Omega_{22}$ is the noncentrality parameter.

We have used the sum of relative mean absolute deviations, $SRMAD$, as an aggregate criterion to evaluate how close each estimator of a set of structural coefficients is to the true values of these coefficients. The MAD or MSE criteria, however, are not relevant in the evaluation of estimators that do not possess finite sample moments. Moreover, some biased estimators may have a higher probability concentration around the true parameter value than unbiased estimators. Moreover, the MAD or MSE criteria do not provide information on the distribution of an estimator as to its shape or the probability of concentration. Using such measures as quartiles, median, and interquartile range, Anderson, Kunitomo, and Sawa (1982) and Morimune (1985) show that the $LIML$, which does not possess finite moment, is approximately median unbiased whereas the $2SLS$ estimator is not approximately median unbiased. Fiebig (1985), in comparing also the $LIML$ to $2SLS$ estimator, employs Rao's (1980) suggestion to use the probability of concentration.

Table 8 presents means, medians, quartiles, skewnesses and probabilities of concentration of the nine limited information estimators of γ_{12}, the coefficient associated with the right hand side endogenous variable for the case of $\mathbf{R}^2 = .9$, $\rho_{12}^2 = .7$ and with multicollinearity and for the sample sizes of $n = 20$, 100, and 300. In each case, we made 400 replications. A few noteworthy points emerge:

(1) When the sample size is small, $n = 20$, all the estimators' distributions are negatively skewed, but skewness disappears as the sample size increases. The inconsistency of the OLS estimator is clear: with $n = 300$, it is tightly distributed around $-.76$.

(2) For $n = 100$, and $n = 300$, the medians of the $LIML$ estimator are .216 and .255, respectively, indicating the approximate median unbiasedness property of the $LIML$. However, $LIML$ has very large dispersion.

(3) The ZEM estimator, on the other hand, has a tighter distribution, indicating that its bias tends to be compensated by a tighter distribution near the true parameter. This may be the reason why the ZEM tends to dominate the other estimators in many cases on using a MAD or MSE criterion.

Table 9 examines the impact on the $SRMAD$ an increase in the number of exogenous variables excluded from the equation of interest, k_2, for the case of $\mathbf{R}^2 = .9$, $n = 100$, $\rho_{12}^2 = .7$ and with multicollinearity. Column (1) is the same as in Table 4.

In Columns (2)–(5), k_2 is set at 9, but the degrees of multicollinearity among the 6 newly added exogenous variables and the existing exogenous variables, $x_{i2}, \cdots, x_{i7}$ are varied. Also the values of coefficients, β_{ij}'s of the newly added exogenous variables in the second and third equations of the model are varied. In the exact small sample distributions of *K-class* estimators (e.g. Anderson, Kunitomo and Sawa (1982) and Morimune (1985)), the number of exogenous variables excluded from the equation, k_2

Table 8: Means, Medians, Quartiles, and Skewness of Various Estimators of $\gamma_{12} = .222$ ($R^2 = .9$, $\rho_{12}^2 = .7$, with Multicollinearity)

	LQ	UQ	IQ	Med	Mean	S.D.	Skew	PC
				$n = 20$				
OLS	-.856	-.617	.239	-.742	-.748	.176	-.032	0
2SLS	-.842	-.177	.664	-.482	-.492	.701	-.015	.083
LIML	-.906	.411	1.317	-.222	-.560	7.020	-.046	.140
FLIM	-.847	-.476	.371	-.664	-.675	.264	-.043	.005
ZM	-.859	-.589	.270	-.727	-.732	.186	-.024	0
ZEM	-.384	-.193	.190	-.271	-.296	.148	-.172	.010
CLIB	-.863	-.039	.824	-.360	-.457	.812	-.119	.143
ULIB	-.694	-.162	.532	-.404	-.408	.379	-.012	.125
PMOD	-.839	-.327	.512	-.543	-.605	.421	-.149	.050
				$n = 100$				
OLS	-.808	-.717	.091	-.761	-.763	.066	-.022	0
2SLS	-.316	.193	.509	-.094	-.005	.534	.168	.263
LIML	-.108	.800	.908	.216	1.051	9.988	.084	.278
FLIM	-.421	-.180	.242	-.315	-.311	.202	.018	.404
ZM	-.589	-.386	.203	-.481	-.487	.148	-.043	0
ZEM	-.266	-.051	.215	-.149	-.163	.154	-.095	.123
CLIB	-.215	.290	.504	.031	.084	.553	.114	.350
ULIB	-.143	.437	.580	.128	.163	.415	.083	.375
PMOD	-.300	.077	.377	-.111	-.112	.344	-.001	.283
				$n = 300$				
OLS	-.791	-.735	.056	-.764	-.764	.040	.018	0
2SLS	-.080	.290	.370	-.091	.156	.348	.186	.490
LIML	.047	.539	.492	.255	.421	1.429	.117	.473
FLIM	-.119	.122	.242	-.002	-.002	.175	-.0003	.480
ZM	-.288	-.105	.184	-.200	-.199	.129	.007	.065
ZEM	-.113	.109	.222	.005	.005	.157	-.0006	.255
CLIB	-.014	.356	.370	.141	.205	.342	.186	.563
ULIB	.047	.507	.460	.243	.070	.172	.169	.495
PMOD	-.084	.239	.323	.077	.105	.257	.108	.528

Notes: 400 replications are made for each sample size. LQ = lower quartile; UQ = upper quartile; IQ = interquartile range; Med = median; S.D. = standard deviation; Skew = measure of skewness, calculated by Skew = (mean-median)/S.D.; PC = probability of concentration given by PC = $Pr\left[0 < \hat{\gamma}_{12}^{(i)} \leq .444\right]$ where $\hat{\gamma}_{12}^{(i)}$ is the estimate of γ_{12} by the estimation procedure i.

(or more precisely, $k_2 - m_1$) is treated as a parameter independent of the noncentrality parameter, η. If k_2 increases, however, η tends to increase and the extent to which η increases depends also on the new values of Π_{22}, the dimension of which increases with k_2. From Table 9, we see that an increase in k_2 (accompanied by an increase in η) tends to improve the relative performances of the *LIML*, *FLIM*, and *CLIB* estimators. Improvement of the *LIML* estimator as k_2 increases is also demonstrated in Anderson, Kunitomo and Sawa (1982) and Morimune (1985).

Table 9: Increase in the Number of Exogenous Variables Excluded from the Equation; R= .9, $n = 100$, $\rho_{12}^2 = .7$, With Multicollinearity

	(1) $k_2 = 3$ *SRMAD*	(2) $k_2 = 9$ *SRMAD*	(3) $k_2 = 9$ *SRMAD*	(4) $k_2 = 9$ *SRMAD*	(5) $k_2 = 9$ *SRMAD*
OLS	12.682(8)	14.577(9)	12.483(9)	13.888(9)	15.021(9)
2SLS	7.159(5)	6.233(6)	6.291(6)	5.705(6)	5.782(6)
LIML	19.495(9)	6.183(5)	6.759(7)	5.536(3)	5.686(4)
FLIM	7.556(6)	5.314(1)	5.201(1)	5.476(1)	5.433(1)
ZM	9.429(7)	9.585(8)	9.147(8)	6.848(8)	7.964(8)
ZEM	6.044(1)	6.683(7)	6.360(6)	5.870(7)	6.120(7)
CLIB	6.900(4)	5.397(2)	5.275(2)	5.537(4)	5.466(2)
ULIB	6.198(2)	5.640(3)	5.612(3)	5.525(2)	5.532(3)
PMOD	6.362(3)	6.102(4)	6.065(4)	5.641(5)	5.740(5)
η	185.7	753.1	502.2	3940.2	1796.5

Notes: Number of replacation is 400. *SRMAD* = sum of relative mean absolute deviation; $\ddot{\eta}$ = noncentrality parameter.

Col. (1)= Equation (21)

Col. (2)= 6 exogenous variables are added to the model: 3 each to the second and third equations of the model (see equations (30 and (31)). These 6 exogenous variables are correlated among themselves but uncorrelated with those that are originally in the system. $\beta_{ij} = 0.1$ in equations (30) and (31).

Col. (3)= Same as in Col. (2) except that the 6 new exogenous variables are correlated with those that are originally in the system. $\beta_{ij} = 0.1$.

Col. (4)= Same as in Col. (3), except $\beta_{ij} = 0.5$ in (30) and (31).

Col. (5)= Same as in Col. (2), except that the 6 new exogenous variables are uncorrelated among themselves as well as with those that are originally in the system.

The diffuse prior ZEM estimate is derived by minimizing the posterior expectation loss of the following loss function:

$$L = \left(X\pi_1 - \bar{Z}_1\hat{\delta}_1\right)' \left(X\pi_1 - \bar{Z}_1\hat{\delta}_1\right) + g\left(y_1 - \bar{Z}_1\hat{\delta}_1\right)' \left(y_1 - \bar{Z}_1\hat{\delta}_1\right) \tag{26}$$

where $\bar{Z}_1 = (X\pi_1, X_1)$, $\pi_1 = (\Pi'_{11}, \Pi'_{21})'$, $\delta'_1 = (\gamma'_1, \beta'_1)'$, $g \geq 0$, and $\hat{\delta}_1$ is an estimate of δ_1. The first term in (26) measures the extent to which the restrictions on the reduced form parameters fail to be satisfied and the second term represents lack of fit. If we modify equation (26) as

$$L = \mu_1\left(X\pi_1 - \bar{Z}_1\hat{\delta}_1\right)' \left(X\pi_1 - \bar{Z}_1\hat{\delta}_1\right) + \mu_2\left(y_1 - \bar{Z}_1\hat{\delta}_1\right)' \left(y_1 - \bar{Z}_1\hat{\delta}_1\right),$$
$$\mu_1 + \mu_2 = 1$$

then the ZEM becomes the *double K class* estimator with $\mathbf{k}_1 = 1 - k/(n-k-m_1-2)$ and $\mathbf{k}_2 = 1 - \mu_1 k/(n-k-m_1-2)$, and hence $\mu_1 = 1/(1+g)$. In our sampling experiments we set $g = 3$ or $\mu_1 = .25$, and $\mu_2 = .75$. If the researcher has a clear preference on how to weight μ_1 and μ_2, he should choose the values of μ_1 and μ_2 to satisfy his preference. For example, in Table 9, k_2, the number of exogenous variables that are excluded from the equation of interest, is raised from 3 to 9. The researcher will take into account such an increase in k_2 that results from a model specification change, and he will adjust g (or μ_1 and μ_2).

4. Concluding Remarks

In this paper we derived conditional and unconditional limited information Bayesian (LIB) estimators. The conditional LIB estimator is the mean of the conditional posterior pdf for the structural coefficients given the three-stage least squares ($3SLS$) estimates of the reduced form coefficients. The unconditional LIB estimators are from the marginal posterior pdf for an individual structural coefficient. We suggest the posterior mean (if loss is quadratic), modal (if loss is $0-1$ type), or median (if loss is absolute deviation) as the unconditional LIB estimators.

Using sampling experiments we compared the performances of nine limited information estimators: OLS, $2SLS$, $LIML$, $FLIM$, ZM, ZEM, $CLIB$, $ULIB$ and $PMOD$. The results of the experiments show that the performance rankings of these estimators vary across the degrees of correlation between the stochastic regressor and the error term, the reduced form coefficient of determination, multicollinearity, and sample sizes. The $LIML$ estimator is quite sensitive to multicollinearity which reduces the noncentrality parameter, η. When η becomes large either due to lack of multicollinearity or to the increase in the number of exogenous variables excluded from the equation of interest, the performance of the *LIML* estimator noticably improves. When the correlation between the stochastic regressor and the error term is high, the performance of such Bayesian estimators as the *PMOD*, *ULIB*, or *CLIB* improve. In our experiments, Zellner's extended *MELO* often ranked first and certainly among the top three in all but the case of $\mathbf{R}^2 = .9$, $n = 100$, and without multicollinearity.

Park (1982) compared, using the direct Monte Carlo method, the *OLS*, 2*SLS* and *ZM* estimators. The multiple correlation coefficient between the endogenous variables included in the right hand side and the error term of the equation of interest for each of the three equations Park considered varies from .001 in the third equation in Experiment No. 16 to as high as .578 in the first equation in Experiment No. 20. Our results are generally in agreement with those of Park.

We used the sum of relative mean absolute deviations (*SRMAD*'s) as the aggregate measure of evaluating how close the estimates are to the true values of structural coefficients. We also evaluated the lower quartiles, upper quartiles, skewnesses and probabilities of concentration (PC) of the various estimators of γ_{12} and we found that the *ZEM* estimator tended to be biased but it had a tighter distribution near the true value of the parameter.

Although the conditional and unconditional *LIB* and Zellner's *MELO* estimators are derived as Bayes' estimators, we treated them as if they are sampling estimators so that we can make sampling comparisons with the existing sampling estimators.

From our sampling experiments, we observe that given a particular sample theoretic criterion (such as *MAD*, *PC*, or *median unbiasedness*), there is no one estimator that dominates all other estimators in all cases, although frequently such Bayesian estimators as *ZEM*, *ULIB*, *PMOD*, and *CLIB* tend to outperform the other estimators as the degree of simultaneity increases. Even within a given set of parameters that influence performance of various estimators, the ranking of estimators varies if one changes the performance criterion. For example, we see that given $\rho_{12}^2 = .7$, $\mathbf{R}^2 = .9$, and multicollinearity, the top ranked estimators vary as shown in Table 10.

If *MAD* is chosen as a performance criterion, *ZEM* will be the preferred estimator for $n = 20$, 100 and 300. If *median unbiasedness* (*MUB*) is chosen as a criterion, then *LIML* is a good choice. Criteria such as *MAD*, *MUB*, and *PC* rely on the assumption that we can repeat the sample of a fixed size infinitely many times, but in actual economic data analysis, we often have non-repeatable and non-experimental data.

Table 10: Top 3 Ranked Estimators of γ_{12} for Different Performance Criteria ($R^2 = .9$, $\rho_{12}^2 = .7$, With Multicollinearity)

Performance Criterion	*MAD*	*MUB*	*PC*
$n = 20$	(1)ZEM (2)ULIB (3)CLIB	(1)LIML (2)ZEM (3)CLIB	(1)CLIB (2)LIML (3)ULIB
$n = 100$	(1)ULIB (2)ZEM (3)PMOD	(1)LIML (2)ULIB (3)CLIB	(1)ULIB (2)CLIB (3)PMOD
$n = 300$	(1)ZEM (2)PMOD (3)FLIM	(1)ULIB (2)LIML (3)CLIB	(1)CLIB (2)PMOD (3)ULIB

Notes: Experiments are the same as those given in Table 8. *MAD* = Mean Absolute Deviation; *MUB* = Median unbiasedness; *PC* = Probability of concentration.

What our sampling experiments demonstrate is that a researcher who exercises his good subjective judgement based on the knowledge he has should follow Bayesian inferential principles (for example, those given in Zellner (1988)) and choose an estimator that is optimal given a particular loss function. Bayesian estimates and estimators are tailored to be optimal for the particular loss function considered. For example, posterior mean for squared loss, and median for absolute error loss, and modal value for $0 - 1$ loss. We can even interpret sampling estimators such as the *LIML* as a Bayesian estimator based on a $0 - 1$ loss function, since given a diffuse prior, the mode of the posterior pdf is close to the *LIML*. Also, if one follows Bayesian inferential procedures, one can examine the entire posterior pdf of a parameter, not just a point estimate with its estimated standard error.

Acknowledgment

The author is grateful to Professor Zellner and three anonymous referees for their stimulating comments that helped improve the paper.

References

Anderson, T.W. (1983). "Some Recent Developments on the Distributions of Single-Equation Estimators," in *Advances in Econometrics*, W. Hilderbrand, ed., Amsterdam: North-Holland.

Anderson, T.W., N. Kunitomo and T. Sawa (1982). "Evaluation of the Distribution Function of the Limited Information Maximum Likelihood Estimator," *Econometrica*, Vol. 50, 1009-1027.

Anderson, T.W., K. Morimune and T. Sawa (1983). "The Numerical Values of Some Key Parameters in Econometric Models," *Journal of Econometrics*, Vol. 21, 229-243.

Anderson, T.W., N. Kunitomo and K. Morimune (1986). "Comparing Single Equation Estimators in a Simultaneous Equation System," *Econometric Theory*, Vol. 2, 1-32.

Anderson, T.W. and H. Rubin (1949). "Estimation of the Parameters of a Single Equation in a Complete System of Stochastic Equations," *Annals of Mathematical Statistics*, Vol. 20, 46-63.

Basmann, R.L. (1957). "A Generalized Classical Method of Linear Estimation of Coefficients in a Structural Equation," *Econometrica*, Vol. 25, 77-83.

Cragg, J.G. (1967). "On the Relative Small-Sample Properties of Several Structural-Equation Estimators," *Econometrica*, Vol. 35, 89-110.

Drèze, J. (1976). "Bayesian Limited Information Analysis of the Simultaneous Equations Model," *Econometrica*, Vol. 44, 1045-1075.

Fiebig, D.G. (1985). "Evaluating Estimators Without Moments," *Review of Economics and Statistics*, Vol. 67, 529-534.

Fuller, W.A. (1977). "Some Properties of a Modification of the Limited Information Estimator," *Econometrica*, Vol. 45, 939-953.

Klein, L.R. and M. Nakamura (1962). "Singularity in the Estimation Systems of Econometrics: Some Aspects of the Problem of Multicollinearity," *International Economic Review*, Vol. 3, 274-299.

Koopmans, T.C. and W.C. Hood (1953). "The Estimation of Simultaneous Economic Relationships," in *Studies in Econometric Methods*, W.C. Hood and T.C. Koopmans, eds., New York: Wiley.

Mariano, R. (1982). "Analytical Small-Sample Distribution Theory in Econometrics: The Simultaneous-Equations Case," *International Economic Review*, Vol. 23, 503-533.

Morimune, K. (1985). *Keizai Model No Suitei To Kentei*, (Estimation and Testing of Economic Models), Tokyo: Kyoritsu Shuppan.

Park, S.B. (1982). "Some Sampling Properties of Minimum Expected Loss (MELO) Estimators of Structural Coefficients," *Journal of Econometrics*, Vol. 18, 295-311.

Phillips, P.C.B. (1983). "Exact Small Sample Theory," in *Handbook of Econometrics*, Vol. 1, Z. Grilliches and M.D. Intrilligator, eds., Amsterdam: North-Holland, 449-516.

Raj, B. (1980). "A Monte Carlo Study of Small-Sample Properties of Simultaneous Equation Estimators with Normal and Nonnormal Disturbances," *Journal of American Statistical Association*, Vol. 75, 221-229.

Rao, C.R. (1980). "Some Comments on the Minimum Mean Squared Error as a Criterion of Estimation," in *Statistics and Related Topics*, M. George et. al., eds., Amsterdam: North-Holland.

Summers, R. (1965). "A Capital Intensive Approach to the Small Sample Properties of Various Simultaneous Equation Estimators," *Econometrica*, Vol. 33, 1-41.

Theil, H. (1958). *Economic Forecast and Policy*, lst edition. Amsterdam: North Holland.

Tsurumi, H. (1985). "Limited Information Bayesian Analysis of a Structural Coefficient in a Simultaneous Equations System," *Communications in Statistics*, Vol. 14, 1103-1120.

Zellner, A. (1971). *Introduction to Bayesian Inference in Econometrics*, New York: Wiley.

Zellner, A. (1978). "Estimation of Functions of Population Means and Regression Coefficients Including Structural Coefficients: A Minimum Expected Loss (MELO) Approach," *Journal of Econometrics*, Vol. 8, 127-158.

Zellner, A. (1986). "Further Results in Bayesian Minimum Expected Loss (MELO) Estimates and Posterior Distributions for Structural Coefficients," *Advances in Econometrics*, Vol. 5, 171-182.

Zellner, A. (1988). "Bayesian Analysis in Econometrics," *Journal of Econometrics*, Vol. 37, 27-50.

Appendix: The Design of the Sampling Experiments

We set up a linear simultaneous equations system consisting of three structural equations for our direct Monte Carlo method of sampling experiments. The model is a modification of Cragg's model (1967):

$$Y\Gamma = XB + U \tag{27}$$

where

$$\Gamma = \begin{pmatrix} 1.0 & -.267 & -.087 \\ -.222 & 1.0 & 0 \\ 0 & -.046 & 1.0 \end{pmatrix} \qquad B = \begin{pmatrix} .62 & 4.4 & 4.0 \\ 0 & .74 & 0 \\ .7 & 0 & .53 \\ 0 & 0 & .11 \\ .96 & .13 & 0 \\ 0 & 0 & .56 \\ .06 & 0 & 0 \end{pmatrix}$$

We shall use the first equation of interest:

$$y_{t1} = \gamma_{12} y_{t2} + \beta_{11} x_{t3} + \beta_{15} x_{t5} + \beta_{17} x_{t7} + u_{t1} \tag{28}$$

$$\gamma_{12} = .222,\ \beta_{11} = 6.2,\ \beta_{13} = .7,\ \beta_{15} = .96\ \beta_{17} = .06\ .$$

The variance-covariance matrix of the row of U, $Cov(u_t)$, is specified as

$$\text{Cov}\,(u_t) = 36.0 \begin{pmatrix} 1.0 & \rho_\omega & .25\ \rho_\omega \\ \rho_\omega & 1.0 & 0 \\ .25\ \rho_\omega & 0 & 1.0 \end{pmatrix} \tag{29}$$

where the values of the parameter, ρ_ω, are chosen so that the correlation between y_{t2} and u_{t1}, ρ_{12} will be controlled. The exogenous variables, other than the constants, $x_{t2}, \cdots, x_{t7}$, are drawn from uniform distribution over the interval $[0, \mathbf{a}]$, where $\mathbf{a}$ is a scalar whose value will be specified so that $\mathbf{R}^2$ (the coefficient of determination of the reduced form for y_{t2}) is controlled. The combination of the values of ρ_ω, $\mathbf{a}$, and $\mathbf{R}^2$ are given in Table 11. The value of $\mathbf{a}$ is affected by the sample size, $\mathbf{R}^2$, and multicollinearity whereas the values of ρ_ω are only influenced by the degrees of simultaneity, ρ_{12}^2.

As the structure of the correlation among the exogenous variables, we use the following correlation matrices, one for almost nonexistence of multicollinearity and the other for high multicollinearity:

Correlation Matrix of X,Corr(X)

	No Multicollinearity					Multicollinearity				
	x_3	x_4	x_5	x_6	x_7	x_3	x_4	x_5	x_6	x_7
x_2	-.15	-.11	-.18	-.37	-.18	.99	.93	.87	.69	.60
x_3		-.18	-.01	-.27	-.03		.93	.88	.73	.61
x_4			-.15	-.37	.02			.87	.67	.64
x_5				-.09	.05				.75	.69
x_6					.56					.85
	Det(Corr(X)) = .3074					Det(Corr(X)) = .0000393				

The determinant of the correlation matrix for the case of no multicollinearity is .3074, indicating a low degree of multicollinearity among x_{ti}'s, whereas that for the case of multicollinearity is .0000393, which shows a high degree of multicollinearity.

To examine how the performances of the different estimators are influenced by an increase in the number of exogenous variables excluded from the equation of interest, we added three exogenous variables each to the second and third structural equations in (27):

$$\begin{aligned} y_{t2} &= .267y_{t1} + .046y_{t3} + 4.4 + .74x_{t2} + .13x_{t5} \\ &\quad + \beta_{82}x_{t8} + \beta_{92}x_{t9} + \beta_{10,2}x_{t,10} + u_{t2} \qquad (30) \\ y_{t3} &= .087y_{t1} + 4.0 + .53x_{t3} + .11x_{t4} + .56x_{t6} \\ &\quad + \beta_{11,3}x_{t,11} + \beta_{12,3}x_{t,12} + \beta_{13,3}x_{t,13} + u_{t3} \qquad (31) \end{aligned}$$

The six newly added exogenous variables are drawn from uniform distributions over the interval [0, **a**]. In the Column (1) of Table 9 these six exogenous variables are made to be correlated among themselves by the correlation matrix for multicollinearity, but they are uncorrelated with those exogenous variables that are originally in the system, $x_{t1}, \cdots, x_{t7}$. In the experiment that is reported in Column 3 of Table 9, the six new exogenous variables, $z_{t1} = (x_{t8}, \cdots, x_{t,13})$ that are correlated among themselves by the correlation matrix for multicollinearity are added to $z_{t2} = (x_{t2}, \cdots, x_{t7})$: $.5z_{t1} + .5z_{t2}$, to form six newly added exogenous variables in equations (30) and (31). In the experiment that is reported in Column 5 of Table 9, the six newly added variables that are drawn independent uniform distributions over the interval [0, **a**] are used without making them correlated among themselves.

Table 11: Combinations of $\mathbf{R}^2$, Sample Size, ρ_ω, and a Used for Sampling Experiments

Without Multicollinearity					With Multicollinearity				
$\mathbf{R}^2$	n	ρ_{12}^2	ρ_ω	a	$\mathbf{R}^2$	n	ρ_{12}^2	ρ_ω	a
.9	100	0	-.25	83.39	.9	100	0	-.25	45.04
		.3	-.71	72.64			.3	-.71	39.49
		.5	-.823	69.43			.5	-.823	37.74
		.7	-.90	67.02			.7	-.90	36.39
		.9	-.963	65.00			.9	-.936	35.18
.9	20	0	-.25	111.14	.9	20	0	-.25	53.70
		.3	-.71	94.49			.3	-.71	47.08
		.5	-.823	89.76			.5	-.823	45.07
		.7	-.90	86.39			.7	-.90	43.58
		.9	-.963	83.96			.9	-.963	42.41
.35	100	0	-.25	21.57	.35	100	0	-.25	12.49
		.3	-.71	19.51			.3	-.71	11.59
		.5	-.823	18.76			.5	-.823	11.14
		.7	-.90	18.14			.7	-.90	10.72
		.9	-.963	17.55			.9	-.963	10.24
.35	20	0	-.25	37.82	.35	20	0	-.25	19.05
		.3	-.71	34.46			.3	-.71	19.23
		.5	-.823	32.39			.5	-.823	18.57
		.7	-.90	30.35			.7	-.90	17.72
		.9	-.963	28.03			.9	-.963	16.54

Notes: $\mathbf{R}^2$ = the coefficient of determination of the reduced form equation for Y_1.

n = sample size.

ρ_{12}^2 = the multiple correlation coefficient between Y_1 and u_1.

ρ_ω = parameter in $\mathrm{Cov}(u_t)$ in (29).

a = upper limit of the uniform distribution, [0,**a**] from which the exogenous variables, x_{ti}, are drawn.

V. SELECTED INFERENCE PROBLEMS

Bayesian and Likelihood Methods in Statistics and Econometrics
S. Geisser, J.S. Hodges, S.J. Press and A. Zellner (Editors)

SELECTION MODELS AND SELECTION MECHANISMS

M. J. Bayarri and M. H. DeGroot
University of Valencia and Carnegie Mellon University

Abstract

In many situations, experimenters are not able to draw a random sample from the population in which they are interested, and statistical models that incorporate the nonrandomness or bias in their observations must be developed. We consider problems in which observations can be obtained only from certain selected portions of the population. Bayesian methods for the analysis of such selection, or truncation, models are discussed. Selection mechanisms that could lead to selected data of this type are studied and represented as binomial sampling plans. Examples are given in which the selection mechanism can be ignored in the sense that the data can be analyzed as a random sample of fixed size from a selection model.

Keywords and Phrases – Bayesian analysis, binomial sampling plans, improper distributions, reporting, statistical significance, stopping rules, truncation.

1. Introduction

Suppose that the distribution of some real-valued characteristic X in a given population is represented by the (generalized) density $g(x \mid \theta)$, and that individuals for whom the value of X is greater than or equal to an unknown threshold τ manifest a certain disease. Suppose also that an experimenter is interested in the distribution of X in the general population, but that the value of X is only measured for individuals who manifest the disease because the observation of X is expensive, painful or even dangerous (involving, perhaps, some kind of surgery). In this situation, inferences about θ and τ must be based on data obtained only from persons who have the disease because X will be observed for these persons in the course of their treatment. The density of X for such a person is

$$f(x \mid \theta) = \frac{g(x \mid \theta)}{Pr(X \geq \tau \mid \theta)} \quad \text{for} \quad x \geq \tau, \tag{1.1}$$

and $f(x \mid \theta) = 0$ for $x < \tau$. This problem is just an example of a wider class of problems arising in statistical practice in which observations are obtained only from some "selected" portions of the population of interest, either because experimental conditions make it impossible to obtain data from the whole population or because the experimenter chooses to restrict the observations in this way.

Throughout the paper we will assume that X is a random variable (or random vector) which is distributed over the population of interest according to the density $g(x \mid \theta)$ where θ is unknown and belongs to some fixed parameter space Ω. Observations however are restricted to lie in some fixed subset S of the sample space. In many problems S is unknown and belongs to a specified class $\mathcal{S}$ of subsets. We may be interested in making inferences about either θ or S or both. In these problems data consist of observations of the random variable (which, to avoid cumbersome notation we will denote also by X) with density

$$f(x \mid \theta, S) = \frac{g(x \mid \theta)}{Pr(X \in S \mid \theta)} \quad \text{for} \quad x \in S \tag{1.2}$$

and $f(x \mid \theta, S) = 0$ otherwise. It is assumed that $Pr(X \in S \mid \theta) > 0$ for all $\theta \in \Omega$ and $S \in \mathcal{S}$. The model (1.2) is called a *selection model* or *truncation model* and it is also referred to as the selected or truncated version of $g(x \mid \theta)$. A random sample from (1.2) will be called a *selection sample*, and the set S, the *selection set.* Moreover, if S is unknown and can be characterized by some parameter τ, then τ will be called the *selection parameter.* For a given selection sample $\mathbf{x} = (x_1, \ldots, x_n)$ let

$$H(\mathbf{x}) = \bigcap_{S \in \mathcal{S}} \{S \mid x_i \in S \quad \text{for} \quad i = 1, \ldots, n\}. \tag{1.3}$$

It can be shown (see, e.g., Fraser, 1952, and Smith, 1957) that if $M(\mathbf{x})$ is a minimal sufficient statistic for the family $\{g(x \mid \theta); \theta \in \Omega\}$ then the pair (M, H) is a jointly minimal sufficient statistic for the selection model $\{f(x \mid \theta, S); \theta \in \Omega, S \in \mathcal{S}\}$.

Experiments yielding selection samples are quite different from experiments yielding censored samples. In a censored sample it is known which observations and how many of them are censored (that is, lie outside of S), even though the exact values of these observations are unknown. In a selection sample there is no information whatsoever about observations outside of S.

Selection models occur frequently in practice and some examples will be given in this paper. Several examples and references are also given in Bayarri and DeGroot (1987a) together with a Bayesian analysis of these models. The question of whether a selection sample can provide more information about the parameter θ than an unrestricted random sample from the original density $g(x \mid \theta)$ is addressed, for some particular selection models, in Bayarri and DeGroot (1987b). In some problems, this comparison is carried out by studying the Fisher information in each type of experiment. In certain cases, stronger results are obtained based on the theory of the comparison of statistical experiments as developed originally by Blackwell (1951, 1953). The name "selection models" in this context is due to Fraser (1952, 1966), although the term "selection" was used in a more general setting by Tukey (1949).

In this paper we will present some of the striking statistical features that appear when analyzing selection models. Also, by explicitly considering some selection mechanisms that can give rise to the selection sample, we will discuss how what is not observed by the experimenter can in fact modify a conditional analysis (of either Bayesian or likelihood type) of the selection sample. We will also show that in some cases a Bayesian analysis can ignore the effect of the selection mechanism. Unless otherwise stated, consideration is restricted to univariate problems in which S can be characterized by a single parameter τ.

In Section 2 we show that many selection models represent simple situations in which improper prior distributions for the selection parameter are inappropriate because they *always* yield improper posterior distributions, regardless of how many observations are taken and what their values are. It is also shown that the usual conjugate prior distributions in these models are improper. In Section 3 we consider the opposite situation in which, when sampling from a uniform distribution over an unknown set S in the plane, the effect of the prior distribution becomes completely erased by the data.

In Section 4 we discuss the selection effect that is present in the reporting and publication of only statistically significant experimental results.

In Section 5 we present a discussion of general selection mechanisms by which selection samples might be generated and observed. In Section 6 these mechanisms are represented as binomial sampling plans. Finally, in Section 7, some explicit examples are given in which the selection mechanism and the data that were not observed can be ignored, and the observed data can be analyzed simply as a random sample from the selection model (1.2).

2. Improper Posterior Distributions

Selection models provide very simple examples where improper priors *always* lead to improper posteriors, no matter which observations or how many observations have been obtained. Suppose that the underlying density g of x, is completely specified and that the selection set S is the lower tail of the distribution containing all the values $x \leq \tau$. In this case the model (1.2) reduces to $f(x \mid \tau)$ and the joint density of a selection sample $\mathbf{x} = (x_1, \ldots, x_n)$ is

$$f(\mathbf{x} \mid \tau) = \frac{\Pi_{i=1}^n g(x_i)}{[G(\tau)]^n} \quad \text{for} \quad \max(x_1, \ldots, x_n) \leq \tau \tag{2.1}$$

and $f(\mathbf{x} \mid \tau) = 0$ otherwise, where $g(x)$ is a known density and $G(x)$ is the corresponding d.f. Since the likelihood function $L(\tau)$ is of the form

$$L(\tau) = [G(\tau)]^{-n} \quad \text{for} \quad \tau \geq \max(x_1, \ldots, x_n) \tag{2.2}$$

and $L(\tau) = 0$ otherwise, the natural choice for a conjugate prior density for τ is of the form

$$\xi_1(\tau) \propto [G(\tau)]^a \quad \text{for} \quad \tau \geq b \tag{2.3}$$

where the hyperparameters a and b are constants to be specified. A distribution of this form may be appropriate in problems in which the support of g is bounded on the right, so the possible values of τ are also bounded on the right. In this case, the prior densitty $\xi_1(\tau)$ can be normalized for any value of a so that it is a proper probability distribution. However, if the possible values of τ are unbounded on the right then the density $\xi_1(\tau)$ will be improper for every real number a, regardless of whether a is positive, negative, or zero. Although improper prior distributions are widely used in Bayesian statistics, the density $\xi_1(\tau)$ seems to be of no use whatsoever because it follows from the statement just given that the posterior distribution of τ will also be improper for all values of n and all values of $x_1, \ldots, x_n$.

Moreover, $\xi_1(\tau)$ is just a particular case of a wide class of improper prior distributions that yield improper posterior distributions for all $\mathbf{x}$. Indeed, let $\xi_u(\tau)$ be *any* improper density for which the area under the upper tail is infinite, that is

$$\int_c^\infty \xi_u(\tau) d\tau = \infty \quad \text{for all} \quad c \quad . \tag{2.4}$$

Since the posterior density $\xi(\tau \mid \mathbf{x})$ for any given $\mathbf{x}$ and prior density $\xi(\tau)$ is

$$\xi(\tau \mid \mathbf{x}) \propto \frac{\xi(\tau)}{[G(\tau)]^n} \quad \text{for} \quad \tau \geq \max(x_1, \ldots, x_n) \quad , \tag{2.5}$$

it follows that (2.4) will also hold for the posterior density $\xi_u(\tau \mid \mathbf{x})$ corresponding to $\xi_u(\tau)$ and therefore it will be improper.

It should be noticed that no assumptions have been made about $g(x)$ nor $\xi_u(\tau)$ apart from (2.4); the support of $g(x)$ may or may not be bounded and the area under the lower tail of $\xi_u(\tau)$ may or may not be finite: As long as the impropriety of $\xi_u(\tau)$ is due to (2.4), then $\xi_u(\tau \mid \mathbf{x})$ will also be improper. Furthermore, if $\xi(\tau)$ is a proper prior for which exactly the first k moments exist ($k = 0, 1 \ldots$) and for which higher order moments do not exist because the integral $\int_c^\infty |\tau|^{k+1} \xi(\tau) d\tau$ is not finite, then it follows that the corresponding posterior distribution $\xi(\tau \mid \mathbf{x})$ will also have exactly k moments for all $\mathbf{x}$. (For a description of another Bayesian model in a different context that exhibits similar features, see Kahn, 1987).

In light of this discussion it seems desirable to use a different form for the conjugate prior distribution other than (2.3) for carrying out a Bayesian analysis of the selection model (2.1). There are many different families of proper prior distributions for τ that are closed under sampling. For example, any density of the form

$$\xi(\tau) \propto h(\tau)[G(\tau)]^a \quad \text{for} \quad \tau \geq b \tag{2.6}$$

will have this property, where h is any function for which the right-hand side of (2.6) is integrable over τ. One convenient choice for h is $h = g$. It can be verified that the resulting prior density

$$\xi_2(\tau) \propto g(\tau)[G(\tau)]^a \quad \text{for} \quad \tau \geq b \tag{2.7}$$

is proper for every real number a. The Bayesian analysis of this selection model is then straightforward.

When G is absolutely continuous, so g is a p.d.f., the problem can be reduced to making inferences about the parameter of a uniform distribution in the following way: Let

$$Y = G(X) \quad \text{and} \quad \beta = G(\tau) \quad . \tag{2.8}$$

Then the transformed selection sample $Y_1, \ldots, Y_n$ is simply a random sample from the uniform distribution on the interval $(0, \beta)$ where $\beta < 1$. Furthermore, β has the usual conjugate prior p.d.f. of the form

$$\xi_3(\beta) \propto \beta^a \quad \text{for} \quad \beta \geq b \quad , \tag{2.9}$$

if and only if τ has the p.d.f. ξ_2 given by (2.7).

3. The Case of the Disappearing Prior

In Section 2 we have seen examples of selection models in which the data have little influence on the posterior distribution in the sense that they cannot make proper an improper prior. In this section we will encounter the opposite situation in which the posterior distribution is not influenced by the hyperparameters of the prior. Although it is quite frequent in Bayesian analysis that the influence of the prior on the posterior decreases as the number of observations increases, there are few examples in which the prior has no effect whatsoever on the posterior after just a finite number of observations are obtained.

Consider the following alternative expression of a selection model as given by (1.2):

$$f(x \mid \theta, S) = \frac{g(x \mid \theta)}{\int_S g(x \mid \theta) dx} \quad \text{for} \quad x \in S, \tag{3.1}$$

and $f(x \mid \theta, S) = 0$ otherwise. It is interesting to note that there is nothing in (3.1) that forces $g(x \mid \theta)$ to be a proper probability density. Indeed, we can remove this restriction and consider g to be any nonnegative function for which the integral in (3.1) is positive and finite for all $\theta \in \Omega$ and $S \in \mathcal{S}$.

A simple example in which g is not a proper density is the following: Suppose that it is desired to estimate the boundary of an unknown bounded set S in the plane, and suppose that a random sample can be drawn from a uniform distribution over that set. In this example, the density f of each observation can be represented in the form (3.1) where g is a positive constant over the entire plane and hence is not a probability density. Although there is no parameter θ in this example because g is completely specified, it can easily be extended to any nonnegative function $g(x \mid \theta)$ over the plane. Problems of this type have been studied in DeGroot and Eddy (1983).

In the example just presented, let $z_i = (x_i, y_i)$ denote an observation in the plane $\Re^2$ and let $\mathbf{z} = (z_1, \ldots, z_n)$ be a selection sample. Then its joint density is given by:

$$f(\mathbf{z} \mid S) = \begin{cases} \dfrac{1}{[Area(S)]^n} & \text{for} \quad z_i \in S, i = 1, \ldots, n, \\ 0 & \text{otherwise.} \end{cases} \tag{3.2}$$

For the purposes of this section, we will now restrict S to belong to $\mathcal{S}$ = {all circles with unit radius}. Thus, each $S \in \mathcal{S}$ can conveniently be represented by its center and we shall let $S(s)$ or $S(\alpha, \beta)$ denote the circle with unit radius and center at $s = (\alpha, \beta)$. It can be shown from (1.3) that a minimal sufficient statistic in this case is the convex hull $< \mathbf{z} >$ of $\mathbf{z}$, because $H(\mathbf{z})$ can be determined from $< \mathbf{z} >$ and conversely (except in the degenerate situation where $< \mathbf{z} >$ is a line segment of length 2).

The likelihood function of the unknown center $s = (\alpha, \beta)$ is

$$f(\mathbf{z} \mid \alpha, \beta) = 1 \quad \text{if} \quad (x_i - \alpha)^2 + (y_i - \beta)^2 \leq 1 \quad \text{for} \quad i = 1, \ldots, n \tag{3.3}$$

and $f(\mathbf{z} \mid \alpha, \beta) = 0$ otherwise. For any arbitrary finite collection $\mathbf{s} = \{s_1, \ldots, s_m\}$ of points $s_i \in \Re^2$, let $\Sigma_{<\mathbf{s}>}$ denote the intersection of all the circles $S(s_i), i = 1, \ldots, m$.

Notice that $\Sigma_{<\mathbf{s}>}$ depends on $\mathbf{s}$ only through $<\mathbf{s}>$, so the notation is appropriate. Then, (3.3) can be rewritten as

$$f(\mathbf{z} \mid \alpha, \beta) = 1 \quad \text{if} \quad (\alpha, \beta) \in \bigcap_{i=1}^{n} S(z_i) = \Sigma_{<\mathbf{z}>}, \tag{3.4}$$

and $f(\mathbf{z} \mid \alpha, \beta) = 0$ otherwise. From (3.4) the form of a conjugate prior for (α, β) can be easily deduced: Prior information is quantified by choosing n_0 points $a_1, \ldots, a_{n_0}$ in $\Re^2$ which are known to lie in $S(\alpha, \beta)$. Let $\mathbf{a} = \{a_1, \ldots, a_{n_0}\}$. Then the prior density for (α, β) is taken to be

$$\xi(\alpha, \beta) \propto 1 \quad \text{if} \quad (\alpha, \beta) \in \bigcap_{i=1}^{n_0} S(a_i) = \Sigma_{<\mathbf{a}>}, \tag{3.5}$$

and $\xi(\alpha, \beta) = 0$ otherwise. The posterior density is then simply

$$\xi(\alpha, \beta \mid \mathbf{z}) \propto 1 \quad \text{if} \quad (\alpha, \beta) \in \Sigma_{<\mathbf{a} \cup \mathbf{z}>} \tag{3.6}$$

and $\xi(\alpha, \beta \mid \mathbf{z}) = 0$ otherwise. The posterior (3.6) is the example we were looking for. Indeed, as soon as $<\mathbf{z}> \supset <\mathbf{a}>$, which with probability one will occur sooner or later after a finite number of observations z_i, the posterior set $\Sigma_{<\mathbf{a} \cup \mathbf{z}>}$ will not depend on $\mathbf{a}$ and all traces of this aspect of the prior distribution will have disappeared from the posterior. (The uniform nature of the prior density will be reflected in the uniform nature of the posterior.) This example is an extension to $\Re^2$ of the standard problem of sampling from a uniform distribution on the real line of known length and unknown location, in which a similar phenomenon occurs.

4. What You Didn't See

The analysis of selection samples forces us to consider the issue of unobserved variables. In building a model and carrying out an analysis, when must we consider variables which have not been observed and when can they safely be ignored? This question is addressed in a more general setting in Bayarri, DeGroot and Kadane (1988). In this section we will discuss a simple yet interesting example in which the consideration of variables that are not observed can in fact dramatically change the conclusions of a statistical analysis of the data.

Assume that you, as the reader of some scientific journal, are learning about the results of an experiment that was recently carried out and suppose that the conclusions reported are based on a statistical analysis which includes a test of hypotheses.

This is not an unrealistic assumption; the unfortunate misuse of statistical methods in many areas of application through the almost-exclusive reliance on tests of hypotheses is well known (Zellner, 1980; DeGroot and Mezzich, 1985). Assume also that only experimental results that are found to be "statistically significant" are published. This situation is very common both because the editors of some journals will publish only articles in which statistical significance has been obtained, and because many experimenters themselves regard their results as being useless unless the results are statistically significant and will not even submit them for publication. Suppose, finally, that in the paper you are reading, the null hypothesis H_0 is rejected and the data are declared significant because they yield a small p-value. Do these significant results provide strong evidence against H_0?

For simplicity we will present our discussion in the context of a one-sided test of hypotheses on the mean of a normal distribution with known variance. Accordingly, suppose that $y_1, \ldots, y_n$ form a random sample from a normal distribution with unknown mean μ and known variance σ^2, and it is desired to test the hypotheses

$$\begin{aligned} H_0 &: \mu \leq 0, \\ H_1 &: \mu > 0. \end{aligned} \tag{4.1}$$

The uniformly most powerful test for any specified value of α is to reject H_0 whenever

$$U = \frac{n^{\frac{1}{2}}}{\sigma} \bar{y}_n \geq \Phi^{-1}(1-\alpha) \tag{4.2}$$

and the distribution $g(u \mid \theta)$ of the test statistic U is normal with mean $\theta = n^{\frac{1}{2}}\mu/\sigma$ and variance 1. But, as readers of the journal, we will not learn the results of this experiment unless they lead to the rejection of H_0; that is, unless $U \geq \Phi^{-1}(1-\alpha)$. In this situation, the density of any value of U that we will actually get to observe is not simply $g(u \mid \theta)$ but is given by the selection model

$$f(u \mid \theta) = \frac{\phi(u-\theta)}{1-\Phi(\tau-\theta)} \quad \text{for} \quad u \geq \tau \tag{4.3}$$

where $\phi(\cdot)$ and $\Phi(\cdot)$ are the p.d.f and d.f. of the standard normal distribution, respectively, and $\tau = \Phi^{-1}(1-\alpha)$. When a statistically significant result is analyzed using the model (4.3) it may very well provide strong *support* for the null hypothesis. To see this, it is enlightening to calculate the maximum likelihood estimate (MLE) of θ for an observed value $U = u$.

It can be shown from (4.3) that the MLE $\hat{\theta}$ is the unique solution of the following equation:

$$(u-\hat{\theta})M(\tau-\hat{\theta}) = 1 \tag{4.4}$$

where $M(x)$ denotes Mills' ratio $[1 - \Phi(x)]/\phi(x)$ for $-\infty < x < \infty$. The values of $\hat{\theta}$ for $\alpha = 0.01, 0.05$, and 0.10 and various observed values of u are given in Table 1. Of course, the most widely used criterion for statistical significance is $\alpha = 0.05$. In Table 1 we also give the p-values corresponding to the observed values of u as they would be reported in the literature, that is, calculated from the standard normal distribution. Since only "significant" values of U can be observed, all the p-values must be less than α.

TABLE 1

	u	p	$\hat{\theta}$
$\alpha = 0.01$	3.424	0.0003	3
	2.792	0.0026	1
	2.665	0.0038	0
	2.588	0.0048	-1
	2.503	0.0062	-3
$\alpha = 0.05$	3.175	0.0008	3
	2.249	0.0124	1
	2.063	0.0195	0
	1.955	0.0253	-1
	1.844	0.0326	-3
$\alpha = 0.10$	3.095	0.0010	3
	1.985	0.0236	1
	1.755	0.0397	0
	1.625	0.0526	-1
	1.495	0.0675	-3

Since $\hat{\mu} = \sigma\hat{\theta}/n^{\frac{1}{2}}$, it follows from (4.1) that negative values of $\hat{\theta}$ support the null hypothesis H_0. The basic conclusion to be drawn from the discussion in this section is that even observed values of u that appear to be highly significant can yield MLE's that are very negative and, therefore, give strong support to H_0. For example, it can be seen from Table 1 that when $\alpha = 0.05$, a p-value as small as 0.033 yields $\hat{\theta} = -3$, an estimate that is at least 3 standard deviations away from the parameter values in H_1. Similar behavior is found for $\alpha = 0.01$ and $\alpha = 0.10$.

These results emphasize the fact that under current statistical practice, a published p-value only slightly smaller than the effective α can only be regarded as supporting H_0 rather than rejecting H_0. In fact, as the p-value approaches α in this selection model, $\hat{\theta} \to -\infty$.

Now let us consider our example further. When we used the selection model (4.3) to conclude that the published statistically significant result actually supported H_0, we were implicitly assuming that the experiment under consideration was performed repeatedly, either by the same experimenter or by different experimenters in different locations around the world, until a significant result was obtained. Suppose, on the other hand, that you knew that only one experiment had been performed in the entire world. In this case, when you read about the experiment in the journal because the result turned out to be significant and got published, you know that you are seeing the only data that had ever been obtained in an experiment of this type. Hence, the data and their associated p-value should be taken at face value.

Conclusions intermediate to these two extreme ones can be reached by explicitly incorporating into the analysis the selection rule that actually produced the selection sample to be analyzed. Thus, in this example, what you don't see, the number of experiments that were performed, can have a decisive influence on the statistical analysis of the data.

Problems of this type, dubbed the "file drawer problem" by Rosenthal (1979), are also discussed by Sterling (1959), Dawid and Dickey (1977), Hedges and Olkin (1985, Chapter 14), and Iyengar and Greenhouse (1988), among others. In the remainder of this paper we will present a general discussion of the selection mechanisms that might generate an observed selection sample, and the conditions under which what you don't see can be ignored in the analysis.

5. Selection Mechanisms

In this section we will again consider the general problem in which our observations $x_1, \ldots, x_n$ form a selection sample of size n from some underlying density $g(x \mid \theta)$ restricted to a selection set S. In general, both the parameter θ and the selection set S are unknown, with $\theta \in \Omega$ and $S \in \mathcal{S}$. By assuming that the joint density of the selection sample $\mathbf{x} = (x_1, \ldots, x_n)$ is given by

$$f(\mathbf{x} \mid \theta, S) = \frac{\Pi_{i=1}^{n} g(x_i \mid \theta)}{[Pr(S \mid \theta)]^n} \quad \text{for} \quad x_i \in S, i = 1, \ldots, n, \tag{5.1}$$

we have been implicitly assuming that the number n of observations was fixed in advance or, in any event, that the observed value of n contained no information about θ or S.

In general, however, depending on the mechanism that was used to obtain the selection sample, the value of n itself may contain useful information. For example, in many problems, a small number of observations in the selection sample may indicate that they were relatively difficult to obtain, which in turn may indicate that $Pr(S \mid \theta)$ is small. In this section we will begin our study of selection mechanisms and the totality of the information in a selection sample.

In recognition of the fact that n will now be regarded as an observable random variable, we will explicitly include it in our notation for densities. Thus, our objective now becomes to determine the posterior joint density $p(\theta, S \mid \mathbf{x}, n)$, which can be represented in the useful form

$$p(\theta, S \mid \mathbf{x}, n) \propto p(\mathbf{x} \mid n, \theta, S)p(n \mid \theta, S)p(\theta, S). \tag{5.2}$$

Of course, once we have observed $\mathbf{x}$ we have automatically also observed n, the number of components of $\mathbf{x}$, but it will nevertheless be convenient to denote our observation as the pair $(\mathbf{x}, n)$.

In order to present our discussion of selection mechanisms, it will also be convenient to consider both an investigator (possibly fictional) who somehow generates a selection sample from the original underlying density $g(x \mid \theta)$, and a statistician who observes only the selection sample $(\mathbf{x}, n)$ reported by the investigator. Our general model for the mechanism that the investigator uses to generate the selection sample is a sequential stopping rule of the following type:

Let $y_1, y_2, \ldots$ denote a sequential random sample from the underlying density $g(x \mid \theta)$, and for $i = 1, 2, \ldots$, let

$$z_i = I_S(y_i) = \begin{cases} 1 & \text{if} \quad y_i \in S, \\ 0 & \text{if} \quad y_i \in S^c. \end{cases} \tag{5.3}$$

The investigator observes the values of y_i (and, of course, of z_i) sequentially, and uses a sequential stopping rule δ that depends only on the observed values of $z_1, z_2, \ldots$ to determine when to stop sampling. We assume that there is probability 0 that sampling will continue forever. In other words, for each value of $\theta \in \Omega$ and $S \in \mathcal{S}$, we assume that the stopping rule δ is such that, with probability 1, sampling will eventually terminate. Let T denote the total number of observations that were obtained when sampling terminated, let $n = \sum_{i=1}^{T} z_i$ and let $x_1, \ldots, x_n$ denote the values of the n observations y_i for which $z_i = 1$. After sampling has terminated, the statistician observes only $\mathbf{x} = (x_1, \ldots, x_n)$, and in general does not know the stopping rule δ used by the investigator or all the values $y_1, \ldots, y_T$ that were obtained.

A critical assumption in this model is that the stopping rule δ used by the investigator depends on each observation y_i only through z_i. For given θ and S, the sequence $z_1, z_2, \ldots$ is a sequence of Bernoulli trials, so we are assuming, in effect, that δ is a binomial sampling plan of the general type described by Girshick, Mosteller, and Savage (1946) and DeGroot (1959). Indeed, binomial sampling plans provide the general framework for representing the different mechanisms by which a selection sample can be obtained from a fixed set S.

We will now evaluate each of the first two factors on the right-hand side of (5.2) by explicitly considering the stopping rule δ. Because δ is a binomial sampling plan, it follows that the conditional density $p(\mathbf{x} \mid n, \theta, S, \delta)$ is given by the right-hand side of (5.1). In other words, given the observed value of n, the observations $\mathbf{x}$ form a selection sample of size n regardless of the particular stopping rule δ that was used. Thus, this conditional density does not depend on δ and we may write simply

$$p(\mathbf{x} \mid n, \theta, S) = \frac{\Pi_{i=1}^{n} g(x_i \mid \theta)}{[Pr(S \mid \theta)]^n} \quad \text{for} \quad x_i \in S, i = 1, \ldots, n. \tag{5.4}$$

6. Binomial Sampling Plans

The second factor on the right-hand side of (5.2) will be evaluated by considering first, for a given stopping rule δ, the probability $p(n \mid \theta, S, \delta)$ that exactly n observations in the selection set S will have been obtained at the time that sampling is terminated. It is important to recognize that the usual consequence of the likelihood principle, which states that the likelihood function of θ and S generated by the data does not depend on the stopping rule that was used, is *not* valid here. The reason is simply that we do not observe all the data that were generated in the sampling process; we observe the number n of observations that belonged to S but not the number $T - n$ that belonged to S^c. To make this statement precise and to derive an explicit expression for $p(n \mid \theta, S, \delta)$ we must review the notation and terminology of binomial sampling plans (see, e.g., DeGroot, 1959).

We shall refer to $\Re^2$ as the $\alpha\beta-$plane and in what follows we restrict attention to points γ in the $\alpha\beta-$plane both of whose coordinates $\alpha(\gamma)$ and $\beta(\gamma)$ are non-negative integers.

A binomial sampling plan or stopping rule is a function δ defined on the points γ, taking only the values 0 and 1, and such that $\delta(\mathbf{0}) = 1$ for the point $\mathbf{0} = (0, 0)$. A *path* to γ is a sequence of points $\mathbf{0} = \gamma_0, \gamma_1, \ldots, \gamma_t = \gamma$ such that $\delta(\gamma_i) = 1$ for $i = 0, 1, \ldots, t-1$ and either

$$\begin{aligned} \alpha(\gamma_{i+1}) &= \alpha(\gamma_i) + 1, \\ \beta(\gamma_{i+1}) &= \beta(\gamma_i), \end{aligned} \tag{6.1}$$

or

$$\begin{aligned} \alpha(\gamma_{i+1}) &= \alpha(\gamma_i), \\ \beta(\gamma_{i+1}) &= \beta(\gamma_i) + 1. \end{aligned} \tag{6.2}$$

Thus each point in the path is either one unit to the right or one unit above its predecessor.

Under a given sampling plan δ, a point γ is a ***boundary point*** if there exists at least one path to γ and $\delta(\gamma) = 0$. It is a continuation point if there exists at least one path to γ and $\delta(\gamma) = 1$. The ***boundary*** B_δ is the set of all boundary points. The ***sample size*** $T(\gamma)$ of any point γ is the sum of its coordinates $\alpha(\gamma) + \beta(\gamma)$.

In this framework, a sequential random sample $y_1, y_2, \ldots$ from the density $g(x \mid \theta)$ can be visualized as generating a path in the $\alpha\beta$ – plane as follows: The path starts at **0** and is extended at each stage of the sequential process either one unit in the horizontal direction or one unit in the vertical direction. If the observation y_i at a given stage lies in the selection set S, which occurs with probability $P = Pr(S \mid \theta)$, then the path is extended in the horizontal direction. On the other hand, if y_i lies in S^c, which occurs with probability $Q = 1 - P$, then the path is extended in the vertical direction. For a given stopping rule δ, the sequential sampling process continues as long as the point γ reached by the path at each stage is a continuation point, and the process stops as soon as a point γ in the boundary B_δ is reached.

The stopping rule δ is characterized by its boundary B_δ. For each point $\gamma \in B_\delta$, let $K_\delta(\gamma)$ denote the number of distinct paths to γ. Then the probability of reaching a particular boundary point γ is

$$Pr(\gamma \mid \theta, S, \delta) = K_\delta(\gamma) P^{\alpha(\gamma)} Q^{\beta(\gamma)}. \tag{6.3}$$

The assumption that there is probability 0 that sampling will continue forever can now be expressed as the assumption that

$$\Sigma_{\gamma \in B_\delta} K_\delta(\gamma) P^{\alpha(\gamma)} Q^{\beta(\gamma)} = 1 \tag{6.4}$$

for all $\theta \in \Omega$ and $S \in \mathcal{S}$.

Now let us again recall that the stopping rule δ is being used as a selection mechanism to generate a selection sample. Hence, when sampling terminates at some boundary point γ, the statistician learns only the value of $n = \alpha(\gamma)$ (along with the actual values of the n corresponding y_i's). It now follows from (6.3) that

$$p(n \mid \theta, S, \delta) = \Sigma_{\gamma \in B_\delta, \alpha(\delta) = n} K_\delta(\gamma) P^{\alpha(\gamma)} Q^{\beta(\gamma)}. \tag{6.5}$$

Moreover, the statistician may not know the exact stopping rule δ that was used by the investigator. Let $p(\delta \mid \theta, S)$ denotes the statistician's prior distribution for δ over some available set Δ of possible sampling plans. (As our notation indicates, the statistician's beliefs about δ may not be independent of his or her prior beliefs about θ and S.) Then

$$p(n \mid \theta, S) = \Sigma_{\delta \in \Delta} p(n \mid \theta, S, \delta) p(\delta \mid \theta, S). \tag{6.6}$$

In general, $p(n \mid \theta, S)$ as given by (6.6) will depend on θ and S, which means that merely learning the number of observations n in the selection sample will give the statistician information about θ and S. On the other hand, if (6.6) does not actually depend on θ and S, then the statistician can analyze the data in the selection sample conditionally on the observed value of n and can ignore the mechanism which generated the data without losing any information. In this situation, the posterior density of θ and S as given by (5.2) reduces to the simple form

$$p(\theta, S \mid \mathbf{x}, n) \propto p(\mathbf{x} \mid n, \theta, S)p(\theta, S), \tag{6.7}$$

where $p(\mathbf{x} \mid n, \theta, S)$ is given by the right-hand side of (5.1).

We now conclude this paper with a discussion of a few situations in which $p(n \mid \theta, S)$ does not actually depend on θ and S. Other conditions under which the selection mechanism can be ignored, in a different but related context, are studied by Dawid and Dickey (1977).

7. Ignoring the Selection Mechanism

We will consider just two special types of sampling plans: inverse binomial plans in which sampling is continued until a fixed number n of observations lying in S have been obtained, and plans with a fixed sample size in which the total number T of observations is fixed in advance.

Consider first an inverse binomial plan δ in which sampling continues until exactly n_0 observations from the selection set S have been obtained, where n_0 is a known positive integer. In this case, $p(n \mid \theta, S, \delta)$ as defined by (6.5) is a degenerate distribution concentrated on the single value $n = n_0$. The selection mechanism can therefore be ignored and the data in the selection sample can be analyzed directly from (6.7). Moreover, even if the value of n_0 to be used in the sampling plan is not known with certainty by the statistician until the selection sample has actually been observed, the analysis can still be based on (6.7) provided that n_0 is independent of (θ, S) in the statistician's prior distribution.

Next consider a sampling plan δ in which the total number T of observations to be drawn from the density $g(x \mid \theta)$ is fixed in advance. In other words, suppose that the investigator draws a random sample of fixed size T from $g(x \mid \theta)$ and reports to the statistician only the values that belong to S. Then

$$p(n \mid \theta, S, \delta) = \binom{T}{n} P^n Q^{T-n} \tag{7.1}$$

and the observed value of n provides the statistician with information about $P = Pr(S \mid \theta)$.

Now suppose that the statistician is uncertain about the value of T used by the investigator. If the statistician's prior probability function for T is $p(T \mid \theta, S)$, then it follows from (6.6) and (7.1) that

$$p(n \mid \theta, S) = \sum_{T=n}^{\infty} \binom{T}{n} P^n Q^{T-n} p(T \mid \theta, S). \tag{7.2}$$

We will present two different examples of prior distributions of T for which $p(n \mid \theta, S)$ does not actually depend on θ and S, so the selection mechanism is ignorable.

First, suppose that the conditional distribution of T given (θ, S) is a Poisson distribution with mean λ/P, where λ is a fixed and known positive constant. Then it is well-known, and follows easily from (7.2), that $p(n \mid \theta, S)$ will be a Poisson distribution with mean λ.

In the second example, $p(T \mid \theta, S)$ itself does not depend on θ and S, but is an improper prior distribution. To be specific, suppose that

$$p(T \mid \theta, S) = \frac{1}{T} \quad \text{for} \quad T = 1, 2, \ldots \tag{7.3}$$

Then for any positive integer n, it follows from (7.2) that

$$\begin{aligned} p(n \mid \theta, S) &= \frac{1}{n} \sum_{T=n}^{\infty} \binom{T-1}{n-1} P^n Q^{T-n} \\ &= \frac{1}{n}. \end{aligned} \tag{7.4}$$

Thus, again, the (improper) distribution of n does not depend on θ and S, so the selection mechanism is ignorable.

Now consider again the example described in Section 4 in which you read the published report of an experiment in which the results were statistically significant at some prescribed level of significance α. We assume that there is only one published report of an experiment of this type, so the size of the selection sample in this example is $n = 1$. A wide variety of possible selection mechanisms that might have led to this published report can easily be described in terms of the concepts that have been developed in this section.

For example, if you believe that the experiment under consideration was performed repeatedly, possibly by different experimenters in different laboratories, until a significant result was obtained and published, then the selection mechanism is based on an inverse binomial sampling plan with $n_0 = 1$. Hence, as explained at the beginning of this section, the selection mechanism can be ignored and the published result can be analyzed directly from the selection model (4.3).

On the other hand, if you believe that the published significant result represents the outcome from the only experiment of this type that had been performed, then the selection mechanism is based on a sampling plan of fixed total sample size $T = 1$. It follows from (7.1) that the selection mechanism cannot be ignored, and the selection model (4.3) is not appropriate for analyzing the published results. In fact, in this case, if we combine (4.3) and (7.1) (with appropriate changes of notation), we find that $P = 1 - \Phi(\tau - \theta)$ and

$$p(u, n = 1 \mid \theta) = \varphi(u - \theta) \quad \text{for} \quad u > \tau. \tag{7.5}$$

Hence, the analysis of the observed outcome u should be based on the original underlying density $\varphi(u - \theta)$, rather than on the selection model (4.3).

As this example and the preceding discussion show, when analyzing data that come just from a selected portion of the population, it is important both to model carefully the distribution of the observables and to incorporate explicitly the selection mechanism. In practice, it is only when the mechanism is ignorable that the number of observations in the selected sample provides no information about the parameters of interest.

Acknowledgment

This research was supported in part by the National Science Foundation under grant DMS-8701770.

References

Bayarri, M.J. and M.H. DeGroot (1987a). "Bayesian Analysis of Selection Models," *The Statistician*, Vol. 36, 137-146.

Bayarri, M.J. and M.H. DeGroot (1987b). "Information in Selection Models," in *Probability and Bayesian Statistics*, R. Viertl, ed., New York: Plenum Press, 39-51.

Bayarri, M.J., M.H. DeGroot, and J.B. Kadane (1988). What is the Likelihood Function? (with discussion) in *Statistical Decision Theory and Related Topics IV*, S. S. Gupta and J. O. Berger, eds., (Springer-Verlag, New York) Vol. 1, 3-27.

Blackwell, D. (1951). "Comparison of Experiments," in *Proceedings of the Second Berkeley Symposium on Mathematical Statistics and Probability*, Berkeley, California: University of California Press, 93-102.

Blackwell, D. (1953). "Equivalent Comparison of Experiments," *Annals of Mathematical Statistics*, Vol. 24, 265-272.

Dawid, A.P. and J.M. Dickey (1977). "Likelihood and Bayesian Inference from Selectively Reported Data," *Journal of the American Statistical Association*, Vol. 72, 845-850.

DeGroot, M.H. (1959). "Unbiased Sequential Estimation for Binomial Populations," *Annals of Mathematical Statistics*, Vol. 30, 80-101.

DeGroot, M.H. and W.F. Eddy (1983). "Set-Valued Parameters and Set-Valued Statistics," in *Recent Advances in Statistics*, M.H. Rizvi, J.S. Rustagi, and D. Siegmund, eds., New York: Academic Press, 175-195.

DeGroot, M.H. and J.E. Mezzich (1985). "Psychiatric Statistics," in *A Celebration of Statistics: The ISI Centenary Volume*, A.C. Atkinson and S.E. Fienberg, eds., New York: Springer-Verlag, 145-165.

Fraser, D.A.S. (1952). "Sufficient Statistics and Selection Depending on the Parameters," *Annals of Mathematical Statistics*, Vol. 23, 417-425.

Fraser, D.A.S. (1966). "Sufficiency for Selection Models," *Sankhyā A*, Vol. 28, 329-334.

Girshick, M.A., F. Mosteller, and L.J. Savage (1946). "Unbiased Estimates for Certain Binomial Sampling Problems with Applications," *Annals of Mathematical Statistics*, Vol. 17, 13-23.

Hedges, L.V. and I. Olkin (1985). *Statistical Methods for Meta-Analysis*, Orlando, Florida: Academic Press.

Iyengar, S. and J.B. Greenhouse (1988). "Selection Models and the File Drawer Problem" (with discussion), *Statistical Science*, Vol. 3, 109-135.

Kahn, W.D. (1987). "A Cautionary Note for Bayesian Estimation of the Binomial Parameter n," *American Statistician*, Vol. 41, 38-40.

Rosenthal, R. (1979). "The 'File Drawer Problem' and Tolerance for Null Results," *Psychological Bulletin*, Vol. 86, 638-641.

Smith, W.L. (1957). "A Note on Truncation and Sufficient Statistics," *Annals of Mathematical Statistics*, Vol. 28, 247-252.

Sterling, T.D. (1959). "Publication Decisions and Their Possible Effects on Inferences Drawn from Tests of Significance — Or Vice Versa," *Journal of the American Statistical Association*, Vol. 54, 30-34.

Tukey, J.W. (1949). "Sufficiency, Truncation, and Selection," *Annals of Mathematical Statistics*, Vol. 20, 309-311.

Zellner, A. (1980). "Statistical Analysis of Hypothesis in Economics and Econometrics," *Proceedings of the Business and Economic Statistics Section, American Statistical Association*, 199-203.

Bayesian and Likelihood Methods in Statistics and Econometrics
S. Geisser, J.S. Hodges, S.J. Press and A. Zellner (Editors)

A BAYESIAN SOLUTION TO THE BEHRENS-FISHER PROBLEM

Lyle D. Broemeling, Mun S. Son, and Hosny I. Hamdy

University of Texas Medical Branch, University of Vermont, and University of Vermont

Abstract

A *Bayesian solution* to the *Behrens-Fisher problem* is proposed. The solution is to approximate the posterior distribution of a set of $k-1$ independent contrasts among the k means of normal populations by a multivariate normal distribution. The approximation is accurate to the extent that the mean vector and covariance matrix of the approximation is the same as the mean vector and covariance matrix of the true posterior distribution. Using *Jeffreys vague prior density*, the *approximate posterior distribution* is easily found, and an *HPD region* for the contrasts is constructed in order to test for equality of means by using a chi-square test. The approximation is compared to the true posterior density and is illustrated with two examples.

1. Introduction

This chapter proposes a Bayesian solution to the Behrens-Fisher problem of testing the equality of means of k normal populations, without the assumption of the equality of covariance matrices. The proposed solution is to approximate the posterior distribution of $k-1$ independent contrasts among the means by a multivariate normal distribution. The mean vector and covariance matrix of the approximation are the same as the mean vector and covariance matrix of the true posterior distribution. By using either a Jeffreys vague prior density or a conjugate type density for prior information, the posterior density is determined without numerical integration and the HPD region for the $k-1$ contrasts is easily found using the chi-square distribution. As a by-product, simultaneous confidence regions for the $k-1$ contrasts are also found.

The solution is derived for comparing the means of multivariate normal populations and several examples demonstrate that the proposed solution behaves in a reasonable manner.

The literature devoted to the Behrens-Fisher problem is immense and there appears no satisfactory solution within classical statistics. Two recent references are Barnard (1984) and Nel and Van der Merwe (1986). On the other hand, there is little controversy within the Bayesian community and the Bayes solution for comparing two scalar normal means is reported by Box and Tiao (1973), and a t-approximation by Patil (1964) is also reported by Box and Tiao. In a recent study, Johnson and Weerahandi (1988) have proposed a Bayesian solution for the two-sample case. Their test statistic is a quadratic form that is similar to the one that we use. They have determined the posterior distribution of the test statistic by one-dimensional numerical integration, but we have not compared their solution to ours. The problem for the Bayesian is implementation. Since the exact posterior density is not a standard distribution, one is forced to use either numerical integration as was done by Box and Tiao (1973) or analytical approximation which is what Patil (1964) did. In this chapter, we propose the latter approach.

There is no unique way to test hypotheses from a Bayesian point of view and what is proposed here is a chi-square credibility region. Other Bayesian testing procedures are based on odds ratios which require a non zero prior probability for the null hypothesis. For additional information, DeGroot (1970, Chapter 11) gives the foundation, while Berger and Deely (1988) provide a recent application of the method.

2. The Model and Prior Information

Consider k independent p-dimensional normal populations $N(\theta_i, P_i^{-1})$, where θ_i is the mean vector and P_i the precision matrix of the i-th population, where $i = 1, 2, \ldots, k$. The Behrens-Fisher problem is to test the hypothesis

$$H: \quad \theta_1 = \theta_2 = \cdots = \theta_k, \tag{2.1}$$

where all the parameters are unknown, and without assuming $P_1 = P_2 = \cdots = P_k$.

If $\{X_{ij} : j = 1, 2, \ldots, n_i\}$ is a random sample of size n_i from the i-th population, the likelihood function for $\theta_1, \ldots, \theta_k; P_1, \ldots, P_k$ is

$$L(\theta_1, \ldots, \theta_k; P_1, \ldots, P_k \mid X) \propto \prod_{i=1}^{k} |P_i|^{n_i/2}$$

$$\times \exp -\frac{T_r}{2}[(n_i - p)S_i + n_i(\theta_i - \bar{X}_i)(\theta_i - \bar{X}_i)']P_i \tag{2.2}$$

where

$$(n_i - p)S_i = \sum_{j=1}^{n_i}(X_{ij} - \bar{X}_i)(X_{ij} - \bar{X}_i)', \tag{2.3}$$

$$\bar{X}_i = n_i^{-1}\sum_{j=1}^{n_i} X_{ij}, \tag{2.4}$$

and T_r denotes the trace operation.

If one has little prior information, Jeffreys prior density

$$g(\theta_1, \ldots, \theta_k; P_1, \ldots, P_k) \propto \prod_{i=1}^{k} |P_i|^{-(p+1)/2} \tag{2.5}$$

is appropriate. The conjugate prior density for this problem is a sequence of k independent normal-gamma densities, which may be appropriate if one is confident of his prior information. The following posterior analysis is done with Jeffreys prior density (2.5), however the conjugate density does not introduce any extra complication.

3. The Posterior Analysis

Our goal is to find the marginal posterior distribution of the k means $\theta_1, \ldots, \theta_k$. Combining the prior density (2.5) with the likelihood function (2.2) and eliminating the k precision matrices $P_1, \ldots, P_k$ via integration (using the normalizing constant of the Wishart density), one may show the marginal posterior density of the means is

$$g(\theta_1, \ldots, \theta_k \mid X) \propto \prod_{i=1}^{k}\left[1 + \frac{n_i(\theta_i - \bar{X}_i)' S_i^{-1}(\theta_i - \bar{X}_i)}{(n_i - p)}\right]^{-n_i/2}. \tag{3.1}$$

Thus the k means are independent and θ_i is distributed

$$\theta_i \mid X \sim T_p[n_i - p, \bar{X}_i, S_i^{-1}n_i], \tag{3.2}$$

that is θ_i has a p-dimensional t-distribution with $n_i - p$ degrees of freedom, mean vector $\bar{X}_i$ and precision matrix $n_i S_i^{-1}$. Furthermore, the dispersion matrix of θ_i is

$$D_i = \frac{S_i(n_i - p)}{n_i(n_i - p - 2)}, \quad n_i > p + 2. \tag{3.3}$$

Our proposed solution to the Behrens-Fisher problem is based on approximating the t-density (3.1) of θ_i with the normal approximation

$$\theta_i \mid X \mathrel{\dot\sim} N(\bar{X}_i, D_i), \tag{3.4}$$

Thus the approximate and true (3.2) posterior densities of θ_i have the same mean vector and covariance matrix. This also implies that the approximate marginal posterior distribution of the k means is the kp-dimensional normal distribution

$$\begin{pmatrix} \theta_1 \\ \theta_2 \\ \vdots \\ \theta_k \end{pmatrix} \dot{\sim} N \left[\begin{matrix} \bar{X}_1 \\ \bar{X}_2 \\ \vdots \\ \bar{X}_k \end{matrix} \middle| \begin{pmatrix} D_1 & & & \\ & D_2 & & 0 \\ & 0 & \ddots & \\ & & & D_k \end{pmatrix} \right] \tag{3.5}$$

which has the same mean vector and covariance matrix as the actual posterior density (3.1) of the means.

4. The Solution to the Behrens-Fisher Problem

In order to test the hypothesis H of the equality of means, the marginal posterior distribution of $k-1$ independent contrasts

$$\gamma_i = \theta_i - \theta_{i+1}, \quad i = 1, 2, \ldots, k-1 \tag{4.1}$$

will be used. Transforming from the $k-$ means distribution (3.5) to the distribution of the $k-1$ contrasts gives

$$\gamma \dot{\sim} N(\hat{\gamma}, D) \tag{4.2}$$

for the approximate posterior distribution of the contrast vector γ, where

$$\gamma = \begin{pmatrix} \gamma_1 \\ \gamma_2 \\ \vdots \\ \gamma_{k-1} \end{pmatrix},$$

$$\hat{\gamma} = \begin{pmatrix} \bar{X}_1 - \bar{X}_2 \\ \bar{X}_2 - \bar{X}_3 \\ \vdots \\ \bar{X}_{k-1} - \bar{X}_k \end{pmatrix}, \tag{4.3}$$

and the $p(k-1)$ order dispersion matrix

$$D = \begin{bmatrix} D_1 + D_2 & -D_2 & 0 & \cdot & \cdot & \cdot & 0 \\ -D_2 & D_2 + D_3 & -D_3 & 0 & \cdot & \cdot & 0 \\ 0 & -D_3 & \cdot & & & & \\ \vdots & & & \cdot & & & \vdots \\ & & & & \cdot & & -D_{k-1} \\ 0 & \cdot & \cdot & \cdot & \cdot & -D_{k-1} & D_{k-1} + D_k \end{bmatrix} \tag{4.4}$$

A $(1-a)$ posterior credible region for the vector γ of contrasts is given by

$$\{\gamma : (\gamma - \hat{\gamma})' D^{-1} (\gamma - \hat{\gamma}) \leq \chi^2_{\alpha,p(k-1)}\} \tag{4.5}$$

where $\chi^2_{\alpha,p(k-1)}$ is the upper $100\alpha\%$ point of the chi-square distribution with $p(k-1)$ degrees of freedom, and the hypothesis

$$H : \theta_1 = \theta_2 = \cdots = \theta_k$$

is rejected whenever

$$\hat{\gamma}' D^{-1} \hat{\gamma} > \chi^2_{\alpha,p(k-1)}. \tag{4.6}$$

For the scalar version of the Behrens-Fisher problem, $p = 1$, and $k = 2$, thus (4.6) is equivalent to rejecting H whenever

$$|\bar{X}_1 - \bar{X}_2| > z_{\alpha/2} \left(\frac{S_1^2(n_1-1)}{n_1(n_1-3)} + \frac{S_2^2(n_2-1)}{n_2(n_2-3)} \right)^{1/2} \tag{4.7}$$

where

$$(n_i - 1)S_i^2 = \sum_{j=1}^{n_i} (X_{ij} - \bar{X}_i)^2, \quad i = 1, 2.$$

We also see that the test procedure (4.6) gives an approximate Bayesian solution to the multivariate one-way analysis of variance problem and that (4.5) produces simultaneous confidence regions for the $k-1$ contrasts γ_i, $i = 1, 2, \ldots, k-1$. It should be emphasized again that the approximate posterior distribution (4.2) has the same mean vector and covariance matrix as the exact posterior distribution of the $k-1$ contrasts.

5. Some Examples

The Bayesian solution to the Behrens-Fisher problem is illustrated by Box and Tiao (1973, pp. 107-110) where they give the exact posterior density of $\theta_1 - \theta_2$ of the difference in two normal means, and they also provide an approximate Bayesian posterior distribution which was proposed by Patil (1964). It is shown that the exact and approximate posterior densities are virtually the same and that the 95% HPD interval for $\theta_1 - \theta_2$ is given by

$$(.673, \; 9.327) \tag{5.1}$$

The following graph and table compare the exact posterior density of Box and Tiao (1973, p. 109) and our approximate $N(5, \; 4.74466)$ posterior density (4.2) for

$\theta_2 - \theta_1$ given the data:

$$n_1 = 20, \quad n_2 = 12$$
$$\bar{X}_1 = 50, \quad \bar{X}_2 = 55$$
$$S_1^2 = 12, \quad S_2^2 = 40.$$

FIGURE 1
EXACT AND APPROXIMATE DENSITIES

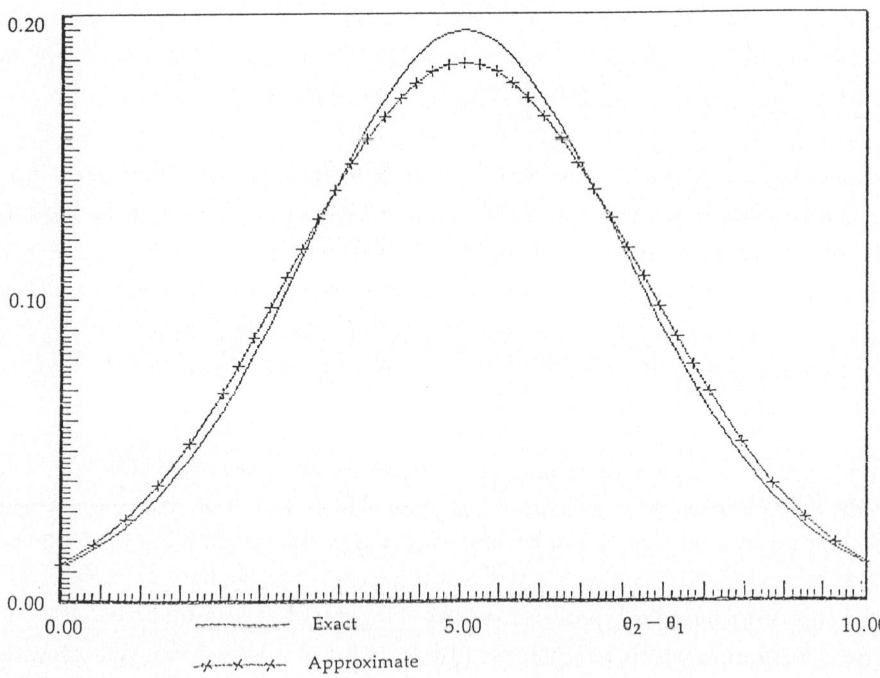

TABLE 1: Exact and Approximate Posterior Densities

$\theta_2 - \theta_1$	Exact	Approximate
5.0	.19411	.18315
5.2	.19310	.18237
5.4	.19010	.18008
5.6	.18523	.17620
5.8	.17863	.17100
6.0	.17054	.16471
6.2	.16112	.15724
6.4	.15089	.14886
6.6	.13990	.13974
6.8	.12851	.13007
7.0	.11700	.12015
7.2	.10561	.10997
7.4	.09453	.09974
7.6	.08395	.08976
7.8	.07398	.08005
8.0	.06474	.07089
8.4	.04859	.05412
8.8	.03562	.03995
9.2	.02558	.02852
9.6	.01805	.01968
10.0	.01257	.01313

We see that the approximate and exact densities are almost the same and that the approximate density is 'below' the exact for values of $\theta_2 - \theta_1$ between 5.0 and 6.8, but after that stays above. It should be noted that Patil's approximate t-density is not listed in the above table but that it does appear in the Table 2.5.1 of Box and Tiao.

The 95% HPD interval for $\theta_2 - \theta_1$ computed from the approximate density is given by

$$\bar{X}_2 - \bar{X}_1 \pm 1.96 \left(\frac{S_1^2(n_1 - 1)}{n_1(n_1 - 3)} + \frac{S_2^2(n_2 - 1)}{n_2(n_2 - 3)} \right)^{1/2} \tag{5.2}$$

or

$$(.731,\ 9.269). \tag{5.3}$$

Compared to the Box-Tiao interval of (.673, 9.327), the approximate *HPD* interval is shorter by .116, and that in both cases, the hypothesis $\theta_1 = \theta_2$ would be rejected. Although there is no substitute for the exact procedure, the approximate solution is easier to compute than either the exact or Patil's approximation and is easily extended to more than two multivariate populations.

As another example of the proposed solution, we examine problem 7.6 of Press (1982, p. 201) which gives $n_1 = 10,\ n_2 = 20$,

$$\bar{X}_1 = \begin{pmatrix} 4 \\ 7 \end{pmatrix}$$

$$\bar{X}_2 = \begin{pmatrix} 6 \\ 6 \end{pmatrix}$$

$$(n_1 - 2)S_1 = \begin{pmatrix} 10 & 3 \\ 3 & 8 \end{pmatrix}$$

and

$$(n_2 - 2)S_2 = \begin{pmatrix} 6 & 2 \\ 2 & 4 \end{pmatrix}$$

The approximate posterior density of $\theta_1 - \theta_2$ is a bivariate normal with mean vector $X_1 - X_2$ and dispersion matrix

$$D = \frac{(n_1 - 2)S_1}{n_1(n_1 - 4)} + \frac{(n_2 - 2)S_2}{n_2(n_2 - 4)} = \begin{pmatrix} .18541 & .05625 \\ .05625 & .14583 \end{pmatrix}$$

The hypothesis of interest is

$$H:\ \theta_1 = \theta_2,$$

that is, doubling the number of patrolling police does not have an effect on the average number of robberies (X_1) and on the average number of car thefts (X_2). The hypothesis is rejected at the 5% level if

$$(\bar{X}_1 - \bar{X}_2)' D^{-1} (\bar{X}_1 - \bar{X}_2) > \chi^2_{.05,2}.$$

In this case, the calculated value of χ^2 is 41.9015 and the table value is 5.99, thus the equality of means hypothesis is rejected. The usual test, assuming the equality of the dispersion matrices would also reject the hypothesis.

6. Some Related Inference Problems with Linear Models

Consider a sequence of k independent linear models

$$Y_i = X_i \theta_i + E_i \tag{6.1}$$

where Y_i is a $n_i \times 1$ observation vector, X_i a $n_i \times p$ known full-rank matrix, θ_i a $p \times 1$ unknown parameter vector, and E_i a $N(0, \tau_i^{-1} I_{n_i})$ random vector. Assuming Jeffreys vague prior density

$$g(\theta_1, \ldots, \theta_k; \tau_1, \ldots, \theta_k) \propto \prod_{i=1}^{k} \tau_i^{-1}$$

it can be shown that the marginal posterior density of the k parameter vectors is

$$g(\theta_1, \ldots, \theta_k \mid Y) \propto \prod_{i=1}^{k} \left[1 + \frac{(\theta_i - \hat{\theta}_i)' X_i X_i (\theta_i - \hat{\theta}_i)}{(n_i - P_i) S_i^2} \right]^{-n_i/2} \tag{6.2}$$

and that θ_i has a p-dimensional t-distribution with $n_i - p$ degrees of freedom, location vector

$$\hat{\theta}_i = (X_i' X_i)^{-1} X_i' Y_i,$$

and dispersion matrix

$$D_i = S_i^2 (X_i' X_i)^{-1} (n_i - p)(n_i - p - 2)^{-1},$$

with $(n_i - p) S_i^2 = (Y_i - X_i \hat{\theta}_i)'(Y_i - X_i \hat{\theta}_i)$, $n_i > p + 2$.

The normal approximation to (6.2) is

$$N \left[\begin{array}{c} \hat{\theta}_1 \\ \hat{\theta}_2 \\ \vdots \\ \hat{\theta}_k \end{array} \middle| \begin{pmatrix} D_1 & & & \\ & D_2 & & 0 \\ & 0 & \ddots & \\ & & & D_k \end{pmatrix} \right]$$

and the hypothesis

$$H: \quad \theta_1 = \cdots = \theta_k$$

is rejected if

$$\hat{\gamma}' D^{-1} \hat{\gamma} > \chi^2_{\alpha, p(k-1)}$$

where

$$\hat{\gamma} = \begin{pmatrix} \hat{\theta}_1 & - & \hat{\theta}_2 \\ \hat{\theta}_2 & - & \hat{\theta}_3 \\ & \vdots & \\ \hat{\theta}_{k-1} & - & \hat{\theta}_k \end{pmatrix}$$

and the $p(k-1)$-th order dispersion matrix is

$$D = \begin{bmatrix} D_1 + D_2 & -D_2 & 0 & \cdot & \cdot & \cdot & 0 \\ -D_2 & D_2 + D_3 & -D_3 & 0 & \cdot & \cdot & 0 \\ 0 & -D_3 & \cdot & & & & \cdot \\ \vdots & & & \cdot & & & \vdots \\ & & & & \cdot & & -D_{k-1} \\ 0 & \cdot & \cdot & \cdot & \cdot & -D_{k-1} & D_{k-1} - D_k \end{bmatrix}$$

The proposed solution will work just as well with multivariate normal linear models. See Zellner (1971, chapter VIII) and Box and Tiao (1973, chapter 8) for the Bayesian analysis of multivariate linear models.

A related problem in linear models is to consider a two-way classification model with possibly unequal cell variance τ_{ij}^{-1}, that is, let the k-th observation in the ij-th cell be

$$y_{ijk} = \mu_{ij} + e_{ijk}$$

where $i = 1, 2, \ldots, a$; $j = 1, 2, \ldots, b$, and $k = 1, 2, \ldots, n_{ij} > 3$ where $n_{ij} > 3$. The cell means are the μ_{ij}, thus the likelihood function is

$$L(\mu, \tau \mid Y) \propto \prod_{i,j,k} \tau_{ij}^{n_{ij}/2} \exp -\frac{\tau_{ij}}{2}(y_{ijk} - \mu_{ij})^2$$

and when combined with Jeffreys vague prior density will produce a ab- dimensional multivariate t-density for the cell means. When the marginal posterior distribution of the cell means is approximated by the appropriate normal density, it is possible to test for the interaction and main effects using a chi-square test. In a similar way, the proposed solution can be applied to other crossed and nested models.

7. Some Comments

The proposed solution has several advantages, namely it is easy to apply because it is based on the standard distributions (normal, t, and chi-square), and it avoids numerical integration, which is necessary for the exact Bayesian analysis. The solution not only solves the usual formulation (the scalar two-population case) of the Behrens-Fisher problem but also applies to any number of multivariate populations, thus giving a generalized analysis of variance. When no more than one of the samples is 'small', we would expect the approximation to be reasonable, because then the normal distribution is a good approximation to the t-distribution. The approximation should even be better when it is used for contrasts, since contrasts among means are more normal than the individual means.

Of course the disadvantage is that the proposed solution is an approximation, however as we have seen, the first two moments of the exact and true densities are the same. To this extent, the small-sample properties should be good. One may also give an asymptotic justification to the approximation. Indeed this was done by Box and Tiao (1973, p. 447) for multivariate linear models.

Acknowledgment

We are grateful to the Vermont Experimental Program to Stimulate Competitive Research (VT EPSCoR) for its support and to the referees who made many useful suggestions for improving the manuscript.

References

Barnard, G.A. (1984). "Comparing the Means of Two Independent Samples, *Applied Statistics*, Vol. 33, 266-271.

Berger, J.O. and J. Deely (1988). "A Bayesian Approach to Ranking and Selection of Related Means with Alternatives to Analysis-of-Variance Methodology," *Journal of the American Statistical Association*, Vol. 83, No. 42, 364-373.

Box, G.E.P. and G.C. Tiao (1973). *Bayesian Inference in Statistical Analysis.* Reading, Massachusetts: Addison-Wesley Publishing Company.

DeGroot, M.H. (1970). *Optimal Statistical Decisions.* New York: McGraw-Hill, Inc.

Johnson, R.A. and S. Weerahandi (1988). "A Bayesian Solution to the Multivariate Behrens-Fisher Problem," *Journal of the American Statistical Association*, Vol. 83, No. 401, 145-149.

Nel, D.G. and C.A. Van der Merwe (1986). "A Solution to the Behrens-Fisher Problem," *Communications in Statistics*, Vol. 15, No. 12, 3719-3736.

Patil, V.H. (1964). "The Behrens-Fisher Problem and Its Bayesian Solution," *Journal of the Indian Statistical Association*, Vol. 2.

Press, J.S. (1982). *Applied Multivariate Analysis: Using Bayesian and Frequentist Methods of Inference.* Malabar, Florida: Robert E. Krieger Publishing Company.

Zellner, A. (1971). *An Introduction to Bayesian Inference in Econometrics.* New York: John Wiley and Sons, Inc.

Bayesian and Likelihood Methods in Statistics and Econometrics
S. Geisser, J.S. Hodges, S.J. Press and A. Zellner (Editors)
© Elsevier Science Publishers B.V. (North-Holland), 1990

EXTENDING FAMILIES OF DENSITIES BY SPECIAL-FUNCTION FACTORS

Chong-Hong Chen and James M. Dickey
National Cheng-Kung University and University of Minnesota

Abstract

A variety of parameterized univariate and vector-variate distributions are generalized by multiplying the usual density by a generalized hypergeometric function or a multiple hypergeometric function with an appropriately transformed argument variable. A benefit of our choice of the new special-function factor is that the normalizing constant has a closed functional form. Also, the moments and characteristic functions of the generalized distributions are averages of the moments and characteristic functions of the usual distributions. The new random quantities, in the cases of the multiple gamma, Dirichlet, and inverted Dirichlet, obey functional relations similar to those of the usual random quantities. The new generalized chi-squared distribution includes, as a special case, the usual noncentral chi-squared distribution. The method preserves the property of being a Bayesian conjugate prior density.

1. Introduction

The multitude of ways to extend a family of distributions or, otherwise, develop a new parameterized family include the following:

(i) Develop the nonnull distribution of a test statistic. [For example, the noncentral chi-squared, noncentral t, noncentral F, double noncentral t, and double noncentral F distributions.]

(ii) Directly introduce a further parameter. [For example, the generalized gamma distribution, with density function $P(x \mid \alpha, \beta, c) = cx^{c\alpha-1}e^{-x^c/\beta}/(\beta^x\Gamma(\alpha)), \quad x \geq 0.$]

(iii) Imbed a familiar distribution as a univariate margin of a new multivariate or matrix-variate distribution. [The beta distribution, relative to the Dirichlet distribution.]

(iv) Work with a function of the coordinates of a multivariate distribution. [The Rayleigh distribution.]

(v) Mix over a random parameter. That is, using a mixing distribution $H(\theta)$, one obtains a probability density function as $\int g(x;\theta)dH(\theta)$ or $\sum_1^c g(x;\theta_1)\cdot H(\theta_i)$, where $g(x;\theta)$ is the parameterized mixand density. [The mixed normal, compound Poisson, Bayesian prior-predictive or posterior-predictive distributions such as the Dirichlet-Bernoulli distribution or Dirichlet-multinomial distribution.]

(vi) Multiply by a further factor. [Truncated distributions, Bayesian posterior distributions.]

In this article, we will extend several distributions by method (vi), using a generalized hypergeometric function or a multiple hypergeometric function as the new factor, following a suitable transformation in its argument. We will find that distributions obtained by other methods often can be rewritten in the form of method (vi). For example, the noncentral chi-squared distribution, a mixture of central chi-squared distributions with Poisson mixing, has a density expressible as the product of the central chi-squared density and a further factor—a generalized hypergeometric function. For another example, noncentral F and double noncentral F, both defined as mixtures of central F distributions, have densities expressible as the product of a central F density and, respectively, a generalized hypergeometric function, and a multiple hypergeometric function. As a benefit of our general choice of factor and its argument, the normalizing constant will have a closed functional form, and the new moments and characteristic functions will be averages of the familiar quantities. In particular cases, the new random variables will be seen to satisfy relations similar to familiar relations holding between the usual random variables.

A Bayesian conjugate prior family is useful in that the updating to obtain a posterior distribution proceeds merely by transforming the prior parameters to posterior parameters according to the formulas for sample-pooling of sufficient statistics. Any summary quantity describing the posterior distribution can then be computed by using exactly the same formula or technique as for describing the prior distribution. When extended by method (vi), a conjugate family retains its property of being closed under sampling. The various simple forms for complete integrals of our extended densities, then, apply to both prior and posterior distributions. A related method of extension was introduced by Roux (1971) for densities of a positive-definite symmetric matrix, using as a further factor a particular invariant hypergeometric function of matrix argument, expressible in terms of zonal polynomials. (See Press (1980) for a conjugate prior family based on an extended density by Roux.)

We begin, Section 2, by providing a brief review of generalized hypergeometric functions and multiple hypergeometric functions. These special functions are well studied as solutions to differential equations. Their series representations and recursive integral representations will be useful for our purposes. After giving the definitions and relevant properties of the further factors, we develop the method of generalization. In Section 3, we extend several families of univariate and multivariate distributions. Their properties and possible applications are given. We conclude, in Section 4, by discussing the effects of the further factors and the problem of estimating the parameters.

2. A Method of Generalization

2.1 The Hypergeometric Functions

The special functions called generalized hypergeometric functions play an important role in statistics, as characteristic functions of distributions, moment-generating functions, probability-generating functions, cumulative distribution functions (Johnson and Kotz, 1970), and even probability density function themselves (Steyn, 1951). Their series representations can provide methods for calculating entities of statistical interest. This is increasingly so with the availability of computers.

The *generalized hypergeometric function* extends the Gauss hypergeometric function by increasing the number of numerator and denominator parameters. Its series representation is

$$ {}_AF_B[(a);(b);x] = \sum_{k=0}^{\infty} \frac{((a),k)x^k}{((b),k)k!}, \tag{2.1} $$

where the symbol (a) denotes the sequence of parameters $a_1, \ldots a_A$, the multiple form $((a),k)$ is the product $(a_1,k)\ldots,(a_A,k)$, and the Pochhammer symbol (a,r) is defined by the relations

$$ \begin{aligned} (a,r) &= (a+r)/(a) \\ &= a(a+1)\ldots(a+r+1) \\ (a,0) &= 1 \\ (a,-r) &= (-1)/(1-a,r). \end{aligned} \tag{2.2} $$

The B many denominator parameters $b_1, b_2, \ldots, b_B$ are required to take values other than negative integers. The series converges for all finite values of its argument if $A < B$, and for $|x| < 1$ if $A = B+1$. The case $A > B+1$ is considered only if one or more of the numerator parameters is a negative integer, for which the series terminates and the question of convergence does not arise.

Many familiar functions are lower-order special cases of the generalized hypergeometric function. For example, the exponential function, ${}_0F_0(\cdot;\cdot;x) = e^x$, the binomial function, ${}_1F_0(a;\cdot;x) = (1-x)^{-a}$, the Bessel function, ${}_0F_1(\cdot;b;x) = \Gamma(b)(ix)^{(b-1)}J_{b-1}(2ix^{1/2})$, the incomplete gamma function, ${}_1F_1(a;a+1;-x) = ax^{-a}r(a,x) = ax^{-a}\int_0^x y^{a-1}e^{-y}dy$ and the incomplete beta function, ${}_2F_1(a,1-b;x) = ax^{-a}B_x(a,b) = ax^{-a}\int_0^x y^{a-1}(1-y)^{b-1}dy$.

The *multiple hypergeometric function* extends the generalized hypergeometric function by an increase in the number of argument variables. It has the following multiple series representation,

$$F^{A:B^{(1)};\ldots;B^{(n)}}_{C:D^{(1)};\ldots;D^{(n)}}\begin{pmatrix}(a):(b^{(1)});\ldots;(b^{(n)}) & \\ & ;x_1,\ldots,x_n \\ (c):(d^{(1)});\ldots;(d^{(n)}) & \end{pmatrix}$$

$$=\sum\frac{((a),m_1+\ldots+m_n)((b^{(1)}),m_1)\ldots((b^{(n)}),m_n)x_1^{m_1}\ldots x_n^{m_n}}{((c),m_1+\ldots+m_n)((d^{(1)}),m_1)\ldots((d^{(n)}),m_n)m_1!\ldots m_n!} \tag{2.3}$$

where the summation is over all integers m_i from 0 to infinity for each $i = 1,\ldots,n$. The most important hypergeometric functions of several variables are the Lauricella functions F_A, F_B, F_C, F_D which are the special cases defined by

(i) $F_A(a,b_1,\ldots,b_n;c_1,\ldots,c_n;x_1,\ldots,x_n)$

$$=F^{1:1;\ldots;1}_{0:1;\ldots;1}\begin{pmatrix}a:b_1;\ldots;b_n & \\ & ;x_1,\ldots,x_n \\ \cdot:c_1;\ldots;c_n & \end{pmatrix} \tag{2.4}$$

(ii) $F_B(a_1,\ldots,a_n,b_1,\ldots,b_n;c;x_1,\ldots,x_n)$

$$=F^{0:2;\ldots;2}_{1:0;\ldots;0}\begin{pmatrix}\cdot:(a_1,b_1);\ldots;(a_n,b_n) & \\ & ;x_1,\ldots,x_n \\ c:\ \cdot\ ;\ldots;\ \cdot & \end{pmatrix} \tag{2.5}$$

(iii) $F_C(a,b;c_1,\ldots,c_n;x_1,\ldots,x_n) = F^{2:0;\ldots;0}_{0:1;\ldots;1}\begin{pmatrix}(a,b):\ \cdot;\ldots;\cdot & \\ & ;x_1,\ldots,x_n \\ \cdot\ :c_1;\ldots;c_n & \end{pmatrix}$

(2.6)

(iv) $F_D(a,b_1,\ldots,b_n;c;x_1,\ldots,x_n) = F^{1:1;\ldots;1}_{1:0;\ldots;0}\begin{pmatrix}a:b_1;\ldots;b_n & \\ & ;x_1,\ldots,x_n \\ c:\cdot;\ldots;\cdot & \end{pmatrix}$

(2.7)

If the number n of variables, is taken equal to two, these four functions reduce to the Appell functions F_2, F_3, F_4 and F_1, respectively; and if $n = 1$, all four functions become the Gauss function ${}_2F_1$. The convergence of multiple hypergeometric series was examined by Horn (1889). In case $A + B^{(i)} < C + D^{(i)} + 1$ for all i, the series (2.3) converges for all finite values of $x_i, x_2, \ldots, x_n$.

The Lauricella functions have a wide diversity of applications in statistics, mathematics, and theoretical physics, such as: an expectation associated with the multiple beta distribution; a Dirichlet distribution function associated with a transition density function in the field of genetics (for F_A); the distribution of the ratio of two Dirichlet variates and the expectation associated with the multiple gamma distribution (for F_D) (Exton, 1976).

Relevant properties of the multiple hypergeometric function include the following:

1. $$F^{A:0;\ldots;0}_{C:0;\ldots;0}\left(\begin{matrix}(a);\ \cdot\ ;\ldots;\ \cdot \\ (c);\ \cdot\ ;\ldots;\ \cdot\end{matrix}\ ;x_1,\ldots,x_n\right) = {}_AF_C\big((a);(c);x_1,\ldots,x_n\big) \tag{2.8}$$

2. $$F^{0:B^{(1)};\ldots;B^{(n)}}_{0:D^{(1)};\ldots;D^{(n)}}\left(\begin{matrix}\cdot\ :(b^{(1)});\ldots;(b^{(n)}) \\ \cdot\ :(d^{(1)});\ldots;(d^{(n)})\end{matrix}\ ;x_1,\ldots,x_n\right)$$
$$= \prod_{j=1}^{n} {}_{B^{(j)}}F_{D^{(j)}}\big((b^{(j)});(d^{(j)});x_j\big) \tag{2.9}$$

3. $$F^{A:1;\ldots;1}_{C:0;\ldots;0}\left(\begin{matrix}(a):b_1;\ldots,\ b_n \\ (c):\ \cdot\ ;\ldots;\ \cdot\end{matrix}\ ;x,\ldots,x\right)$$
$$= {}_{A+1}F_C\big((a),b_1+\ldots+b_n;(c);x\big) \tag{2.10}$$

4. $$F_A(a,b_1,\ldots,b_n;c_1,\ldots,c_n;x_1,\ldots,x_n) = E(1-\sum_{j=1}^{n}U_ix_i)^{-a} \tag{2.11}$$

where U_i has the beta distribution $U_i \sim \text{Beta}(b_i, c_i - b_i)$ for $i = 1, 2, \ldots, n$, independently.

5. $$F_B(a_1,\ldots,a_n,b_1,\ldots,b_n;c;x_1,\ldots,x_n) = E\sum_{j=1}^{n}(1-x_jU_j)^{-b_j} \tag{2.12}$$

where $\underset{\sim}{U} = (U_1,\ldots,U_n)^T \sim \text{Dirichlet}(a_1,\ldots,a_n,c-\sum_{j=1}^{n}a_j)$.

$$6.\quad F_D(a, b_1, \ldots, b_n; c; x_1, \ldots, x_n) = E(1 - \sum_1^n U_j x_j)^{-a} \tag{2.13}$$

where $\underset{\sim}{U}$ has the Dirichlet distribution as in (2.12). Furthermore, the function F_D may also be represented by a single integral:

$$F_D(a, b_1, \ldots, b_n; c; x_1, \ldots, x_n) = E \sum_{j=1}^n (1 - x_j U)^{-b_j} \tag{2.14}$$

where U has the beta distribution $U \sim \text{Beta}(a, c - a)$.

2.2 The Method of Generalization

By multiplying a familiar probability density function by a further factor and then renormalizing, the probability density function of a generalized distribution can be obtained. This is analogous to the operation performed in Bayes' Theorem. Let $p(x \mid \underset{\sim}{\theta})$ be such a generalized density function,

$$p(x \mid \underset{\sim}{\theta}) = p_0(x \mid \underset{\sim}{\theta}_0) \cdot g(z(x) \mid \underset{\sim}{\theta}_1)/c, \tag{2.15}$$

where $p_0(x \mid \underset{\sim}{\theta}_0)$ is a known probability density function having the parameter vector $\underset{\sim}{\theta}_0$ and $g(z(x) \mid \underset{\sim}{\theta}_1)$ is some factor function with argument $z(x)$ and parameter vector $\underset{\sim}{\theta}_1$. The parameter vector $\underset{\sim}{\theta}$ is a function $\underset{\sim}{\theta} = \underset{\sim}{\theta}(\underset{\sim}{\theta}_0, \underset{\sim}{\theta}_1)$ of $\underset{\sim}{\theta}_0$d and $\underset{\sim}{\theta}_1$ and c is the renormalizing constant. Here, x is a general notation for a single random variable, or a random vector. The argument $z(x)$ can be a real-valued or a vector-valued function of x.

It is important that the further factor selected be positive. For cases where the factor is a multiple hypergeometric function and the argument variables are nonnegative, the parameters can be allowed to be negative. If the number of negative parameters is even, the coefficients of the hypergeometric function are positive, and so is the function itself. But if the number of negative parameters is odd, the signs of the coefficients of the series alternate, and then the positivity of the function is not guaranteed if the arguments can be any real value. Gasper (1975) studied the positivity of special functions by reducing them to a simpler function involving fewer parameters. The positivity of our further factor will imply the positivity of the renormalizing constant. This will allow a recurrence-relation method to prove the positivity of a special function having a few more parameters than a simple one.

2.3 Initial Examples

1. *Truncated distributions:* the multiplier is an indicator function parameterized by the maximum and minimum possible values of the random variable, for example, the truncated normal distribution with density function,

$$p(x \mid \mu, \sigma, a, b) = p_0(x \mid \mu, \sigma) \cdot I_{[a,b]}(x) \Big/ \left[\Phi\left(\frac{b-\mu}{\sigma}\right) - \Phi\left(\frac{a-\mu}{\sigma}\right)\right] \tag{2.16}$$

where $p_0(x \mid \mu, \sigma)$ is the normal density with mean μ and standard deviation σ, and Φ is the standard normal c.d.f.

2. *Rayleigh distribution:* Let $\underset{\sim}{X}$ be normally distributed in n dimensions with mean vector $\underset{\sim}{\mu}$ and covariance matrix σI_n, where σ is a positive constant and I_n is the identity matrix. Then $r = |\underset{\sim}{X}|$, the norm of $\underset{\sim}{X}$, follows the Rayleigh distribution with density function

$$p(r) = \left(\frac{\mu}{\sigma}\right)\left(\frac{r}{\mu}\right)^{n/2} e^{-(r^2+\mu^2)/2\sigma} \cdot I_{\frac{1}{2}(n-2)}\left(\frac{r\mu}{\sigma}\right), \quad r \geq 0 \tag{2.17}$$

where $\mu = |\underset{\sim}{\mu}|$ and $I_\nu(\cdot)$ is the modified Bessel function of the first kind of order ν.

3. *Generalized beta and gamma distributions:* Exton (1978) used a product of generalized hypergeometric functions of form ${}_AF_B(\ldots\ldots;z)$ as a multiplier to the density functions of a special beta distribution and the gamma distribution, respectively, to obtain new densities,

$$k^{-1}x^{d-1}{}_{A^{(1)}}F_{B^{(1)}}\big((a^{(1)});(b^{(1)});h_1x\big)\ldots{}_{A^{(n)}}F_{B^{(n)}}\big((a^{(n)});(b^{(n)});h_nx\big), \quad 0 \leq x \leq 1,$$

where

$$k = d^{-1}F^{1:A^{(1)};\ldots;A^{(n)}}_{1:B^{(1)};\ldots;B^{(n)}}\left(\begin{matrix} d & :(a^{(1)});\ \ldots;\ (a^{(n)}) \\ d+1 & :(b^{(1)});\ \ldots;\ (b^{(n)}) \end{matrix} ;h_1,\ldots,h_n\right) \tag{2.18}$$

and the density

$$k^{-1}e^{-px}x^{d-1}{}_{A^{(1)}}F_{B^{(1)}}\big((a^{(1)});(b^{(1)});h_1x\big)\ldots{}_{A^{(n)}}F_{B^{(n)}}\big((a^{(n)});(b^{(n)});h_nx\big) \tag{2.19}$$

where

$$k = \frac{\Gamma(d)}{pd} \cdot F^{1:A^{(1)};\ldots;A^{(n)}}_{0:B^{(1)};\ldots;B^{(n)}}\left(\begin{matrix} d:\ (a^{(1)});\ldots;(a^{(n)}) \\ \cdot\ :\ (b^{(1)});\ldots;(b^{(n)}) \end{matrix} ;h_1/p,\ldots,h_n/p\right).$$

Properties and Requirements: 1. The normal distribution, the beta distribution, and the gamma distribution are known as the conjugate prior distributions, respectively, for the normal sampling model, binomial sampling model and Poisson sampling process. There is a natural sense in which this relation extends to their generalized distributions as defined by this method. Generally, if $p_0(x \mid \underset{\sim}{\theta}_0)$ is a conjugate prior density for a sampling model, then the new generalized density $p(x \mid \underset{\sim}{\theta})$ is a family similarly closed under sampling from the same sampling process.

2. If the further-factor function g can be expressed in series form or merely a polynomial by special choice of the numerator parameter in the generalized hypergeometric function, then the generalization can be considered as a mixture distribution. For a mixture, the moments and characteristic function are simple linear combinations of the moments or characteristic functions of the mixand distributions if such moments and characteristic functions exit. Examples include the mixed normal and the compound Poisson distributions.

3. In case the function g is a series having positive coefficients, and the argument $z(x)$ is proportional to the kernel of the original distribution, then the generalized distribution is completely monotonic, that is, a scale-mixture of normal distributions, provided that the parent distribution is completely monotonic. This follows from the fact that the complete monotone class is closed under such summation.

4. Generally, it may be difficult to find the renormalizing constant, even though the factor function $g(z(x) \mid \underset{\sim}{\theta}_1)$ is simple. To avoid such difficulty, the choice of the function g and the argument $z(x)$ should be made with care. In Example 3, Exton used the generalized hypergeometric function as the function g to extend the beta and gamma distributions with the argument $z(x)$ chosen proportional to the kernel of a respective beta or gamma density. Of course, a renormalizing constant may also be computed by numerical integration, as is often done in connection with posterior distributions.

5. To form the new density (2.15), the further factor should be positive. Such a restriction can reduce the variety in the effect of the further factor. However, the numerous parameters of the special functions can allow important differences from the usual distribution. We shall consider the special functions to have positive coefficients throughout, except for cases of particular mention.

3. Specific Parameterized Families

3.1 Generalized Multiple Gamma Distribution

We extend the multiple gamma distribution, a joint distribution of independent gamma variables with vectors of parameters $\underset{\sim}{\alpha}$, $\underset{\sim}{\beta}$ by multiplying the density by the generalized multiple hypergeometric function and then renormalizing, to yield

$$\left(\prod_{j=1}^{k}\frac{\beta_j^{\alpha_j}x_j^{\alpha_j-1}e^{-\beta_j x_j}}{\Gamma(\alpha_j)}\right)F^{A:B^{(1)};\ldots;B^{(k)}}_{C:D^{(1)};\ldots;D^{(k)}}\left(\begin{matrix}(a):(b^{(1)});\ldots;(b^{(k)})\\ \\(c):(d^{(1)});\ldots;(d^{(k)})\end{matrix};\rho_1x_1,\ldots,\rho_kx_k\right)$$

$$\Bigg/F^{A:B^{(1)}+1;\ldots;B^{(k)}+1}_{C:D^{(1)};\ldots;D^{(k)}}\left(\begin{matrix}(a):(b^{(1)}),\alpha_1;\ldots;(b^{(k)}),\alpha_k\\ \\(c):(d^{(1)})\quad;\ldots;(d^{(k)})\end{matrix};\rho_1/\beta_1,\ldots,\rho_k/\beta_k\right) \tag{3.1}$$

As a generalized multiple gamma distribution, this comprises a Bayesian conjugate prior family for the multiple Poisson sampling process, the joint distribution of K independent Poisson variables. With such a sampling model, if we denote the prior distribution by $\underset{\sim}{x} \sim MG(\underset{\sim}{\alpha},\underset{\sim}{\beta},F)$, where F denotes the new factor, then the posterior distribution takes the form,

$$\underset{\sim}{x} \mid \underset{\sim}{n} \sim MG(\underset{\sim}{\alpha}+\underset{\sim}{n},\underset{\sim}{\beta}+\underset{\sim}{1};F) \tag{3.2}$$

The corresponding Bayesian prior predictive distribution, the marginal distribution of vector $\underset{\sim}{n}$, is obtained as a generalized multiple gamma mixture of the multiple Poisson distribution,

$$p(\underset{\sim}{n}) = E_{\underset{\sim}{x}|\underset{\sim}{\alpha},\underset{\sim}{\beta},F}(p(\underset{\sim}{n}\mid\underset{\sim}{x})) = \prod_{j=1}^{k} n_j^{-1}B(n_j,\alpha_j)^{-1}\left(\frac{\beta_j}{1+\beta_j}\right)^{\alpha_j}\left(\frac{1}{\beta_j+1}\right)^{n_j}$$

$$\cdot\frac{F^{A:B^{(1)}+1;\ldots;B^{(k)}+1}_{C:D^{(1)}\quad;\ldots;D^{(k)}}\left(\begin{matrix}(a):(b^{(1)}),\alpha_1+n_1;\ldots;(b^{(k)}),\alpha_k+n_k\\ \\(c):(d^{(1)})\qquad;\ldots;(d^{(k)})\end{matrix};\frac{\rho_1}{\beta_1+1},\ldots,\frac{\rho_k}{\beta_k+1}\right)}{F^{A:B^{(1)}+1;\ldots;B^{(k)}+1}_{C:D^{(1)}\quad;\ldots;D^{(k)}}\left(\begin{matrix}(a):(b^{(1)}),\alpha_1;\ldots;(b^{(k)}),\alpha_k\\ \\(c):(d^{(1)})\quad;\ldots;(d^{(k)})\end{matrix};\frac{\rho_1}{\beta_1},\ldots,\frac{\rho_k}{\beta_k}\right)} \tag{3.3}$$

The vector $\underset{\sim}{n}$ no longer has independent coordinates, due to the dependent coordinates in the generalized multiple gamma distribution. Such a prior predictive distribution can be viewed as a generalized multiple negative binomial distribution if the α_is are integers.

The characteristic function of (3.1) has the form,

$$E(e^{i\underset{\sim}{t}^T \underset{\sim}{x}}) = \left(\prod_{j=1}^{k} \left(\frac{\beta_j}{\beta_j - it_j} \right)^{\alpha_j} \right) \cdot \frac{F^{A:B^{(1)}+1;\ldots;B^{(k)}+1}_{C:D^{(1)} \quad ;\ldots;D^{(k)}} \left(\begin{matrix} (a):(b^{(1)}),\alpha_1;\ldots;(b^{(k)}),\alpha_k \\ \\ (c):(d^{(1)}) \quad ;\ldots;(d^{(k)}) \end{matrix} ;\left(\frac{\rho_1}{\beta_1 - it_1}\right),\ldots,\left(\frac{\rho_k}{\beta_k - it_k}\right) \right)}{F^{A:B^{(1)}+1;\ldots;B^{(k)}+1}_{C:D^{(1)} \quad ;\ldots;D^{(k)}} \left(\begin{matrix} (a):(b^{(1)}),\alpha_1;\ldots;(b^{(k)}),\alpha_k \\ \\ (c):(d^{(1)}) \quad ;\ldots;(d^{(k)}) \end{matrix} ;\left(\frac{\rho_1}{\beta_1}\right),\ldots,\left(\frac{\rho_k}{\beta_k}\right) \right)} \tag{3.4}$$

This is a product of the characteristic function of the multiple gamma distribution and a ratio of multiple hypergeometric functions. The moment generating function has a similar form.

Univariate Case: For the univariate case, somewhat more generally than merely letting $k = 1$ in (3.1), we can allow the further factor to be a multiple hypergeometric function. Then the generalized gamma distribution will have the following density,

$$\frac{\beta^\alpha x^{\alpha-1} e^{-\beta x}}{\Gamma(\alpha)} \cdot F^{A:B^{(1)};\ldots;B^{(k)}}_{C:D^{(1)};\ldots;D^{(k)}} \left(\begin{matrix} (a):(b^{(1)});\ldots;(b^{(k)}) \\ (c):(d^{(1)});\ldots;(d^{(k)}) \end{matrix} ;\rho_1 x,\ldots,\rho_k x \right)$$

$$\Big/ F^{A+1:B^{(1)};\ldots;B^{(k)}}_{C \quad :D^{(1)};\ldots;D^{(k)}} \left(\begin{matrix} (a),\alpha;\ (b^{(1)});\ldots;(b^{(k)}) \\ \\ (c):(d^{(1)});\ldots;(d^{(k)}) \end{matrix} ;\frac{\rho_1}{\beta_1},\ldots,\frac{\rho_k}{\beta_k} \right) \tag{3.5}$$

This yields, as special cases, Exton's generalized gamma distribution as in (2.19) for all B_is and D_is equal to zero, and also the noncentral chi-squared distribution, for which the density function can be expressed as the mixture of central chi-square density functions,

$$P_{\chi^2_{\delta(\lambda)}}(x) = \sum_{j=0}^{\infty} \frac{\left(\frac{\lambda}{2}\right)^j e^{-\frac{1}{2}\lambda}}{j!} \; P_{\chi^2_{\delta+2j}}(x)$$

$$= \frac{\left(\frac{1}{2}\right)^{\delta/2} x^{\frac{1}{2}\delta-1} e^{-\frac{1}{2}x}}{\Gamma\left(\frac{\delta}{2}\right)} \cdot \sum_{j=0}^{\infty} \frac{\left(\left(\frac{\lambda}{4}\right)x\right)^j}{\left(\frac{\delta}{2},j\right)j!} \Big/ e^{\lambda/2}$$

$$= P_{\chi^2_\delta(x)} \cdot {}_0F_1\left(\cdot;\frac{\delta}{2};\frac{\lambda}{4}x\right) \Big/ {}_1F_1\left(\frac{\delta}{2};\frac{\delta}{2};\frac{\lambda}{2}\right). \tag{3.6}$$

We define the generalized chi-squared distribution having the density

$$\frac{x^{\frac{1}{2}\sigma-1}e^{-\frac{1}{2}x}}{2^{\delta/2}\Gamma\left(\frac{\delta}{2}\right)} \; {}_AF_B\big((a);(b);\rho x\big) \big/ {}_{A+1}F_B\big((a),\frac{\delta}{2};(b);2\rho\big), \tag{3.7}$$

This contains (3.6) as a special case. For $\delta = 1$ in (3.7), we can define a generalized normal distribution by the transformation $y = \pm\sqrt{x}$,

$$P(y) = \frac{1}{\sqrt{2\pi\sigma^2}} e^{-\frac{1}{2\sigma^2}y^2} \cdot {}_AF_B\left((a);(b);\frac{\mu}{\sigma}y\right) \Big/$$

$${}_{2A}F_{2B}\left(\left(\frac{a}{2}\right),\left(\frac{a+1}{2}\right);\left(\frac{b}{2}\right),\left(\frac{b+1}{2}\right);2^{2A-2B-1}\frac{\mu^2}{\sigma^2}\right). \tag{3.8}$$

When $A = B = 0$, this reduces to the usual normal distribution with mean μ and variance σ^2. A sum of independent generalized chi-squared variables, each distributed as (3.7), will have a generalized gamma density, as in (3.1).

For application to Bayesian inference, in addition to the new conjugate prior family for the Poisson process, let us consider, as a sampling model, one of the generalized gamma distributions introduced by Stacy (1962). Such a conditional distribution has the density,

$$P(x \mid p) = \frac{cp^{w/c}}{\Gamma\left(\frac{w}{c}\right)} x^{w-1} \exp\{-px^c\}, \qquad \text{for } x > 0, \tag{3.9}$$

(w, c, and $p > 0$). This distribution includes as special cases the exponential ($p = c = 1$), the gamma ($c = 1$), and the Weibull ($w = c$). The log-normal is also a limiting special case as $w \longrightarrow \infty$. The distribution (3.9) is easily visualized in terms of $y = \log x$, the log survival time. We let the parameter p have the prior density $\pi(p)$ as follows,

$$\pi(p) = \frac{\beta^\alpha p^{\alpha-1} e^{-\beta p} \prod_{j=1}^k {}_1F_1(b_j;d_j;\rho_j p)}{\Gamma(\alpha)\cdot F_A\left(\alpha,b_1,\ldots,b_k;d_1,\ldots,d_k;\frac{\rho_1}{\beta},\ldots,\frac{\rho_k}{\beta}\right)}, \tag{3.10}$$

a special case of (3.1). The posterior density, $\pi(p \mid x)$, then has the closed form,

$$\frac{(\beta+x^c)^{\frac{w}{c}+\alpha} p^{(\frac{w}{c}+\alpha)-1} e^{-(\beta+x^c)p} \prod_{j=1}^{k} \{{}_1F_1(b_j; d_j; \rho_j p)\}}{\Gamma\left(\frac{w}{c}+\alpha\right) \cdot F_A\left(\beta+x^c, b_1, \ldots, b_k; d_1, \ldots, d_k; \frac{\rho_1}{\beta+x^c}, \ldots, \frac{\rho_k}{\beta+x^c}\right)}. \tag{3.11}$$

The prior predictive density is given by

$$P(x) = \frac{c x^{w-1} (\beta+x^c)^{-(\frac{w}{c}+\alpha)} F_A\left(\beta+x^c, b_1, \ldots, b_k; d_1, \ldots, d_k; \frac{\rho_1}{\beta+x^c}, \ldots, \frac{\rho_k}{\beta+x^c}\right)}{B\left(\frac{w}{c}, \alpha\right) \cdot F_A\left(\beta, b_1, \ldots, b_k; d_1, \ldots, d_k; \frac{\rho_1}{\beta}, \ldots, \frac{\rho_k}{\beta}\right)}. \tag{3.12}$$

3.2 Extended Dirichlet Distributions

The moments of the Dirichlet distribution have the form of a ratio of two generalized beta functions. It is easy to derive from this the Dirichlet expectation of a multiple hypergeometric function.

THEOREM 3.1: *If $\underset{\sim}{u} \sim D(\underset{\sim}{b})$, then*

$$E^{(k-1)}_{\underset{\sim}{u}|\underset{\sim}{b}} F^{A:B^{(1)};\ldots;B^{(k)}}_{C:D^{(1)};\ldots;D^{(k)}} \begin{pmatrix} (a):(b^{(1)});\ldots;(b^{(k)}) \\ & ;\rho_1 u_1, \ldots, \rho_k u_k \\ (c):(d^{(1)});\ldots;(d^{(k)}) \end{pmatrix}$$

$$= F^{A:B^{(1)}+1;\ldots;B^{(k)}+1}_{C+1:D^{(1)};\ldots;D^{(k)}} \begin{pmatrix} (a) \quad :(b^{(1)}), b_1; \ldots; (b^{(k)}), b_k \\ & ;\rho_1, \ldots, \rho_k \\ (c), \Sigma b_i : (d^{(1)}) \quad ; \ldots; (d^{(k)}) \end{pmatrix} \tag{3.13}$$

again a multiple hypergeometric function.

We define the generalized Dirichlet distribution, to have the density,

$$P(\underset{\sim}{u} \mid \underset{\sim}{b}, \underset{\sim}{\omega})$$

$$= B(\underset{\sim}{b})^{-1} \left(\prod_{i=1}^{k} u_i^{b_i-1}\right) F^{A:B^{(1)};\ldots;B^{(k)}}_{C:D^{(1)};\ldots;D^{(k)}} \begin{pmatrix} (a):(b^{(1)});\ldots;(b^{(k)}) \\ & ;\rho_1 u_1, \ldots, \rho_k u_k \\ (c):(d^{(1)});\ldots;(d^{(k)}) \end{pmatrix}$$

$$\Big/ F^{A:B^{(1)}+1;\ldots;B^{(k)}+1}_{C+1:D^{(1)};\ldots;D^{(k)}} \begin{pmatrix} (a) \quad :(b^{(1)}), b_1; \ldots; (b^{(k)}) \\ & ;\rho_1, \ldots, \rho_k \\ (c), b. : \quad (d^{(1)}); \ldots; (d^{(k)}) \end{pmatrix} \tag{3.14}$$

where $\underset{\sim}{\omega} = \{(a),(c),(b^{(1)}),\ldots,(b^{(k)}),(d^{(1)}),\ldots,(d^{(k)}),\rho_1,\ldots,\rho_k\}$. The density (3.14) can also be obtained from the generalized multiple gamma distribution (3.1), by the transformation, $u_i = x_i/\Sigma x_i$, $i = 1,2,\ldots,k$. This generalization gives Janardan (1973) and Dickey (1983) as special cases, as follows.

(i) Janarden's extended Dirichlet distribution has the density function,

$$B(\underset{\sim}{b})^{-1}\left(\prod_{i=1}^{k-1} u_i^{b_i-1}\right)\left(1-\sum_{i=1}^{k-1} u_i\right)^{b_k-1} e^{\sum_{i=1}^{k-1}\rho_i u_i}$$

$$\Big/ \Phi_2^{(k-1)}(b_1,\ldots,b_{k-1};b.;-\rho_1,\ldots,-\rho_{k-1}), \tag{3.15}$$

where $\Phi_2^{(n)}(b_1,\ldots,b_n;c;x_1,\ldots,x_n) = \lim_{\epsilon\to 0} F_D\left(\frac{1}{\epsilon},b_1,\ldots,b_n;c;\epsilon x_1,\ldots,\epsilon x_n\right)$, where

$$0 < u_i < 1 \text{ for all } i, \sum_1^{k-1} u_i < 1,$$

$$b_i > 0 \text{ for all } i,\ b. = \left(\sum_{i=1}^{k-1} b_i\right) + b_k,$$

$$\rho_i > 0 \text{ for all } i.$$

If $A = C = B^{(1)} = \ldots = B^{(k)} = D^{(1)} = \ldots = D^{(k)} = 0$, by properties (2.9), (2.2), ${}_0F_0(\cdot;\rho_1 u_1,\ldots,\rho_{k-1}u_{k-1},0)$ reduces to $\exp\left(\sum_1^{k-1}\rho_i u_i\right)$.

(ii) Dickey's extended Dirichlet distribution has the density function,

$$B(\underset{\sim}{b})^{-1}\left(\prod_{i=1}^{k} u_i^{b_i-1}\right)(\underset{\sim}{u}\cdot\underset{\sim}{z})^{-\beta}/R_{-\beta}(\underset{\sim}{b},\underset{\sim}{z}) \tag{3.16}$$

where $R_a(\underset{\sim}{b},\underset{\sim}{z})$ is B.C. Carlson's (1963, 1977) symmetrized multiple hypergeometric function, a special case of the Lauricella F_D function: $R_a(\underset{\sim}{b},z) = F_D^{(k)}(a,b_1,\ldots,b_k; b_1+\ldots+b_k;1-z_1,\ldots,1-z_k)$. For $A=1$, $C = B^{(1)} = \ldots = B^{(k)} = D^{(1)} = \ldots = D^{(k)} = 0$, $a=\beta$ and property (2.8),

$$F^{1:0;\ldots;0}_{0:0;\ldots;0}\left(\begin{matrix}\beta :\cdot \\ \\ \cdot :\cdot\end{matrix};(1-z_1)u_1,\ldots,(1-z_k)u_k\right)$$

reduces to $(\underset{\sim}{u}\cdot\underset{\sim}{z})^{-\beta}$. We note that Dickey's other extensions, involving several linear forms and a double average, are not special cases of (3.14).

From Theorem (3.1), we have the following expectation associated with the Lauricella functions.

COROLLARY:

$$\underset{\underset{\sim}{u}|\underset{\sim}{b}}{E}{}^{(k-1)} F_A^{(k)}(c, d_1, \ldots, d_k; b_1, \ldots, b_k; \rho_1 u_1, \ldots, \rho_k u_k)$$

$$= F_D^{(k)}(c, d_1, \ldots, d_k; b_1 + \ldots + b_k; \rho_1, \ldots, \rho_k) \tag{3.17}$$

$$\underset{\underset{\sim}{u}|\underset{\sim}{b}}{E}{}^{(k-1)} F_D^{(k)}(b_1 + \ldots + b_k; d_1, \ldots, d_k; e; \rho_1 u_1, \ldots, \rho_k u_k)$$

$$= \underset{\underset{\sim}{u}|\underset{\sim}{b}}{E}{}^{(k-1)} \underset{\underset{\sim}{v}|(\underset{\sim}{d}, e)}{\overset{(k)}{E}} \left(1 - \sum_1^k \rho_i u_i v_i\right)^{-(b_1 + \ldots + b_k)}$$

$$= F_B^{(k)}(d_1, \ldots, d_k, b_1, \ldots, b_k; e; \rho_1, \ldots, \rho_k) \tag{3.18}$$

COROLLARY: *The moment of the generalized Dirichlet distribution is the product of a moment of the ordinary Dirichlet distribution and a ratio of multiple hypergeometric functions; that is,*

$$\underset{\underset{\sim}{u}|\underset{\sim}{b},\underset{\sim}{\omega}}{E}\left(\prod_{i=1}^{k} u_i^{n_i}\right) = \frac{B(\underset{\sim}{b} + \underset{\sim}{n})}{B(\underset{\sim}{b})}$$

$$\cdot \frac{F^{A:B^{(1)}+1;\ldots;B^{(k)}+1}_{c+1:D^{(1)};\ldots;D^{(k)}}\left(\begin{matrix}(a) & :(b^{(1)}), b_1 + n_1; \ldots; (b^{(k)}), b_k + n_k \\ (c), b + n : & (d^{(1)}); \ldots; (d^{(k)})\end{matrix} ; \rho_i, \ldots, \rho_k\right)}{F^{A:B^{(1)}+1;\ldots;B^{(k)}+1}_{c+1:D^{(1)};\ldots;D^{(k)}}\left(\begin{matrix}(a) & :(b^{(1)}), b_1; \ldots; (b^{(k)}), b_k \\ (c), b. : & (d^{(1)}); \ldots; (d^{(k)})\end{matrix} ; \rho_1, \ldots, \rho_k\right)}. \tag{3.19}$$

Marginal and Conditional Distributions: In general, the marginal density of a joint extended Dirichlet distribution with density function (3.14) can be derived as a series form, but not a closed (or named) functional form. However, if we restrict discussion to the special case $B^{(1)} = \ldots = B^{(k)} = D^{(1)} = \ldots = D^{(k)} = 0$, the density (3.14) reduces by (2.8) to

$$B(\underset{\sim}{b})^{-1}\left(\prod_{i=1}^{k} u_i^{b_i-1}\right) {}_AF_C((a);(c);\rho_1 u_1+\ldots+\rho_k u_k)$$

$$\Big/ F^{A:1;\ldots;1}_{C+1:0;\ldots;0}\left(\begin{matrix}(a) & : b_1;\ldots;b_k & \\ & & \rho_1,\ldots,\rho_k \\ (c),b. & : \ \cdot\ ;\ldots;\ \cdot & \end{matrix}\right). \tag{3.20}$$

Conformably partition $\underset{\sim}{u} = (\underset{\sim}{u}_{(1)}^T, \underset{\sim}{u}_{(2)}^T)^T, \underset{\sim}{b} = (\underset{\sim}{b}_{(1)}^T, \underset{\sim}{b}_{(2)}^T)^T$ and $\underset{\sim}{\rho} = (\rho_1,\ldots,\rho_k)^T = (\underset{\sim}{\rho}_{(1)}^T, \underset{\sim}{\rho}_{(2)}^T)^T$, where $\underset{\sim}{u}_{(i)} : k_i \times 1 \quad (i = 1,2)$ and $k_1 + k_2 = k$. Then we have the following theorem concerning the marginal and conditional densities of (3.20).

THEOREM 3.2: *Suppose $\underset{\sim}{u}$ has the density function (3.20), and let $\underset{\sim}{v} = (\underset{\sim}{u}_{(1)}^T, u_{(2)\cdot})$, $\underset{\sim}{w} = \underset{\sim}{u}_{(2)}/u_{(2)\cdot}$. Then the conditional distribution of $\underset{\sim}{w}$ given $\underset{\sim}{v}$ has the density function,*

$$p(\underset{\sim}{w} \mid \underset{\sim}{v}) = B(\underset{\sim}{b}_{(2)})^{-1}\left(\prod_{i=1}^{k_2} w_i^{b_{(2)i}-1}\right) \cdot {}_AF_C((a);(c);\tilde{\rho}_1 w_1,\ldots,\tilde{\rho}_{k_2} w_{k_2})$$

$$\Big/ F^{A\ :1;\ldots;1}_{C+1:0;\ldots;0}\left(\begin{matrix}(a) & :b_{(2)1};\ldots;\ b_{(2)k_2} & \\ & & ;\tilde{\rho}_1,\ldots,\tilde{\rho}_{k_2} \\ (c),b_{(2)} & : \ \cdot\ ;\ldots;\ \cdot & \end{matrix}\right), \tag{3.21}$$

and the marginal distribution of v has the density function

$$p(\underset{\sim}{v}) = B(\underset{\sim}{b}_{(1)}, b_{(2)\cdot})^{-1}\left(\prod_{i=1}^{k_1} v_i^{b_{(1)i}-1}\right)$$

$$\cdot\, v_{k_1+1}^{b_{(2)\cdot}-1} F^{A\ :1;\ldots;1}_{C+1:0;\ldots;0}\left(\begin{matrix}(a) & :b_{(2)1};\ldots;\ b_{(2)k_2} & \\ & & ;\rho_1,\ldots,\rho_k \\ (c),b_{(2)\cdot} & : \ \cdot;\ldots;\cdot & \end{matrix}\right)$$

$$\Big/ F^{A\ :1;\ldots;1}_{C+1:0;\ldots;0}\left(\begin{matrix}(a) & :b_1;\ldots;b_k & \\ & & ;\rho_1,\ldots,\rho_k \\ (c),b & : \ \cdot;\ldots;\cdot & \end{matrix}\right) \tag{3.22}$$

where $\tilde{\rho}_i = \rho_1 u_1 + \ldots + \rho_{k_1} u_{k_1} + u_{(2)\cdot}\rho_{k_1+i}$, $i = 1,2,\ldots,k_2$.

PROOF: Indicate by $(*)$ the different distribution for the same variables, $u \sim * D(\underset{\sim}{b})$ with density $p^*(\underset{\sim}{u})$. Then the given distribution for (3.20) satisfies

$$p(\underset{\sim}{u}) = p^*(\underset{\sim}{u}) \cdot {}_AF_C\big((a);(c); \underset{\sim}{u}_{(1)}^T \underset{\sim}{\rho}_{(1)} + \underset{\sim}{u}_{(2)}^T \underset{\sim}{\rho}_{(2)}\big)/\text{constant}$$

where the constant denominator is

$$F^{A\ \ :1;\ldots;1}_{C+1:0;\ldots;0}\left(\begin{matrix}(a):b_1;\ldots;b_k\\ \\ (c),b.;.;..;.\end{matrix}\ ;\rho_1,\ldots,\rho_k\right).$$

Change variables to $\underset{\sim}{w}$, $\underset{\sim}{v}$,

$$p(\underset{\sim}{w},\underset{\sim}{v}) = p^*(\underset{\sim}{w},\underset{\sim}{v})_AF_C\big((a);(c); \underset{\sim}{w}^T \underset{\sim}{1}_{k_2} \underset{\sim}{u}_{(1)}^T \underset{\sim}{\rho}_{(1)} + \underset{\sim}{w}^T u_{(2)\cdot} \underset{\sim}{\rho}_{(2)}\big)/\text{const.}$$

Since $\underset{\sim}{w} \mid \underset{\sim}{v} \sim * D(\underset{\sim}{b}_{(2)})$ and $\underset{\sim}{v} \sim * D(\underset{\sim}{b}_{(1)}, b_{(2)\cdot})$, then

$$\begin{aligned}p(\underset{\sim}{w}\mid\underset{\sim}{v})p(\underset{\sim}{v}) = \Bigg(&p^*(\underset{\sim}{w}\mid\underset{\sim}{v})_AF_C\big((a);(c);\tilde\rho_1,\ldots,\tilde\rho_{k_2}\big)\\ &\Big/F^{A\ \ :1;\ldots;1}_{C+1:0;\ldots;0}\left(\begin{matrix}(a):b_{(2)1};\ldots;b_{(2)k_2}\\ \\ (c),b_{(2)}:.;..;.\end{matrix}\ ;\tilde\rho_1,\ldots,\tilde\rho_{k_2}\right)\Bigg)\\ &\cdot p^*(\underline{v})F^{A\ \ :1;\ldots;1}_{C+1:0;\ldots;0}\left(\begin{matrix}(a):b_{(2)1};\ldots;b_{(2)k_2}\\ \\ (c),b_{(2)\cdot}:.;..;.\end{matrix}\ ;\tilde\rho_1,\ldots,\tilde\rho_{k_2}\right)\\ &\Big/\text{constant.}\end{aligned}$$

A special case of this theorem was obtained by Dickey for his generalizations of the Dirichlet family.

In addition to the convenient use of such a generalization for the prior and posterior distributions for multiple Bernoulli sampling, this class can also be used to model the variability of personal posterior probabilities over a population of Bayesian scientists (Dickey (1968) and Dickey and Freeman (1975)).

Univariate Case (Generalized Family of Beta Distributions): The beta distribution is the univariate case of the Dirichlet ($k = 2$). A generalization of the beta distribution can be obtained as the univriate case of our extended Dirichlet distribution. Consider the following expectations of the beta distribution, $u \sim \text{Beta}(b_1, b_2)$,

$$E_{u|b_1,b_2} F^{A:B^{(1)};\ldots;B^{(k)}}_{C:D^{(1)};\ldots;D^{(k)}} \left(\begin{matrix} (a):(b^{(1)});\ldots;\ (b^{(k)}) \\ (c):(d^{(1)});\ldots;(d^{(k)}) \end{matrix} ;\rho_1 u^n, \ldots, \rho_k u^n \right)$$

$$= F^{A+n:B^{(1)};\ldots;B^{(k)}}_{C+n:D^{(1)};\ldots;D^{(k)}} \left(\begin{matrix} (a),\ \frac{b_1}{n}\ ,\ldots,\ \frac{b_1+n-1}{n}\ :\ (b^{(1)});\ldots;(b^{(k)}) \\ (c),\ \frac{b_1+b_2}{n},\ldots,\ \frac{b_1+b_2+n-1}{n}:\ (d^{(1)});\ldots;(d^{(k)}) \end{matrix} ;\rho_1,\ldots,\rho_k \right) \tag{3.23}$$

and

$$E_{u|b_1,b_2} F^{A:B^{(1)};\ldots;B^{(k)}}_{C:D^{(1)};\ldots;D^{(k)}} \left(\begin{matrix} (a):(b^{(1)});\ldots;(b^{(k)}) \\ (c):(d^{(1)});\ldots;(d^{(k)}) \end{matrix} ;\rho_1 u^n(1-u)^n, \ldots . \rho_k u^n(1-u)^n \right)$$

$$= F^{A+n:B^{(1)};\ldots;B^{(k)}}_{C+n:D^{(1)};\ldots;D^{(k)}} \left(\begin{matrix} (a),\ \frac{b_1}{n}\ ,\ldots,\ \frac{b_1+n-1}{n}\ :\ (b^{(1)});\ldots;(b^{(k)}) \\ (c),\frac{b_1+b_2}{n},\ldots,\frac{b_1+b_2+n-1}{n}:\ (d^{(1)});\ldots;(d^{(k)}) \end{matrix} ;\ \frac{\rho_1}{4^n},\ldots,\frac{\rho_k}{4^n} \right). \tag{3.24}$$

We can use the functions under these expectations as further factors to extend the beta distributions, and their expectations become the renormalizing constants. The generalization by (3.23) contains Exton's generalized beta distribution (2.18) as a special case, and also the noncentral beta distribution. The latter is defined as the distribution of the ratio, involving independent random variables,

$$\chi'^2_{\nu_1}(\lambda)/[\chi^2_{\nu_2} + \chi'^2_{\nu_1}(\lambda)], \tag{3.25}$$

and has the density function

$$e^{-\lambda/2} \sum_{j=0}^{\infty} \left(\frac{\lambda}{2}\right)^j (j!)^{-1} u^{\frac{1}{2}\nu_1+j-1}(1-u)^{\frac{\nu_2}{2}-1} / B\left(\frac{\nu_1}{2}+j, \frac{\nu}{2}\right)$$

$$= B\left(\frac{\nu_1}{2},\frac{\nu_2}{2}\right)^{-1} u^{\frac{\nu_1}{2}-1}(1-u)^{\frac{\nu_2}{2}-1} \sum_{j=0}^{\infty} \frac{\left(\frac{\nu_1+\nu_2}{2},j\right)\left(\frac{\lambda u}{2}\right)}{\left(\frac{\nu_1}{2},j\right) j!} \Big/ e^{\lambda/2}$$

$$= p_{u|b_1,b_2}(u) \cdot {}_1F_1\left(\frac{\nu_1+\nu_2}{2}, \frac{\nu_1}{2}; \frac{\lambda u}{2}\right) \Big/ e^{\lambda/2}.$$

If we use the generalized chi-squared random variable instead of noncentral chi-squared in the representation (3.25), we obtain a generalized beta distribution which is a special case of (3.14) for $k = 1$ or (3.23) for $k = 1,\ n = 1$.

For interest, let us consider the very special case with density function

$$p(x) = \frac{x^{b_1-1}(1-x)^{b_2-1}\prod_{i=1}^n(1-\alpha_i x)^{-a_i}}{B(b_1,b_2)F_D^{(n)}(b_1,a_1,\dots,a_n;b_1+b_2;\alpha_1,\dots,\alpha_n)}; \; 0 \le x \le 1. \tag{3.26}$$

The cumulative distribution function then has the following form,

$$F(x) = B(b_1,b_2)^{-1}\frac{x_1^{b_1}F_D^{(n+1)}(b_1,a_1,\dots,a_n,1-b_2;1+b_1;\alpha_1,\dots,a_n,x)}{b_1F_D^{(n)}(b_1,a_1,\dots,a_n,b_1+b_2;\alpha_1,\dots,\alpha_n)} \tag{3.27}$$

Of course, when we let the a's equal zero, the well-known special case of the incomplete beta function is obtained,

$$B(b_1,b_2)^{-1}\frac{x^{b_1}}{b_1}{}_2F_1(b_1,1-b_2;1+b_1;x). \tag{3.28}$$

The moments of (3.26) are easy to derive as ratios of the Lauricella function F_D. We also have the expectation of a particular function in the same form as the further factor,

$$E\prod_{i=1}^m(1-\beta_i x)^{c_i} = \frac{F_D^{(n+m)}(b_1,a_1,\dots,a_n,c_1,\dots,c_m;b_1+b_2;\alpha_1,\dots,\alpha_n,\beta_1,\dots,\beta_m)}{F_D^{(n)}(b_1,a_1,\dots,a_n;b_1+b_2;\alpha_1,\dots,\alpha_n)} \tag{3.29}$$

For Bayesian inference, similarly to the Dirichlet family, the distributions of our generalized family are conjugate prior for Bernoulli sampling trials or the Binomial model. We can also consider a sampling model having the density function,

$$p(x \mid \beta) = e^{-\beta x}/[\beta(1-\rho^{-\beta})], \quad 0 \le x \le 1, \tag{3.30}$$

the density of a truncated exponential distribution. Let the unknown parameter have the prior density $\pi(\beta)$ as in (3.26). Then the prior-predictive density is as given by Exton (1976)

$$p(x) = \sum_{m=0}^{\infty}\Phi_D^{(n+1)}(b_1-1,a_1,\dots,a_n,\cdot;b_1+b_2-1;\alpha_1,\dots,\alpha_n,-x-m), \tag{3.31}$$

where $\Phi_D^{(n+1)}$ is the confluent form of the Lauricella Function F_D.

3.3 The Extended Multiple F and Inverted Dirichlet Distributions

Let $\underset{\sim}{u} = (u_1,u_2,\dots,u_k)^T$ have the extended Dirichlet distribution (3.14). Consider the usual transformation,

$$T: x_i = u_i/(1-u_1-\dots-u_{k-1}), \quad i=1,2,\dots,k-1. \tag{3.32}$$

Then we have the extended inverted Dirichlet distribution with density function

$$p(\underset{\sim}{x}) = B(\underset{\sim}{b})^{-1}\left(\prod_{j=1}^{k-1} x_j^{b_j-1}\right)\left(1+\sum_{j=1}^{k-1} x_j\right)^{-b\cdot}$$

$$\cdot F^{A:B^{(1)};\ldots;B^{(k)}}_{C:D^{(1)};\ldots;D^{(k)}}\left(\begin{matrix}(a):(b^{(1)});\ldots;\ (b^{(k)})\\ \\ (c):(d^{(1)});\ldots;\ (d^{(k)})\end{matrix}\ ;\ \frac{\rho_1 x_1}{1+\sum x_i},\ldots,\frac{\rho_{k-1}x_{k-1}}{1+\sum x_i},\frac{\rho_k}{1+\sum x_i}\right)$$

$$\Big/ F^{A:B^{(1)}+1;\ldots;B^{(k)}+1}_{C+1:D^{(1)};\ldots;D^{(k)}}\left(\begin{matrix}(a)\quad :(b^1),b_1;\ldots;(b^{(k)}),b_k\\ \\ (c),b.:\quad (d^1)\quad;\ldots;(d^{(k)})\end{matrix}\ ;\ \rho_1,\ldots,\rho_k\right). \tag{3.33}$$

The density (3.33) can also be obtained by the transformation

$$X_i = Y_i/Y_k, \quad i = 1,2,\ldots,k-1 \tag{3.34}$$

where $(Y_1,\ldots,Y_k)$ has the generalized multiple gamma distributions (3.1). The generalized multiple F distribution $(z_1,z_2,\ldots,z_k)$ can be obtained by the scale transformation of (3.33), $z_i = (b_k/b_1)x_i$, $i = 1,2,\ldots,k$. The moments are ratios of multiple hypergeometric functions, similarly to (3.19).

Univariate Case: The univariate case of the extended multiple F is of particular interest. The usual noncentral and double noncentral F distributions are both mixtures of central F distributions (Johnson and Kotz, 1970). These are both special cases of (3.33), as follows.

The extended noncentral F density is

$$p_{F'}(x) = p_F(x)\sum_{j=0}^{\infty}\frac{\left(\frac{\nu_1+\nu_2}{2},j\right)\left(\frac{\frac{1}{2}\nu_1\lambda_1 x}{\nu_2+\nu_1 x}\right)^j}{\left(\frac{\nu_1}{2},j\right)j!}\Big/ e^{\lambda/2}$$

$$= p_F(x)\ {}_1F_1\left(\frac{\nu_1+\nu_2}{2},\frac{\nu_1}{2};\frac{\frac{1}{2}\lambda_1\nu_1 x}{\nu_2+\nu_1 x}\right)\Big/ e^{\lambda/2} \tag{3.35}$$

The double noncentral F has density

$$p_{F''}(x) = \sum_{j=0}^{\infty}\sum_{k=0}^{\infty}\left[\exp\left(\frac{-\lambda_1}{2}\right)\left(\frac{\lambda_1}{2}\right)^j/j!\right]\left[\exp\left(\frac{-\lambda_2}{2}\right)\left(\frac{\lambda_2}{2}\right)^k/k!\right]$$

$$\cdot B\left(\frac{\nu_1}{2}+j,\frac{\nu_2}{2}+k\right)^{-1}\nu_1^{\frac{1}{2}\nu_1+j}\nu_2^{\frac{1}{2}\nu_2+k}x^{\frac{1}{2}\nu_1+j-1}(\nu_2+\nu_1 x)^{-\frac{1}{2}(\nu_1+\nu_2)-j-k}$$

$$= p_F(x)\cdot F^{1:0;0}_{0:1;1}\left(\begin{matrix}\frac{\nu_1+\nu_2}{2}:\cdot\,;\ \cdot\\ \\ \cdot\ :\frac{\nu_1}{2};\ \frac{\nu_2}{2}\end{matrix}\ ;\ \frac{\frac{1}{2}\nu_1\lambda_1 x}{\nu_2+\nu_1 x},\frac{\frac{1}{2}\lambda_2\nu_2}{\nu_2+\nu_1 x}\right)\Big/ e^{\frac{1}{2}(\lambda_1+\nu_2)}. \tag{3.36}$$

We began this section by defining the generalized multiple gamma distribution, which has dependent coordinates. Its univariate case contains as a special case the noncentral chi-squared distribution. By transforming independent chi-squared random variables, one obtains the ordinary Dirichlet and inverted Dirichlet distributions. The generalized Dirichlet and generalized inverted Dirichlet distributions, as we defined them, can also be derived by the same transformations from the generalized multiple gamma distribution. Moreover, using the generalized chi-squared distribution instead of the noncentral chi-squared distribution, the transformed variables associated with noncentral beta, noncentral F, and double noncentral F turn out to be special cases of our generalized beta and generalized F distributions.

4. Conclusion

The method of section 2 successfully extends various distributions of multivariate form by introducing, as a further factor, a transformed special function, a generalized hypergeometric function, or multiple hypergeometric function. In this article, we restrict our attention to the special functions having positive coefficients. Such special functions can still have a large effect on the ordinary distributions. Firstly, the effect can be a change in the parameters of the ordinary distribution. For example, in the beta distribution, Beta(α, β), the choice of ${}_1F_0(a; \cdot; x) = (1-x)^{-a}$ will change β to $\beta - a$, or the choice of ${}_1F_0(a; \cdot; (1-x))$ will change α to $\alpha - a$; and in the gamma distribution, Gamma(α, β), the choice of ${}_0F_0(-; -; \rho x)$ will change β to $\beta - \rho$. Secondly, after choosing a particular special function, if we still need a further change, we can introduce two further parameters in the function, e.g., ${}_2F_1(a, b; c; \cdot)$ to ${}_3F_2(a, b, d_1; c, d_2; \cdot)$. The greater the difference is between d_1 and d_2, the greater the effect of the factor will be. To allow an even greater effect by a wider choice of factor functions, we have the following remedy: Use a linear combination of special functions as a further factor. This will allow one to choose a function having a lower bound. For example, by working with the cosine function $\cos\sqrt{z} = {}_0F_1(\frac{1}{2}, -\frac{1}{4}z) \geq -1$, or the Legendre polynomials $P_n(x) = {}_2F_1\left(-n, n+1; 1; \frac{1-x}{2}\right) \geq \frac{1}{2}$, we may use $\cos\sqrt{z} + 1$ and $P_n(x) - \frac{1}{2}$ as further factors. Other examples include the Hermite polynomials and Laguerre polynomials. The renormalizing constant will then be just a linear combination of the renormalizing constants for each individual term as a factor. In general, we may consider the following weighted average

$$p(x \mid \xi) = \sum_i p_0(x \mid \theta_i(\xi)) \cdot \tilde{w}_i(x \mid \eta_i(\xi)) \tag{4.1}$$

where the ith weight is the ith linear combination,

$$\tilde{w}_i(x \mid \eta_i) = \sum_j c_{ij} f_j(x \mid \psi_j(\eta_i)) \geq 0, \quad \text{and} \quad \sum \tilde{w}_i \equiv 1,$$

and the f_js are particular special functions. The weights $\tilde{w}(x \mid \eta_i)$ would be independent of the variable x in the case of constant f_js, yielding an ordinary mixture of distributions.

Perhaps, the most conspicuous problem when using a parameterized distribution to fit data is the estimation of the parameters. Many methods have being developed for estimation of the "ordinary" distributions, such as the frequency substitution principle, the method of moments, least squares, maximum likelihood, and Bayesian posterior mean, mode, etc. The method of moments might seem to offer a promising way of estimating the parameters of the "generalized" distributions, since the moments are available in closed form. However, it turns out to be difficult to solve the resulting simultaneous equations for many parameters.

Acknowledgment

Research by Professor Dickey was supported by the U.S. National Science Foundation (Research Grant DMS-8614793).

References

Carlson, B.C. (1963). "Lauricella's Hypergeometric Function F_D," *Journal of Mathematical Analysis and Applications*, Vol. 7, 452-470.

Carlson, B.C. (1977). *Special Functions of Applied Mathematics.* New York: Academic Press.

Dickey, J.M. (1980). "Beliefs About Beliefs, A Theory for Stochastic Assessments of Subjective Probabilities," in *Bayesian Statistics*, J.M. Bernardo, M.H. DeGroot, D.V. Lindley, and A.F.M. Smith, eds., Valencia, Spain: Valencia University Press, 1980.

Dickey, J.M. (1983). "Multiple Hypergeometric Functions: Probabilistic Interpretations and Statistical Uses," *Journal of the American Statistical Association*, Vol. 78, 628-637.

Dickey, J.M. and C.H. Chen (1983). "Direct Subjective-Probability Modeling Using Ellipsoidal Distributions," in *Bayesian Statistics 2*,

J.M. Bernardo, M.H. DeGroot, D.V. Lindley and A.F.M. Smith, eds., Amsterdam: North-Holland Publishing.

Dickey, J.M. and P. Freeman (1975). "Population-Distributed Personal Probabilities," *Journal of the American Statistical Association*, Vol. 70, No. 350, 362-364.

Exton, H. (1976). *Handbook of Hypergeometric Integrals: Theory, Applications, Computer Programs.* New York: John Wiley & Sons.

Exton, H. (1978). *Multiple Hypergeometric Functions and Applications.* New York: John Wiley & Sons.

Gasper, G. (1975). "Positivity and Special Functions," in *Theory and Application of Special Functions*, R.A. Askey, ed., New York: Academic Press, 375-433.

Horn, J. (1889). "Ueber die Convergenz der Hypergeometrische Reihen Zweier und Reier Veränderlichen," *Math. Ann.*, Vol. 34, 544-600.

Janardan, K.G. (1973). "Chance Mechanisms for Multivariate Hypergeometric Models," *Sankhyā, A*, Vol. 35, 465-478.

Johnson, N.L. and S. Kotz (1970). *Distributions in Statistics: Continuous Univariate Distributions-2.* New York: John Wiley.

Johnson, N.L. and S. Kotz (1972). *Distributions in Statistics: Continuous Multivariate Distributions.* New York: John Wiley.

Press, S.J. (1980). "Bayesian Inference in Group Judgment Formulation and Decision Making Using Qualitative Feedback," in *Bayesian Statistics*, J.M. Bernardo, M.H. DeGroot, D.V. Lindley and A.F.M. Smith, eds., Valencia, Spain: Valencia University Press, 397-414.

Roux, J.J. (1971). "On Generalized Multivariate Distributions," *South African Journal of Statistics*, Vol. 5, 91-100.

Stacy, E.W. (1962). "A Generalization of the Gamma Distribution," *Annals of Mathematical Statistics*, Vol. 33, 1187-1192.

Steyn, H.S. (1951). "On Discrete Multivariate Probability Functions," *Akad. Van Wetensch.*, Vol. A58, No. 5, 23-30.

Bayesian and Likelihood Methods in Statistics and Econometrics
S. Geisser, J.S. Hodges, S.J. Press and A. Zellner (Editors)

BAYES ESTIMATION IN THE MARKOV RENEWAL PROCESS

George T. Duncan
Carnegie Mellon University

Abstract

The form of Bayes estimators for the parameters of a finite state Markov renewal process observed for an arbitrary time T are developed, both for parameters of the waiting time distributions and for the transition probabilities. An approximation is presented that eases integration problems. Explicit approximate Bayes estimators are presented for the transition probabilities in the case of a matrix beta prior distribution.

Key Words – Markov renewal process, Bayesian estimation, waiting times, transition probabilities.

1. Introduction

This article develops Bayesian statistical estimation procedures for the finite state Markov renewal process, a flexible model for stochastic phenomena that has found considerable application. The Building Fire Safety Model of Berlin (1980), for example, is a Markov renewal model. It has six states of fire development identified as fire "realms." For residential rooms these realms are defined as in Table 1.

Based on a literature review of eight published reports, Berlin assessed a transition probability matrix $\mathbf{P}$ for a single-family dwelling couch fire to be

$$\mathbf{P} = \begin{bmatrix} 1 & 0 & 0 & 0 & 0 & 0 \\ .33 & 0 & .67 & 0 & 0 & 0 \\ 0 & .25 & 0 & .75 & 0 & 0 \\ 0 & 0 & .25 & 0 & .75 & 0 \\ 0 & 0 & 0 & .08 & 0 & .92 \\ 1 & 0 & 0 & 0 & 0 & 0 \end{bmatrix}$$

TABLE 1: Residential Room Fire Realm Definitions
(Derived from Berlin (1980))

REALM I	Non-fire State. Situations prior to ignition and following termination of a fire.
REALM II	Sustained Burning. There is sustained burning of the fuel which includes smoldering.
REALM III	Vigorous Burning. The heat release rate exceeds 2 kW, the flame height is approximately 25 cm, or the upper room gas temperature has increased 15°C above the ambient temperature.
REALM IV	Interactive Burning. The heat release rate exceeds 50 kW, the flame height is approximately 120 cm, or the upper room gas temperature exceeds 150°C.
REALM V	Remote Burning. The external heat flux returning to the fuel surface exceeds 5 kW/sq. m or the upper room gas temperature exceeds 450°C.
REALM VI	Full Room Involvement.

The waiting time distributions were assessed as in Table 2.

TABLE 2: Parameter Estimates of Waiting Time Distributions for a Smoldering Fire in a Couch with Cotton Cushions

Realm Transition From	To	Type	Mean	Standard Deviation
II	I	Uniform	2	5
II	III	Lognormal	8.45	0.78
III	II	Uniform	1	2
III	IV	Normal	5.55	3.22
IV	III	Uniform	1.5	9
IV	V	Uniform	0.5	3.5
V	IV	Uniform	0.6	6.0
V	VI	Lognormal	5.18	4.18

The general case is treated here where uncertainty exists about both the waiting time distributions and the transition probabilities. This work extends the Bayesian results of Martin (1967), who only considers the Markov chain case, and Brock (1971), who assumes that the waiting time distributions are known. Moore and Pyke (1968) deal only with non-Bayesian estimation methods. The sampling schemes considered here are either (1) to observe n transitions and their associated waiting times, or, more generally, (2) to observe the process (transitions and associated waiting times) for some time T, where T is not necessarily a transition time.

The Markov renewal process is a convenient and workable generalization of both a Markov process and a renewal process, incorporating the essential features of each. Essentially, a finite state semi-Markov process is a stationary stochastic process modeling moves among a finite number of states. The successive states visited form a Markov chain. The length of stay in a given state is a random variable, the distribution of which may depend on the origin state as well as on the destination state. The finite state semi-Markov process can be thought of as a Markov chain for which the time scale has been randomly transformed.

As noted by Pyke (1961), the semi-Markov process is equivalent to the Markov renewal process which records at each time t the number of times an entity has visited each of the possible states up to time t. The entity moves from state to state according to a Markov chain, and the time required for each successive move is a random variable whose distribution function depends only on the two states between which the move is being made.

An early application of the semi-Markov process was in psychology by Cane (1959). Its application to social mobility and migration was outlined by Ginsberg (1971, 1972) and it was reviewed briefly by Bartholomew (1973). Weiss and Zelen (1965) applied it to clinical trials. Duncan and Job (1978) have applied it to short-term probability forecasting in international relations. Fienberg (1978) has applied it in crime victimization studies. Other applications have treated inventory, reliability, and maintenance problems. Hershey, Weiss, and Cohen (1981) and Weiss, Cohen, and Hershey (1982) have applied it to hospital patient flow. Shonick (1973) has used it to analyze multi-stage queueing models.

2. A Likelihood Function for the Markov Renewal Process

Statistical inference questions about the transition probabilities of a Markov renewal process are considered by Moore and Pyke (1968). In this section we establish notation, present the likelihood function, and discuss some issues in estimating the waiting time distributions as well as the transition probabilities.

We deal with a Markov renewal process having a finite state space, $S = \{1, 2, \ldots, s\}$. For each time $t \geq 0$, and each $i, j \in S$, we let $Q_{ij}(t)$ denote the probability that after making a transition into state i, the process next makes a transition into state j in an amount of time less than or equal to t. The Markov renewal model allows for the possibility of transitions from state i to state i. Such transitions may or may not have substantive meaning in a particular application. For example, in migration studies a move within a particular geographic region (state i) might signal a transition. On the other hand, in reliability studies, states usually reflect the operating status of a system, and a "within state" transition may have no physical meaning. In this latter case, $Q_{ii}(t) = 0$ for each $i \in S$. Since each $Q_{ij}(t)$ is increasing in t, we may define transition probabilities $P_{ij} = Q_{ij}(\infty) = \lim Q_{ij}(t)$. Note that $\sum_j P_{ij} = 1$ for each i. With this notation, P_{ij} represents the probability that the next transition will be into state j, given that the process has just entered state i. If $P_{ij} > 0$, we define the conditional probability $F_{ij}(t)$ that a transition will take place within an amount of time t, given that the process has just entered i and will next enter j, as

$$F_{ij}(t) = \frac{Q_{ij}(t)}{P_{ij}}. \tag{1}$$

When i is entered, the next state is chosen according to the transition probabilities P_{ij}; then, given that the state chosen is j, the time until transition has a distribution function $F_{ij}(\cdot)$. These quantities P_{ij} and $F_{ij}(t)$ are estimable from data through the transition frequencies and observed waiting times between the various transitions.

If the Markov renewal process is observed through time T, during which period n transitions take place, the data will have the form

$$(X_0, T_1, X_1, \ldots, T_n, X_n, W) = (x_0, t_1, x_1, \ldots, t_n, x_n, w),$$

where X_0 is the state initially occupied, X_i is the state occupied after the ith transition, T_i is the waiting time between the $(i-1)$st transition and ith transition and $W = T - \sum_{i=1} T_i$. Assuming continuous waiting time distributions and noting the stationarity of the process, the likelihood function is proportional to

$$L = \left\{\prod f_{x_{i-1},x_i}(t_i)\right\}\left\{\prod P_{x_{i-1},x_i}\right\}\left\{1 - \sum F_{x_n,j}(w)P_{x_n,j}\right\}, \tag{2}$$

where $f_{ij}(t)$ is the probability density function corresponding to $F_{ij}(t)$. This is the likelihood function derived by Moore and Pyke (1968).

In most practical applications there will not be enough data to adequately estimate the waiting time distribution functions F, unless restrictions are placed on F. Therefore often it is not feasible to estimate F through the empirical distribution function. Instead, it will be necessary to restrict the class of distribution functions F to some parametric family and then estimate the indexing parameters θ of this family. Adopting this approach, we write as F_{ij}^{θ} the distribution function of the waiting times between transitions from state i to state j, with f_{ij}^{θ} denoting the corresponding density function. Then letting n_{ru} denote the number of observed transitions from state r to state u, the likelihood function $L(\theta, P)$ can be rewritten as

$$L = L_\theta L_P - \sum L_\theta F^\theta_{x_n,j}(w) L_P P_{x_n,j}, \tag{3}$$

where

$$L_\theta = \prod_{r,u} \prod_k f^\theta_{ru}(t_{ru(k)}),$$

with $t_{ru(k)}$ being the observed waiting time from state r to state u on the kth occasion that this transition occurs, and

$$L_P = \prod P_{ru}^{n_{ru}}.$$

Thus a sufficient statistic for this model is a vector $\mathbf{Z} = (W, \mathbf{Z}^{(1)}, \mathbf{Z}^{(2)})$, where $\mathbf{Z}^{(1)}$ denotes the vector of observed waiting times, $t_{ru}(k)$, $1 \leq k \leq n_{ru}$, between each of the pairs of states for which transitions are observed, and $\mathbf{Z}^{(2)}$ denotes the s^2-vector of observed transition counts n_{ru}.

3. The Form of Bayes Estimators

In certain applications, Bayes estimation is desirable because of its flexibility in incorporating varying amounts of prior information and its success in handling the problem of limited amounts of relevant sampling data. Building on the work of Martin (1967), Brock (1971) examined a Bayesian procedure for inference about transition probabilities. In this section, we develop Bayes estimators with respect to squared error loss for both the transition probabilities $\mathbf{P}$ and the parameters θ of the waiting time distributions.

Use of a squared error loss criterion often gives results which are reasonably approximate to what would be obtained with other smooth symmetric loss functions, implements as an analytically tractable procedure, and yields interpretable estimators. The Bayes estimators of $\mathbf{P}$ and θ using squared error loss are given by the mean vector of the joint posterior distribution of $\mathbf{P} = [P_{ij}]$ and θ. Denoting the prior joint density of $\mathbf{P}$ and θ by π, the Bayes estimator of $\mathbf{P}$ is

$$E(\mathbf{P} \mid \mathbf{Z}) = \frac{\int \mathbf{P} L\pi d\mathbf{P} d\theta}{\int L\pi d\mathbf{P} d\theta}, \tag{4a}$$

and the Bayes estimator of θ is

$$E(\theta \mid \mathbf{Z}) = \frac{\int \theta L\pi d\mathbf{P} d\theta}{\int L\pi d\mathbf{P} d\theta}. \tag{4b}$$

Bayes estimators could be developed directly from Equations (4a) and (4b), but the form simplifies substantially with independence of $\mathbf{P}$ and θ in the prior distribution, since then

$$\pi = \pi(\mathbf{P}) \times \pi(\theta).$$

Such an independence assumption may not be unduly restrictive, since, in many applications, analysts tend to think of the transition probabilities and waiting time distributions separately. Indeed, it may be that one of the major strengths of the Markov renewal model is the way it isolates the role of waiting time distributions and transition probabilities in the mind of the analyst. For the rest of this paper we assume prior independence of $\mathbf{P}$ and θ.

With prior independence of $\mathbf{P}$ and θ, the calculation of the Bayes estimators of $\mathbf{P}$ and θ simplifies to separate evaluations of integrals with respect to θ and integrals with respect to $\mathbf{P}$. We now develop the form of the Bayes estimators in this prior independence case. First, we derive the Bayes estimator $\hat{\mathbf{P}}$ of $\mathbf{P}$, where

$$\hat{\mathbf{P}} = E(\mathbf{P} \mid \mathbf{Z}) = \frac{\int \mathbf{P} L\pi d\mathbf{P} d\theta}{\int L\pi d\mathbf{P} d\theta}.$$

Using Equation (3), the denominator can be rewritten as follows:

$$\begin{aligned}\int L\pi d\mathbf{P} d\theta &= \int L_\theta L_{\mathbf{P}} \pi(\mathbf{P})\pi(\theta) d\mathbf{P} d\theta \\ &\quad - \sum_{j=1}^{s} \int L_\theta F_j^\theta L_{\mathbf{P}} P_j \pi(\mathbf{P})\pi(\theta) d\mathbf{P} d\theta \\ &= \left[\int L_\theta \pi(\theta) d\theta\right]\left[\int L_{\mathbf{P}} \pi(\mathbf{P}) d\mathbf{P}\right] \\ &\quad - \sum_{j=1}^{s} \left[\int L_\theta F_j^\theta \pi(\theta) d\theta\right]\left[\int L_{\mathbf{P}} P_j \pi(\mathbf{P}) d\mathbf{P}\right],\end{aligned}$$

where

$$F_j^\theta = F_{x_n,j}^\theta(w) \quad \text{and} \quad P_j = P_{x_n,j}.$$

Similarly, the numerator can be rewritten as follows:

$$\int \mathbf{P} L\pi d\mathbf{P} d\theta =$$

$$\left[\int L_\theta \pi(\theta) d\theta\right]\left[\int \mathbf{P} L_{\mathbf{P}}(\mathbf{P})\pi(\mathbf{P}) d\mathbf{P}\right] - \sum_{j=1}^{s} \left[\int L_\theta F_j^\theta \pi(\theta) d\theta\right]\left[\int \mathbf{P} L_{\mathbf{P}} P_j \pi(\mathbf{P}) d\mathbf{P}\right].$$

Second, we derive the Bayes estimator of θ,

$$\hat{\theta} = E(\theta \mid \mathbf{Z}) = \frac{\int \theta L\pi d\mathbf{P} d\theta}{\int L\pi d\mathbf{P} d\theta}.$$

Using Equation (3), the numerator can be rewritten as follows:

$$\int \theta L\pi d\mathbf{P} d\theta =$$

$$\left[\int \theta L_\theta \pi(\theta) d\theta\right]\left[\int L_\mathbf{P} \pi(\mathbf{P}) d\mathbf{P}\right] - \sum_{j=1}^{s}\left[\int \theta L_\theta F_j^\theta \pi(\theta) d\theta\right]\left[\int L_\mathbf{P} P_j \pi(\mathbf{P}) d\mathbf{P}\right].$$

A further simplification occurs when the terminal time T is a transition time, i.e., when $w = 0$. In this special case, the Bayes estimators are just

$$E(\mathbf{P} \mid \mathbf{Z}) = E(\mathbf{P} \mid \mathbf{Z}^{(2)}) = \frac{\int \mathbf{P} L_P \pi(\mathbf{P}) d\mathbf{P}}{\int L_P \pi(\mathbf{P}) d\mathbf{P}} = \mathbf{P}^*, \quad \text{say},$$

and

$$E(\theta \mid \mathbf{Z}) = E(\theta \mid \mathbf{Z}^{(1)}) = \frac{\int \theta L_\theta \pi(\theta) d\theta}{\int L_\theta \pi(\theta) d\theta} = \theta^*, \quad \text{say}.$$

The case $w = 0$ is important because it occurs under any sampling plan calling for stopping at a transition time.

In the general case where $w \geq 0$, the Bayes estimators take a form adjusted away from $\mathbf{P}^*$ and θ^*. Specifically, we can write the Bayes estimator $\hat{\mathbf{P}}$ as

$$\hat{\mathbf{P}} = \frac{\mathbf{P}^* - \sum \alpha_j E(\mathbf{P} P_j \mid \mathbf{Z}^{(2)})}{1 - \sum \alpha_j \beta_j}, \tag{5}$$

where

$$\alpha_j = E(F_j^\theta \mid \mathbf{Z}^{(1)}),$$

and

$$\beta_j = E(P_j \mid \mathbf{Z}^{(2)}).$$

The Bayes estimator $\hat{\theta}$ can be written as

$$\hat{\theta} = \frac{\theta^* - \sum \beta_j E(\theta F_j^\theta \mid \mathbf{Z}^{(1)})}{1 - \sum \alpha_j \beta_j}. \tag{6}$$

The practical use of the Bayes estimators $\hat{\mathbf{P}}$ and $\hat{\theta}$ is limited by the fact that integration of functions of θ involving F^{θ} is required, and F^{θ} is generally a complicated function of θ. These difficulties motivate developing an alternative pair of estimators $P^{\#}$ and $\theta^{\#}$ that can be derived by replacing F^{θ} in Equations (5) and (6) with the constant (over θ), $F^{\theta *}$. Writing then $F_j^{\theta *}$ as just α_j^*, we have the simplified, and workable, estimators,

$$\mathbf{P}^{\#} = \frac{\mathbf{P}^* - \sum \alpha_j^* E(\mathbf{P}P_j \mid \mathbf{Z}^{(2)})}{1 - \sum \alpha_j^* \beta_j}, \tag{7}$$

and

$$\theta^{\#} = E(\theta \mid \mathbf{Z}^1) = \theta^*. \tag{8}$$

These approximate Bayes estimators $P^{\#}$ and $\theta^{\#}$ are intuitively appealing and are easily interpreted. Their adequacy in particular settings has not yet been demonstrated. If the number of transitions is large, the dependence introduced by sampling which does not stop at a transition time is minimal and both P^* and $P^{\#}$ are presumably satisfactory.

4. Approximate Bayes Estimators of the Transition Probabilities

In order to get calculable approximate Bayes estimators of the transition probabilities, we can specialize Equation (7) using the framework of Martin (1967). Martin noted that the observed transition counts in each row of the frequency count matrix $\mathbf{N}$ come from a multinomial distribution with cell probabilities given by the associated row of $\mathbf{P}$. He took what he called a *matrix beta* distribution as appropriate for a prior distribution on $\mathbf{P}$. This is just the distribution of s independent Dirichlet random vectors (see Johnson and Kotz (1972; p. 233)). The matrix beta distribution is, like the Dirichlet distribution, *closed under sampling* so that the posterior distribution is also a member of the matrix beta family. The Dirichlet distribution is not without criticism as a prior distribution (see, Good (1965) and Aitchison and Shen (1980)) but its use does lead to a tractable analysis and a beginning has been made on developing subjective elicitation procedures; see Chaloner and Duncan (1983, 1987).

The matrix beta distribution is determined by a matrix parameter $\mathbf{M}$. If $\mathbf{M}_0$ is taken to be the matrix parameter specifying the prior distribution and a matrix of frequency counts $\mathbf{N}$ is observed, then the posterior distribution of $\mathbf{P}$ is matrix beta with matrix parameter $\mathbf{M} = \mathbf{M}_0 + \mathbf{N}$. Under quadratic loss the Bayes estimator of $\mathbf{P}$ is given by the posterior expectation of $\mathbf{P}$. But the expectation of a random matrix

P having a matrix beta distribution with matrix parameter $\mathbf{M} = [m_{ij}]$ is simply,

$$P^* = EP = \left[\frac{m_{ij}}{m_{i+}}\right]$$

where $m_{i+} = \sum_j m_{ij}$; see Martin (1967; p. 17). This same approach is followed by Brock (1971) for the Markov renewal process when the waiting time distributions are assumed known.

Our goal is Bayesian estimation for the transition probabilities when the parameters of the waiting time distributions are *not* known. We now derive approximate Bayes estimates $\bar{\mathbf{P}}$ that are readily computed in practice.

For Bayesian estimation in the general Markov renewal process, we require, according to Equation (7), the conditional expectation, $E(\mathbf{P}P_{x_n,j} \mid \mathbf{Z}^{(2)})$. It can be computed as follows: We basically require, for **P** having a matrix beta distribution with matrix parameter $\mathbf{M} = [m_{ru}]$, the expectation

$$E(\mathbf{P}P_{ij}) = [EP_{ru}P_{ij}].$$

Based on the moments of the Dirichlet distribution, we have

$$\begin{aligned} E(P_{ru}P_{ij}) &= \frac{m_{ru}m_{ij}}{m_{r+}(m_{i+}+1)}, \quad i=r,\ j\neq u, \\ &= \frac{m_{ru}(m_{ij}+1)}{m_{r+}(m_{i+}+1)}, \quad i=r,\ j=u, \\ &= \frac{m_{ru}m_{ij}}{m_{r+}m_{i+}}, \quad i\neq r. \end{aligned} \tag{9}$$

We obtain after simplification,

$$\begin{aligned} \bar{P}_{ru} &= \frac{m_{ru}}{m_{r+}}, \quad r\neq x_n \\ &= a_u\frac{m_{ru}}{m_{r+}}, \quad r = x_n, \end{aligned} \tag{10}$$

where

$$a_u = \left\{1-(m_{x_n+}+1)^{-1}(\alpha_u^* + \sum \alpha_j^* m_{x_n j})\right\} \Big/ \left\{1 - m_{x_n+}^{-1}\sum \alpha_j^* m_{x_n j}\right\}.$$

Both the numerator and denominator of a_u are in the form of 1 minus a weighted average of the α_j^*. Since the α_j^* are less than or equal to 1, both numerator and denominator are positive. So we take $\bar{\mathbf{P}} = [\bar{P}_{ru}]$.

It is easy to verify directly that with the matrix beta prior distribution either the Bayes estimates $\mathbf{P}^{\#}$ or their approximation $\bar{\mathbf{P}}$ have row sums equal to one and have non-negative entries. Hence either $\mathbf{P}^{\#}$ or $\bar{\mathbf{P}}$ can be used as a transition probability matrix.

5. Conclusion

It is possible to develop relatively simple approximate Bayes estimators for the parameters of a Markov renewal model. The parameters of the waiting time distributions need not be assumed known.

References

Aitchison, J. and S.M. Shen (1980). "Logistic-Normal Distributions: Some Properties and Uses," *Biometrika*, Vol. 67, No. 62, 261-272.

Bartholomew, D.J. (1973). *Stochastic Models for Social Processes.* New York: Wiley.

Berlin, G.N. (1980). "Managing the Variability of Fire Behavior," unpublished report to U.S. Department of Housing and Urban Development.

Brock, D.B. (1971). "Bayesian Analysis for Markov Renewal Processes," *Metron*, Vol. 39, 185-198.

Cane, V.R. (1959). "Behavior Sequences as Semi-Markov Chains," *Journal of Royal Statistical Society, Series B*, Vol. 21, 36-58.

Chaloner, K.M. and G.T. Duncan (1983). "Assessment of a Beta Prior Distribution: PM Elicitation," *The Statistician, Journal of the Institute of Statisticians*, Vol. 32, 174-180.

Chaloner, K.M. and G.T. Duncan (1987). "Some Properties of the Dirichlet-Multinomial Distribution and Its Use in Prior Elicitation," *Communications in Statistics: Theory and Methods*, Vol. 16, 511-523.

DeGroot, M.H. (1970). *Optimal Statistical Decisions.* New York: McGraw-Hill.

Duncan, G.T. and B.L. Job (1978). "Crisis Forecasting Using the Markov Renewal Model," *International Interactions*, Vol. 4, 225-246.

Duncan, G.T. and B.L. Job (1981). " Expert-Based International Crisis Forecasting," in *Expert Generated Data: Applications in International Affairs*, G. W. Hopple, ed., Boulder: Westview Press, 233-268.

Fienberg, S.E. (1978). "Victimization and the National Crime Survey: Problems of Design and Analysis," in *Survey Sampling and Measurement*, N. K. Namboodiri, ed., New York: Academic Press.

Ginsberg, R.B. (1971). "Semi-Markov Processes and Mobility," *Journal of Mathematical Sociology*, July, Vol. 1, 223-262.

Ginsberg, R.B. (1972). "Critique of Probabilistic Models: Application of the Semi-Markov Model to Migration," *Journal of Mathematical Sociology*, Vol. 1, No. 2, 63-82.

Good, I.J. (1965). *The Estimation of Probabilities: An Essay on Modern Bayesian Methods*. Cambridge, MA: MIT Press.

Hershey, J.C., E.N. Weiss and M.A. Cohen (1981). "A Stochastic Service Network Model With Application to Hospital Facilities," *Operations Research*, Vol. 29, 1-22.

Johnson, N.L. and S. Kotz (1972). *Distributions in Statistics: Continuous Multivariate Distributions*. New York: Wiley.

Martin, J.J. (1967). *Bayesian Decision Problems and Markov Chains*. New York: Wiley.

Moore, E.H. and R. Pyke (1968). "Estimation of the Transition Distributions of a Markov Renewal Process," *Annals of the Institute of Statistical Mathematics*, Vol. 20, 411-424.

Pyke, R. (1961). "Markov Renewal Processes: Definitions and Preliminary Properties," *Annals of Mathematical Statistics*, Vol. 32, 1231-1242.

Shonick, W.J. (1973). "An Improved Stochastic Model for Occupancy Related Random Variables in General-Acute Hospitals," *Operations Research*, Vol. 21, 952-965.

Weiss, E., M.A. Cohen and J.C. Hershey (1982). "An Iterative Estimation and Validation Procedure for Specification of Semi-Markov Models with Application to Hospital Patient Flow," *Operations Research*, December, Vol. 30, No. 6, 1082-1104.

Weiss, G.H. and M. Zelen (1965). "A Semi-Markov Model for Clinical Trials," *Journal of Applied Probability*, Vol. 2, 269-285.

Bayesian and Likelihood Methods in Statistics and Econometrics
S. Geisser, J.S. Hodges, S.J. Press and A. Zellner (Editors)

A HIERARCHICAL BAYES ANALYSIS OF CROSS-CLASSIFICATION MODELS

Malay Ghosh, Leone Y. Low and Glen Meeden
University of Florida, Wright State University and Iowa State University

Abstract

The paper discusses a hierarchical Bayes approach to certain cross-classification situations. The Bayesian analysis provides a unified treatment of the so-called fixed, random or mixed model examples. Interestingly, the Bayes approach can also handle certain nested model situations as special cases.

1. Introduction

This article provides a new Bayesian approach to a classical mixed model situation. As an example, consider the following problem in quality assurance. It is well-known that the success of any quality assurance program depends in part on the measuring process generating the data which it uses. For this reason, an accurate assessment of the amount of variation present in the measuring process is crucial. For definiteness, consider a factory which produces a given material. During the manufacturing process, a gauge is used to measure some quality of the material. Suppose there are t possible gauges that can be used by any of the p possible workers where t is considerably smaller than p. For a given batch of material, let M_{ij} denote the true average value assessed by worker i using gauge j. To assess the variation in the measuring process, the following experiment is conducted. Out of the p possible workers, $q(< p)$ are chosen, and each of these workers makes n measurements with each of the t gauges on the given batch of material. Some characteristics of interest of the measuring system are (i) repeatability (the error variance) or (ii) reproducibility (the variation among the workers).

If observations on all the p workers were taken, a classical approach to this problem would be to use the representation $M_{ij} = \mu + \alpha_i + \beta_j + \gamma_{ij}$ $(i = 1, \ldots, p; j = 1, \ldots, t)$, where μ is the general effect, α_i is the effect of the ith worker, β_j is the effect of the jth gauge, and γ_{ij} is the interaction effect due to the ith worker and the jth gauge. This is the so-called fixed effects model where one tests $H_0 : \alpha_1 = \cdots = \alpha_p$ to assess the variation among workers. Instead, if one collects observations on a subset q of workers, one assumes the α_i's as well as the γ_{ij}'s to be random, but β_j's are assumed to be fixed. In this case α_i's are assumed to have zero means, and common variance σ_α^2, where one tests $H_0 : \sigma_\alpha^2 = 0$ to assess the variation among workers.

The present article considers instead a Bayesian approach to assess the variation among workers. Based on observations on q workers, we find the posterior distribution of $\sum_{i=1}^{p}(\overline{M}_{i.} - \overline{M})^2$, where $\overline{M}_{i.} = t^{-1}\sum_{j=1}^{t} M_{ij}$ and $\overline{M} = p^{-1}\sum_{i=1}^{p}\overline{M}_{i.}$. Thus, from a Bayesian viewpoint, there is no difference in the analyses between the fixed and the mixed model situations, the former being a special case of the latter with $q = p$.

Next, we present a mathematical formulation of the problem considered in this article. Observations are collected for a subset of workers of size q, the selected workers being labelled $1, 2, \ldots, q$. Let X_{ijk} denote the measurement obtained by the ith worker operating on the jth machine for the kth day ($i = 1, \ldots, q; j = 1, \ldots, t; k = 1, \ldots, n$). In the familiar two-way mixed model, one assumes that $X_{ijk} = \mu + \alpha_i + \beta_j + \gamma_{ij} + e_{ijk}$, where the general effect μ is fixed, while the α_i's, β_j 's, γ_{ij}'s and e_{ijk}'s are mutually independent with α_i's iid $N(0, \sigma_\alpha^2)$, β_j's iid $N(0, \sigma_\beta^2)$, γ_{ij}'s iid $N(0, \sigma_\gamma^2)$, and e_{ijk}'s iid $N(0, \sigma^2)$. Recall $M_{ij} = \mu + \alpha_i + \beta_j + \gamma_{ij}$. We shall write $r^{-1} = \sigma^2, \eta_1 = \sigma_\gamma^2/\sigma^2, \eta_2 = \sigma_\alpha^2/\sigma^2$ and $\eta_3 = \sigma_\beta^2/\sigma^2$. Note that minimal sufficiency reduction allows one to consider only $(Y_{11}, \ldots, Y_{1t}, \ldots, Y_{q1}, \ldots, Y_{qt}, S)$ where $Y_{ij} = n^{-1}\sum_{k=1}^{n} X_{ijk}$ and $S = \sum_{i=1}^{q}\sum_{j=1}^{t}\sum_{k=1}^{n}(X_{ijk} - Y_{ij})^2$. We shall denote by $\underset{\sim}{Y}$ the column vector of all the Y_{ij}'s. The general Bayesian formulation of the problem is as follows:

A) Conditional on $M_{ij} = m_{ij}(1 \le i \le q, 1 \le j \le t)$ and $R = r$, Y_{ij}'s and S are mutually independent with $Y_{ij} \sim N\left(m_{ij}, (nr)^{-1}\right)$ and $S \sim r^{-1}\chi^2_{qt(n1)}$.

B) Conditional on $U = \mu$, $R = r$, $G_i = \eta_i$ $(i = 1, 2, 3)$, $\underset{\sim}{M} = (M_{11}, \ldots, M_{1t}, \ldots, M_{p1}, \ldots, M_{pt})^T \sim N\left(\mu \underset{\sim}{1}_{pt}, r^{-1}(\eta_1 \underset{\sim}{I}_p \otimes \underset{\sim}{I}_t + \eta_2 \underset{\sim}{I}_p \otimes \underset{\sim}{J}_t + \eta_3 \underset{\sim}{J}_p \otimes \underset{\sim}{I}_t)\right)$.

C) U, R, G_1, G_2 and G_3 have a joint pdf (proper or improper) $h(\mu, r, \eta_1, \eta_2, \eta_3)$.

Throughout this paper, we use the notations $\underset{\sim}{1}_m$ for a m-component column vector with each element equal to 1, $\underset{\sim}{I}_m$ for the identity matrix of order m, $\underset{\sim}{J}_{m,n}$ for a $m \times n$ matrix with each element equal to 1, $\underset{\sim}{J}_m = \underset{\sim}{J}_{m,m}$, and $\otimes$ for a Kronecker product.

First, in Section 2 of this article, we consider Lindley's (1972) prior according to which G_1, G_2 and G_3 are degenerate at η_1, η_2 and η_3 (which amounts to known ratios of the variance components), while U and R are marginally independent with $U \sim$ uniform $(-\infty, \infty)$ and $R \sim$ Gamma$(\frac{1}{2}a, \frac{1}{2}b)$. In this section we first find the posterior distribution of $\underset{\sim}{M}$ given $\underset{\sim}{Y}$ and S. In the case when $q = p$, our result agrees with that of Lindley (1972). Our method of proof, however, makes numerous uses of the inverse of a matrix of the form $b_1 \underset{\sim}{I}_m \otimes \underset{\sim}{I}_n + b_2 \underset{\sim}{I}_m \otimes \underset{\sim}{J}_n + b_3 \underset{\sim}{J}_m \otimes \underset{\sim}{I}_n + b_4 \underset{\sim}{J}_m \otimes \underset{\sim}{J}_n$ which is not directly used in Lindley. Next in this section, we find posterior distributions of $\sum_{i=1}^{p}(\overline{M}_{i.} - \overline{M})^2$, $\overline{M}_{i.} = t^{-1}\sum_{j=1}^{t} M_{ij}, \overline{M} = p^{-1}\sum_{i=1}^{p}\overline{M}_{i.}$, the variation among the workers, and of $\sum_{j=1}^{t}(\overline{M}_{.j} - \overline{M})^2(\overline{M}_{.j} = p^{-1}\sum_{i=1}^{p} M_{ij}; j = 1, \ldots, t)$, the variation among the gauges. We also find the posterior means and variances of these parameters and interpret the results.

Next, in Section 3, we consider the more general situation when the variance ratios are not known, and assign general noninformative priors to U, R, G_1, G_2 and G_3. We use Jeffreys' (1961) noninformative prior as well as the noninformative limiting gamma prior given in Box and Tiao (1973). The posterior distributions, means and variances of $\sum_{i=1}^{p}(\overline{M}_{i.} - \overline{M})^2$ and $\sum_{j=1}^{t}(\overline{M}_{.j} - \overline{M})^2$ are obtained. A numerical example is considered in this section, and it is shown that the results are not very sensitive to the choice of the noninformative prior. The proofs of all the technical results are deferred to the Appendix.

There are certain important consequences of our findings in Sections 2 and 3. Instead of the above cross-classificatory model, consider the nested model where one measures the performance of q out of p operators, each working on several shifts, and for each operator working on a particular shift, there are n replications. The usual nested ANOVA model in this situation is $X_{ijk} = \mu + \alpha_i + \gamma_{ij} + e_{ijk}$ $(i = 1, \ldots, q; j = 1, \ldots, t; k = 1, \ldots, n)$. In the above, X_{ijk} denotes some measurement corresponding to the kth replication of the ith worker working on the jth shift. In our framework, conditional on $M_{ij} = m_{ij}$'s and $R = r$, it is assumed that X_{ijk}'s are independent and X_{ijk}'s $\sim N(m_{ij}, r^{-1})$. The assumed prior for $\underset{\sim}{M}$ is the same as before, except that one puts $\eta_3 = 0$ in the variance-covariance matrix introduced in (B). Hence, the posterior distribution of $\sum_{i=1}^{p}(\overline{M}_{i.} - \overline{M})^2$ is obtained from the earlier result by putting $\eta_3 = 0$.

Our methods are also applicable to certain subsampling problems where sampling is done in two stages. In a typical subsampling situation, a number of primary units is selected at the first stage, and then from each selected primary unit, a number of subunits is selected at the second stage. Suppose, the number of primary units is not very large, so that they can all be used, but we take a random sample of subunits from the selected primary units. For example, suppose a company which manufactures a certain instrument has factories in two different locations, and we want to measure the variation in the output of workers in these two locations. We select a number of workers from each location, and record their output for the same number n of days, and study the total variation among the workers. In this case, let X_{ijk} denote the output of the ith worker in the jth location for the kth day.

Once again, assume that conditional on $M_{ij} = m_{ij}$'s and $R = r$, X_{ijk}'s are independent, and $X_{ijk} \sim N(m_{ij}, r^{-1})$. The prior distribution for $\underset{\sim}{M}$ is similar as before, but this time one puts $\eta_2 = 0$ in the variance-covariance matrix given in (B). The total variation among the workers can be obtained from the posterior distribution of $\sum_{i=1}^{p}\sum_{j=1}^{t}(\overline{M}_{ij} - \overline{M})^2$.

A Bayesian analysis in one way random effects models by putting prior distributions on the variance components was initiated by Hill (1965) see also Hill (1977). Box and Tiao (1973) provide a very comprehensive account of the use of noninformative priors in many balanced designs.

2. Hierarchical Bayes Analysis with Known Ratios of Variance Components

Consider the model given in (A)-(C) of Section 1, where η_1, η_2 and η_3 are known, while U and R are marginally independently distributed with $U \sim$ uniform $(-\infty, \infty)$ and $R \sim$ Gamma$(\frac{1}{2}a, \frac{1}{2}b)$. This prior is similar to the one given in Lindley (1972). Under this model, we shall first find the posterior distribution of $\underset{\sim}{M}$ given $\underset{\sim}{Y} = \underset{\sim}{y}$ and $S = s$, and subsequently the posterior distributions of $\sum_{i=1}^{p}(\overline{M}_{i.} - \overline{M})^2$ and $\sum_{j=1}^{t}(\overline{M}_{.j} - \overline{M})^2$ given $\underset{\sim}{Y} = \underset{\sim}{y}$ and $S = s$.

Before stating the results, we need to develop a few notations. Use the standards ANOVA notations

$$SSE = s;$$

$$\overline{y}_{i.} = t^{-1}\sum_{j=1}^{t} y_{ij},\ y_{.j} = q^{-1}\sum_{i=1}^{q} y_{ij},\ \overline{y} = (tq)^{-1}\sum_{i=1}^{q}\sum_{j=1}^{t} y_{ij}; \tag{2.1}$$

$$SSB = nt\sum_{i=1}^{q}(\overline{y}_{i.} - \overline{y})^2; \tag{2.2}$$

$$SST = nq\sum_{j=1}^{t}(\overline{y}_{.j} - \overline{y})^2; \tag{2.3}$$

$$SSI = n\sum_{i=1}^{q}\sum_{j=1}^{t}(y_{ij} - \overline{y}_{i.} - \overline{y}_{.j} + \overline{y})^2. \tag{2.4}$$

Also, we shall use the notations SSB, SST, SSI and SSE for the corresponding random variables, whenever no confusion occurs. Next define

$$\nu_1 = 1 + n\eta_1, \ \nu_2 = 1 + n\eta_1 + nt\eta_2 \text{ and } \nu_3 = 1 + n\eta_1 + nq\eta_3; \tag{2.5}$$

$$\underset{\sim}{A}_1 = \nu_1^{-1}\left[\eta_1 \underset{\sim}{I}_q \otimes \underset{\sim}{I}_t + \nu_2^{-1}\eta_2 \underset{\sim}{I}_q \otimes \underset{\sim}{J}_t + \nu_3^{-1}\eta_3 \underset{\sim}{J}_q \otimes \underset{\sim}{I}_t \right.$$

$$\left. + \left\{-t^{-1}\nu_3^{-1}\eta_3 + (nqt)^{-1}\nu_2^{-1}\nu_1\right\} \underset{\sim}{J}_q \otimes \underset{\sim}{J}_t\right]; \tag{2.6}$$

$$\underset{\sim}{A}_2 = \nu_3^{-1}\underset{\sim}{J}_{p-q,q} \otimes \left[\eta_3 \underset{\sim}{I}_t + (nqt)^{-1}\nu_1 \underset{\sim}{J}_t\right]; \tag{2.7}$$

$$\underset{\sim}{A}_3 = \eta_1 \underset{\sim}{I}_{p-q} \otimes \underset{\sim}{I}_t + \eta_2 \underset{\sim}{I}_{p-q} \otimes \underset{\sim}{J}_t + \nu_3^{-1}\eta_3\nu_1 \underset{\sim}{J}_{p-q} \otimes \underset{\sim}{I}_t$$

$$+ \left[(nqt)^{-1}\nu_2 - t^{-1}\nu_3^{-1}\eta_3\nu_1\right] \underset{\sim}{J}_{p-q} \otimes \underset{\sim}{J}_t; \tag{2.8}$$

$$\underset{\sim}{d}_1^T = \nu_1^{-1}\left[n\eta_1 \underset{\sim}{y}^T + \nu_2^{-1}\left((nt\eta_2\overline{y}_{1.} + \nu_1\overline{y})\underset{\sim}{1}_t^T \ldots (nt\eta_2\overline{y}_{q.} + \nu_1\overline{y})\underset{\sim}{1}_t^T\right)\right.$$

$$\left. + \nu_3^{-1}(nq\eta_3)(\overline{y}_{.1} - \overline{y} \ldots \overline{y}_{.t} - \overline{y} \ldots \overline{y}_{.1} - \overline{y} \ldots \overline{y}_{.t} - \overline{y})\right]; \tag{2.9}$$

$$\underset{\sim}{d}_2^T = \nu_3^{-1}\underset{\sim}{1}_{p-q}^T \otimes (nq\eta_3\overline{y}_{.1} + \nu_1\overline{y} \ldots nq\eta_3\overline{y}_{.t} + \nu_1\overline{y}). \tag{2.10}$$

The first main result of this section is as follows.

THEOREM 1: *(I) Conditional on* $\underset{\sim}{Y} = \underset{\sim}{y}$, $R = r$ *and* $S = s$,

$$\underset{\sim}{M} \sim N\left[\begin{pmatrix}\underset{\sim}{d}_1 \\ \underset{\sim}{d}_2\end{pmatrix}, \begin{pmatrix}\underset{\sim}{A}_1 & \underset{\sim}{A}_2^T \\ \underset{\sim}{A}_2 & \underset{\sim}{A}_3\end{pmatrix}\right]. \tag{2.11}$$

(II) The conditional distribution of R *given* $\underset{\sim}{Y} = \underset{\sim}{y}$ *and* $S = s$ *is*

$$\text{Gamma}\left[\frac{1}{2}(s + \nu_1^{-1}SSI + \nu_2^{-1}SSB + \nu_3^{-1}SST + a), \ \frac{1}{2}(nqt + b - 1)\right]. \tag{2.12}$$

REMARK 1: Note that the posterior mean of $\underset{\sim}{M}$ as given in Theorem 1 depends only on $\underset{\sim}{y}$ and not on r and s. In the special case when $q = p$, the expression agrees with the one given in Lindley (1972). To see this, first identify p, t, n, η_1, η_2 and η_3 of ours with m, n, r, σ_c^2/σ^2, σ_a^2/σ^2 and σ_b^2/σ^2 respectively of Lindley (1972). Now observe from (2.9) and (2.11) that the posterior mean of M_{ij} is

$$\begin{aligned}
&\nu_1^{-1}\left[n\eta_1 y_{ij} + \nu_2^{-1}(nt\eta_2\overline{y}_{i.} + \nu_1\overline{y}) + \nu_3^{-1}(nq\eta_3)(\overline{y}_{.j} - \overline{y})\right] \\
&= (1-\nu_1^{-1})(y_{ij} - \overline{y}_{i.} - \overline{y}_{.j} + \overline{y}) + (1-\nu_2^{-1})(\overline{y}_{i.} - \overline{y}) \\
&\quad + (1-\nu_3^{-1})(\overline{y}_{.j} - \overline{y}) + \overline{y}
\end{aligned} \tag{2.12}$$

which agrees with the equation (7) of Lindley (1972).

Recall that $\overline{M}_{i.} = p^{-1}\sum_{j=1}^{t} M_{ij}$. Hence, writing $\underset{\sim}{M}_{.}^{T} = (\overline{M}_{10}\ldots\overline{M}_{p0})$, it follows that

$$\underset{\sim}{M}_{.} = (\underset{\sim}{I}_p \otimes t^{-1}\underset{\sim}{1}_t^T)\,\underset{\sim}{M} = \begin{bmatrix} \underset{\sim}{I}_q \otimes t^{-1}\underset{\sim}{1}_t^T & \underset{\sim}{0} \\ \underset{\sim}{0} \;\ldots & \underset{\sim}{I}_{p-q} \otimes t^{-1}\underset{\sim}{1}_t^T \end{bmatrix} \underset{\sim}{M}.$$

Then using the multivariate normal theory, Part (I) of Theorem 1, and much algebraic manipulations, one gets the following theorem whose proof is omitted. The details may be found in Ghosh, Low and Meeden (1987).

THEOREM 2: *Conditional on* $\underset{\sim}{Y} = \underset{\sim}{y}$, $R = r$ *and* $S = s$,

$$\underset{\sim}{M}_{.} \sim N\left[\left(\begin{matrix} (n+\lambda)^{-1}\begin{pmatrix} n\overline{y}_{1.} + \lambda\overline{y} \\ \vdots \\ n\overline{y}_{q.} + \lambda\overline{y} \end{pmatrix} \\ \overline{y}\underset{\sim}{1}_{p-q} \end{matrix}\right),\right.$$

$$\left.(rt)^{-1}\begin{bmatrix} (n+\lambda)^{-1}(\underset{\sim}{I}_q + \lambda n^{-1}q^{-1}\underset{\sim}{J}_q)(nq)^{-1} & \underset{\sim}{J}_{q,p-q} \\ (nq)^{-1}\underset{\sim}{J}_{p-q,q}\lambda^{-1} & (\underset{\sim}{I}_{p-q} + (nq)^{-1}(n+\lambda)\underset{\sim}{J}_{p-q}) \end{bmatrix}\right],$$

where $\lambda = \{(\nu_2 - 1)/n\}^{-1} = (\eta_1 + \eta_2 t)^{-1}$.

It is important to identify the above posterior distribution of $\underset{\sim}{M}_{.}$ with the posterior distribution of $\underset{\sim}{M}$ as given in Theorem 1 of Ghosh and Meeden (1984). The present $\overline{y}_{i.}$'s should be identified with y_i's of the latter. The only point of difference to note is that the present posterior variance matrix involves an extra factor t^{-1} as compared to the other paper. Noting that $\sum_{i=1}^{p}(M_{i.} - \overline{M})^2 = \underset{\sim}{M}_{.}^{T}(\underset{\sim}{I}_p - p^{-1}\underset{\sim}{J}_p)\underset{\sim}{M}_{.}$, a direct application of Theorem 2 of Ghosh and Meeden (1984) now leads to the following theorem.

THEOREM 3: *Conditional on* $\underset{\sim}{Y} = \underset{\sim}{y}$, $R = r$ and $S = s$,

$$\sum_{i=1}^{p}(\overline{M}_{i.} - \overline{M})^2 \sim (rt)^{-1}[(n+\lambda)^{-1}W_1 + \lambda^{-1}W_2],$$

where W_1 *and* W_2 *are independently distributed with* $W \sim \chi^2_{q-1}(nr(n+\lambda)^{-1}SSB)$ *and* $W_2 \sim \chi^2_{p-q}$. *Note that the above distribution does not depend on* $S = s$.

Using the above theorem, it follows that writing $\gamma(\underset{\sim}{M}) = (p-1)^{-1}\sum_{i=1}^{p}(\overline{M}_{i.} - \overline{M})^2$,

$$E[\gamma(\underset{\sim}{M}) \mid \underset{\sim}{y}, r, s] = (p-1)^{-1}[(rt)^{-1}\{(q-1)(n+\lambda)^{-1} + (p-q)\lambda^{-1}\} + n(n+\lambda)^{-2}t^{-1}SSB]. \tag{2.13}$$

Using Part (II) of Theorem 1,

$$E(R^{-1} \mid \underset{\sim}{y}, s) = (nqt + b - 3)^{-1}(s + \nu_1^{-1}SSI + \nu_2^{-1}SSB + \nu_3^{-1}SST + a). \tag{2.14}$$

Now from (2.13) and (2.14) it follows that the posterior mean of $\gamma(\underset{\sim}{M})$ is

$$\begin{aligned} e_1 &= E[\gamma(\underset{\sim}{M}) \mid \underset{\sim}{y}, s] \\ &= (p-1)^{-1}(\lambda t)^{-1}\left[\{(q-1)\lambda(n+\lambda)^{-1} + p - q\}(nqt + b - 3)^{-1}\right. \\ &\quad \times (s + \nu_1^{-1}SSI + \nu_2^{-1}SSB + \nu_3^{-1}SST + a) \\ &\quad \left. + n\lambda(n+\lambda)^{-2}SSB\right]. \end{aligned} \tag{2.15}$$

Note that $\lambda(n+\lambda)^{-1} = \nu_2^{-1}$. Write $w = (p-1)^{-1}[(q-1)\nu_2^{-1} + p - q]$, $z_1 = \nu_1^{-1}SSI/((q-1)(t-1))$, $z_2 = \nu_2^{-1}SSB/(q-1)$, $z_3 = \nu_3^{-1}SST/(t-1)$, $z_4 = s/(qt(n-1))$, $z_5 = a/(b-2)$(a constant), $c_1 = (nqt+b-3)^{-1}(q-1)(t-1)$, $c_2 = (nqt + b - 3)^{-1}(q-1)$, $c_3 = (nqt+b-3)^{-1}(t-1)$, $c_4 = (nqt+b-3)^{-1}qt(n-1)$ and $c_5 = (nqt+b-3)^{-1}(b-2)$. Then from (2.15), it follows that

$$\begin{aligned} e_1 &= (\lambda t)^{-1}w\sum_{i=1}^{5}c_i z_i + (1-w)z_2 \\ &= (\lambda t)^{-1}[(wc_2 + 1 - w)z_2 + w(c_1z_1 + c_3z_3 + c_4z_4 + c_5z_5). \end{aligned} \tag{2.16}$$

REMARK 2: Note that $\sum_{i=1}^{5} c_i = 1$. Hence, e_1 is $(\lambda t)^{-1}$ times a weighted average of $z_1, \ldots, z_5$. Writing Z_i's as the *rv*'s corresponding to the z_i's ($i = 1, \ldots, 4$) ($z_5 = a/(b-2)$ is a constant), it follows that conditional on $R = r$, $E(Z_i) = r^{-1}$ $(1 \le i \le 4)$.

Further, $E(R^{-1}) = a/(b-2)$ if $b > 2$. Thus, the posterior mean of $\gamma(\underset{\sim}{M})$ is a weighted average of several quantities each estimating $(\lambda t)^{-1}E(R^{-1}) = (\eta_1 t^{-1} + \eta_2)E(R^{-1})$. For fixed r, $\eta_2 r^{-1}$ can be interpreted as the population variation among workers. Hence, when $\eta_1 = 0$, that is, there is no interaction, the posterior mean does indeed meet its intended objective of measuring the variation among workers.

REMARK 3: Note that the posterior mean of $R^{-1}(\eta_1 + \eta_2 t^{-1})$ is

$$(\lambda t)^{-1}[s + \nu_1^{-1}SSI + \nu_2^{-1}SSB + \nu_3^{-1}SST + a]\Big/(nqt + b - 3)$$

$$= (\lambda t)^{-1}\sum_{i=1}^{5} c_i z_i. \tag{2.17}$$

Although this quantity is also a weighted average of the same four sums of squares as well as the prior mean of R^{-1}, the weights in this case are different from the ones in the earlier case, and unless p is very large compared to q, that is, w is very close to 1, one can indeed get different answers for measuring variation among the workers with a finite p. Indeed, the posterior mean given in (2.16) assigns a greater weight to SSB than the estimate given in (2.17). Note that for fixed r, SSB is the very natural quantity for estimating $\eta_2 r^{-1}$ when $\eta_1 = 0$. Accordingly, for small or moderate p, the estimate given in (2.16) seems more appropriate than the one given in (2.17).

Next we find the expression for $V[\gamma(\underset{\sim}{M}) \mid \underset{\sim}{Y} = \underset{\sim}{y},\ S = s]$. Using the standard formula for iterated conditional expectations,

$$\begin{aligned}
&V[\gamma(\underset{\sim}{M}) \mid \underset{\sim}{Y} = \underset{\sim}{y},\ S = s] \\
&\quad = V[E(\gamma(\underset{\sim}{M})|\underset{\sim}{Y} = \underset{\sim}{y},\ R = r,\ S = s)|\underset{\sim}{Y} = \underset{\sim}{y},\ S = s] \\
&\qquad + E[V(\gamma(\underset{\sim}{M})|\underset{\sim}{Y} = \underset{\sim}{y},\ R = r,\ S = s)|\underset{\sim}{Y} = \underset{\sim}{y},\ S = s] \\
&\quad = (p-1)^{-2}t^{-2}\{(q-1)(n+\lambda)^{-1} + (p-q)\lambda^{-1}\}^2 \\
&\qquad \times 2(s + \nu_1^{-1}SSI + \nu_2^{-1}SSB + \nu_3^{-1}SST + a)^2/((nqt + b - 3)^2(nqt + b - 5)) \\
&\qquad + 2(p-1)^{-2}t^{-2}\left[\{(n+\lambda)^{-2}(q-1) + \lambda^{-2}(p-q)\}\right. \\
&\qquad \times (s + \nu_1^{-1}SSI + \nu_2^{-1}SSB + \nu_3^{-1}SST + a)^2/((nqt + b - 3)(nqt + b - 5)) \\
&\qquad + 2n(n+\lambda)^{-3}SSB(s + \nu_1^{-1}SSI + \nu_2^{-1}SSB + \nu_3^{-1}SST + a)/ \\
&\qquad\quad \left.(nqt + b - 3)\right].
\end{aligned} \tag{2.18}$$

Next we find the posterior distribution of $\sum_{j=1}^{t}(\overline{M}_{.j} - \overline{M})^2$, the variation among the gauges in our first example.

With this end, we find first the posterior distribution of $\overline{\underset{\sim}{M}}_{.(*)} = (\overline{M}_{.1}, \ldots, \overline{M}_{.t})^T$ given $\underset{\sim}{Y} = \underset{\sim}{y}$ and $S = s$. Note that

$$\overline{\underset{\sim}{M}}_{.(*)} = p_1^{-1}(\underset{\sim}{1}_p^T \otimes \underset{\sim}{I}_t)\underset{\sim}{M}. \tag{2.19}$$

Once again, using the multivariate normal theory, and much algebraic manipulations, one finds the posterior distribution of $\overline{\underset{\sim}{M}}_{.(*)}$ given $\underset{\sim}{Y} = \underset{\sim}{y}$ and $S = s$. The result is given in the next theorem whose proof is omitted. The details may be found in Ghosh, Low and Meeden (1987).

THEOREM 4: *Conditional on* $\underset{\sim}{Y} = \underset{\sim}{y}$, $R = r$ *and* $S = s$,

$$\overline{\underset{\sim}{M}}_{.(*)} \sim N\left[k\begin{pmatrix}\overline{y}_{.1}\\ \vdots \\ y_{.t}\end{pmatrix} + (1-k)\overline{y}_{..}\underset{\sim}{1}_t,\ r^{-1}(l\underset{\sim}{I}_t + gt^{-1}\underset{\sim}{J}_t)\right],$$

where

$$k = p^{-1}\nu_3^{-1}\big(p\nu_3 - (p-q)\nu_1 - q\big); \tag{2.20}$$

$$l = p^{-2}\nu_3^{-1}(nq)^{-1}\big(p\nu_3 - (p-q)\nu_1 - q\big)\big((p-q)\nu_1 + q\big); \tag{2.21}$$

$$g = q^{-1}p^{-1}(p-q)\eta_2 + p^{-2}n^{-1}\nu_3^{-1}\big((p-q)\nu_1 + q\big)^2. \tag{2.22}$$

From Theorem 4, it is possible to find the posterior distribution of $\sum_{j=1}^{t}(\overline{M}_{.j} - \overline{M})^2$. The following theorem is proved.

THEOREM 5: *Conditional on* $\underset{\sim}{Y} = \underset{\sim}{y}$, $R = r$ *and* $S = s$,

$$\sum_{j=1}^{t}(\overline{M}_{.j} - \overline{M})^2 \sim r^{-1}l\chi^2_{t-1}(d), \tag{2.23}$$

where

$$d = \nu_3^{-1}\big((p-q)\nu_1 + q\big)^{-1}r\big(p\nu_3 - (p-q)\nu_1 - q\big)SST. \tag{2.24}$$

Using Theorem 5 and (2.14), for $nqt \geq 3$, one gets the posterior mean of $(t-1)^{-1}\sum_{j=1}^{t}(\overline{M}_{.j} - \overline{M})^2$ given by

$$\begin{aligned} e_2 &= E\left[(t-1)^{-1}\sum_{j=1}^{t}(\overline{M}_{.j} - \overline{M})^2 \mid \underset{\sim}{Y} = \underset{\sim}{y},\ S = s\right] \\ &= E\left[R^{-1}l(1 + (t-1)^{-1}d) \mid \underset{\sim}{Y} = \underset{\sim}{y},\ S = s\right] \\ &= l(nqt + b - 3)^{-1}(s + \nu_1^{-1}SSI + \nu_2^{-1}SSB + \nu_3^{-1}SST + a) \\ &\quad + (t-1)^{-1}p^{-2}\nu_3^{-2}(nq)^{-1}\big(p\nu_3 - (p-q)\nu_1 - q\big)^2 SST. \end{aligned} \tag{2.25}$$

It is possible to rewrite the right hand side of (2.25) as

$$\left[w_0 \sum_{i=1}^{5} c_i z_i + (1 - w_0) z_3\right] (\eta_1 p^{-1} + \eta_3), \tag{2.26}$$

where the z_i's and c_i's are defined after (2.15) and $w_0 = p^{-1}\nu_3^{-1}((p-q)\nu_1 + q)$. Note again that conditional on $R = r$, $E(Z_i) = r^{-1}(1 \le i \le 4)$, and $E(R^{-1}) = a/(b-2)$ for $b > 2$. Hence e_2 is the weighted average of several quantities each estimating $(\eta_1 p^{-1} + \eta_3)E(R^{-1})$. If $\eta_1 = 0$, η_2 estimates $\eta_3 E(R^{-1})$, which in our first example is the population variability among gauges. Also, unless p is much larger than q, the estimator given in (2.25) or (2.26) differs from the posterior mean of $R^{-1}(\eta_1 p^{-1} + \eta_3)$.

Next, using the iterative formula for conditional variances, one gets

$$V\left[(t-1)^{-1} \sum_{j=1}^{t} (\overline{M}_{.j} - \overline{M})^2 \mid \underset{\sim}{Y} = \underset{\sim}{y},\ S = s\right]$$

$$= 2l^2(s + \nu_1^{-1}SSI + \nu_2^{-1}SSB + \nu_3^{-1}SST + a)^2 / \{(nqt + b - 3)^2(nqt + b - 5)\}$$

$$+ 2(t-1)^1 l^2 (s + \nu_1^{-1}SSI + \nu_2^{-1}SSB + \nu_3^{-1}SST + a)^2 / \{(nqt + b - 3)(nqt + b - 5)\}$$

$$+ 4(t-1)^{-2}\nu_3^{-3}(nq)^{-2}(p\nu_3 - (p-q)\nu_1 - q)^3((p-q)\nu_1 + q)SST$$

$$\times (s + \nu_1^{-1}SSI + \nu_2^{-1}SSB + \nu_3^{-1}SST + a)^2 / (nqt + b - 3). \tag{2.27}$$

We may note in passing that when $q = p$, writing $\lambda_0 = (\eta_1 + p\eta_3)^{-1}$, it follows from Theorem 5 that the posterior distribution of $\sum_{j=1}^{t}(\overline{M}_{.j} - \overline{M})^2$ given $\underset{\sim}{Y} = \underset{\sim}{y}$, $R = r$ and $S = s$ is $r^{-1}p^{-1}(n + \lambda_0)^{-1}\chi^2_{t-1}\left(\frac{nr}{n+\lambda_0}SST\right)$. In this case, from Theorem 3, the posterior distribution of $\sum_{i=1}^{p}(\overline{M}_{.j} - \overline{M})^2$ given $\underset{\sim}{Y} = \underset{\sim}{y}$, $R = r$ and $S = s$ is $r^{-1}t^{-1}\chi^2_{p-1}\left(\frac{nr}{n+\lambda}SSB\right)$. The symmetry in the two situations is quite anticipated, since this is the so called fixed effects situation.

3. Hierarchical Bayes Analysis with Unknown Ratios of Variance Components

In this section, we consider the more realistic situation when η_1, η_2 and η_3 are unknown. First, we find Jeffreys' noninformative prior for the unknown parameters R, G_1, G_2 and G_3, assuming as before that $U \sim$ uniform$(-\infty, \infty)$ and is distributed independently of (R, G_1, G_2, G_3). We find it convenient to use the reparametrization $\nu_1 = 1 + n\eta_1$, $\nu_2 = 1 + n\eta_1 + nt\eta_2$, $\nu_3 = 1 + n\eta_1 + nq\eta_3$ as done earlier in Section 2.

Note that conditional on r, η_1, η_2 and η_3,

$$\underset{\sim}{Y} \sim N\left(\mu \underset{\sim}{1}_{qt},\ r^{-1}[(n^{-1} + \eta_1)\underset{\sim}{I}_q \otimes \underset{\sim}{I}_t + \eta_2 \underset{\sim}{I}_q \otimes \underset{\sim}{J}_t + \eta_3 \underset{\sim}{J}_q \otimes \underset{\sim}{I}_t]\right). \qquad (3.1)$$

An application of Lemma 1 given in the Appendix now provides

$$\begin{aligned}
&[(n^{-1} + \eta_1)\underset{\sim}{I}_q \otimes \underset{\sim}{I}_t + \eta_2 \underset{\sim}{I}_q \otimes \underset{\sim}{J}_t + \eta_3 \underset{\sim}{J}_q \otimes \underset{\sim}{I}_t]^{-1} \\
&= n\nu_1^{-1}[\underset{\sim}{I}_q \otimes \underset{\sim}{I}_t - n\eta_2\nu_2^{-1}\underset{\sim}{I}_q \otimes \underset{\sim}{J}_t - n\eta_3\nu_3^{-1}\underset{\sim}{J}_q \otimes \underset{\sim}{I}_t \\
&\quad + n^2\eta_2\eta_3(\nu_2 + \nu_3 - \nu_1)^{-1}(\nu_2^{-1} + \nu_3^{-1})\underset{\sim}{J}_q \otimes \underset{\sim}{J}_t].
\end{aligned} \qquad (3.2)$$

Using (3.2), it follows after much simplications that

$$\begin{aligned}
&(\underset{\sim}{y} - \mu \underset{\sim}{1}_{qt})^T[(n^{-1} + \eta_1)\underset{\sim}{I}_q \otimes I_t + \eta_2 \underset{\sim}{I}_q \otimes \underset{\sim}{J}_t + \eta_3 \underset{\sim}{J}_q \otimes \underset{\sim}{I}_t]^{-1}(\underset{\sim}{y} - \mu \underset{\sim}{1}_{qt}) \\
&= n[\nu_1^{-1}SSI + \nu_2^{-1}SSB + \nu_3^{-1}SST + (\nu_2 + \nu_3 - \nu_1)^{-1}qt(\overline{y} - \mu)^2].
\end{aligned} \qquad (3.3)$$

Again, since

$$(n^{-1} + \eta_1)\underset{\sim}{I}_q \otimes \underset{\sim}{I}_t + \eta_2 \underset{\sim}{I}_q \otimes \underset{\sim}{J}_t + \eta_3 \underset{\sim}{J}_q \otimes \underset{\sim}{I}_t$$

$$= \begin{bmatrix}
(n^{-1} + \eta_1 + \eta_3)\underset{\sim}{I}_t + \eta_2 \underset{\sim}{J}_t & \eta_3 \underset{\sim}{I}_t & \cdots & \eta_3 \underset{\sim}{I}_t \\
\eta_3 \underset{\sim}{I}_t & (n^{-1} + \eta_1 + \eta_3)\underset{\sim}{I}_t + \eta_2 \underset{\sim}{J}_t & \cdots & \eta_3 \underset{\sim}{I}_t \\
\cdots\cdots\cdots & \cdots\cdots\cdots & & \cdots\cdots\cdots \\
\eta_3 \underset{\sim}{I}_t & \eta_3 \underset{\sim}{I}_t & \cdots & (n^{-1} + \eta_1 + \eta_3)\underset{\sim}{I}_t + \eta_2 \underset{\sim}{J}_t
\end{bmatrix} \qquad (3.4)$$

using Exercises 1.3, p. 68 and 1.1, p. 67 of Rao (1973), one gets

$$|(n^{-1}+\eta_1)\underset{\sim}{I}_q\otimes\underset{\sim}{I}_t+\eta_2\underset{\sim}{I}_q\otimes\underset{\sim}{J}_t+\eta_3\underset{\sim}{J}_q\otimes\underset{\sim}{I}_t| = n^{-qt}(\nu_2+\nu_3-\nu_1)\nu_3^{t-1}\nu_2^{q-1}\nu_1^{(q-1)(t-1)}. \quad (3.5)$$

Conditional on μ, r, ν_1, ν_2 and ν_3, the joint likelihood function of $\underset{\sim}{Y}$ and S is given by

$$f(\underset{\sim}{y},\ s \mid \mu,\ r,\ \nu_1,\ \nu_2,\ \nu_3)$$

$$\propto \nu_1^{-\frac{1}{2}(q-1)(t-1)}\nu_2^{-\frac{1}{2}(q-1)}\nu_3^{-\frac{1}{2}(t-1)}(\nu_2+\nu_3-\nu_1)^{-\frac{1}{2}}r^{\frac{1}{2}nqt}s^{\frac{1}{2}qt(n-1)-1}$$

$$\times \exp\left[\frac{1}{2}r\{s+\nu_1^{-1}SSI+\nu_2^{-1}SSB+\nu_3^{-1}SST + (\nu_2+\nu_3-\nu_1)^{-1}qt(\overline{y}-\mu)^2\}\right]. \quad (3.6)$$

Since U is marginally uniform $(-\infty,\infty)$, and is distributed independently of $(R,\ G_1,\ G_2,\ G_3)$, integrating with respect to μ over $(-\infty,\infty)$ in (3.6), one gets

$$f(\underset{\sim}{y}, s \mid r, \nu_1, \nu_2, \nu_3)$$

$$\propto \nu_1^{-\frac{1}{2}(q-1)(t-1)}\nu_2^{-\frac{1}{2}(q-1)}\nu_3^{-\frac{1}{2}(t-1)}r^{\frac{1}{2}(nqt-1)}s^{\frac{1}{2}qt(n-1)-1}$$

$$\times \exp\left[-\frac{1}{2}r(s+\nu_1^{-1}SSI+\nu_2^{-1}SSB+\nu_3^{-1}SST)\right]. \quad (3.7)$$

With (3.6) as the starting point, one finds Jeffreys' (1961) noninformative prior for R, ν_1, ν_2 and ν_3. The following theorem is proved.

THEOREM 6: *Jeffreys' noninformative prior for* (R,ν_1,ν_2,ν_3) *is given by*

$$f(r,\nu_1,\nu_2,\nu_3)$$

$$\propto r^{-1}\nu_1^{-1}\nu_2^{-1}\nu_3^{-1}\left[1+(\nu_2+\nu_3-\nu_1)^{-2}\left\{\frac{\nu_1^2}{(q-1)(t-1)}+\frac{\nu_2^2}{q-1}+\frac{\nu_3^2}{t-1}\right\}\right]^{\frac{1}{2}}. \quad (3.8)$$

The proof of this theorem involves routine but tedious calculations, and is omitted.

For numerical computations, we shall find it convenient to reparametrize further from (r,ν_1,ν_2,ν_3) to (r,u_1,u_2,u_3), where $u_1 = \nu_1^{-1}$, $u_1u_2 = \nu_2^{-1}$ and $u_1u_3 = \nu_3^{-1}$. Then, from Theorem 6, Jeffreys' noninformative prior for (R,U_1,U_2,U_3) is given by

$$f(r, u_1, u_2, u_3)$$

$$\propto r^{-1}u_1^{-1}u_2^{-1}u_3^{-1}\left[1 + (u_2 + u_3 - u_2u_3)^{-2}\left\{\frac{u_2^2u_3^2}{(q-1)(t-1)} + \frac{u_2^2}{t-1} + \frac{u_3^2}{q-1}\right\}\right]^{\frac{1}{2}}. \tag{3.9}$$

From (3.7) and (3.9) it follows that the joint posterior pdf of R, U_1, U_2, U_3 given $\underset{\sim}{Y} = \underset{\sim}{y}$ and $S = s$ is

$$f(r, u_1, u_2, u_3 \mid \underset{\sim}{y}, s)$$

$$\propto u_1^{\frac{1}{2}(qt-1)-1}u_2^{\frac{1}{2}(q-1)-1}u_3^{\frac{1}{2}(t-1)-1}r^{\frac{1}{2}(nqt-3)}$$

$$\times \exp[-\frac{1}{2}r(s + u_1SSI + u_1u_2SSB + u_1u_3SST)]$$

$$\times\left[1 + (u_2 + u_3 - u_2u_3)^{-2}\left\{\frac{u_2^2u_3^2}{(q-1)(t-1)} + \frac{u_2^2}{t-1} + \frac{u_3^2}{q-1}\right\}\right]^{\frac{1}{2}}. \tag{3.10}$$

Also, integrating wrt r in (3.10) it follows that

$$f(u_1, u_2, u_3 \mid \underset{\sim}{y}, s)$$

$$\propto u_1^{\frac{1}{2}(qt-1)-1}u_2^{\frac{1}{2}(q-1)-1}u_3^{\frac{1}{2}(t-1)-1}[s + u_1(SSI + u_2SSB + u_3SST)]^{-\frac{1}{2}(nqt-1)}$$

$$\times\left[1 + (u_2 + u_3 - u_2u_3)^{-2}\left\{\frac{u_2^2u_3^2}{(q-1)(t-1)} + \frac{u_2^2}{t-1} + \frac{u_3^2}{q-1}\right\}\right]^{\frac{1}{2}}. \tag{3.11}$$

From (2.15), (2.18), (2.25) and (2.27) (with the necessary reparametrization to u_1, u_2 and u_3) as well (3.10) and (3.11), one gets after using the iterative formulas for conditional expectations and variances,

$$E\left[\sum_{i=1}^{p}(\overline{M}_{i.} - \overline{M})^2 \mid \underset{\sim}{Y} = \underset{\sim}{y},\ S = s\right] = K_1, \tag{3.12}$$

where

$$K_1 = (nt)^{-1} \int_0^1 \int_0^1 \int_0^1 u_2^{-1}(1 - u_1u_2)$$

$$\times \Big[\frac{SSE + u_1(SSI + u_2SSB + u_3SST)}{nqt - 3}\{p - q + (q-1)u_1u_2\}$$

$$+ u_1u_2(1 - u_1u_2)SSB\Big]$$

$$\times f(u_1, u_2, u_3 \mid \underset{\sim}{y}, s)du_1du_2du_3.$$

$$V\left[\sum_{i=1}^{p}(\overline{M}_{i.} - \overline{M})^2 \mid \underset{\sim}{Y} = \underset{\sim}{y},\ S = s\right] = K_2 + K_3 + K_4 - K_1^2 \tag{3.13}$$

where

$$K_2 = 2(nt)^{-2}(nqt-3)^{-1} \int_0^1 \int_0^1 \int_0^1 u_1^{-2}u_2^{-2}[SSE + u_1(SSI + u_2SSB + u_3SST)]$$

$$\times (1 - u_1u_2)^2 \Big[\frac{q-1}{nqt-5}u_1^2u_2^2\{SSE + u_1(SSI + u_2SSB + u_3SST)\}$$

$$+ 2u_1^2u_2^2(1 - u_1u_2)SSB$$

$$+ (p-q)(nqt-5)^{-1}\{SSE + u_1(SSI + u_2SSB + u_3SST)\}\Big]$$

$$\times f(u_1, u_2, u_3 \mid \underset{\sim}{y}, s)du_1du_2du_3; \tag{3.14}$$

$$K_3 = (nt)^{-2} \int_0^1 \int_0^1 \int_0^1 (u_1u_2)^{-2}(1 - u_1u_2)^2$$

$$\times \Big[\frac{SSE + u_1(SSI + u_2SSB + u_3SST)}{nqt - 3}\{q - 1 + u_1u_2(p-q)\}$$

$$+ u_1u_2(1 - u_1u_2)SSB\Big]^2$$

$$\times f(u_1, u_2, u_3 \mid \underset{\sim}{y}, s)du_1du_2du_3; \tag{3.15}$$

$$K_4 = \frac{2}{n^2t^2(nqt-3)^2(nqt-5)} \int_0^1\int_0^1\int_0^1 u_1^{-2}u_2^{-2}$$

$$[SSE + u_1(SSI + u_2SSB + u_3SST)]^2(1-u_1u_2)^2\left(p-q+(p-q)u_1u_2\right)^2$$

$$\times f(u_1,u_2,u_3 \mid \underset{\sim}{y}, s)du_1du_2du_3. \tag{3.16}$$

$$E\left[\sum_{j=1}^{t}(\overline{M}_{.j}-\overline{M})^2 \mid \underset{\sim}{Y} = \underset{\sim}{y},\ S = s\right] = K_5, \tag{3.17}$$

where

$$K_5 = (nqp^2)^{-1}\int_0^1\int_0^1\int_0^1 u_1^{-1}$$

$$\times\ \left(p-(p-q)u_3qu_1u_3\right)\Big[\frac{t-1}{nqt-3}(qu_1+p-q)$$

$$\times\ \left(SSE + u_1(SSI + u_2SSB + u_3SST)\right)$$

$$+\, u_1SST\left(p-(p-q)u_3 - qu_1u_3\right)\Big]$$

$$\times\ f(u_1,u_2,u_3 \mid \underset{\sim}{y}, s)du_1du_2du_3.$$

$$V\left[\sum_{j=1}^{t}(\overline{M}_{.j}-\overline{M})^2\underset{\sim}{Y} = \underset{\sim}{y},\ S = s\right] = K_6 + K_7 + K_8 - K_5^2, \tag{3.18}$$

where

$$K_6 = 2(t-1)p^{-4}(nqt-3)^{-1}(nqt-5)^{-1}(nq)^{-2}\int_0^1\int_0^1\int_0^1 u_1^{-2}$$

$$\times\ [SSE + u_1(SSI + u_2SSB + u_3SST)]^2$$

$$\times\ [p-(p-q)u_3 - qu_1u_3]^2(p-q+qu_1)^2$$

$$\times\ f(u_1,u_2,u_3 \mid \underset{\sim}{y}, s)du_1du_2du_3; \tag{3.19}$$

$$K_7 = 4p^{-4}(nqt-3)^{-1}(nq)^{-2}SST \int_0^1 \int_0^1 \int_0^1 u_1^{-1}$$

$$\times [SSE + u_1(SSI + u_2 SSB + u_3 SST)]$$

$$\times [p - (p-q)u_3 - qu_1u_3]^3 (p - q + qu_1)$$

$$\times [SSE + u_1(SSI + u_2 SSB + u_3 SST)] f(u_1, u_2, u_3 \mid \underset{\sim}{y}, s) du_1 du_2 du_3; \tag{3.20}$$

$$K_8 = \frac{2(t-1)^2}{n^2 q^2 p^4 (nqt-3)^2 (nqt-5)} \int_0^1 \int_0^1 \int_0^1 u_1^{-2}$$

$$\times [SSE + u_1(SSI + u_2 SSB + u_3 SST)]^2 [p - (p-q)u_3 q u_1 u_3]^2$$

$$\times (p - q + qu_1)^2 f(u_1, u_2, u_3 \mid \underset{\sim}{y}, s) du_1 du_2 du_3. \tag{3.21}$$

These formulae look very complicated, but we have been able to carry out the necessary numerical integrations for a given set of data fairly easily.

Next note that instead of using Jeffreys' prior as given in (3.9), one can consider a limiting gamma prior as given in p. 332 of Box and Tiao (1973). With the necessary reparametrization to u_1, u_2 and u_3, this prior is given by

$$f(r, u_1, u_2, u_3) \propto r^{-1} u_1^{-1} u_2^{-1} u_3^{-1}. \tag{3.22}$$

This prior is similar to the one in (3.9) with the factor

$$h = \left[1 + (u_2 + u_3 - u_2 u_3)^{-2} \left\{ \frac{u_2^2 u_3^2}{(q-1)(t-1)} + \frac{u_2^2}{t-1} + \frac{u_3^2}{q-1} \right\}\right]^{\frac{1}{2}}$$

missing. Accordingly, the posterior distributions are similar to the ones given in (3.10) and (3.11) with the same factor h missing. The posterior means and variances of $\sum_{i=1}^{p}(\overline{M}_{i.} - \overline{M})^2$ and $\sum_{j=1}^{t}(\overline{M}_{.j} - \overline{M})^2$ are now obtained with J_i's replacing K_i's, where each J_i is similar as K_i except that the factor h will not appear in the integrands.

We apply the results of this section to a set of data given in Table 16.32, p. 472 of Hald (1952). The data involve measurements of the waterproof quality (logarithms of permeabilities in seconds; the higher the measurement the better the waterproof) of sheets of material manufactured on three different machines on nine different days. The nine days may be treated as being selected from a large (but finite) number of days the three machines are in operation. Thus $q = 9$, whereas we have tried different values of p starting with $p = 10$. Also, for each of the 9 selected days, there are 3 measurements available on each machine so that $n = 3$. In this situation, we are primarily interested in studying the variation among the three machines, that is, the posterior distribution of $\theta_1 = \sum_{j=1}^{3}(\overline{M}_{.j} - \overline{M})^2/2$, but for the sake of illustration, we have discussed also measurements of variation in the quality over the different days, that is the posterior distribution of $\theta_2 = \sum_{i=1}^{p}(\overline{M}_{i.} - \overline{M})^2/(p-1)$.

The following table summarizes our findings. The table provides posterior means and standard deviations of variation among the machines as well as variation among the days. The small numbers in the table can be accounted for by the fact that in this example $SSI = 2.0134$, $SSI = .8708$, $SSB = .5541$ and $SST = .9082$. It is clear from the table that the posterior means and the posterior s.d.'s tend to stabilize for large p. The estimates of the variations among the three machines does not change much with p, about a 3% difference when $p = 10$ and $p = 200$. But, whereas there is a 1% difference between $p = 10$ and $p = 15$, there is only a 0.2% difference between $p = 100$ and $p = 200$.

The difference in the estimates of the variability among the days is much more pronounced for different p. It is clear that the percentage difference of the estimates of the variability among the days is somewhere within the 45 to 46 per cent range between $p = 10$ and $p = 200$. Also, whereas the percentage difference in the estimates between $p = 10$ and $p = 15$ is 16.7, the percentage difference in the estimates between $p = 100$ and $p = 200$ is only 1.4. Thus, while the situation is fairly stable for large values of p, things may be different quite drastically for small p.

The other thing to note is the insensitivity of the estimates of the variability among the machines or the variability among the days for a given p. The difference in these estimates has never exceeded even 1.7%. Other types of noninformative priors can, of course, be used, but the lack of dependence of these estimates on the (noninformative) priors we have tried, appears quite encouraging.

TABLE 1: Posterior Means and S.D.'s for Different p

	Posterior Means of Variation Among the				Posterior S.D.'s of Variation Among the			
	machines		days		machines		days	
p	(under Jeffreys' prior)	(under the Box-Tiao prior)	(under Jeffreys' prior)	(under the Box-Tiao prior)	(under Jeffreys' prior)	(under the Box-Tiao prior)	(under Jeffreys' prior)	(under the Box-Tiao prior)
10	.01547	.01544	.00572	.00577	.00674	.00674	.00383	.00387
15	.01528	.01525	.00668	.00676	.00721	.00721	.00541	.00553
20	.01519	.01516	.00713	.00723	.00744	.00744	.00610	.00625
30	.01510	.01507	.00757	.00768	.00765	.00765	.00673	.00692
40	.01506	.01502	.00778	.00790	.00775	.00775	.00704	.00723
50	.01503	.01500	.00791	.00803	.00781	.00781	.00722	.00742
60	.01501	.01498	.00799	.00812	.00785	.00785	.00733	.00755
100	.01498	.01494	.00816	.00829	.00793	.00793	.00756	.00779
200	.01495	.01492	.00828	.00841	.00799	.00799	.00773	.00797

Acknowledgments

The first author's research supported in part by NSF Grant Numbers DMS-8600666 and DMS-8701814. The third author's research supported in part by NSF Grant Number MCS-8401740.

Thanks are due to the referees for their useful comments which led to the development of Section 3. Thanks are also due to Ms. Li Chu Lee for doing the numerical computations.

References

Box, G.E.P. and G.C. Tiao (1973). *Bayesian Inference in Statistical Analysis.* Reading, Massachusetts: Addison-Wesley.

Ghosh, M. and G. Meeden (1984). "A New Bayesian Analysis of a Random Effects Model," *Journal of the Royal Statistical Society, Series B*, Vol. 46, 474-482.

Ghosh, M., L.Y. Low, and G. Meeden (1987). "A Hierarchical Bayes Analysis of Cross-Classification Models," Technical Report Number 292, University of Florida.

Hald, A. (1952). *Statistical Theory with Engineering Applications.* New York: Wiley.

Hill, B.M. (1965). "Inference About Variance Components in the One-way Model," *Journal of the American Statistical Association*, Vol. 60, 806-825.

Hill, B.M. (1977). "Exact and Approximate Bayesian Solutions for Inference About Variance Components and Multivariate Inadmissibility," in *New Developments in the Applications of Bayesian Methods*, A. Aykac and C. Brumat, eds., Amsterdam: North-Holland, 129-152.

Jeffreys, H. (1961). *Theory of Probability*, 3rd Edition. London: Oxford University Press.

Lindley, D.V. (1972). "A Bayesian Solution for the Two-way Analysis of Variance," in *Progress in Statistics*, J. Gani, K. Sarkadi and L. Vincze, eds., Amsterdam: North Holland, 475-496.

Lindley, D.V. and A.F.M. Smith (1972). "Bayes Estimates for the Linear Model (with discussion)," *Journal of the Royal Statistical Society, Series B*, Vol. 34, 1-41.

Rao, C.R. (1973). *Linear Statistical Inference and its Applications.* New York: Wiley.

Searle, S.R. (1971). *Linear Models.* New York: Wiley.

Appendix: Proofs

PROOF OF THEOREM 1. Write $\underset{\sim}{V} = \eta_1 \underset{\sim}{I}_p \otimes \underset{\sim}{I}_t + \eta_2 \underset{\sim}{I}_p \otimes \underset{\sim}{J}_t + \eta_3 \underset{\sim}{J}_p \otimes \underset{\sim}{I}_t$. Also, partition $\underset{\sim}{M}^T$ as $(\underset{\sim}{M}_1^T \underset{\sim}{M}_2^T)$, where $\underset{\sim}{M}_1^T = (m_{11}, \ldots, m_{1t}, \ldots, m_{q1}, \ldots, m_{qt})$, and $\underset{\sim}{M}_2^T = (m_{q+1,1}, \ldots, m_{q+1,t}, \ldots, m_{p1}, \ldots, m_{pt})$. Then, under the conditions of Theorem 1, the joint pdf of $\underset{\sim}{Y}$, S, $\underset{\sim}{M}$, U and R is given by

$$
\begin{aligned}
&f(\underset{\sim}{y}, s, \underset{\sim}{m}, \mu, r) \\
&\quad \propto r^{\frac{1}{2}qt} \exp\left[-\frac{1}{2}r(\underset{\sim}{y} - \underset{\sim}{m}_1)^T(\underset{\sim}{y} - \underset{\sim}{m}_1)\right] r^{\frac{1}{2}qt(n-1)} s^{\frac{1}{2}qt(n-1)-1} \exp\left(-\frac{1}{2}rs\right) \\
&\quad \times r^{\frac{1}{2}pt} \exp\left[-\frac{1}{2}r(\underset{\sim}{m} - \mu \underset{\sim}{1}_{pt})^T \underset{\sim}{V}^{-1}(\underset{\sim}{m} - \mu \underset{\sim}{1}_{pt})\right] \exp\left(-\frac{1}{2}ar\right) r^{\frac{1}{2}b-1}. \qquad \text{(A.1)}
\end{aligned}
$$

Integrating with respect to μ over $(-\infty, \infty)$, one gets

$$
\begin{aligned}
f(\underset{\sim}{y}, s, \underset{\sim}{m}, r) \propto r^{\frac{1}{2}(pt+nqt+b-3)} \exp\Big[-\frac{1}{2}r\{(\underset{\sim}{y} - \underset{\sim}{m}_1)^T(\underset{\sim}{y} - \underset{\sim}{m}_1) \\
+ \underset{\sim}{m}^T \underset{\sim}{D} \underset{\sim}{m} + s + a\}\Big], \qquad \text{(A.2)}
\end{aligned}
$$

where

$$
\underset{\sim}{D} = \underset{\sim}{V}^{-1} - \left(\underset{\sim}{1}_{pt}^T \underset{\sim}{V}^{-1} \underset{\sim}{1}_{pt}\right)^{-1} \underset{\sim}{V}^{-1} \underset{\sim}{J}_{pt} \underset{\sim}{V}^{-1}. \qquad \text{(A.3)}
$$

The calculation of $\underset{\sim}{V}^{-1}$ is greatly facilitated by the use of the following lemma which provides the inverse of a linear function of matrices involving the Kronecker products of $\underset{\sim}{I}$'s and $\underset{\sim}{J}$'s.

LEMMA 1.:

$$
\begin{aligned}
&[b_1 \underset{\sim}{I}_m \otimes \underset{\sim}{I}_n + b_2 \underset{\sim}{I}_m \otimes \underset{\sim}{J}_n + b_3 \underset{\sim}{J}_m \otimes \underset{\sim}{I}_n + b_4 \underset{\sim}{J}_m \otimes \underset{\sim}{J}_n]^{-1} \\
&= b_1^{-1}[\underset{\sim}{I}_m \otimes \underset{\sim}{I}_n - b_2(b_1 + nb_2)^{-1} \underset{\sim}{I}_m \otimes \underset{\sim}{J}_n \\
&\qquad - b_3(b_1 + mb_3)^{-1} \underset{\sim}{J}_m \otimes \underset{\sim}{I}_n + (b_1 + nb_2 + mb_3 + mnb_4)^{-1} \\
&\qquad \times \{-b_4 + b_2(b_3 + nb_4)(b_1 + nb_2)^{-1} \\
&\qquad\quad + b_3(b_2 + mb_4)(b_1 + mb_3)^{-1} \underset{\sim}{J}_m \otimes \underset{\sim}{J}_n], \qquad \text{(A.4)}
\end{aligned}
$$

where b_i's are constants for which the matrix given in the left hand side of (A.4) is invertible .

Proof. Straightforward.

Since $\underset{\sim}{V} = \eta_1 \underset{\sim}{I}_p \otimes \underset{\sim}{I}_t + \eta_2 \underset{\sim}{I}_p \otimes \underset{\sim}{J}_t + \eta_3 \underset{\sim}{J}_p \otimes \underset{\sim}{I}_t$, applying Lemma 1 with $m = p$, $n = t$, $b_i = \eta_i$ $(i = 1, 2, 3)$ and $b_4 = 0$, one gets after some algebraic simplifications,

$$\underset{\sim}{V}^{-1} = \eta_1^{-1}[\underset{\sim}{I}_p \otimes \underset{\sim}{I}_t - u_2 \underset{\sim}{I}_p \otimes \underset{\sim}{J}_t - u_3 \underset{\sim}{J}_p \otimes \underset{\sim}{I}_t + u_2 u_3 (1+g) \underset{\sim}{J}_p \otimes \underset{\sim}{J}_t \tag{A.5}$$

where $u_2 = \eta_2(\eta_1 + t\eta_2)^{-1}$, $u_3 = \eta_3(\eta_1 + p\eta_3)^{-1}$ and $g = \eta_1(\eta_1 + t\eta_2 + p\eta_3)^{-1}$. Then $\underset{\sim}{1}_{pt}^T \underset{\sim}{V}^{-1} \underset{\sim}{1}_{pt} = \eta_1^{-1} g \underset{\sim}{1}_{pt}$, and after some simplifications, one gets

$$\begin{aligned} \underset{\sim}{D} &= \eta_1^{-1}[\underset{\sim}{I}_p \otimes \underset{\sim}{I}_t - u_2 \underset{\sim}{I}_p \otimes \underset{\sim}{J}_t - u_3 \underset{\sim}{J}_p \otimes \underset{\sim}{I}_t \\ &\quad + \{u_2 u_3 (1+g) - g(pt)^{-1}\} \underset{\sim}{J}_p \otimes \underset{\sim}{J}_t] \\ &= \eta_1^{-1}[\underset{\sim}{I}_p \otimes \underset{\sim}{I}_t - u_2 \underset{\sim}{I}_p \otimes J_t - u_3 \underset{\sim}{J}_p \otimes \underset{\sim}{I}_t \\ &\quad + (p^{-1}u_2 + t^{-1}u_3 - (pt)^{-1}) \underset{\sim}{J}_p \otimes \underset{\sim}{J}_t. \end{aligned} \tag{A.6}$$

We shall find it convenient to express $\underset{\sim}{D}$ as the partitioned matrix

$$\underset{\sim}{D} = \begin{bmatrix} \underset{\sim}{D}_{11} & \underset{\sim}{D}_{12} \\ \underset{\sim}{D}_{21} & \underset{\sim}{D}_{22}, \end{bmatrix} \tag{A.7}$$

where

$$\begin{aligned} \underset{\sim}{D}_{11} &= \eta_1^{-1}\big[\underset{\sim}{I}_q \otimes \underset{\sim}{I}_t - u_2 \underset{\sim}{I}_q \otimes \underset{\sim}{J}_t - u_3 \underset{\sim}{J}_q \otimes \underset{\sim}{I}_t \\ &\quad + \big(p^{-1}u_2 + t^{-1}u_3 - (pt)^{-1}\big) \underset{\sim}{J}_q \otimes \underset{\sim}{J}_t\big]; && \text{(A.8)} \\ \underset{\sim}{D}_{12} &= \underset{\sim}{D}_{21}^T = \eta_1^{-1} \underset{\sim}{J}_{q,p-q} \otimes \big[- u_3 \underset{\sim}{I}_t \\ &\quad + \big(p^{-1}u_2 + t^{-1}u_3 - (pt)^{-1}\big) \underset{\sim}{J}_t\big]; && \text{(A.9)} \\ \underset{\sim}{D}_{22} &= \eta_1^{-1}\big[\underset{\sim}{I}_{p-q} \otimes \underset{\sim}{I}_t - u_2 \underset{\sim}{I}_{p-q} \otimes \underset{\sim}{J}_t - u_3 \underset{\sim}{J}_{p-q} \otimes \underset{\sim}{I}_t \\ &\quad + \big(p^{-1}u_2 + t^{-1}u_3 - (pt)^{-1}\big) \underset{\sim}{J}_{p-q} \otimes \underset{\sim}{J}_t\big]. && \text{(A.10)} \end{aligned}$$

Using (A.5) and (A.7), it follows from (A.2) that

$$\begin{aligned} &f(\underset{\sim}{y}, \underset{\sim}{m}^{(1)}, \underset{\sim}{m}^{(2)}, s, r) \\ &\quad \propto r^{\frac{1}{2}(pt+nqt+b-3)} s^{\frac{1}{2}qt(n-1)-1} \\ &\quad \times \exp[\frac{1}{2} r\{\underset{\sim}{m}^{(1)T} \underset{\sim}{G}_{11} \underset{\sim}{m}^{(1)} + \underset{\sim}{m}^{(2)T} \underset{\sim}{G}_{22} \underset{\sim}{m}^{(2)} \\ &\qquad + 2\underset{\sim}{m}^{(1)T} \underset{\sim}{G}_{12} \underset{\sim}{m}^{(2)} - 2n\underset{\sim}{m}^{(1)T} \underset{\sim}{y} + n\underset{\sim}{y}^T \underset{\sim}{y} + s + a\}] \end{aligned} \tag{A.11}$$

where $\underset{\sim}{G}_{11} = n\underset{\sim}{I}_q \otimes \underset{\sim}{I}_t + \underset{\sim}{D}_{11}$, $utG_{12} = \underset{\sim}{D}_{12}$, $\underset{\sim}{G}_{22} = \underset{\sim}{D}_{22}$. Next we write

$$\begin{aligned} &\underset{\sim}{m}^{(1)T}\underset{\sim}{G}_{11}\underset{\sim}{m}^{(1)} + \underset{\sim}{m}^{(2)T}\underset{\sim}{G}_{22}\underset{\sim}{m}^{(2)} + 2\underset{\sim}{m}^{(1)T}\underset{\sim}{G}_{12}\underset{\sim}{m}^{(2)} - 2n\underset{\sim}{m}^{(1)T}\underset{\sim}{y} + n\underset{\sim}{y}^T\underset{\sim}{y} \\ &= (\underset{\sim}{m}^{(1)} - \underset{\sim}{A}_1\underset{\sim}{y})^T\underset{\sim}{G}_{11}(\underset{\sim}{m}^{(1)} - \underset{\sim}{A}_1\underset{\sim}{y}) + (\underset{\sim}{m}^{(2)} - \underset{\sim}{A}_2\underset{\sim}{y})^T\underset{\sim}{G}_{22}(\underset{\sim}{m}^{(2)} - \underset{\sim}{A}_2\underset{\sim}{y}) \\ &\quad + 2(\underset{\sim}{m}^{(1)} - \underset{\sim}{A}_1\underset{\sim}{y})^T\underset{\sim}{G}_{12}(\underset{\sim}{m}^{(2)} - \underset{\sim}{A}_2\underset{\sim}{y}) + N, \end{aligned} \tag{A.12}$$

where

$$N = \underset{\sim}{y}^T[n\underset{\sim}{I}_q \otimes \underset{\sim}{I}_t - \underset{\sim}{A}_1^T\underset{\sim}{G}_{11}\underset{\sim}{A}_1 - \underset{\sim}{A}_2^T\underset{\sim}{G}_{22}\underset{\sim}{A}_2 - 2\underset{\sim}{A}_1^T\underset{\sim}{G}_{12}\underset{\sim}{A}_2]\underset{\sim}{y}. \tag{A.13}$$

Note that $\underset{\sim}{A}_1$ and $\underset{\sim}{A}_2$ satisfy the equations

$$\begin{aligned} &\underset{\sim}{G}_{11}\underset{\sim}{A}_1 + \underset{\sim}{G}_{12}\underset{\sim}{A}_2 = \underset{\sim}{I}_q \otimes \underset{\sim}{I}_t; \\ &\underset{\sim}{G}_{22}\underset{\sim}{A}_2 + \underset{\sim}{G}_{21}\underset{\sim}{A}_1 = \underset{\sim}{0}. \end{aligned} \tag{A.14}$$

Solving the equations given in (A.14), one gets

$$\underset{\sim}{A}_1 = (\underset{\sim}{G}_{11} - \underset{\sim}{G}_{12}\underset{\sim}{G}_{22}^{-1}\underset{\sim}{G}_{22})^{-1}, \ \underset{\sim}{A}_2 = -\underset{\sim}{G}_{22}^{-1}\underset{\sim}{G}_{21}\underset{\sim}{A}_1.$$

An application of Lemma 1 gives after some simplifications

$$\begin{aligned} \underset{\sim}{G}_{22}^{-1} = \underset{\sim}{D}_{22}^{-1} = \eta_1\big[&\underset{\sim}{I}_{p-q} \otimes \underset{\sim}{I}_t + u_2(1-tu_2)^{-1}\underset{\sim}{I}_{p-q} \otimes \underset{\sim}{J}_t \\ &+ u_3\big(1-(p-q)u_3\big)^{-1}\underset{\sim}{J}_{p-q} \otimes \underset{\sim}{I}_t \\ &+ (qt)^{-1}\{(1-tu_2)^{-1} - qu_3\big(1-(p-q)u_3\big)^{-1}\underset{\sim}{J}_{p-q} \otimes \underset{\sim}{J}_t\big]. \end{aligned} \tag{A.15}$$

Hence, after some simplifications, it follows that

$$\underset{\sim}{G}_{22}^{-1}\underset{\sim}{G}_{21} = \underset{\sim}{D}_{22}^{-1}\underset{\sim}{D}_{21} = \underset{\sim}{J}_{p-q,q} \otimes \left[\frac{u_3}{1-(p-q)u_3}(\underset{\sim}{I}_t - t^{-1}\underset{\sim}{J}_t) + (qt)^{-1}\underset{\sim}{J}_t\right]; \tag{A.16}$$

$$\begin{aligned} \underset{\sim}{G}_{12}\underset{\sim}{G}_{22}^{-1}\underset{\sim}{G}_{21} &= \underset{\sim}{D}_{12}\underset{\sim}{D}_{22}^{-1}\underset{\sim}{D}_{21} \\ &= \eta_1^{-1}(p-q)\underset{\sim}{J}_q \otimes \big[u_3^2\big(1-(p-q)u_3\big)^{-1}(\underset{\sim}{I}_t - t^{-1}\underset{\sim}{J}_t) \\ &\qquad + (pqt)^{-1}(1-tu_2)\underset{\sim}{J}_t\big]; \end{aligned} \tag{A.17}$$

$$\begin{aligned} \underset{\sim}{G}_{11} - \underset{\sim}{G}_{12}\underset{\sim}{G}_{22}^{-1}\underset{\sim}{G}_{21} = \eta_1^{-1}\big[&(1+n\eta_1)\underset{\sim}{I}_q \otimes \underset{\sim}{I}_t - u_2\underset{\sim}{I}_q \otimes \underset{\sim}{J}_t \\ &- u_3\big(1-(p-q)u_3\big)^{-1}\underset{\sim}{J}_q \otimes \underset{\sim}{I}_t \\ &+ t^{-1}\{u_3\big(1-(p-q)u_3\big)^{-1} - q^{-1}(1-tu_2)\}\underset{\sim}{J}_q \otimes \underset{\sim}{J}_t\big]. \end{aligned} \tag{A.18}$$

Noting that $\underset{\sim}{A}_1 = (\underset{\sim}{G}_{11} - \underset{\sim}{G}_{12}\underset{\sim}{G}_{22}^{-1}\underset{\sim}{G}_{21})^{-1}$, an application of Lemma 1 in (A.18) provides after much simplifications the expression for $\underset{\sim}{A}_1$ given in (2.6). Next noting that $\underset{\sim}{A}_2 = -\underset{\sim}{G}_{22}^{-1}\underset{\sim}{G}_{21}\underset{\sim}{A}_1$, the expression for $\underset{\sim}{A}_2$ obtains from (A.16), and the formula for $\underset{\sim}{A}_1$ given in (2.6).

It follows from (A.11) and (A.12) that conditional on $\underset{\sim}{Y} = \underset{\sim}{y}$, $R = r$ and $S = s$,

$$\underset{\sim}{M} \sim N\left[\begin{pmatrix} \underset{\sim}{A}_1\underset{\sim}{y} \\ \underset{\sim}{A}_2\underset{\sim}{y} \end{pmatrix}, \begin{pmatrix} \underset{\sim}{G}_{11.2}^{-1} & -\underset{\sim}{G}_{11.2}^{-1}\underset{\sim}{G}_{12}\underset{\sim}{G}_{22}^{-1} \\ -\underset{\sim}{G}_{22}^{-1}\underset{\sim}{G}_{21}\underset{\sim}{G}_{11.2}^{-1} & \underset{\sim}{G}_{22.1}^{-1} \end{pmatrix}\right],$$

where $\underset{\sim}{G}_{11.2} = \underset{\sim}{G}_{11} - \underset{\sim}{G}_{12}\underset{\sim}{G}_{22}^{-1}\underset{\sim}{G}_{21}$ and $\underset{\sim}{G}_{22.1} = \underset{\sim}{G}_{22} - \underset{\sim}{G}_{21}\underset{\sim}{G}_{11}^{-1}\underset{\sim}{G}_{12}$.

Next using Exercise 2.9, p. 33 of Rao (1973), one gets

$$\begin{aligned} \underset{\sim}{G}_{22.1}^{-1} &= (\underset{\sim}{G}_{22} - \underset{\sim}{G}_{21}\underset{\sim}{G}_{11}^{-1}\underset{\sim}{G}_{12})^{-1} \\ &= \underset{\sim}{G}_{22}^{-1} + \underset{\sim}{G}_{22}^{-1}\underset{\sim}{G}_{21}(\underset{\sim}{G}_{11} - \underset{\sim}{G}_{12}\underset{\sim}{G}_{22}^{-1}\underset{\sim}{G}_{21})^{-1}\underset{\sim}{G}_{12}\underset{\sim}{G}_{22}^{-1} \\ &= \underset{\sim}{G}_{22}^{-1} + \underset{\sim}{G}_{22}^{-1}\underset{\sim}{G}_{21}\underset{\sim}{G}_{11.2}^{-1}\underset{\sim}{G}_{12}\underset{\sim}{G}_{22}^{-1} \\ &= \underset{\sim}{G}_{22}^{-1} - \underset{\sim}{A}_2\underset{\sim}{G}_{12}\underset{\sim}{G}_{22}^{-1}. \end{aligned} \tag{A.19}$$

Now using (A.15), (A.16) and the expression for $\underset{\sim}{A}_2$ as given in (2.7), it follows from (A.19) that $\underset{\sim}{G}_{22.1}^{-1} = \underset{\sim}{A}_3$, where $\underset{\sim}{A}_3$ is given in (2.8). Also, $\underset{\sim}{A}_1\underset{\sim}{y}$ and $\underset{\sim}{A}_2\underset{\sim}{y}$ simplify to $\underset{\sim}{d}_1$ and $\underset{\sim}{d}_2$. This completes the proof of part (I) of the theorem.

To prove (II), first observe that in view of part (I) of the theorem, it follows from (A.11) and (A.12) that the joint marginal pdf of $\underset{\sim}{Y}$, S and R is

$$f(\underset{\sim}{y}, r, s) \propto r^{\frac{1}{2}(nqt+b-3)} s^{\frac{1}{2}qt(n-1)-1} \exp[-\frac{1}{2}r(s+a+N)]. \tag{A.20}$$

Next observe that using (A.15),

$$\begin{aligned} &\underset{\sim}{A}_1^T\underset{\sim}{G}_{11}\underset{\sim}{A}_1 + \underset{\sim}{A}_2^T\underset{\sim}{G}_{22}\underset{\sim}{A}_2 + 2\underset{\sim}{A}_1^T\underset{\sim}{G}_{12}\underset{\sim}{A}_2 \\ &\quad = \underset{\sim}{A}_1^T\underset{\sim}{G}_{11}\underset{\sim}{A}_1 + \underset{\sim}{A}_1^T\underset{\sim}{G}_{12}\underset{\sim}{G}_{22}^{-1}\underset{\sim}{G}_{21}\underset{\sim}{A}_1 - 2\underset{\sim}{A}_1^T\underset{\sim}{G}_{12}\underset{\sim}{G}_{22}^{-1}\underset{\sim}{G}_{21}\underset{\sim}{A}_1 \\ &\quad = \underset{\sim}{A}_1^T(\underset{\sim}{G}_{11} - \underset{\sim}{G}_{12}\underset{\sim}{G}_{22}^{-1}\underset{\sim}{G}_{21})\underset{\sim}{A}_1 = \underset{\sim}{A}_1. \end{aligned} \tag{A.21}$$

Now using the expression for $\underset{\sim}{A}_1$ from (2.6), it follows from (A.13) and (A.21) that

$$N = n(1+n\eta_1)^{-1}\underset{\sim}{y}^T[\underset{\sim}{I}_q \otimes \underset{\sim}{I}_t - n\eta_2(1+n\eta_1+nt\eta_2)^{-1}\underset{\sim}{I}_q \otimes \underset{\sim}{J}_t$$
$$- n\eta_3(1+n\eta_1+nq\eta_3)^{-1}\underset{\sim}{J}_q \otimes \underset{\sim}{I}_t$$
$$+ n(qt)^{-1}\{(1+n\eta_1+nq\eta_3)^{-1}q\eta_3$$
$$- n^{-1}(1+n\eta_1+nt\eta_2)^{-1}(1+n\eta_1)\}\underset{\sim}{J}q \otimes \underset{\sim}{J}_t]\underset{\sim}{y}. \quad \text{(A.22)}$$

Next note that

$$\underset{\sim}{y}^T(\underset{\sim}{I}_q \otimes \underset{\sim}{I}_t)\underset{\sim}{y} = \sum_{i=1}^{q}\sum_{j=1}^{t} y_{ij}^2; \; \underset{\sim}{y}^T(\underset{\sim}{I}_q \otimes \underset{\sim}{J}_t)\underset{\sim}{y} = t^2 \sum_{i=1}^{q} \overline{y}_{i\cdot}^2; \quad \text{(A.23)}$$

$$\underset{\sim}{y}^T(\underset{\sim}{J}_q \otimes \underset{\sim}{I}_t)\underset{\sim}{y} = q^2 \sum_{j=1}^{q} \overline{y}_{\cdot j}^2; \; \underset{\sim}{y}^T(\underset{\sim}{J}_q \otimes \underset{\sim}{J}_t)\underset{\sim}{y} = q^2t^2\overline{y}^2. \quad \text{(A.24)}$$

Using (A.23) and (A.24), the expression N given in (A.22) simplifies to $\nu_1^{-1}SSI + \nu_2^{-1}SSB + \nu_3^{-1}SST$. This completes the proof of Part (II) of Theorem 1.

PROOF OF THEOREM 5. The proof of Theorem 5 hinges crucially on the following lemma regarding characteristic functions of quadratic forms (see e.g., Searle (1971), p. 55, equation (42)).

LEMMA 2: *If $\underset{\sim}{Z} \sim N_t(\underset{\sim}{\mu}, \underset{\sim}{V}_0)$, then for every $t \times t$ real symmetric matrix $\underset{\sim}{A}$, and for every real u,*

$$E[\exp(iu\underset{\sim}{Z}^T\underset{\sim}{A}\underset{\sim}{Z})] = |\underset{\sim}{I}_t - 2iu\underset{\sim}{A}\underset{\sim}{V}_0|^{-\frac{1}{2}} \exp[-\frac{1}{2}\underset{\sim}{\mu}^T\{\underset{\sim}{I}_t - (\underset{\sim}{I}_t - 2iu\underset{\sim}{A}\underset{\sim}{V}_0)^{-1}\}\underset{\sim}{V}_0^{-1}\underset{\sim}{\mu}]. \quad \text{(A.25)}$$

Next write

$$\sum_{j=1}^{t}(\overline{M}_{\cdot j} - \overline{M})^2 = \sum_{j=1}^{t}\left[(\overline{M}_{\cdot j} - \overline{y})^2 - (\overline{M} - \overline{y})\right]^2$$
$$= (\overline{M}_{\cdot(*)} - \overline{y}\underset{\sim}{1}_t)^T(\underset{\sim}{I}_t - t^{-1}\underset{\sim}{J}_t)(\overline{M}_{\cdot(*)} - \overline{y}\underset{\sim}{1}_t). \quad \text{(A.26)}$$

Note that from Theorem 4, conditional on $\underset{\sim}{Y} = \underset{\sim}{y}$, $R = r$ and $S = s$,

$$\underset{\sim}{M}_{\cdot(*)} - \overline{y}\underset{\sim}{1}_t \sim N\left[k\begin{pmatrix}\overline{y}_{\cdot 1} - \overline{y} \\ \vdots \\ \overline{y}_{\cdot t} - \overline{y}\end{pmatrix}, r^{-1}(l\underset{\sim}{I}_t + gt^{-1}\underset{\sim}{J}_t)\right], \quad \text{(A.27)}$$

where k, l and g are defined in (2.20)-(2.22). Put $\underset{\sim}{\mu} = k(\overline{y}_{\cdot 1} - \overline{y} \cdots \overline{y}_{\cdot t} - \overline{y})^T$, $\underset{\sim}{V}_0 = r^{-1}(l\underset{\sim}{I}_t + gt^{-1}\underset{\sim}{J}_t)$ and $\underset{\sim}{A} = \underset{\sim}{I}_t - t^{-1}\underset{\sim}{J}_t$ in Lemma 2. Then, $\underset{\sim}{A}\underset{\sim}{V}_0 = r^{-1}l\underset{\sim}{A} = r^{-1}l(\underset{\sim}{I}_t - t^{-1}\underset{\sim}{J}_t)$ so that

$$\underset{\sim}{I}_t - 2iu\underset{\sim}{A}\underset{\sim}{V}_0 = (1 - 2iur^{-1}l)\underset{\sim}{I}_t + 2iur^{-1}lt^{-1}\underset{\sim}{J}_t. \tag{A.28}$$

Hence, using Exercise 1.1 (p. 67) and 2.8 (p. 33) of Rao (1973), one gets

$$|\underset{\sim}{I}_t - 2iu\underset{\sim}{A}\underset{\sim}{V}_0| = (1 - 2iur^{-1}l)^{t-1}; \tag{A.29}$$

$$(\underset{\sim}{I}_t - 2iu\underset{\sim}{A}\underset{\sim}{V}_0)^{-1} = (1 - 2iur^{-1}l)^{-1}(\underset{\sim}{I}_t - 2iur^{-1}t^{-1}l\underset{\sim}{J}_t). \tag{A.30}$$

Hence, from (A.30),

$$\begin{aligned}
&\underset{\sim}{\mu}^T[\underset{\sim}{I}_t - (\underset{\sim}{I}_t - 2iu\underset{\sim}{A}\underset{\sim}{V}_0)^{-1}]\underset{\sim}{V}_0^{-1}\underset{\sim}{\mu} \\
&= -(2iur^{-1}l)(1 - 2iur^{-1}l)^{-1}\underset{\sim}{\mu}^T(\underset{\sim}{I} - t^{-1}\underset{\sim}{J}_t)rl^{-1}(\underset{\sim}{I}_t - \frac{gt^{-1}}{l-g}\underset{\sim}{J}_t)\underset{\sim}{\mu} \\
&= -(2iur^{-1}l)(1 - 2iur^{-1}l)^{-1}rk^2l^{-1}\sum_{j=1}^{t}(\overline{y}_{.j} - \overline{y})^2.
\end{aligned} \tag{A.31}$$

Hence, noting that $rk^2l^{-1}\sum_{j=1}^{t}(\overline{y}_{.j} - \overline{y})^2 = d$, it follows from (A.29) and (A.31) that

$$\begin{aligned}
&E[\exp(iu\sum_{j=1}^{t}(\overline{M}_{.j} - \overline{M})^2)] \\
&= (1 - 2iur^{-1}l)^{-\frac{1}{2}(t-1)}\exp[\frac{1}{2}(2iur^{-1}l)(1 - 2iur^{-1}l)^{-1}d].
\end{aligned} \tag{A.32}$$

Using the standard formula for characteristic functions of non-central chi-squares, the result follows.

Bayesian and Likelihood Methods in Statistics and Econometrics
S. Geisser, J.S. Hodges, S.J. Press and A. Zellner (Editors)

INFERENCE WITH IMPERFECT SAMPLING FROM A BERNOULLI PROCESS

Robert L. Winkler and Anil Gaba
Duke University

Abstract

When a sample is taken from a dichotomous process, various sources of noise may cause some observations to be classified incorrectly. In this paper, we consider imperfect sampling from a Bernoulli process with a noise parameter that is not known. A likelihood analysis reveals an identification problem, which is avoided under a Bayesian analysis with a joint prior distribution on the noise parameter and the Bernoulli proportion. An example is presented to illustrate the methodology and to provide some flavor of the implications of the noise and our uncertainty about the noise for inferences about a proportion.

1. Introduction

Data are frequently generated from dichotomous processes, with the parameter of interest being a proportion. In practice, inferences about a proportion p are almost always based on the assumption that the data-generating process is Bernoulli with parameter p. The maximum likelihood estimator is then the sample proportion, and a conjugate Bayesian analysis involves beta prior and posterior distributions.

When a sample is taken from a simple Bernoulli process, various sources of noise may cause some observations to be incorrect. For example, responses to dichotomous survey questions may be incorrect because some respondents lie or misinterpret the questions; observations of good and defective items in a quality control setting or positive and negative results in medical tests may be incorrect because of imperfect testing. Even with no lying, no misinterpretations, and perfect testing, errors in recording, coding, and handling the data provide sources of noise. As a result, the process that is actually observed may be somewhat different than the process that is of interest because of such contaminating factors.

The problem of making inferences about a Bernoulli parameter p based on imperfect sampling has received limited attention in the literature. Some work involving measurement error models (e.g., Fuller, 1987) is in the same spirit but emphasizes normal models. Research on classification models is relevant, although the focus is typically on classifying individual items instead of making inferences about proportions (however, see Press, 1968). The area of randomized response sampling (e.g., Warner, 1965) uses models similar to those considered here, but the noise is intentionally introduced to the process and is therefore known and carefully controlled (in an attempt to suppress the unknown and uncontrollable noise that might otherwise be present.)

Inferences for a noisy Bernoulli process are studied in Winkler (1985) under the assumption that the noise parameters are known. The results of likelihood and Bayesian analyses indicate that the noise can have considerable impact on inferences that are made about the Bernoulli parameter. For some values of the noise parameters, the noise leads to large shifts in point estimates and greatly increases the uncertainty about the parameter of interest. In a Bayesian analysis, the reduction in effective sample size due to the noise results in more weight being given to the prior distribution.

In this paper, we consider the case of imperfect sampling from a Bernoulli process with a single noise parameter under the assumption that the noise parameter is not known. The basic model is presented and a likelihood analysis is discussed in Section 2. The presence of the noise parameter and its interactions with the Bernoulli parameter lead to difficulties in the likelihood analysis. A full Bayesian analysis with a joint prior distribution on the noise parameter and the Bernoulli parameter is developed in Section 3. Joint and marginal posterior distributions can be expressed as mixtures and related to our uncertainty about the true status of the elements in the sample. Some examples given in Section 4 illustrate the methodology and provide some flavor of the implications of the noise and our uncertainty about the noise. The prior distribution assumes an even greater role than it does when the noise parameter is known. Section 5 concludes the paper with a brief summary and discussion.

2. A Likelihood Analysis

Suppose that each member of a large population is in either Group A or Group B, but not both, and let p denote the proportion of the population in Group A. For example, Group A might consist of items from a production process that are defective, individuals who will purchase a certain brand, individuals with a particular disease, or voters who will vote for a given candidate. A random sample is taken from the population, and each member of the sample is classified as being in Group A or Group B. The classification is imperfect. Testing of items from a production process or medical patients yields less than perfect information, as do market research surveys or pre-election polls. We denote the probability that a member of the sample is classified incorrectly by λ. Thus, p is the Bernoulli parameter of interest, and λ is the noise parameter in our model.

Our inferences are based on a random sample of size n, with r denoting the number of members of the sample classified as being in Group A. Let $x_i = 1$ if the ith member of the sample is classified as being in Group A and $x_i = 0$ if classified as Group B. Then the process we actually observe is not Bernoulli in p, but is Bernoulli in q, where

$$q = P(x_i = 1 \mid p, \lambda) = p(1-\lambda) + (1-p)\lambda \tag{1}$$

and

$$1 - q = P(x_i = 0 \mid p, \lambda) = p\lambda + (1-p)(1-\lambda). \tag{2}$$

The likelihood function is thus of the form

$$\ell(r, n \mid p, \lambda) = [p(1-\lambda) + (1-p)\lambda]^r [p\lambda + (1-p)(1-\lambda)]^{n-r}. \tag{3}$$

By expanding the terms in (3), we obtain an alternate representation for the likelihood function:

$$\ell(r, n \mid p, \lambda) = \sum_{j=0}^{r} \sum_{t=0}^{n-r} \binom{r}{j} \binom{n-r}{t} p^{n-j-t}(1-p)^{j+t}\lambda^{n+j-r-t}(1-\lambda)^{r-j+t}. \tag{4}$$

In this expression, j can be interpreted as the number of items classified in Group A but really in Group B, and t can be interpreted as the number of items classified in Group B and actually in Group B. Thus, j of the r classified in Group A and $n-r-t$ of the $n-r$ classified in Group B are misclassified. Of course, we do not know the values of j and t, and that is precisely what complicates our inferential problem. If we knew j and t, the likelihood function would be separable in p and λ: a product of Bernoulli likelihoods. With imperfect sampling, we lose this separability and wind up with a mixture of products of Bernoulli likelihoods.

The likelihood function is shown in Figure 1 for $r = 21$ and $n = 104$, the sample statistics for an example discussed in Section 4. This illustrates the identification problem that arises in a likelihood analysis of this noisy process. The likelihood function is unable to distinguish among all points (p, λ) with the same value of q from (1). As a result, the maximum likelihood estimate of (p, λ) is not unique, but

Figure 1: Likelihood Function and Likelihood Contour Plot with $r = 21$ and $n = 104$

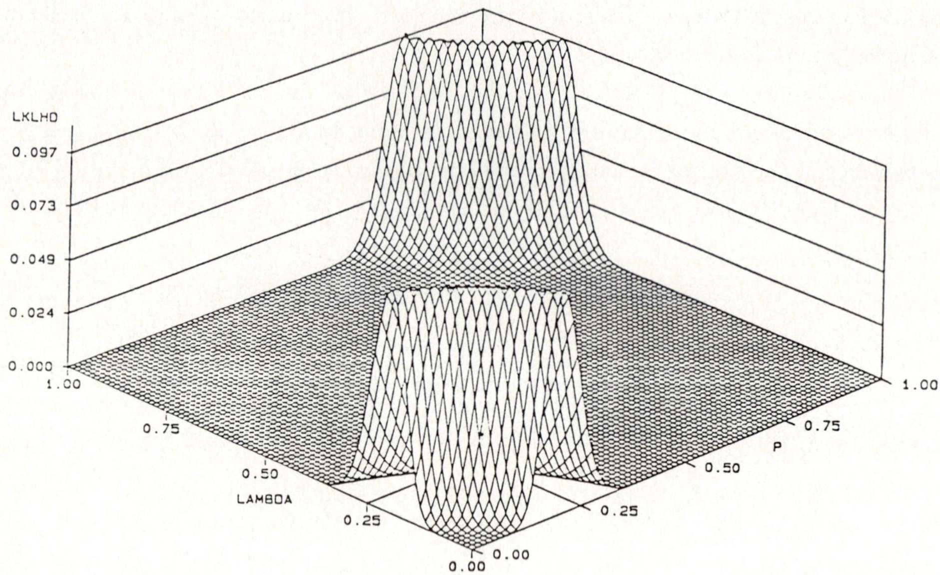

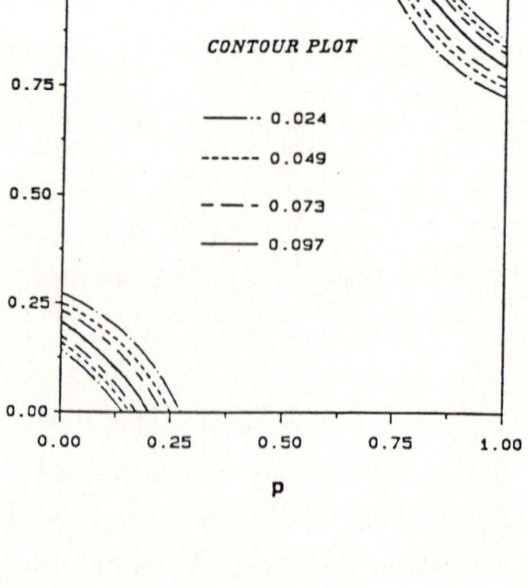

consists of those points (p, λ) such that $p(1-\lambda)+(1-p)\lambda = r/n$. For example, the set of maximum likelihood estimates for $r = 21$, $n = 104$ is shown in the contour plot in Figure 1. This set includes points such as $p = 0$ with a 20% misclassification rate, $p = 0.1$ with a 13% misclassification rate, $p = 0.2$ with no misclassification, $p = 0.8$ with a 100% misclassification rate, and $p = 1$ with an 80% misclassification rate. Without any further information about p or λ, alternative explanations for a given set of sample results seem equally compelling.

3. A Bayesian Analysis

With imperfect sampling from a Bernoulli process, we would expect to have some relevant prior information. In most cases, for instance, we would probably view high values of λ as quite unlikely. Thus, in a survey with noise due to possible lying on the part of the respondents, $(p, \lambda) = (0.8, 0.9)$ and (0.9,0.8) would have the same likelihood as (0.1,0.2) and (0.2,0.1), but we might dismiss the former two pairs because lying rates of 80-90% seem extremely unrealistic. A Bayesian analysis enables us to formalize our prior judgments and can provide the information necessary to avoid identification problems that arise in the likelihood analysis. Furthermore, it is not just a matter of incorporating additional information about p; information about λ is just as valuable in our model with imperfect sampling.

In the case of perfect sampling, the family of beta distributions is conjugate with respect to a Bernoulli process, and beta distributions are viewed as satisfactory approximations for a wide variety of types of prior information concerning a proportion. Both p and λ are proportions in our model with imperfect sampling, and we will assume that the prior density for (p, λ) is a product of beta densities for p and λ:

$$f(p,\lambda) = f_\beta(p)f_\beta(\lambda) \propto p^{\alpha_1-1}(1-p)^{\beta_1-1}\lambda^{\alpha_2-1}(1-\lambda)^{\beta_2-1}, \tag{5}$$

with α_1, α_2, β_1, and $\beta_2 > 0$. The independence assumption is not unreasonable for this symmetric model with a single noise parameter. Any dependence between p and λ might be expected to be negative, but the symmetry tempers this because a given value of λ is associated with both p and $1-p$.

Given the likelihood function in (3) or (4) and the prior distribution in (5), an application of Bayes' theorem yields the following posterior distribution:

$$f(p,\lambda \mid r,n) = \sum_{j=0}^{r}\sum_{t=0}^{n-r} w_{jt} f_\beta(\lambda \mid \alpha_2+n-r-t+j, \beta_2+r-j+t)$$

$$\cdot f_\beta(p \mid \alpha_1+n-j-t, \beta_1+j+t), \tag{6}$$

where

$$w_{jt} = a_{jt} \Big/ \sum_{j=0}^{r} \sum_{t=0}^{n-r} a_{jt} \tag{7}$$

and

$$a_{jt} = \binom{r}{j}\binom{n-r}{t}$$

$$\frac{(\alpha_2 + n - r - t + j - 1)!(\beta_2 + r - j + t - 1)!(\alpha_1 + n - j - t - 1)!(\beta_1 + j + t - 1)!}{(\alpha_2 + \beta_2 + n - 1)!(\alpha_1 + \beta_1 + n - 1)!}. \tag{8}$$

Thus, the posterior distribution is a mixture of products of beta distributions. The interpretation of the weight w_{jt} is that it is the posterior probability that j of the r classified in Group A and $n-r-t$ of the $n-r$ classified in Group B are misclassified.

Our primary interest focuses on p. If we integrate $f(p, \lambda \mid r, n)$ with respect to λ, we arrive at the following marginal posterior density for p:

$$f(p \mid r, n) = \sum_{s=0}^{n} v_s f_\beta(p \mid \alpha_1 + n - s, \beta_1 + s), \tag{9}$$

where

$$v_s = \sum_{(j,t)\ni j+t=s} w_{jt} = P(s \mid r, n) \tag{10}$$

is the posterior probability that there are really s members of Group B in the sample. The posterior mean of p is

$$E(p \mid r, n) = \sum_{s=0}^{n} v_s \left(\frac{\alpha_1 + n - s}{\alpha_1 + \beta_1 + n} \right), \tag{11}$$

and the posterior variance can be found from $E(p \mid r, n)$ and

$$E(p^2 \mid r, n) = \sum_{s=0}^{n} v_s \frac{(\alpha_1 + n - s + 1)(\alpha_1 + n - s)}{(\alpha_1 + \beta_1 + n + 1)(\alpha_1 + \beta_1 + n)}. \tag{12}$$

We can also determine how the sample information modifies the distribution of λ. The marginal posterior density for λ is

$$f(\lambda \mid r, n) = \sum_{q=0}^{n} u_q f_\beta(\lambda \mid \alpha_2 + n - q, \beta_2 + q), \tag{13}$$

where

$$u_q = \sum_{(j,t)\ni r+t-j=q} w_{jt} = P(q \mid r, n) \tag{14}$$

is the posterior probability that there are really q correctly classified elements and $n-q$ misclassifications in the sample. The posterior mean of λ and $E(\lambda^2 \mid r, n)$ are

$$E(\lambda \mid r, n) = \sum_{q=0}^{n} u_q \left(\frac{\alpha_2 + n - q}{\alpha_2 + \beta_2 + n} \right) \tag{15}$$

and

$$E(\lambda^2 \mid r, n) = \sum_{q=0}^{n} u_q \frac{(\alpha_2 + n - q + 1)(\alpha_2 + n - q)}{(\alpha_2 + \beta_2 + n + 1)(\alpha_2 + \beta_2 + n)}. \tag{16}$$

In general, then, the joint and marginal densities can be expressed as mixtures of the possible posterior densities that could arise under perfect knowledge of the exact number of misclassifications of each type. The shapes and moments of these mixtures will of course depend on the prior parameters α_1, β_1, α_2, and β_2 as well as the sample statistics r and n. Numerical examples in the following section illustrate some possible shapes for the posterior distributions and give an indication of possible implications of the noise parameter λ and our uncertainty about λ on inferences about p.

4. Examples

Self-reported incidents of delinquent behavior by youths provide examples of imperfect sampling. Clark and Tifft (1966) administered a questionnaire asking college students whether they had committed certain delinquent acts and then conducted extensive interviews and polygraph tests to check the validity of the answers to the questionnaire. Estimated misclassification probabilities ranged from zero to 0.625, with many questions resulting in equal or almost equal misclassification probabilities (suggesting that our symmetric model with one noise parameter might be appropriate in such cases) and many other questions resulting in substantial discrepancies between overreporting and underreporting (indicating that an asymmetric model with two noise parameters is sometimes more appropriate).

To illustrate the symmetric model, we chose a question with equal estimated misclassification probabilities: "Have you 'beaten up' on someone who hadn't done anything to you?" The same question was asked by Gould (1969) in a later study, and we use Gould's results ($r = 21, n = 104$) as our sample data. Considering the information from Clark and Tifft (1966) for this question and attempting to take into account differences in the populations being sampled in the two studies, we assessed prior distributions for p and λ with $\alpha_1 = 2$, $\beta_1 = 8$, $\alpha_2 = 2$, and $\beta_2 = 18$. The expected proportion who had beaten up someone is 0.20, and the expected misclassification probability is 0.10; the corresponding standard deviations are 0.12 and 0.07, respectively. The joint prior density is shown in Figure 2.

The joint posterior density determined from (6) is also shown in Figure 2. Note that in the posterior distribution, the second hump in the likelihood function (see Figure 1) disappears because most of the prior probability is concentrated on regions of low values of p and λ. The posterior distribution is unimodal despite the multimodality of the likelihood function. The posterior means for p and λ are 0.14 and 0.09, with standard deviations of 0.05 and 0.05. Thus, we estimate that 14% of the members of the population have beaten up on someone and 9% would lie when asked the question.

As noted in Section 1, inferences about a proportion p are usually based on the assumption of a Bernoulli process in p. An interesting question, therefore, is how inferences based on our model recognizing imperfect sampling differ from those based on an assumption of perfect sampling (i.e., $\lambda = 0$). With $\lambda = 0$ and the same prior distribution for p, the posterior distribution for p is a beta distribution with a mean that is 0.06 higher than our posterior mean under imperfect sampling and a standard deviation that is smaller than the posterior standard deviation under imperfect sampling by a factor of 1.45. The impact of the noise is to shift r/n toward 0.5. For a given r/n, then, the model with noise allows for this impact by placing more weight on more extreme values of p (lower values in our example because the prior mean and r/n are less than one-half), thereby shifting the posterior mean further away from 0.5.

The fact that the posterior standard deviation is larger under the imperfect-sampling model suggests that the noise causes a reduction in the information content of the sample regarding p. Another way to investigate the information content is to determine an "effective sample size" for the noisy sample. We generate this value by fitting beta distributions to the marginal posterior distributions of p and λ and defining the effective sample size as

$$n_i^* = (\alpha_i^* + \beta_i^*) - (\alpha_i + \beta_i), \tag{17}$$

where $i = 1$ for p, $i = 2$ for λ, and the parameters of the posterior beta fits are α_i^* and β_i^* for $i = 1, 2$. Such beta fits may be somewhat crude in some cases, since the marginal posterior densities cannot always be closely approximated by beta distributions, but n_i^* still seems useful as a rough measure of effective sample size. In our example, the effective sample sizes are 30.4 for p and 13.0 for λ. When compared with $n = 104$, the effective sample size for p indicates a considerable loss of information due to the noise, and the effective sample size for the noise parameter itself is even smaller.

Figure 2: Prior and Posterior Distributions with
$\alpha_1 = 2,\ \beta_1 = 8,\ \alpha_2 = 2,\ \beta_2 = 18,\ r = 21,$ and $n = 104$

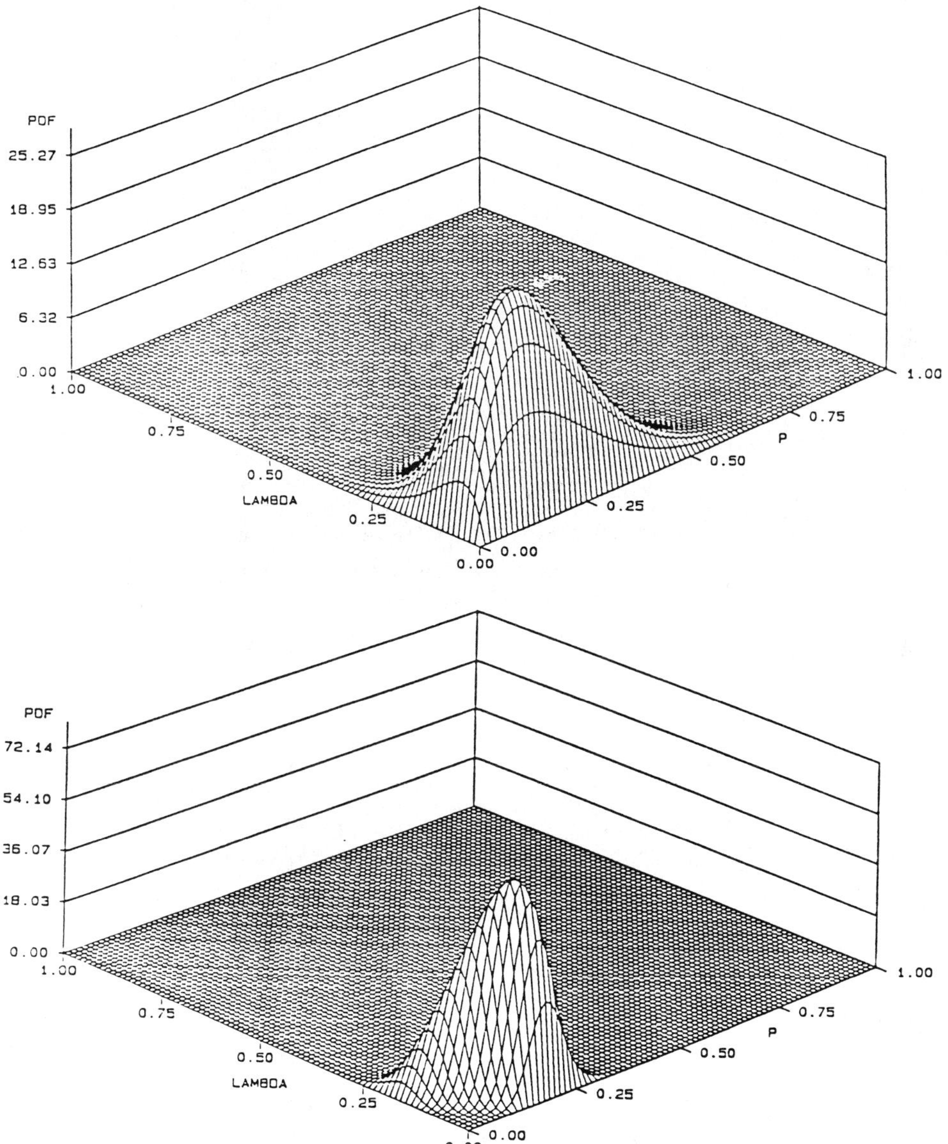

The impact of the noise, then, is substantial in our example: a reduction of about 30% in the posterior mean and an increase of 45% in the standard deviation, as compared with an analysis assuming perfect sampling. Since the prior distribution plays a particularly important role in the imperfect-sampling analysis, the robustness of inferences to variations in the prior distribution is of interest. For p, we considered one tighter distribution ($\alpha_1 = 4$, $\beta_1 = 16$), and for those who prefer to use diffuse prior distributions, we obtained results for $\alpha_1 = \beta_1 = 1$, a uniform distribution. For λ, we varied the prior mean from 0.10 to 0.05 and 0.15 and $\alpha_2 + \beta_2$ from 20 to 10 and 40. (A diffuse prior distribution for λ seems highly unrealistic.)

TABLE 1: Posterior Means (Standard Deviations) for p and λ in "Have you beaten up..." Example

			(α_2, β_2)				
			(1,9)	(1,19)	(2,18)	(3, 17)	(4,36)
(α_1, β_1)	(1,1)	p	0.15 (0.06)	0.17 (0.06)	0.14 (0.06)	0.11 (0.06)	0.13 (0.06)
		λ	0.08 (0.06)	0.05 (0.05)	0.09 (0.05)	0.13 (0.05)	0.10 (0.04)
	(2,8)	p	0.15 (0.06)	0.17 (0.05)	0.14 (0.05)	0.12 (0.05)	0.14 (0.05)
		λ	0.08 (0.05)	0.05 (0.04)	0.09 (0.05)	0.12 (0.05)	0.10 (0.04)
	(4,16)	p	0.16 (0.05)	0.18 (0.04)	0.16 (0.05)	0.14 (0.05)	0.15 (0.05)
		λ	0.07 (0.05)	0.05 (0.04)	0.08 (0.05)	0.11 (0.05)	0.09 (0.04)

The results, which are presented in Tables 1-3, provide some (albeit limited) evidence about the behavior of the imperfect sampling model and the sensitivity of estimates of p and λ as well as the posterior level of uncertainty to variations in

TABLE 2: Differences in Posterior Means for p With and Without Noise (Ratios of Posterior Standard Deviations for p With and Without Noise) in "Have you beaten up..." Example

		(α_2, β_2)				
		(1,9)	(1,19)	(2,18)	(3,17)	(4,36)
(α_1, β_1)	(1,1)	-0.06 (1.65)	-0.04 (1.42)	-0.07 (1.59)	-0.09 (1.59)	-0.07 (1.50)
	(2,8)	-0.05 (1.50)	-0.03 (1.35)	-0.06 (1.45)	-0.08 (1.45)	-0.07 (1.38)
	(4,16)	-0.04 (1.33)	-0.03 (1.23)	-0.05 (1.31)	-0.06 (1.32)	-0.05 (1.27)

TABLE 3: Effective Sample Sizes for $p(\lambda)$ in "Have you beaten up..." Example

		(α_2, β_2)				
		(1,9)	(1,19)	(2,18)	(3,17)	(4,36)
(α_1, β_1)	(1,1)	27.1 (10.1)	42.6 (3.2)	27.8 (10.1)	22.9 (19.4)	30.4 (8.6)
	(2,8)	29.5 (12.1)	44.1 (5.1)	30.4 (13.0)	25.6 (22.0)	33.1 (12.7)
	(4,16)	39.2 (14.0)	53.1 (8.1)	39.1 (15.7)	33.5 (23.5)	40.9 (16.6)

the prior distribution. The general nature of the basic results from the imperfect-sampling model (a reduction in the posterior mean of p and a loss of information as reflected by a larger standard deviation and an effective sample size considerably less than n) seems to hold consistently. The magnitude of the shifts and the balance between p and λ vary somewhat. For example, for a given prior distribution for p, note what happens as we move from $(\alpha_2, \beta_2) = (1, 19)$ through $(\alpha_2, \beta_2) = (2, 18)$ to $(\alpha_2, \beta_2) = (3, 17)$. The prior places more weight on higher values of λ, causing the posterior mean of λ to increase and the posterior mean of p to decrease because of the λ-p tradeoffs inherent in the likelihood function. At the same time, the effective sample size decreases for p and increases for λ.

Further insight into the imperfect-sampling model is provided by shapes of marginal posterior densities for p, which are given in Figure 3 for $r = 21$, $n = 104$, and a variety of prior distributions. Some of these densities are very similar to individual beta distributions, but others are very different. For example, when $\alpha_1 = \alpha_2 = 1$ and $\beta_1 = \beta_2 = 19$, there is a mode near 0.15 and the density flips up on the left to reach a local maximum at $p = 0$. Here the mode corresponds to the sample results being explained by a p near the sample proportion r/n and a small misclassification probability. The rise near $p = 0$ suggests an alternative explanation: a very small proportion combined with a relatively large λ. Explanations with intermediate values of p and λ apparently are less likely.

To demonstrate that the imperfect-sampling model can be extended to the case of unequal misclassification probabilities, we considered another question from Clark and Tifft (1966): "Have you skipped school without a legitimate excuse?" This question, which led to vastly different misclassification probabilities, was also asked by Gould (1969), who reports that $r = 12$ and $n = 104$. Once again considering the information from Clark and Tifft (1966) and attempting to judge differences in the populations used in the two studies, we assessed independent beta distributions for p, λ_1 (the probability that someone who skipped school answers no), and λ_2 (the probability that someone who never skipped school answers yes). The parameters for these distributions are, respectively, (3,7), (5,5), and (1,49). Note that λ_1 is judged to be considerably higher than λ_2, as might be anticipated. We expect half of those who have skipped school to lie, as opposed to only 2% of those who have not skipped school.

To conserve space, equations analogous to (6), (9), and (13) are not presented here, but the posterior distributions for p, λ_1, and λ_2 are mixtures of beta distributions. The posterior mean of p is 0.25, which is almost twice the posterior mean assuming perfect sampling, and the standard deviation of p is higher by a factor of 3.03 due to the noise. The effective sample size for inferences concerning p is 9.6, as

Figure 3: Marginal Posterior Distributions for p for $r = 21$, $n = 104$, and Selected Prior Parameters.

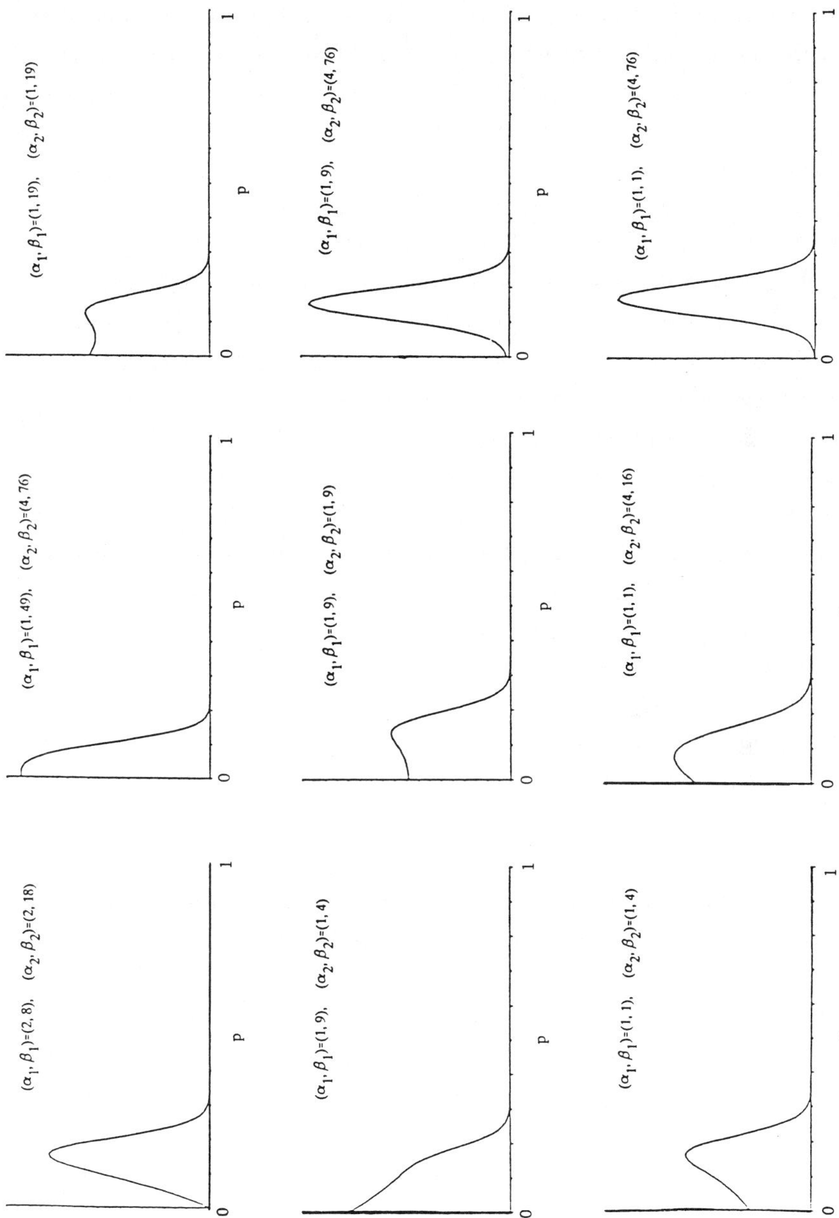

compared with $n = 104$. A uniform prior distribution for p, with the same distributions for λ_1 and λ_2, yields an even higher posterior mean for p, a smaller standard deviation, and an effective sample size of only 2.7.

The results in this section certainly do not cover all of the possible types of cases that might occur in the model with noise. Nonetheless, they illustrate the model and provide some flavor of the implications of noise for inferences about a proportion.

5. Summary and Discussion

We have considered inferences about a Bernoulli parameter p with imperfect, or noisy, sampling from the Bernoulli process with a single, unknown noise parameter λ. The likelihood function is Bernoulli not in p, but in $p(1-\lambda)+(1-p)\lambda$, and it can be represented as a mixture of products of Bernoulli likelihoods in p and λ. Because of the symmetry of the model, an identification problem arises in which the likelihood function is unable to distinguish among certain (p, λ) pairs. The incorporation of prior information via a Bayesian analysis avoids this identification problem. Results have been presented based on a joint prior density that is a product of beta densities for p and λ. The joint and marginal posterior densities can be expressed as mixtures of the possible posterior densities that could arise under perfect knowledge of the exact number of misclassifications of each type. Specific examples in Section 4 indicate some possible shapes for the joint posterior distribution and the marginal posterior distribution for p and provide an illustration involving a slightly more general model with two noise parameters. The examples reveal how inferences based on the imperfect-sampling model can differ (often substantially) from those based on an assumption of perfect sampling.

One important implication of the analysis is that ignoring the noise by treating the sampling process as Bernoulli in p can lead to misguided inferences. Point estimates of p will be shifted toward 0.5, and the degree of uncertainty about p often will be understated. Thus, unless λ is viewed as being very close to zero, it is important to include it in the model. Once it is included, however, the prior distribution of p and λ plays a very important role in the analysis, both in terms of identification and in terms of influencing the ultimate inferences. The assessment of $f(p)$ is identical in principle in the noisy and noise-free situations, although noise in past data may render that data less informative about p. As for the noise parameter, sometimes we may have considerable information about λ (e.g., errors in test results with well-known tests in quality control or medicine); other times our information about λ may be quite vague and imprecise (e.g., lying or misrepresentation in surveys, errors with new tests in quality control or medicine, errors in recording or handling data). In the latter case, the assessment of $f(\lambda)$ may be difficult, but this is exactly the situation in which care should be taken because inferences about p may not be highly robust with respect to variations in $f(\lambda)$.

The results presented here provide a model for dealing with a Bernoulli process with imperfect sampling and a flavor for the implications of the noise. Some generalizations being studied include a model with asymmetric noise involving two noise parameters, one for each type of misclassification (as illustrated briefly in Section 4), and a prior distribution allowing dependence between p and the noise parameters in the asymmetric model. We feel that models of this general type have great potential applicability and deserve more attention because of the number of real-world situations in which one or more sources of noise are present.

Acknowledgments

This research was supported in part by the National Science Foundation under Grant IST8600788. We are grateful for helpful comments from an editor and two referees.

References

Clark, J.P. and L.L. Tifft (1966). "Polygraph and Interview Validation of Self-Reported Deviant Behavior," *American Sociological Review*, Vol. 31, 516-523.

Fuller, W.A. (1987). *Measurement Error Models.* New York: Wiley.

Gould, L.C. (1969). "Who Defines Delinquency: A Comparison of Self-Reported and Officially Reported Indices of Delinquency for Three Racial Groups," *Social Problems*, Vol. 16, 325-336.

Press, S.J. (1968). "Estimating from Misclassified Data," *Journal of the American Statistical Association*, Vol. 63, 123-133.

Warner, S.L. (1965). "Randomized Response: A Survey Technique for Eliminating Evasive Answer Bias," *Journal of the American Statistical Association*, Vol. 60, 63-69.

Winkler, R.L. (1985). "Information Loss in Noisy and Dependent Processes," in *Bayesian Statistics 2*, J.M. Bernardo, M.H. DeGroot, D.V. Lindley, and A.F.M. Smith, eds. Amsterdam: North-Holland.

VI. PREDICTIVE METHODS AND APPLICATIONS

Bayesian and Likelihood Methods in Statistics and Econometrics
S. Geisser, J.S. Hodges, S.J. Press and A. Zellner (Editors)

PREDICTIVE APPROACHES TO DISCORDANCY TESTING

Seymour Geisser
University of Minnesota

Abstract

Often a model has proven useful in the past in a given setting, but in the current application a few observations appear to be discordant from the model or from the rest of the observations. In this paper we propose a Bayesian framework for predictive approaches to testing for such discordant observations. Among other advantages, this framework is more readily applied in situations that are difficult for frequentist or non-predictive approaches, e.g. for censored data, or when prior information is available.

1. Introduction

Simple tests of significance are, by all accounts, the most prevalent of statistical procedures in current use. One need only peruse a journal in which papers using statistical methods are published and one is immediately beset by either varying numbers of stars indicating significance at particular levels or the exact P-values themselves. Why this should be so is in one sense rather curious as significance testing has been roundly denounced by most frequentist and Bayesian statisticians alike as being generally useless under any circumstances. However there have been a few rather distinguished statisticians who have vigorously defended tests of significance in certain limited situations Fisher (1956), Barnard (1962), Box (1980), and Cox (1987).

Clearly, so-called pure significance tests, whether based on frequentist, Bayesian, or predictivist principles, may suffer not only from a suitable choice of a critical region but can also be criticized on the basis that alternative hypotheses have not been taken into account. From the Bayesian viewpoint (one which we generally subscribe to) this requires knowledge of all possible alternatives and their prior probabilities—a formidable if not impossible task in many instances. There are also situations where reasonable modeling alternatives are very difficult to contemplate. At any rate we believe that there are situations where a simple test of significance can be useful. In particular, an area that we believe is most susceptible to tests of significance is discordancy testing, where a particular situation is such that a standard model has proven useful in similar settings in the past but perhaps in the current experiment one or two observations may be discordant from the rest. For example, this situation may have arisen from errors in transcribing data, numbers misread, digits transposed, an incorrect sign before a number; or a stipulated experimental condition that did not actually obtain for several of the observables. Therefore the surprise engendered by a small P-value of an appropriate significance test is useful in detecting these and other anomalies. If it is determined that the "discordant" observation(s) make(s) an appreciable difference in a potential inference or decision then this should provide the stimulus for a thorough review of the protocols and procedures in making the observations. A determination then needs to be made as to whether the apparently discordant observable is really incompatible with the rest of the observations or whether the modeling requires revision or both. In simple univariate situations these discordancies often take the form of outliers in that one or two observations appear to be distantly removed from the others. Hence discordancy tests for outliers have been devised. For a compendium of frequentist tests the reader is referred to Barnett and Lewis (1978). In this paper we propose a Bayesian framework for predictive approaches to testing for discordant observations. Hence this approach is one that a Bayesian can comfortably use and will turn out to be considerably more flexible and more widely applicable than the standard frequentist approach.

We shall present a general framework for such discordancy tests that may simply depend on (a) the identification of a potentially discordant observation because of the intrusion of some untoward event connected with the particular observation (b) taking into account a diagnostic ransacking of the data in search of potentially discordant observations. Several different approaches are devised that depend on the differing circumstances in identifying the suspect observations. Illustrations for certain simple distributional models will be presented.

2. Testing an Observation Under Suspicious Circumstances

Consider the situation where $Y_1, \ldots, Y_N$ are independently and identically distributed with common distribution function $F_Y(y \mid \theta)$. Further assume that $g(\theta)$ is a prior probability function for θ. Hence based on this model, say M, one can compute the predictive distribution for a future value Z by calculating

$$F_Z(z \mid y^{(N)}) = E_\theta F_Y(z \mid \theta)$$

where $y^{(N)} = (y_1, \ldots, y_N)$ is the observed set of values of $Y^{(N)} = (Y_1, \ldots, Y_N)$, and the expectation is over the posterior distribution of θ.

If, when obtaining the experimental value for Y_i, it was noted that some untoward event or suspicious circumstance occurred that may have affected the observable, the experimenter could make a determination regarding the inclusion of the experimental outcome by calculating the predictive distribution of Z based on all the observations except for y_i and denoted $y_{(i)}$,

$$F_Z(y_i \mid y_{(i)}) = \int F_Z(y_i \mid \theta) dP(\theta \mid y_{(i)})$$

where $P(\theta \mid y_{(i)})$ is the posterior distribution of θ given $y_{(i)}$. An assessment of the discordancy of y_i with $y_{(i)}$ can be made by calculating a predictive P-value, Geisser (1980), for some suitably defined region R using the conditional predictive

$$P_i = Pr[Z \in R \mid y_{(i)}].$$

It may be more convenient at times to calculate

$$P_i = Pr[T(Z \mid y_{(i)}) \in R \mid y_{(i)}] \tag{2.1}$$

where $T(Z \mid y_{(i)})$, or simply $T(Z)$, is a statistic which is a function of Z and possibly the observed $y_{(i)}$ for some suitably defined region R using the conditional predictive distribution of Z given $y_{(i)}$. As in frequentist significance testing there are several reasonable principles for implementing the above procedure. One would be to order values of the range of $T(Z)$ as to their departure from compatibility with the rest of the observations assuming the truth of the original model. Another is to simply calculate a probability ordering

$$P_i = Pr[f_Z(Z \mid y_{(i)}) \leq f_Z(y_i \mid y_{(i)})] \tag{2.2}$$

which implies that R consists of the lowest probability region. In problems where the predictive distribution is unimodal and symmetric both principles lead to the same test region. When this is not the case the statistician must decide whether in the particular case so-called "scientifically meaningful" ordering or lowest probability ordering should define the region.

This setting assumes that only y_i is suspect and that the set $y_{(i)}$ concords with the model permitting one to condition on $y_{(i)}$. If there is doubt about whether $y_{(i)}$ concords with model one could calculate the marginal or unconditional distribution of Y_i, from

$$F(y_i) = \int F(y_i \mid \theta) dG(\theta)$$

where $G(\theta)$ is the prior distribution for θ. Computations analogous to (1.1) and (1.2) can be made using the unconditional distribution of Y_i, namely

$$P_i = Pr[T(Y_i) \in R] \tag{2.3}$$

or

$$P_i = Pr[f_Z(Z) \leq f_Z(y_i)]. \tag{2.4}$$

These latter tests have the advantage of not depending on any assumption about $Y_{(i)}$, but they can only be used with a proper prior distribution otherwise P_i would not exist. Of course when the assumption that $Y_{(i)}$ concords can confidently be made, the more sensitive test previously described would be available using this additional information. Further this test exists for certain useful improper prior distributions as well. At any rate as we shall see more clearly later on there are generally two kinds of predictive significance tests that one can apply, namely conditional and unconditional ones.

Often the statistician may be unaware of the experimental procedures and only when examining the data by plotting or applying analytic diagnostics does the possibility of outliers or discordant observations for the presumed model become evident. Checking with the investigator could reveal some experimental mishap that might have occurred when taking those observations so that appropriate remedies may be undertaken. However, in cases when no such information from the experimenter is available it would certainly be wise to take account of the diagnostic checking that data are often subject to in computing a significance test lest one put too severe a strain on a potentially adequate model. The burden of the rest of this paper is to propose general predictive testing procedures, when the assumption is made that at most there are only a few discordant observations and the data have been ransacked in some predetermined manner. In the next section we shall discuss some ways of ransacking or searching the data for potentially discordant observations from a predictivistic viewpoint. In section 3, we propose more formally and discuss the two kinds of tests, Unconditional and Conditional Predictive Discordancy Tests. In section 3, we illustrate these ideas for the simple exponential and normal models.

3. Diagnostics and Discordancy Indices

Before a random sample of independent and identically distributed random variables is taken, all of the potential observables are "created equal" in terms of making inferences or decisions. However, after the observations are in hand this is no longer the case and each can have a different effect on the inference e.g. on the predictive distribution of a future observable (or on the estimation of the set of parameters θ or a subset of them).

Various functions have been proposed to rank the varying influence of observations. For purposes of this paper we shall adhere to those that concern themselves solely with prediction. Johnson and Geisser (1982, 1983) proposed predictive influence functions (PIF) using the Kullback-Leibler numbers between the predictive distribution of a future value based on $y_{(i)}$ and $y^{(N)} = (y_i, y_{(i)})$. This can also be used in the random sample case to yield an ordering of the discordancy of the observations, i.e., in the absence of other information the observation yielding the largest Kullback-Leibler number is the foremost candidate for a discordancy test. Thus the PIF, which can be interpreted as a discordancy index here, is

$$I_i = E[\ln f(Z \mid y_{(i)}) - \ln f(Z \mid y^{(N)})] \tag{3.1}$$

where the expectation is taken over the predictive distribution $F_Z(z \mid y_{(i)})$, i.e., with y_i deleted.

Another such diagnostic or index,

$$d_i = f(y_i \mid y_{(i)}) \tag{3.2}$$

called the Conditional Predictive Ordinate (CPO) Geisser (1980,1985), ranks the discordancy of observations—the smaller the value of d_i the more discrepant is y_i from $y_{(i)}$, see also Smith and Pettit (1985). Other indices are the unmodified tests (1.1)-(1.4) of section 1 which are useful for testing an observable which is initially under suspicion for reasons other than its actual value.

The one given by (1.4) is somewhat similar to the Aitchison and Dunsmore (1975) atypicality index—the difference being that here each observable y_i is related to its predictive distribution conditioned on the other observations while their index uses the single predictive distribution which includes y_i.

These and other diagnostics of the data can be graphed to yield a visual description of the possibilities of faulty data but our main focus in this paper is the calculation of significance test assessments associated with potential discordancies.

4. Predictive Discordancy Tests

In the predictive framework an overall test for model adequacy has been suggested by Box (1980) by calculating for a model M the prior predictive density of the entire set of observables $Y^{(N)} = (Y_1, \ldots, Y_N)$,

$$f(y^{(N)} \mid M) = \int f(y^{(N)} \mid \theta, M) g(\theta \mid M) d\theta, \tag{4.1}$$

where $g(\theta \mid M)$ is the prior density of θ under model M.

He tests the adequacy of M by computing the significance level for the observed value $y^{(N)}$

$$P_y = Pr[f(Y^{(N)} \mid M) \leq f(y^{(N)} \mid M)] \tag{4.2}$$

or for some "checking function" $H(Y^{(N)})$

$$P_h = Pr[f_H(H) \leq f_H(h)] \tag{4.3}$$

where h is the observed value of H. Small values of P_y or P_h will serve to call model M into question where by model M he means the joint density of $Y^{(N)}$ and θ

$$f(y^{(N)}, \theta \mid M) = f(y^{(N)} \mid \theta, M) g(\theta \mid M). \tag{4.4}$$

Sometimes a model may be called into question because of one or more faulty observations. Suppose some checking function, diagnostic criterion, or discordancy index is used to identify possibly faulty or discordant observations where say H_C chooses y_C as potentially most discordant i.e., the observed value

$$h_C > h_i \quad \text{for all} \quad i \neq C$$

where $>$ stands for "more discordant." One can check for discordance by calculating a significance level from h_C, the observed value of H_C,

$$Pr[f(H_C \mid M, C) \leq f(h_C \mid M, C)] \tag{4.5}$$

or from

$$Pr[H_C > h_C \mid C, M] \tag{4.6}$$

where $f(H \mid M, C)$ is calculated from $f(y^{(N)} \mid M)$ under choice C i.e., that h_C was the observed value of the most discrepant diagnostic.

We note that if the distribution of the index or diagnostic H is independent of θ and the hyperparameters then the prior distribution has no effect whatever on the significance value. This is the usual case for those diagnostics (statistics) that are currently in use and purport to test discordancy for the sampling part of the model rather than the entire model, i.e., including the prior distribution.

Another possible way of looking at this problem is to consider the conditional predictive density

$$f(y_i \mid y_{(i)}) = \frac{f(y^{(N)} \mid M)}{f(y_{(i)} \mid M)} \tag{4.7}$$

As indicated previously, this is appropriate when suspicion is aroused that something untoward occurred that may have affected the value of a particular observation.

A simple predictive significance test as indicated earlier, conditional on $y_{(i)}$ satisfying the model, is

$$P_i = [Pr[T(Z) \in R_i \mid y_{(i)}] \tag{4.8}$$

where R_i is some specified region. If on the other hand the observation is to be examined solely because its observed value was flagged by the diagnostic H as ranked most discordant and no discernible reason for it adduced, than an adjustment to the previous significance test to take account of this fact, at the very least, is in order. One possible way is to reject only at a P that decreases as N increases. How this might be sensibly implemented is not clear. Another possible way of adjusting the test of significance for the ransacking is to calculate

$$P_C = Pr[H > h_C \mid H > h_{C-1}; y_{(C)}] \tag{4.9}$$

where H_{C-1} is the value for H which ranks y_{C-1} as the second most discordant observation. For example, assume that the discordancy diagnostic is a monotone increasing function of the observation's value. In this case we suggest, as a simple alternative to finding the significance level from the distribution of the maximum, using the original predictive distribution of the observable and conditioning it on being greater than the observed value y_{C-1},

$$P_C = Pr[Z > y_C \mid Z > y_{C-1}, y_{(C)}]. \tag{4.10}$$

As indicated discordancy testing procedures can be developed from both the conditional and unconditional approach. There are several aspects of the conditional approach that many statisticians may find appealing. The first one is that the test procedures depend more on the conditional distribution (likelihood) of the observables than the unconditional procedures which depend more heavily on the prior for θ. The second is that the unconditional approach in general can not be used with an improper prior, while the conditional procedure can be used with certain useful improper priors. A third aspect is that the conditional approach will often lead to much simpler computations than the unconditional approach.

It is to be noted that all these tests are essentially subjective assessments and are not grounded in a frequency theory as there is no attempt to tie this to a class of repetitions. However, for certain combinations of sampling and prior distributions, some of the CPD tests proposed have conditional and unconditional frequency analogues as we shall see in the next section.

5. Simple Exponential Examples

We now illustrate some of the previous proposals in the simple exponential case.

Consider a random sample $Y_1, \ldots, Y_N$ from $\theta e^{-\theta y}$ with prior density $g(\theta)$. Suppose the index H happened to lead to choosing as most discordant

$$y_C = \max_i \ y_i.$$

Then we calculate

$$\begin{aligned} Pr[Y_C \leq y_C \mid M, \theta] &= (1 - e^{-y_C \theta})^N \\ Pr[Y_C \leq y_c \mid M] &= \int (1 - e^{-y_C \theta})^N g(\theta) d\theta \end{aligned} \tag{5.1}$$

and

$$P_C = Pr[Y_C > y_c \mid M] = 1 - P(Y_C \leq y_C \mid M). \tag{5.2}$$

If we have a conjugate prior density for θ with hyperparameters N_0 and $\overline{y}_0$,

$$g(\theta) \ \propto \ \theta^{N_0 - 1} e^{-\theta N_0 \overline{y}_0}, \quad \overline{y}_0 > 0, \ N_0 > 0 \tag{5.3}$$

then the CPO index, for example, will still rank y_C the maximal observation, as relatively most discordant and

$$P_C = \sum_{j=1}^{N} \binom{N}{j} (-1)^{j+1} \left[\frac{N_0 \overline{y}_0}{N_0 \overline{y}_0 + j y_C} \right]^{N_0} \tag{5.4}$$

Although in this case the exact calculation is fairly simple, in most cases it is quite complex. Note also that P_C depends only on N, y_C and the hyperparameters of $g(\theta)$, $\overline{y}_0$ and N_0. Incidentally, if we were suspicious of an observation y_i for reasons other than its value as indicated in section (3), an unconditional test, irrespective of $y_{(i)}$, would merely set $N = 1$ and substitute y_i for y_C in (4) yielding

$$P_i = Pr[Z > y_i \mid M] = \left[\frac{N_0 \overline{y}_0}{N_0 \overline{y}_0 + y_i} \right]^{N_0} \tag{5.5}$$

This unconditional test which assumes nothing about the observation set $y_{(i)}$ may be compared to a conditional test derived using (1)

$$P_i = Pr[Z > y_i \mid M] = \left[\frac{N_0 \overline{y}_0 + N \overline{y}}{N_0 \overline{y}_0 + N \overline{y} + y_i} \right]^{N + N_0} \tag{5.6}$$

which assumes that the set $y_{(i)}$ concords with the model. Another possible unconditional discordancy test alternative to (1.2) would be to calculate

$$\begin{aligned} P_C &= P[Y_C - Y_{C-1} > y_C - y_{C-1} \mid M] \\ &= \left[\frac{N_0 \overline{y}_0}{N_0 \overline{y}_0 + y_C - y_{C-1}} \right]^{N_0}, \end{aligned} \tag{5.7}$$

where y_{C-1} is the second largest observation. This test is completely independent of N but now depends on y_{C-1}. It also appears, as it does in (4.4) to put relatively heavy stress on the prior.

The significance level for the proposed adjusted or Conditional Predictive Discordancy (CPD) test for the exponential problem previously discussed is

$$P_C = \left[\frac{N_0\overline{y}_0 + N\overline{y} + y_{C-1} - y_C}{N_0\overline{y}_0 + N\overline{y}}\right]^{N_0+N+1} \tag{5.8}$$

which depends not only on the prior hyperparameters N_0 and $\overline{y}_0$ but on the observables y_C, y_{C-1}, $\overline{y}$ and N. We also note that the test exists as $N_0 \to 0$ which yields the useful improper prior which purports to reflect little prior information relative to the likelihood. For example in (4.8), $N_0 = 0$ results in the usual improper prior $g(\theta) \propto \theta^{-1}$ and

$$\lim_{N_0 \to 0} P_C = \left[\frac{N\overline{y} + y_{C-1} - y_C}{N\overline{y}}\right]^{N-1} \tag{5.9}$$

This, it is easy to show, leads to the same significance level one would obtain from the classical frequentist statistic

$$T = \frac{Y_C - Y_{C-1}}{Y_C} \tag{5.10}$$

used to test for discordance, Dixon (1950, 1951) if one conditioned on

$$U = \frac{Y_C}{N\overline{Y}} \tag{5.11}$$

so that equivalent to (4.9),

$$Pr[T > t \mid U = u] = P_C. \tag{5.12}$$

However the usual frequentist approach is to calculate the unconditional significance level

$$Pr[T > t] = N(N-1)B\left(\frac{2-t}{1-t}, N-1\right) \tag{5.13}$$

where $B(\cdot,\cdot)$ is the beta function, Likes (1966).

The application of this approach to translated exponential distributions in the presence of censoring is being extensively treated in a future paper.

6. Normal Samples–CPD Tests

For normal samples, Unconditional Predictive Discordancy (UPD) tests lead to quite involved calculation. Hence we shall focus here on CPD tests.

First we present a rather simple illustration. Let $Y_1, \ldots, Y_N$ be $N(\theta, 1)$ and assume θ is $N(\beta, \tau^2)$ where β and τ^2 are presumed known. Then the predictive distribution of Z is $N(a, b_0^2)$ where

$$a = \frac{\tau^2 \bar{y} + \beta N^1}{\tau^2 + N^{-1}} \quad \text{and} \quad b_0^2 = 1 + \tau^2 (N\tau^2 + 1)^{-1} \tag{6.1}$$

Hence for Z based on $y_{(i)}$ rather than $y_{(N)}$ we have $Z - N(a_i, b^2)$ where

$$a_i = \frac{\tau^2 \bar{y}_{(i)} + \frac{1}{N-1}\beta}{\tau^2 + \frac{1}{N-1}}, \; b^2 = 1 + \tau^2[1 + (N-1)\tau^2]^{-1}, \tag{6.2}$$

where $(N-1)\bar{y}_{(i)} = \sum_{j \neq i} y_j$.

In this instance all the discordancy indices previously discussed select that y_i which maximize the checking function

$$v_i^2 = \left(\frac{y_i - a_i}{b}\right)^2 \tag{6.3}$$

To implement the unconditional test we note that $X' = (X_1, \ldots, X_N)$ for

$$X_i = Y_i - a_i$$

is a multivariate $N(\mu, \sum \mid \theta)$ where

$$\mu = \frac{(\theta - \beta)e}{(N-1)\tau^2 + 1}, \qquad e' = (1, \ldots, 1)$$

and $\sum$ is known matrix with equal diagonal elements and equal off diagonal elements. One then would find the distribution of the max $|Y_i - a_i|$ or max $\left(\frac{Y_i - a_i}{b}\right)^2 = V_C^2$ conditional on θ and integrate this over the distribution of θ to obtain the unconditional distribution of V. The calculation of

$$Pr[V_C^2 \geq v_C^2] = P \tag{6.6}$$

would then result in the significance level for a UPD test. It is clear that even for such a simple normal case, the calculation can be quite arduous especially in cases where the variance is also unknown.

We suggest instead using the CPD approach to test discordancy here and for the subsequent normal sampling that we shall discuss. Now conditional on $y_{(i)}$ the checking function

$$V^2 = \left(\frac{Z - a_i}{b}\right) \sim X_1^2. \tag{6.7}$$

A conditional significance test level can be calculated as

$$Pr[V^2 \geq v_C^2 \mid V^2 \geq v_{C-1}^2, y_{(C)}] = P_C \tag{6.8}$$

or

$$P_C = \frac{1 - F(v_C^2)}{1 - F(v_{C-1}^2)} \tag{6.9}$$

where $F(\cdot)$ is the distribution function of a χ_1^2 random variable. The result then is immediately available from the χ_1^2 distribution.

A particular case of greater interest is when the random sample is from a $N(\mu, \sigma^2)$ and μ and σ^2 are unknown. To make this situation closely correspond to a conventional frequency analysis we let the prior density be

$$g(\mu, \sigma^2) \propto 1/\sigma^2, \tag{6.10}$$

noting that now the UPD test is unavailable. Again computation of any of the reasonable discordancy indices leads to the selection of that y_i which maximizes $(y_i - \overline{y}_{(i)})^2$ or the checking function

$$\frac{(y_i - \overline{y}_{(i)}^2(N-1)}{Ns_{(i)}^2} \tag{6.11}$$

which will yield the same y_C, since $s_{(C)}^2 \leq s_{(i)}^2$ where $(N-1)\overline{y}_{(i)} = \sum_{j \neq 1} y_j$ and $(N-2)s_{(i)}^2 = \sum_{j \neq 1} (y_j - \overline{y}_{(i)})^2$.

We note that the predictive distribution of the checking function

$$W^2 = \frac{(Z - \overline{y}_{(i)})^2(N-1)}{Ns_{(i)}^2} \tag{6.12}$$

conditional on $y_{(i)}$ has an $F(1, N-2)$ distribution. Hence to test for discordancy of y_C we can calculate the conditional or adjusted significance level

$$P_C = Pr[W^2 > w_C^2 \mid W^2 > w_{C-1}^2, y_{(C)}] = \frac{1 - F(w_C^2)}{1 - F(w_{C-1}^2} \tag{6.13}$$

where w_{C-1}^2 is the observed second largest standardized deviate and $F(\cdot)$ is the distribution of an F variate with 1 and $N-2$ degrees of freedom.

As an application of this consider the following experiment, Barnett and Lewis (1979), on crushing strength of cement made up of 10 test cubes from a particular mix. After a suitable hardening period the observed strengths of the cubes in pressure per square inch is determined to be 790, 750, 910, 650, 990, 630, 1290, 820, 860, 710. We assume, as the forementioned authors, that previous experience has indicated that the normal distribution is adequate for crushing strength observables. The maximum discordancy index occurs obviously for the value 1290 resulting in $w_c^2 = 15.873$, with the second largest for the value 630, yielding $w_{C-1}^2 = 1.827$. Hence $P_C = .02$

This can be compared with a much more complex calculation for a significance test due to Grubbs (1969). Interpolation in his tables yields for the two sided case a P-value of approximately .04. For another application of this procedure to a data set that gives results indistinguishable from the frequency significance level, see Geisser (1987). The fact that the results are often negligibly different may allow frequentists to use the CPD test as a quick approximation for their own test.

In situations where discordancy indices for the largest y_C and second largest y_{C-1} appear to be appreciably different from the rest, a joint test will depend on the predictive distribution of Z_1 and Z_2, a pair of future observations. This is easily shown to be bivariate student with density function

$$f(z_1, z_2 \mid y_{(i,j)}) \propto \left(1 + \frac{(N-2)t'S^{-1}t}{(N-2)^2-1}\right)^{-(N-1)/2} \tag{6.14}$$

where $y_{(i,j)}$ is the set of observation with y_i and y_j omitted and

$$t' = (z_1 - \overline{y}_{(i,j)}, z_2 - \overline{y}_{(i,j)})$$

$$(N-2)\overline{y}_{(i,j)} = \sum_{k \neq i,j} y_k$$

$$(N-3)s_{(i,j)}^2 = \sum_{k \neq i,j} (y_k - \overline{y}_{(i,j)})^2$$

$$S = s_{(i,j)}^2 \begin{pmatrix} 1 & (N-1)^1 \\ (N-1)^{-1} & 1 \end{pmatrix}.$$

Further we find that the predictive distribution for

$$W = \frac{(N-2)}{(2N-2)} T'S^{-1}T - F(2, N-3) \tag{6.15}$$

is an F-variate with 2 and $N-3$ degrees of freedom. Hence if there were reason to suspect 2 observations prior to calculating discordancy indices we calculate

$$Pr[W \geq w] = P_{ij} \tag{6.16}$$

where w is the observed value of W for y_i and y_j.

When a discordancy index is calculated here for two observations it is natural to choose the most discordant pair as that y_i and y_j that minimizes P_{ij}. Another would be the one that minimized the bivariate CPO, $d_{ij} = f(y_i, y_j \mid y_{(i,j)})$. Both will lead to the same most discordant pair.

A possible adjusted test then is

$$P_C = Pr[W \geq w_C \mid W \geq w_{C-1}] = \frac{1 - F(w_C, 2, N-3)}{1 - F(w_{C-1}, 2, N-3)} \tag{6.17}$$

where $F(\cdot, 2, N-3)$ is the distribution function of an $F(2, N-3)$ variate and W_{C-1} is the second largest observed value of W where both arguments involve two observations other than those contained in w_C. Clearly other regions may also be used to make significance calculations but none are likely to be as simple as the preceding one.

The extension of these CPD tests to regression analysis, Geisser (1987), and to conjugate priors presents no intrinsic difficulty.

7. Remarks

The case of detecting discordant or flawed observations within the predictive framework, where there are no easily discernible model alternatives[1] can be addressed in several different ways. The simplest solution is the case where one observation is suspect because of a known circumstance that could affect the observed value. The simple CPD tests discussed in section 1 are appropriate in that case. In other cases where diagnostics, checking functions, or discordancy indices are used to direct one's attention to the possibility of discordancies that cannot be verified as due to a known circumstance, the statistician can construct either a UPD or CPD test that will provide guidance in concluding whether to consider the observation as flawed. However, except for the simplest cases, an index, which basically is a test in the known circumstance discordancy situation, can be adjusted into a CPD test that in many instances is conveniently calculable and provides an informative guide to the determination of discordancy. Other advantages of these predictive tests are their much greater applicability then standard frequency tests to censoring, inclusion of prior information and in particular the existence of CPD tests for certain useful improper priors distributions.

[1]When there are such alternatives then the methods suggested here will be superseded by taking a known alternative into account. In such a case the posterior odds can be obtained for the alternative as opposed to the null.

As has been pointed out even within the same circumstance there may be a variety of possible tests one could use. Since these tests are designed for situations where model alternatives are not assumed, power like considerations that may aid in discriminating between tests are not available here. What should follow indicators of discordancy, reflected by small P-values, are attempts to find out what possibly could have gone wrong and in the failure of such a determination, a sensitivity analysis should be made to ascertain the effect that the possibly flawed values in the sample have future predictions.

Acknowledgment

This work was supported by National Science Foundation Grant DMS-8601314 and National Institutes of Health Grant GM 25271.

References

Aitchison, J. and I.R. Dunsmore (1975). *Statistical Prediction Analysis*, Cambridge: University Press.

Barnard, G.A. (1962). Prepared Contribution in *The Foundations of Statistical Inference*. L.J. Savage and Other Contributors, G.A. Barnard and D.R. Cox, eds., London: Methuen & Company, 39-49.

Barnett, V. and T. Lewis (1978). *Outliers in Statistical Data*. New York: John Wiley and Sons.

Box, G.E.P. (1980). "Sampling and Bayes' Inference in Scientific Modelling and Robustness," *Journal of the Royal Statistical Society, Series B*, Vol. 143, 383-430.

Cox, D.R. (1987). "Comment on 'Testing Precise Hypotheses' by J.O. Berger and M. Delampady," *Statistical Science*, Vol. 2, No. 3, 335-336.

Dixon, W.J. (1950). "Analysis of Extreme Values," *Annals of Mathematical Statistics*, Vol. 21, 488-506.

Dixon, W.J. (1951). "Ratios Involving Extreme Values," *Annals of Mathematical Statistics*, Vol. 22, 68-78.

Fisher, R.A. (1956). *Statistical Methods and Scientific Inference*, Edinburgh: Oliver and Boyd.

Geisser, S. (1980). "Discussion on 'Sampling and Bayes' Inference in Scientific Modelling and Robustness,' by G.E.P. Box," *Journal of the Royal Statistical Society, Series A*, Vol. 143, 416-417.

Geisser, S. (1985). "On the Predicting of Observables: A Selective Update," in *Bayesian Statistics 2*, with discussion, J.M. Bernardo, M.H. DeGroot, D.V. Lindley, and A.F.M. Smith, eds., Amsterdam: North-Holland, 353-358.

Geisser, S. (1987). "Influential Observations, Diagnostics and Discordancy Tests," *Journal of Applied Statistics*, Vol. 14, No. 2, 133-142.

Grubbs, F.E. (1969). "Procedures for Detecting Outlying Observations in Samples," *Technometrics*, Vol. 11, 1-21.

Johnson, W. and S. Geisser (1982). "Assessing the Predictive influence of Observations," in *Statistics and Probability Essays in Honor of C.R. Rao*, G. Kallianpur, P.R. Krishnaiah, and J. Ghosh, eds., Amsterdam: North Holland, 353-358.

Johnson, W. and S. Geisser (1983). "A Predictive View of the Detection and Characterization of Influential Observations in Regression Analysis," *Journal of the American Statistical Association*, Vol. 78, 137-144.

Likes, J. (1966). "Distribution of Dixon's Statistics in the Case of an Exponential Population," *Metrika*, Vol. 11, 46-54.

Smith, A.F.M. and L.I. Pettit (1985). "Outliers and Influential Observations in Linear Models," *Bayesian Statistics 2*, J.M. Bernardo, M.H. DeGroot, D.V. Lindley, and A.F.M. Smith, eds., Amsterdam: North-Holland, 473-494.

Bayesian and Likelihood Methods in Statistics and Econometrics
S. Geisser, J.S. Hodges, S.J. Press and A. Zellner (Editors)
© Elsevier Science Publishers B.V. (North-Holland), 1990

BAYESIAN CONTEXTUAL CLASSIFICATION WITH NEIGHBORS CORRELATED WITH TRAINING DATA

Ruben Klein and S. James Press

IMPA, Rio de Janeiro, Brazil and the University of California, Riverside

Abstract

We consider the problem of Bayesian contextual classification with spatial correlation structure in which the neighbors of an unclassified observation vector, z_0, and z_0 itself, are correlated with the training data set. We use the neighbors of z_0 to improve the classification procedure. In order to classify z_0 into one of K normal populations we choose that population which maximizes the predictive classification probability, which is shown to be proportional to a matrix T-density in z_0 and its neighbors. Permitting the training data to be correlated with z_0 and its neighbors generalizes the results from that where they are uncorrelated, a case in which they are physically far apart. In the current case, when we classify an observation whose neighbors and itself are correlated with observations coming from known populations, we are taking advantage of this existing correlation and a substantial improvement in probability of correct classification results. We illustrate the results with a variety of examples that are based upon simulated data.

Key Words – Bayesian classification, spatial correlation, remote sensing, correlation with training data.

1. Introduction

We are concerned with the problem of contextual classification from a Bayesian point of view, and in a context involving spatial structure. Electromagnetic signals in p frequencies reflected from a ground plot are received by remote sensing in a satellite; these signals constitute a p-variate observation, z_0. The problem is to classify z_0 as having come from one of K populations representing, say, in an agricultural context, high quality rice fields, low quality rice fields, high quality wheat fields, etc. There are some neighboring observations in the immediate vicinity of z_0 which will be used to improve the classification procedure over the results obtainable from classical approaches (where no neighbors are used, and no spatial correlation is assumed to be present). There is also a set of training data available, *i.e.*, observations from each of the populations which have been classified with certainty on the ground. These observations are referred to as ground truth.

In earlier research (Klein and Press, 1987) we assumed that the training data, and z_0 and its neighbors, were far enough away from each other so that the spatial correlation between them was small enough to be ignorable. In this paper we describe the results of research in which we assume that z_0 and its neighbors are sufficiently close to the training data set so that the spatial correlation between them must be taken into account. This approach, which follows naturally from the Bayesian development, can take advantage of the existing correlation to substantially improve the classification results. As far as we know this is a new idea.

The Bayesian classification procedure we develop is based upon the predictive classification distribution for z_0 and its neighbors, given all of the data, and given that with probability close to one, z_0 and its neighbors all belong to the same population (this is the assumption of local spatial continuity used by Switzer, 1980, and Mardia, 1984). We will classify z_0 into that population for which the predictive classification density multiplied by the prior classification probability is maximum, over all populations. This procedure is due to Geisser (1964, 1966).

The plan of the paper is to introduce the notation in Section 2. The model is set out in Section 3. It is there that we develop the basic result that the predictive classification distribution is matrix T in z_0 and its neighbors. Section 4 contains some examples based upon extensive simulation of data to evaluate the effectiveness of the new procedure in various situations.

2. Notation

$\{Z(s), s \in \mathcal{R}^2\}$ denotes a p-vector valued spatial stochastic process on the plane; s is a 2×1 column vector representing a point in the plane of the ground, say latitude and longitude. A realization of this process at the point s_0 is denoted by the $p \times 1$ column vector $Z(s_0) \equiv z_0$.

Suppose we have n_1 observations from population π_1; n_2 observations from $\pi_2; \ldots; n_k$ observations from π_K. We denote them by

$$\begin{array}{llll} x_1 \equiv [X_1'(s_{11}), & \ldots, & X_1'(s_{1n_1})]' & \in \quad \pi_1, \\ \vdots & & & \quad\ \vdots \\ x_K \equiv [X_K'(s_{K1}), & \ldots, & X_K'(s_{Kn_K})]' & \in \quad \pi_K. \end{array}$$

Note that $X_k(s_{kj})$ denotes a realization from the $Z(s)$ spatial process which arises from π_k; the observation emanates from the point s_{kj} on the ground. The prime denotes transpose. Let $D \equiv (x'_1, \ldots, x'_K)'$ denote the vector of x_j's, the observations known to come from particular populations. D (for "data") represents the ground truth, the training data set.

We assume that $\{X(s), s \in \mathcal{R}^2\}$ is a Gaussian, covariance stationary stochastic process. (This appears to be a reasonable approximation for data obtained from sensors in satellites; see, *e.g.*, Kittler and Pairman, 1985; and Elphinstone, *et al.*, 1985.) Thus, its mean vector doesn't depend upon s, and its covariances depend only upon the vector difference between the points at which we take the observations. So

$$\operatorname{cov}[X_j(s_1), X_j(s_2)] = \underset{(p \times p)}{\Sigma_j} (s_1 - s_2),$$

$j = 1, \ldots, K$. We take $\Sigma_j(0) \equiv \Sigma_j$.

Let ψ denote the cross-covariance function for observation vectors from different populations at different points. Thus, for all $s_1 \neq s_2$,

$$\psi_{ij}(s_1 - s_2) \equiv \operatorname{cov}[X_i(s_1), X_j(s_2)], \quad i \neq j.$$

In this paper we assume $\psi_{ij}(s) = 0$ for all i, j, $i \neq j$, and all $s \neq 0$ ($\psi_{ij}(s)$ is not defined for $s = 0$). That is, signals from two different populations are likely to be approximately uncorrelated.

The elements of each x_i are joint normally distributed, so that if e_r denotes an r-vector of "ones",

$$\mathcal{L}[\underset{(pn_i \times 1)}{x_i} \mid \theta_i, \Sigma_i(s)] = N\left[e_{n_i} \otimes \theta_i, \Sigma_i^*\right], \tag{1}$$

where

$$\Sigma_i^* \equiv \operatorname{var}[x_i \mid \theta_i, \Sigma_i(s)] = (\Sigma_i^*)'$$

$$= \begin{pmatrix} \Sigma_i, \Sigma_i(s_{i1} - s_{i2}), & \ldots, & \Sigma_i(s_{i1} - s_{in_i}) \\ \Sigma_i & \ddots & \vdots \\ & & \Sigma_i \end{pmatrix},$$

where $\mathcal{L}(\cdot)$ denotes probability law, and $\otimes$ denotes the direct or Kronecker product of two matrices. Denote the K populations by $\pi_k \equiv N(\theta_k, \Sigma_k)$, $k = 1, \ldots, K$. The parameters (θ_k, Σ_k) are unknown.

Denote the neighbors of z_0 by the vector $\Delta_{(pr\times 1)} \equiv (z'_1, \ldots, z'_r)'$. For example, if we assume the neighbors of z_0 are located at the interstices of a lattice, an 8-neighbor configuration might be envisioned as shown in Figure 1.

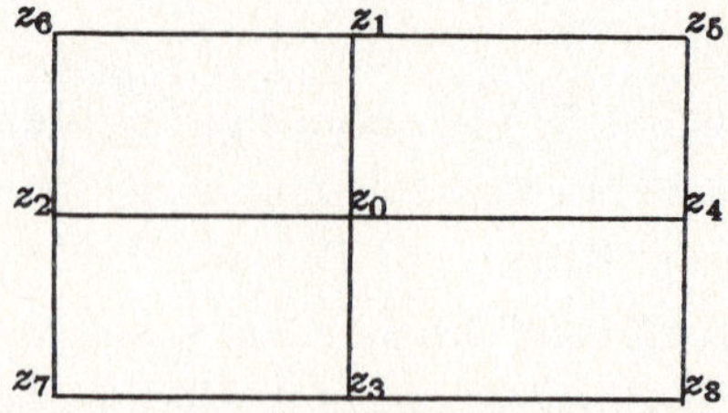

Figure 1: Lattice Neighbor Configuration

Note that z_0 is at the center, as the observation to be classified, and $z_1, \ldots, z_8$ are located on the grid around z_0. A 4-neighbor cross-configuration would consist of the neighbors $z_1, \ldots, z_4$.

The prior probability for classifying z_0 into π_k is denoted by

$$p_k \equiv P\{z_0 \in \pi_k\}.$$

The predictive classification probability is denoted by

$$\tau_k \equiv P\{z_0 \in \pi_k \mid \Delta, z_0, D\}.$$

Defining $\delta \equiv (\Delta', z'_0)'$, gives, for $\delta : p(r+1) \times 1$,

$$\tau_k = P\{z_0 \in \pi_k \mid \delta, D\}. \tag{2}$$

We will use $p(t)$ generically to denote the density of a random variable T, at $T = t$; the density might be taken with respect to Lebesgue measure, or counting measure, whichever is appropriate.

3. Model

Application of Bayes theorem to eqn. (2) gives the predictive classification probability

$$\tau_k = \frac{p(\delta, D \mid z_0 \in \pi_k) p_k}{\sum_{j=1}^{K} p(\delta, D \mid z_0 \in \pi_j) p_j},$$

or if we use just π_k as a shorthand notation for the event, $\{z_0 \in \pi_k\}$, and if we note that the denominator is a constant not depending upon k, we can rewrite τ_k as

$$\tau_k \propto p_k p(\delta, D \mid \pi_k). \tag{3}$$

The symbol "$\propto$" denotes proportionality. Conditioning upon D, and the neighbors gives

$$\tau_k \propto p_k \sum_{\rho_1=1}^{K} \cdots \sum_{\rho_r=1}^{K} p(\delta \mid D, \pi_k; \pi_{\rho_1}, \ldots, \pi_{\rho_r})$$

$$\cdot\, p(D \mid \pi_k; \pi_{\rho_1}, \ldots, \pi_{\rho_r}) p(\pi_{\rho_1}, \ldots, \pi_{\rho_r} \mid \pi_k.)$$

In this equation we have used the notation π_{ρ_j} in place of "$z_j \in \pi_{\rho_j}$", with $\rho_0 \equiv k, j = 0, 1, \ldots, r$. Since $p(D \mid \pi_k; \pi_{\rho_1}, \ldots, \pi_{\rho_r})$ depends only upon the known data, D, we incorporate this term into the proportionality constant to get

$$\tau_k \propto p_k \sum_{\rho_1=1}^{K} \cdots \sum_{\rho_r=1}^{K} p(\delta \mid D, \pi_k; \pi_{\rho_1}, \ldots, \pi_{\rho_r}) p(\pi_{\rho_1}, \ldots, \pi_{\rho_r} \mid \pi_k).$$

In this research we assume along with Switzer (1980) and Mardia (1984) that z_0 and its neighbors all belong to the same population (local spatial continuity), so if $z_0 \in \pi_k,\ \rho_1 = \rho_2 = \ldots = \rho_r = k,\ p(\pi_{\rho_1}, \ldots, \pi_{\rho_r} \mid \pi_k) = p(\pi_k, \ldots, \pi_k \mid \pi_k) = 1$, and the predictive classification probability becomes

$$\tau_k \propto p_k p(\delta \mid D, \pi_k; \pi_k, \ldots, \pi_k). \tag{4}$$

This assumption is realistic when the true map contains large regions of homogeneous populations, as in agricultural applications where the populations might be large fields of wheat, corn, and other crops. We now evaluate $p(\delta \mid D, \pi_k; \pi_k, \ldots, \pi_k)$.

We know that all observations are normally distributed, so the joint distribution of z_0, its neighbors, and the training data set, given the parameters, and that $\delta \in \pi_k$, is for some (t, T),

$$\mathcal{L}(\delta, D \mid \pi_k; \pi_k, \ldots, \pi_k; P) = N(t, T), \tag{5}$$

where P denotes the unknown population parameters:

$$P \equiv (\theta_1, \ldots, \theta_K; \Sigma_1, \ldots, \Sigma_K).$$

We next evaluate the parameters (t, T). Define

$$E(\delta \mid \pi_k; \pi_k, \ldots, \pi_k; P) \equiv \underset{p(r+1)\times 1}{\tilde{\phi}_k} = (e'_{r+1} \otimes \theta'_k)', \tag{6a}$$

$$E(D \mid \pi_k; \pi_k, \ldots, \pi_k; P) = E(D \mid P) \equiv \underset{\left(p\sum_1^K n_j\right)\times 1}{\omega} = (e'_{n_1} \otimes \theta'_1, \ldots, e'_{n_K} \otimes \theta'_K)'. \quad \text{(6b)}$$

Therefore,

$$t = (\tilde{\phi}'_k, \omega')'. \tag{7}$$

The covariance matrix is given by

$$T = \left(\begin{array}{c|c} \text{var}(\delta) & \text{cov}(\delta, D) \\ \hline [\text{cov}(\delta, D)]' & \text{var}(D) \end{array} \right). \tag{8}$$

Since $\delta = (\Delta', z'_0)'$,

$$\text{var}(\delta) \equiv \text{var}(\delta \mid \pi_k; \pi_k, \ldots, \pi_k; P) \equiv \tilde{\Lambda}_k = \left(\begin{array}{c|c} \text{var}(\Delta) & \text{cov}(\Delta, z_0) \\ \hline [\text{cov}(\Delta, z_0)]' & \text{var}(z_0) \end{array} \right),$$

or

$$\tilde{\Lambda}_k = \left(\begin{array}{cccc|c} \Sigma_k, \Sigma_k(s_1 - s_2), & \ldots, & \Sigma_k(s_1 - s_r) & & \Sigma_k(s_1 - s_0) \\ \Sigma_k, & \ldots, & \Sigma_k(s_2 - s_r) & & \Sigma_k(s_2 - s_0) \\ & \ddots & \vdots & & \vdots \\ & & \Sigma_k & & \Sigma_k(s_r - s_0) \\ \hline & & & & \Sigma_k \end{array} \right).$$

Now assume that

$$\Sigma_i(s) = \alpha_i(s)\Sigma_i, \qquad i = 1, \ldots, K, \tag{10}$$

where Σ_i is a constant matrix not depending upon s, and $\alpha_i(s)$ is a correlation function; so it is a monotone, non-increasing, positive definite function of s's, with $\alpha_i(0) = 1$. The actual functional form $\alpha_i(s)$ should take depends upon the data. This could be determined by variogram analysis (see, e.g., Journel and Huijbregts, 1978). We assume for simplicity in the examples (Section 4) for this paper, that

$$\alpha_i(s) = \exp\{-c_i(s's)\}, \qquad c_i > 0. \tag{11}$$

The constants c_i can be assessed directly from variogram analysis or by methods described in Klein and Press (1987), Section 7. Use of eqn. (10) in $\tilde{\Lambda}_k$ yields

$$\tilde{\Lambda}_k = B_k \otimes \Sigma_k \tag{12}$$

where the correlation matrix B_k for (Δ, z_0) is given by

$$\underset{(r+1)\times(r+1)}{B_k} = \begin{pmatrix} 1, \alpha_k(s_1 - s_2), & \ldots, & \alpha_k(s_1 - s_r) & \alpha_k(s_1 - s_0) \\ 1 \cdot & & & \vdots \\ & \cdot & \cdot & \\ & & \cdot \;\cdot\; 1 & \alpha_k(s_r - s_0) \\ & & & 1 \end{pmatrix}.$$

Next, note that since $\text{cov}(x_i, x_j) = 0, \quad i \neq j$,

$$\text{var}(D) \equiv \text{var}(D \mid \pi_k; \pi_k, \ldots, \pi_k; P) = \text{var}(D \mid P) = \begin{pmatrix} \Sigma_1^* & & 0 \\ & \ddots & \\ 0 & & \Sigma_K^* \end{pmatrix},$$

where: $\Sigma^* \equiv \text{var}(x_j)$. But

$$\text{var}(x_j) = \begin{pmatrix} \Sigma_j, \; \Sigma_j(s_{j1} - s_{j2}), & \ldots, & \Sigma_j(s_{j1} - s_{jn_j}) \\ \ddots & & \vdots \\ & & \Sigma_j \end{pmatrix}.$$

Since $\Sigma_j(s) = \alpha_j(s)\Sigma_j$,

$$\text{var}(x_j) = A_j \otimes \Sigma_j, \qquad j = 1, \ldots, K,$$

where:

$$\underset{(n_j \times n_j)}{A_j} \equiv \begin{pmatrix} 1, \; \alpha_j(s_{j1} - s_{j2}), & \ldots, & \alpha_j(s_{j1} - s_{jn_j}) \\ \cdot \quad \cdot & & \vdots \\ & \cdot \quad \cdot & \cdot \; 1 \end{pmatrix}.$$

So

$$\underset{\left(p\sum_1^K n_j\right)\times\left(p\sum_1^K n_j\right)}{\Omega} \equiv \text{var } D = \begin{pmatrix} A_1 \otimes \Sigma_1 & & 0 \\ & \ddots & \\ 0 & & A_K \otimes \Sigma_K \end{pmatrix}. \tag{13}$$

Define

$$\text{Cov}(\delta, D) \equiv \text{Cov}(\delta, D \mid \pi_k; \pi_k, \ldots, \pi_k; P) \equiv \tilde{C}_k,$$

or

$$\tilde{C}_k = [\text{cov}(\delta, x_1), \ldots, \text{cov}(\delta, x_K)]. \tag{14}$$

Substituting eqns. (12), (13), and (14) into eqn. (8) gives

$$T = \begin{pmatrix} B_k \otimes \Sigma_k & \tilde{C}_k \\ \tilde{C}_k' & \Omega \end{pmatrix}. \tag{15}$$

We must still evaluate $\tilde{C}_k$. First note that $\mathrm{cov}(\delta, x_j) = 0$, for $j \neq k$. So from eqn. (14),

$$\tilde{C}_k = [0, \ldots, 0, \mathrm{cov}(\delta, x_k), 0, \ldots, 0].$$

Define $C_k^* \equiv \mathrm{cov}(\delta, x_k)$, so

$$\tilde{C}_k = (0, \ldots, 0, C_k^*, 0, \ldots, 0). \tag{16}$$

But if $x_k \equiv (x'_{k1}, \ldots, x'_{kn_k})'$,

$$C_k^* = \begin{pmatrix} \mathrm{Cov}(z_1, x_{k1}), & \ldots, & \mathrm{Cov}(z_1, x_{kn_k}) \\ \vdots & & \vdots \\ \mathrm{Cov}(z_0, x_{k1}), & \ldots, & \mathrm{Cov}(z_0, x_{kn_k}) \end{pmatrix}.$$

Since z_j and x_k both belong to π_k, for all $j = 0, 1, \ldots, r$,

$$C_k^* = \begin{pmatrix} \Sigma_k(s_1 - s_{k1}), & \ldots, & \Sigma_k(s_1 - s_{kn_k}) \\ \vdots & & \vdots \\ \Sigma_k(s_0 - s_{k1}), & \ldots, & \Sigma_k(s_0 - s_{kn_k}) \end{pmatrix},$$

or,

$$C_k^* = C_k \otimes \Sigma_k, \tag{17}$$

where:

$$\underset{(r+1)\times n_k}{C_k} \equiv \begin{pmatrix} \alpha_k(s_1 - s_{k1}), & \ldots, & \alpha_k(s_1 - s_{kn_k}) \\ \vdots & & \vdots \\ \alpha_k(s_0 - s_{k1}), & \ldots, & \alpha_k(s_0 - s_{kn_k}) \end{pmatrix}$$

Finally,

$$\tilde{C}_k = (0, \ldots, 0, (C_k \otimes \Sigma_k), 0, \ldots, 0), \tag{18}$$

which completes the evaluation of T.

To evaluate the predictive classification probability in eqn. (4) we must still evaluate $p(\delta \mid D, \pi_k; \pi_k, \ldots, \pi_k)$. Conditioning on D in eqn. (5) gives

$$\mathcal{L}(\delta \mid D, \pi_k; \pi_k, \ldots, \pi_k; P) = N[\tilde{\phi}_k + \tilde{C}_k \Omega^{-1}(D - \omega), \tilde{\Lambda}_k - \tilde{C}_k \Omega^{-1} \tilde{C}'_k]. \tag{19}$$

We may now evaluate the predictive classification probability by integrating the density of δ in eqn. (19) with respect to P. Thus,

$$p(\delta \mid D, \pi_k; \pi_k, \ldots, \pi_k) = \int p(\delta \mid D, \pi_k; \pi_k, \ldots, \pi_k; P) p(P \mid D) dP, \tag{20}$$

where $p(P \mid D)$ denotes the posterior density of the population parameters given the data. First evaluate $p(P \mid D)$.

The likelihood function for the data, given the parameters, is readily obtained from the marginal distribution of D in eqn. (5), *viz.*,

$$\mathcal{L}(D \mid P) = N(\omega, \Omega), \tag{21}$$

where ω and Ω are given in eqns. (6b), and (13).

We adopt a vague, Jeffreys invariant, prior distribution for $P \equiv (\theta_1, \ldots, \theta_K; \Sigma_1, \ldots, \Sigma_K)$ (see Jeffreys, 1961, pp. 181, 182 where he shows the prior to be proportional to the square root of the information matrix, which results in our prior; see also, Geisser and Cornfield, 1963, or Press, 1982, Sect. 3.6.2),

$$p(P) = p(\theta_1) \ldots p(\theta_K) p(\Sigma_1) \ldots p(\Sigma_K),$$

with $p(\theta_j) \propto$ constant, and

$$p(\Sigma_j) \propto |\Sigma_j|^{-(p+1)/2}.$$

Thus,

$$p(P) \propto \prod_{j=1}^{K} |\Sigma_j|^{-(p+1)/2}. \tag{22}$$

(Note that we could also adopt alternatively a natural conjugate prior distribution family for P. It is straightforward to check, using normal priors for the $(\theta_j \mid \Sigma_j)$'s and inverted Wishart priors for the Σ_j's, that the results would then be of exactly the same form as we'll find for the vague prior case, but the parameters of the resulting distribution would be modified accordingly.) Combining eqns. (21) and (22) by means of Bayes theorem gives the posterior density for P,

$$p(P \mid D) \propto \frac{1}{\prod_{j=1}^{K} |\Sigma_j|^{(p+1)/2} |\Omega|^{1/2}} \exp\{(-1/2)(D-\omega)'\Omega^{-1}(D-\omega)\}. \tag{23}$$

Now substitute eqns. (19) and (23) into (20) to get

$$\begin{aligned} p(\delta \mid D, \pi_k; \pi_k, \ldots, \pi_k) \propto & \int_\theta \int_\Sigma \prod_{j=1}^{K} |\Sigma_j|^{-(p+1)/2} |\Omega|^{-1/2} \\ & \cdot \exp\{(-1/2)(D-\omega)'\Omega^{-1}(D-\omega)\} \left| \tilde{\Lambda}_k - \tilde{C}_k \Omega^{-1} \tilde{C}_k' \right|^{-1/2} \\ & \cdot \exp\{(-1/2)[\delta - \tilde{\phi}_k - \tilde{C}_k \Omega^{-1}(D-\omega)]'(\tilde{\Lambda}_k - \tilde{C}_k \Omega^{-1} \tilde{C}_k)^{-1} \\ & \cdot [\delta - \tilde{\phi}_k - \tilde{C}_k \Omega^{-1}(D-\omega)]\} d\theta_1 \ldots d\theta_k d\Sigma_1 \ldots d\Sigma_K. \end{aligned} \tag{24}$$

Note that $\int_\theta$ denotes the iterated integral over $\theta_1, \ldots, \theta_K$; and $\int_\Sigma$ denotes the iterated integral over $\Sigma_1, \ldots, \Sigma_K$. To evaluate the integrals in eqn. (24), we first simplify. Note from eqns. (6a,b), (13), (16) and (17) that

$$\tilde{\phi}_k + \tilde{C}_k \Omega^{-1}(D-\omega) = (e_{r+1} \otimes \theta_k) + (C_k A_k^{-1} \otimes I_p)(x_k - e_{n_k} \otimes \theta_k). \tag{25}$$

In a similar way we can find that

$$\tilde{\Lambda}_k - \tilde{C}_k \Omega^{-1} \tilde{C}_k' = (B_k - C_k A_k^{-1} C_k') \otimes \Sigma_k = H_k \otimes \Sigma_k, \tag{26}$$

where: $H_k \equiv B_k - C_k A_k^{-1} C_k'$. Finally, from eqns. (6b) and (13),

$$\begin{aligned}(D-\omega)'\Omega^{-1}(D-\omega) &= \sum_{j=1}^{K} (x_j - e_{n_j} \otimes \theta_j)'(A_j^{-1} \otimes \Sigma_j^{-1})(x_j - e_{n_j} \otimes \theta_j) \\ &= \sum_{j=1}^{K} \sum_{\ell=1}^{n_k} \sum_{m=1}^{n_k} a_j^{\ell m} (x_{j\ell} - \theta_j)' \Sigma_j^{-1} (x_{jm} - \theta_j), \end{aligned} \tag{27}$$

where $A_j^{-1} \equiv (a_j^{\ell m})$.

Since, except for the kth term, the integrand in eqn. (24) doesn't involve δ, we can integrate over all $\theta_1, \ldots, \theta_{k-1}, \theta_{k+1}, \ldots, \theta_K$, and $\Sigma_1, \ldots, \Sigma_{k-1}, \Sigma_{k+1}, \ldots, \Sigma_K$ and get constants that can be incorporated into the proportionality constant. The result is

$$\begin{aligned} p(\delta \mid D, \pi_k; \pi_k, \ldots, \pi_k) \propto & \int_{\theta_k} \int_{\Sigma_k} |\Sigma|^{-(p+n_k+r+2)/2} \\ & \cdot \exp\{(-1/2) \sum_{\ell=1}^{n_k} \sum_{m=1}^{n_k} a_k^{\ell m} (x_{k\ell} - \theta_k)' \Sigma_k^{-1} (x_{km} - \theta_k)\} \\ & \cdot \exp\{(-1/2)[(\delta - \bar{\delta})'(H_k \otimes \Sigma_k)^{-1}(\delta - \bar{\delta})]\} d\theta_k d\Sigma_k, \end{aligned} \tag{28}$$

where: $\bar{\delta} \equiv e_{r+1} \otimes \theta_k + (C_k A_k^{-1} \otimes I_p)(x_k - e_{n_k} \otimes \theta_k)$. Now simplify $\bar{\delta}$. Combining terms in $\bar{\delta}$ gives

$$\bar{\delta} = (C_k A_k^{-1} \otimes I_p) x_k + (e_{r+1} - C_k A_k^{-1} e_{n_k}) \otimes \theta_k. \tag{29}$$

The quadratic form involving δ in eqn. (28) may now be written

$$(\delta - \bar{\delta})'(H_k \otimes \Sigma_k)^{-1}(\delta - \bar{\delta}) = \sum_{i=1}^{r+1} \sum_{j=1}^{r+1} H_k^{ij} (z_i - u_{ki} - v_{ki}\theta_k)' \Sigma_k^{-1} (z_j - u_{kj} - v_{kj}\theta_k), \tag{30}$$

where: $H_k^{-1} \equiv (H_k^{ij})$, $u_k \equiv (u_{ki}) \equiv (C_k A_k^{-1} \otimes I_p) x_k$,

$$v_k \equiv (v_{ki}) \equiv e_{r+1} - C_k A_k^{-1} e_{n_k}, \text{ and } z_{r+1} \equiv z_0.$$

Substituting eqn. (30) into eqn. (28) gives

$$p(\delta \mid D, \pi_k; \pi_k, \ldots, \pi_k) \propto \int_{\theta_k} \int_{\Sigma_k} |\Sigma|^{-(p+n_k+r+2)/2}$$

$$\cdot \exp\{(-1/2) \sum_{\ell=1}^{n_k} \sum_{m=1}^{n_k} a_k^{\ell m} (\theta_k - x_{k\ell})' \Sigma_k^{-1} (\theta_k - x_{km})\}$$

$$\cdot \exp\left\{(-1/2) \sum_{i=1}^{r+1} \sum_{j=1}^{r+1} H_k^{ij} (v_{k\ell}\theta_k - z_i + u_{ki})' \Sigma_k^{-1}\right.$$

$$\left. \cdot (v_{kj}\theta_k - z_j + u_{kj})\right\} d\theta_k d\Sigma_k. \tag{31}$$

Note from eqn. (31) that there are two quadratic forms in θ_k in the exponent. They may be combined to form a single quadratic form. The result then looks like a normal density in θ_k, which is readily integrable. The remaining integral then looks like the kernel of an inverted Wishart density, which is also readily integrable. Combining terms and carrying out the algebra shows that the density of δ (the predictive classification density) is matrix T, and is given by

$$\begin{aligned} &p(\delta \mid D, z_0 \in \pi_k; z_1 \in \pi_k, \ldots, z_r \in \pi_k) \\ &\quad \propto |V_k + (Z - \overset{=}{Z}{}_k^*) Q_k^{-1} (Z - \overset{=}{Z}{}_k^*)'|^{-(n_k+r)/2}, \end{aligned} \tag{32}$$

where:

$$\underset{p\times(r+1)}{Z} \equiv (z_1, \ldots, z_r; z_0), \qquad \underset{p(r+1)\times 1}{\delta} = \text{vec}(Z),$$

where the "vec" operator denotes the stacking of the columns of Z, and we have used a semicolon to separate z_0 from its r neighbors,

$$\underset{(p\times n_k)}{\Xi_k} \equiv (\underset{(p\times 1)}{x_{k1}}, \ldots, \underset{(p\times 1)}{x_{kn_k}}), \quad x_k = (x_{ki}) = \text{vec } \Xi_k,$$

$$\underset{(p\times p)}{V_k} \equiv \sum_{i=1}^{n_k} \sum_{j=1}^{n_k} a_k^{ij} (x_{ki} - \overset{=}{x}_k)(x_{kj} - \overset{=}{x}_k)'$$

denotes the generalized sum-of-squares matrix,

$$\underset{(p\times 1)}{\overset{=}{x}_k} \equiv \frac{\sum_{i=1}^{n_k} \sum_{j=1}^{n_k} a_k^{ij} x_{kj}}{\sum_{i=1}^{n_k} \sum_{j=1}^{n_k} a_k^{ij}},$$

denotes the generalized mean with $A^{-1} \equiv (a^{ij})$, and

$$\underset{(r+1)\times(r+1)}{Q_k^{-1}} \equiv H_k^{-1} - \frac{H_k^{-1} v_k v_k' H_k^{-1}}{v_k' H_k^{-1} v_k + e_{n_k}' A_k^{-1} e_{n_k}},$$

denotes the inverse of the scale matrix of the matrix T-density; the location of the distribution is

$$\underset{p\times(r+1)}{\bar{\bar{Z}}_k^*} \equiv \bar{\bar{Z}}_k + (\Xi_k - \bar{\bar{x}}_k\, e'_{n_k})A_k^{-1}C'_k,$$

where: $\bar{\bar{Z}}_k \equiv (\,\bar{\bar{x}}_k\,, \ldots, \,\bar{\bar{x}}_k\,)$. Note that $\bar{\bar{Z}}_k^*$ is the estimated conditional mean of $(\delta \mid D)$. We also note that when the training data is not correlated with z_0 and its neighbors, $C_k = 0$, which implies that $\bar{\bar{Z}}_k^* = \bar{\bar{Z}}_k$, which is the location matrix in that case (see Klein and Press, 1987). It is also true that when $C_k = 0$, $H_k = B_k$, and $v_k = e_{r+1}$; so Q_k^{-1} reduces to its value for that case as well. Thus, for $C_k = 0$, the distribution of δ reduces to its value computed in Klein and Press (1987); therefore, this result is more general.

In summary, we classify z_0 into that population for which

$$\tau_k = \frac{p_k p(\delta \mid D, \pi_k; \pi_k, \ldots, \pi_k)}{\sum_{k=1}^{k} p_k p(\delta \mid D, \pi_k; \pi_k, \ldots, \pi_k)}, \qquad k = 1, \ldots, K$$

is maximum, where $p(\delta \mid D, \pi_k; \pi_k, \ldots, \pi_k)$, the predictive classification density is given by the matrix T-density in eqn. (32).

4. Examples and Discussion

Since we have developed a new procedure for classifying spatially correlated, remotely sensed, data we would like to evaluate the procedure's efficacy. But we cannot evaluate the efficacy of the procedure by applying it to real, remotely sensed, data unless we have complete "ground truth"; that is, unless we know the true map on the ground from which the data came. Unfortunately, it is difficult to obtain both remotely sensed data and the accompanying ground truth. Moreover, we cannot study special kinds of problems with real data, such as how far off will a classification analyst be if he assumes the data are independent when in fact they exhibit spatial correlation. Many analysts have done this without indicating the nature of the accompanying degradation in correct classification probability. Alternatively, we applied the procedure to simulated data in various contexts. In both examples, we generated spatially correlated normal data. Both populations were one-dimensional, with unknown parameters, and in both cases, $K = 2$. The correlation parameters, c_i, were assumed in both examples to be known and equal; *i.e.*, $c_1 = c_2$. To understand the nature of the variability in the results attributable to sampling, and to see the procedure applied to other examples, see Klein and Press (1988).

The entire classification procedure was repeated 100 times independently for each set of circumstances. At that point, results were stable. Results were averaged, and the averages were compared. In both cases, we have examined two extreme cases, that with zero neighbors, and that with 8 neighbors lying on a lattice (as shown in Figure 1, earlier).

Since the training data is known, and since we condition on the training data, we can make no mistakes in classifying the observations in the training data. For this reason, we only classify the observations not belonging to this training data. At the edge of the lattice, and adjacent to the training data, we have adapted the procedure to use as many neighbors as possible, according to the neighbor scheme. In both cases we have adopted a vague prior for classification, *i.e.*, we have taken $p_1 = p_2 = 1/2$. It is assumed throughout that z_0 and its neighbors are correlated with the training data. We have in both cases also studied the effect of ignoring spatial correlation (which many classification analysts have done) when in fact it is present. This type of analysis sheds new light on an important issue. For these examples, we denote the spatial correlation by $\rho \equiv \alpha(s) = \exp\{-c(s's)\}$.

4.1 Example (1)

In this example we classify unknown observations into $\pi_1 = N(0,1)$ *vs.* $\pi_2 = N(1,1)$. The true map is the 20×20 lattice shown in Figure 2, where it may be seen that the population consisting of the dark points is sectioned into two disjoint regions by the population of light points.

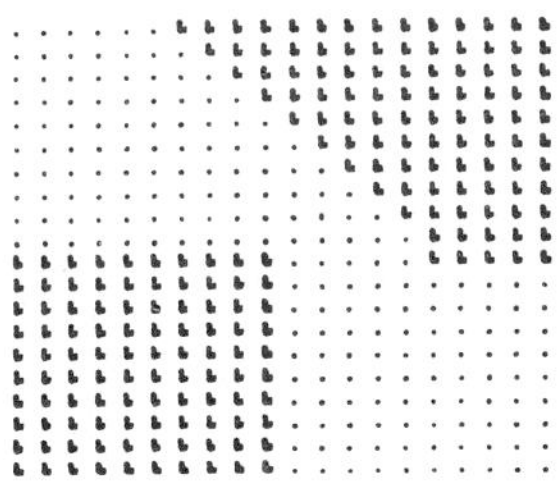

Figure 2: True Map for Examples 1, 2

For training data we have used observations from all of the points in columns 2, 7, 12, 17. We then classified all remaining points. Numerical results of the simulations are summarized in Table 1. Pictorial results are presented in Figure 3.

4.2 Example (2)

This example employs the same lattice configuration for the unknown observa-

Table 1: Simulated Data Showing Mean Percentage
of Correct Classifications $N(0,1)$ vs. $N(1,1)$

c	0 Neighbors		8 Neighbors	
.3 moderate correlation** ($\rho \simeq .75$)	76.29	69.07*	84.22	70.74*
1.0 low correlation ($\rho \simeq .36$)	70.48	68.83*	77.67	76.61*
∞ independent data	69.23		83.84	

*Numbers in lower right hand corner of each box indicate result when analysis assumes data is independent, when in fact it is correlated.

**Correlation values assume s's = 1, $\rho = \exp(-c)$.

Figure 3: Typical Cases with $N(0,1)$ vs. $N(1,1)$ — Example 1
[Percent correctly classified overall, percent correctly classified into $N(0,1)$, percent correctly classified into $N(1,1)$]

(a) zero neighbors

[68.4, 58.9, 77.8]

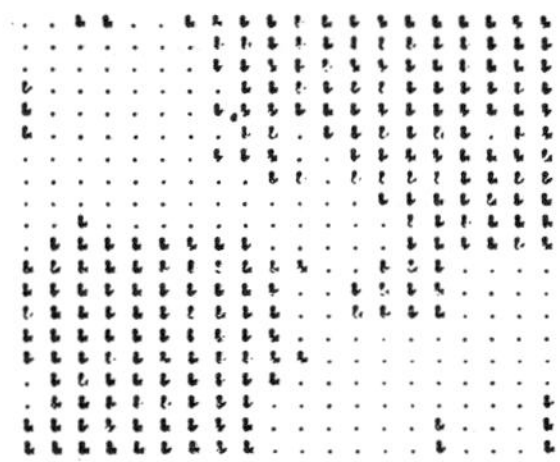

(b) 8 neighbors

[85.0, 74.7, 95.1]

Independent Data

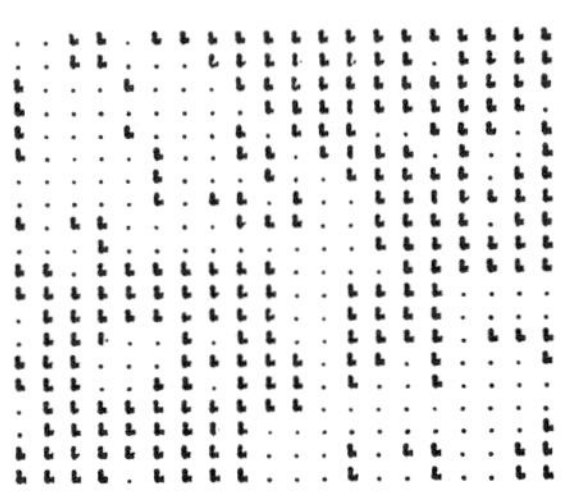

(a) zero neighbors

[70.6, 58.2, 82.7]

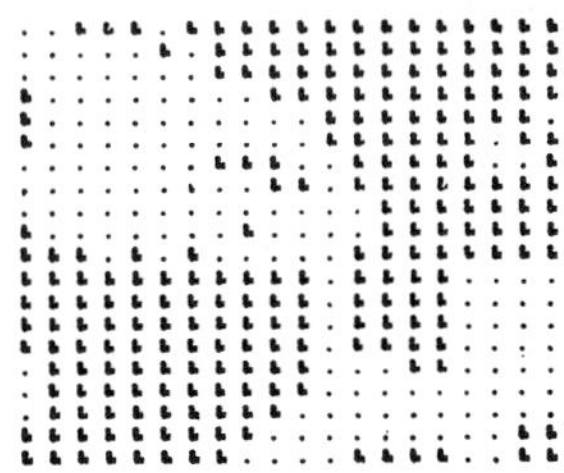

(b) 8 neighbors

[78.1, 65.8, 90.1]

Correlated Data $-c_1 = c_2 = 1.0$

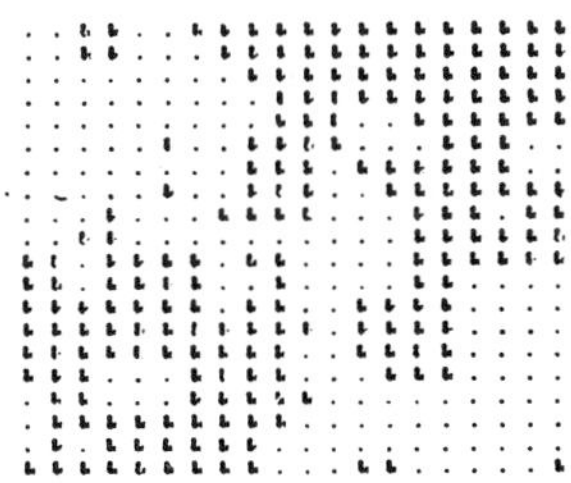

(a) zero neighbors

[76.6, 70.3, 82.7]

(b) 8 neighbors

[84.4, 76.6, 92.0]

Correlated Data $-c_1 = c_2 = 0.3$

Table 2: Simulated Data Showing Mean Percentage of Correct Classifications $N(0,1)$ vs. $N(2,1)$

c	0 Neighbors		8 Neighbors	
.3 moderate correlation** ($\rho \simeq .75$)	87.53	84.40*	88.02	85.94*
1.0 low correlation ($\rho \simeq .36$)	84.67	84.16*	89.74	90.47*
∞ independent data	84.12		94.10	

*Numbers in lower right hand corner of each box indicate result when analysis assumes data is independent, when in fact it is correlated.

**Correlation values assume s's = 1, $\rho = \exp(-c)$.

Figure 4: Typical Cases with $N(0,1)$ vs. $N(2,1)$ — Example 2
[Percent correctly classified overall, percent correctly classified into $N(0,1)$, percent correctly classified into $N(2,1)$]

(a) zero neighbors
[84.4, 84.2, 84.6]

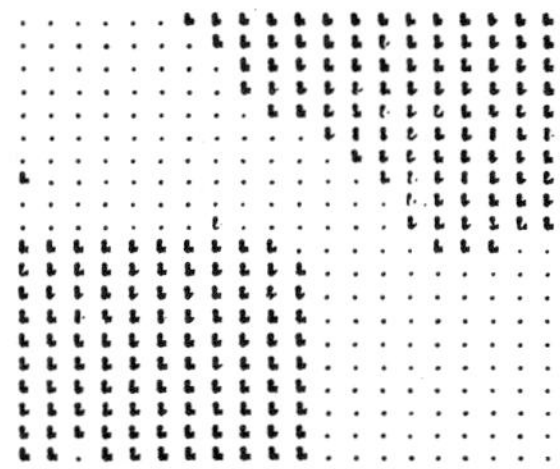

(b) 8 neighbors
[94.7, 91.1, 98.1]

Independent Data

(a) zero neighbors
[84.1, 86.7, 81.5]

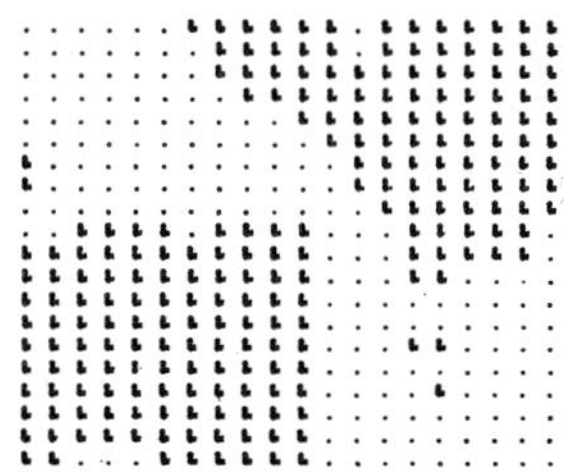

(b) 8 neighbors
[88.1, 80.4, 95.7]

Correlated Data $-c_1 = c_2 = 1.0$

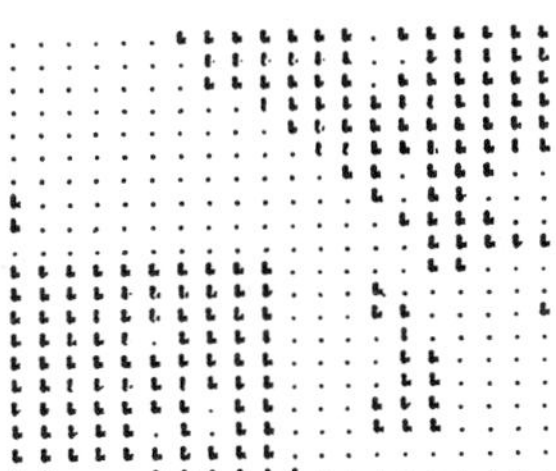

(a) zero neighbors
[86.9, 88.0, 85.8]

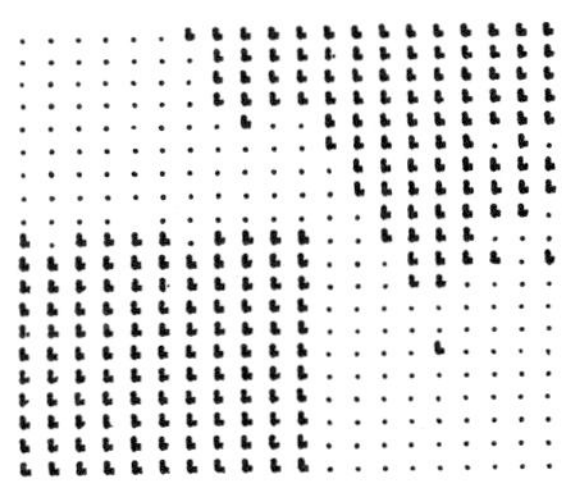

(b) 8 neighbors
[87.8, 80.4, 95.1]

Correlated Data $-c_1 = c_2 = 0.3$

tions to be classified, and the same columns of training data as in Example (1) (see Figure 2 for the true map). In this example, however, we have changed the populations to $\pi_1 = N(0,1)$, and $\pi_2 = N(2,1)$. Thus, it should now be easier to classify than in Example (1). Numerical results for this case are shown in Table 2. It may be seen by comparing Table 2 with Table 1 that as expected, the percentage of correct classifications is much greater when the centers of the populations are further apart. Pictorial results are presented in Figure 4.

Acknowledgments

The first author would like to gratefully acknowledge the support of CAPES (Brazil) proc: 6183/85-4 and of CNPq (Brazil) proc: 20.0213/87.3

We would like to thank Telma S. Klein for helping us to design the simulation programs used to obtain the results in Section 4.

References

Elphinstone, C.D., A.T. Lonergan, L.P. Fatti, and D.M. Hawkins (1985). "An Empirical Investigation into the Application of Some Statistical Techniques to the Classification of Remotely Sensed Data," Special Report SWISK 40, National Research Institute for Mathematical Sciences, CSIR, Pretoria.

Geisser, S. (1964). "Posterior Odds for Multivariate Normal Classifications," *Journal of the Royal Statistical Society (B)*, Vol. 26, 69-76.

Geisser, S. (1966). "Predictive Discrimination." in *Multivariate Analysis*, P.R. Krishnaiah, ed. New York: Academic Press, 149-163.

Geisser, S. and J. Cornfield (1963). "Posterior Distributions for Multivariate Normal Parameters," *JRSS (B)*, Vol. 25, 368-376.

Jeffreys, H. (1961). *Theory of Probability*, 3rd ed., Clarendon Press.

Journel, A.G. and C.J. Huijbregts (1978). *Mining Geostatistics*. New York: Academic Press.

Kittler, J. and D. Pairman (1985). "Contextual Pattern Recognition Applied to Cloud Detection and Identification," *IEEE Transactions on Geoscience and Remote Sensing*, Vol. GE-23, No. 6, 855-863.

Klein, R. and S.J. Press (1987). "Spatial Structure in Bayesian Classification," Technical Report No. 157, Department of Statistics, University of California, Riverside.

Klein, R. and S.J. Press (1988). "Bayesian Contextual Classification with Neighbors Correlated with Training Data," Technical Report No. 167, Department of Statistics, University of California, Riverside.

Mardia, K.V. (1984). "Spatial Discrimination and Classification Maps," *Comm. in Stat., Theory and Methods*, Vol. 13, No. 18, 2181-2197.

Press, S.J. (1982). *Applied Multivariate Analysis: Using Bayesian and Frequentist Methods of Inference*, Malabar, Florida: Krieger Publishing Company.

Switzer, P. (1980). "Extension of Linear Discriminant Analysis for Statistical Classification of Remotely Sensed Satellite Imagery," *Mathematical Geology, Vol. 12, No. 4, 367-376.*

Bayesian and Likelihood Methods in Statistics and Econometrics
S. Geisser, J.S. Hodges, S.J. Press and A. Zellner (Editors)

BAYESIAN PREDICTIVE DISTRIBUTIONS UNDER MULTINOMIAL SAMPLING

Peter J. Lenk
New York University

Abstract

This paper presents a general method of constructing families of distributions for the probabilities of the multinomial distribution such that these families are closed in the prior to posterior Bayesian analysis. The special case of the generalized, logistic normal distribution is studied in detail. The logistic normal distribution is developed by a logistic transform of a multivariate normal random vector. This model for the probabilities of the multinomial distribution has three features which make it a convenient family to express ones prior opinions:

1. The prior means of the probabilities are a logistic transform of the means of the normal vector.
2. The variances of the components of the normal vector determine the tradeoff between the prior distribution and the data.
3. The covariances between the components of the the normal vector determine the amount of smoothing of the sample information. For the logistic normal family, the predictive distribution is approximated with a cubic spline.

1. Introduction

The most widely used family of prior distributions on the simplex for multinomial sampling is the Dirichlet (Good 1965). A major deficiency of the Dirichlet is that it does not have a natural method of introducing a "strong" correlational structure between the probabilities of events. Its correlation is due merely to the fact that the probabilities must sum to one. This "weak" correlation often does not adequately describe the phenomenon. For example, there may exist a natural ordering of the events so that their probabilities are related. If the data is sparse, then observations in neighboring cells could be used to evaluate the probability of a cell that has few or no observations.

An alternative model that has a flexible covariance structure is constructed by the logistic transformation of a multivariate normal random vector. Lindley (1964) noted that the logarithm of a contrast of cell probabilities under the Dirichlet-multinomial model is asymptotically normal. Leonard (1973, 1975), Aitchison and Shen (1980), Aitchison (1981), Simonoff (1983), and Nazaret (1987) proposed logistic normal models. However, they did not compute the predictive distribution of a future observation due to the intractability of the model. In particular, Leonard, Simonoff, and Nazaret used the prior as a penalty term to obtain a penalized, maximum likelihood estimator. Aitchison and Shen developed some properties of the logistic normal family but did not compute the predictive distribution for multinomial sampling.

This paper presents a method of constructing distributions on the simplex of probabilities such that the family of distributions is closed in the prior to posterior analysis. This construction is specialized to a generalization of the logistic normal family, and the predictive distribution is approximated. The logistic normal model offers three features:

1. The mean of the normal random vector determines the prior mean of the multinomial probabilities.
2. The variance of the normal vector determines the influence of the data in the posterior distribution.
3. The covariance of the normal vector determines the amount of smoothing of the sample information.

The analysis for the generalized, logistic normal model is applied to the data of Maguire, Pearson, and Wynn (1952) in Table 1 of section 4. The data is the number of days between fatal explosions in mines. The time interval was discretized into 30 day intervals so that an explosion was recorded as being in the first cell if the previous explosion was 1 to 30 days before it; it was recorded as being in the second cell if the previous explosion was 31 to 60 days before it, etc. Two features of the data are sparse observations in the tails of the distribution and a "bulge" in the early intervals.

2. The Model

$Y_1, Y_2 \ldots$ are discrete, exchangeable random variables with K possible outcomes $\{1, 2, \ldots, K\}$. S_K is the $K-1$ dimensional simplex

$$S_K = (\underline{p} = (p_1, \ldots, p_K)' : 0 \le p_k \text{ and } \Sigma p_k = 1).$$

Using de Finetti's representation theorem (de Finetti, 1964 and Hewitt and Savage, 1955), we suppose that given a parameter θ, the Y's are i.i.d. with common distribution $Pr(Y = i|\theta) = p_i(\theta)$ for $i = 1, \ldots, K$ where $\underline{p}(\theta) \in S_K$. Let θ be a random element with distribution Υ. The marginal distribution of $\underline{Y}_n = (Y_1, \ldots, Y_n)'$ is

$$Pr(\underline{Y}_n = \underline{y}_n) = \int \prod_{i=1}^{K} p_i(\theta)^{n(i)} d\Upsilon(\theta)$$

where $n(i)$ is the number of observations of type i. The predictive distribution of Y_{n+1} given $\underline{n} = (n_1, \ldots, n_K)'$ is

$$Pr(Y_{n+1} = y|\underline{n}) = \int p_y(\theta) \prod_{i=1}^{K} p_i(\theta)^{n(i)} d\Upsilon(\theta) / Pr(\underline{Y}_n = \underline{y}_n)$$

There are at least two critical features of this model. First, no amount of evidence to the contrary will enlarge the family of distributions $\{\underline{p}(\theta)\}$ parameterized by θ. The family must be sufficiently broad to adequately describe the range of anticipated behavior of the phenomenon. Yet an unnecessarily large family inefficiently utilizes the observations. Second, the parameter θ or, more correctly, its distribution Υ should be selected judiciously to represent critical aspects of the phenomenon and, in particular, the source of uncertainty that the modeler wishes to reduce through observation. For example, the number of explosions in each time interval of the mine explosion data could be modelled as a Poisson distribution and the parameter of interest is the mean. However, since the data does have a long tail, a negative binomials distribution provides a better description.

At one extreme, Υ is a point mass at some θ_0. Then the Y's are unconditionally independent with common distribution $\underline{p}(\theta)$. No learning from observation occurs since the marginal and predictive distributions of an observation are equal. At the other extreme, θ is an element of the simplex S_K, and $p_i(\theta) = \theta_i$. Then Υ is a distribution on the simplex and completely determines the joint distribution of the Y's. I will follow the terminology of Bishop, Fienberg, and Holland (1975, p. 9) and call the model "saturated." The saturated model is the widest possible parameterization and is capable of detecting any distribution in the support of Υ in that the posterior distribution of $\underline{P}$ is asymptotically a point mass at the "true" distribution (Freedman, 1963). For convenience the "θ" will be suppressed in the remainder of the paper.

The mine explosion data has a bulge in the early months which is inconsistent with the Poisson or negative binomial models. Saturated models are able to detect this feature while the effect of using the Poisson or negative binomial models is to smooth the bulge to the other cells. The maximum likelihood estimators for the saturated model, i.e., the relative frequencies of the cells, are capable of detecting the bulge but are inadequate in describing the probabilities of the tails of the distribution since the data is sparse. Since there does not exist any structural explanation for having cells with zero probabilities, the relative frequencies provide *a priori* an unreasonable predictive distribution. The Bayesian saturated models are able to adequately describe both of these features.

In the saturated model, the posterior distribution is

$$\Upsilon(A|\underline{n}) = \int_A \prod_{i=1}^{K} p_i^{n(i)} d\Upsilon(\underline{p}) \Big/ E\left[\prod_{i=1}^{K} P_i^{n(i)}\right]$$

where A is a Borel subset of S_K, and the expectation on the right-hand side is taken with respect to Υ. The n_j's are held fixed but not conditioned on, in the latter expectation. The predictive distribution is the posterior mean

$$E[P_i|\underline{n}] = E\left[P_i \prod_{j=1}^{K} P_j^{n(j)}\right] \Big/ \sum_{k=1}^{K} E\left[P_k \prod_{j=1}^{K} P_j^{n(j)}\right].$$

3. Closed Families of Distributions

A general method of constructing priors on the simplex is to normalize positive, continuous random variables. Let $\underline{X}$ be a K dimensional, non-negative random vector with distribution F. Define $\underline{P}$ by the transformation

$$P_i = X_i/S \text{ for } i = 1, \ldots, K \text{ and } S = X_i + \ldots + X_K.$$

The joint distribution of $\underline{P}$ and S is

$$\Upsilon(A, B) = \int_{A\times B} dF(s\underline{p})$$

where A and B are Borel sets of S_K and R^+.

Families of distributions that are closed in the prior to posterior analysis are easily constructed. Let W be a random vector on $(R^+)^K$ with density

$$g(\underline{w}|\underline{\alpha}) = \prod_{i=1}^{K} w_i^{\alpha(i)-1} h(\underline{w})/C(\underline{\alpha}) \tag{1}$$

where h is a non-negative function on $(R^+)^K$ and $\underline{\alpha} \in R^K$ such that

$$C(\underline{\alpha}) = \int_{R^{+K}} \prod_{i=1}^{K} w_i^{\alpha(i)-1} h(\underline{w}) d\underline{w} < \infty.$$

Let $\underline{X}$ be a K dimensional random vector with density

$$f(\underline{x}|\underline{\alpha},\gamma) = (\underline{l}'\underline{x})^{\gamma} g(\underline{x}|\underline{\alpha})/E[S^{\gamma}|\underline{\alpha}] \tag{2}$$

where $\underline{l}$ is a K-vector of ones, $S = \underline{l}'\underline{W}$, and γ is restricted so that $E[S^{\gamma}]$ is finite. The next lemma gives conditions for finite expectations.

LEMMA: *Let $\underline{W}$ be a non-negative, random vector.*

i. If $\gamma < 0$ or $\gamma > 1$ and if $EW_i^{\gamma} < \infty$ for all i, then $E[(\underline{l}'\underline{W}^{\gamma}] < \infty$.

ii. If $0 \leq \gamma \leq 1$ and if $EW_i < \infty$ for all i, then $E[(\underline{l}'\underline{W})^{\gamma}] < \infty$.

PROOF:

i. Fix ω in the sample space. Then $(\underline{l}'\underline{W}(\omega)/K)^{\gamma} \leq \Sigma W_i(\omega)^{\gamma}/K$ by Jensen's inequality. Take expectations.

ii. Follows from Jensen's inequality.

The density on S_K deduced from equation (2) is

$$\tau(\underline{p}) = c\prod_{i=1}^{K} p_i^{\alpha(i)-1}\phi(\underline{p}|\alpha+\gamma)$$

where c is the normalization constant, $\alpha = \underline{l}'\underline{\alpha}$, and

$$\phi(\underline{p}|\alpha+\gamma) = \int_{R+K} s^{\alpha+\gamma-1}h(s\underline{p})ds.$$

Let $\underline{a} \in R^K$ and $a = \underline{l}'\underline{a}$. Assuming that the following expectations are finite, the joint moments of $\underline{P}$ are given by

$$E\left[\prod_{i=1}^{K} P_i^{a(i)}|\underline{\alpha},\gamma\right] = [C(\underline{\alpha}+\underline{a})/C(\underline{\alpha})]\, E[S^{\gamma-a}|\underline{\alpha}+\underline{a}]/E[S^{\gamma}|\underline{\alpha}]$$

where $C(\underline{\alpha}+\underline{a})/C(\underline{\alpha}) = E[\prod W_i^{a(i)}|\underline{\alpha}]$, and the density of $\underline{W}$ is given by equation (1).

The posterior density of $\underline{P}$ is

$$\tau(\underline{p}|\underline{n}) = c\prod_{i=1}^{K} p_i^{\alpha(i)+n(i)-1}\phi(\underline{p}|(\alpha+n)+(\gamma-n))$$

where c is the appropriate normalizing constant. The prior parameters $(\underline{\alpha},\gamma)$ are transformed to the posterior parameters $(\underline{\alpha}+\underline{n},\gamma-n)$. The predictive distribution of Y_{n+1} is

$$E[P_i|\underline{n}] = \frac{E[W_i\prod W_j^{n(j)}]E[S^{\gamma-n-1}|\underline{\alpha}+\underline{n}+\underline{e}_i]}{\sum_{k=1}^{K} E[W_k\prod W_j^{n(j)}]E[S^{\gamma-n-1}|\underline{\alpha}+\underline{n}+\underline{e}_k]} \tag{3}$$

where $\underline{e}_k$ is a vector of zeros with a one in the kth position.

This equation is useful if the joint moments of $\underline{W}$ are easily obtained and if $E[S^{\gamma-n-1}]$ can be approximated. If the distribution of S is known or can be approximated, then the right-hand side of equation (3) requires K, one-dimensional integrals in order to compute the expectations of S, whereas the joint moments of $\underline{P}$ require a $K-1$ dimensional integral. The 'natural' choice of γ is 0 since it is more difficult to envision the nature of the prior distributions for $\underline{P}$ for other choices of γ. In this aspect, γ is introduced merely as a mathematical convenience to obtain families of distributions that are closed in the prior to posterior Bayes' analysis. Nevertheless, the posterior distribution of $\underline{P}$ will, in general, have a non-zero γ, which presents the same problem of understanding the behavior of the posterior distribution that a non-zero γ presents for the prior distribution.

Examples

1. The Dirichlet distribution is constructed by the W's being independent gamma random variables with the same scale parameter. Then $S = \underline{l}'\underline{W}$ is a gamma random variable and depends only on α and not the individual components of $\underline{\alpha}$. The predictive distribution (3) becomes the familiar $(\alpha_i + n_i)/(\alpha + n)$.
2. Consider the F-distribution $\big(F(\alpha,\delta,\beta,\zeta)\big)$ with density

$$g(w) = \frac{\Gamma(\alpha+\delta)}{\Gamma(\alpha)\Gamma(\delta)}(\beta/\zeta)^{\alpha}\frac{w^{\alpha-1}}{\big(1+(\beta/\zeta)w\big)^{\alpha+\delta}}$$

for positive w, α, δ, β, and ζ. Let the W_i's be independent $F(\alpha_i, \delta_i, \beta, \zeta)$. Pick δ such that either (i) $\gamma < 0$ or $\gamma > 1$ and $\delta_i - \gamma > 0$ for all i or (ii) $0 \leq \gamma \leq 1$ and $\delta_i - 1 > 0$. The lemma gives $E[\underline{l}'\underline{W}^{\gamma}] < \infty$. Then

$$\tau(\underline{p}, s) = (\beta/\zeta)^{\alpha}E^{-1}[S^{\gamma}]s^{\gamma+\alpha-1}\prod_{i=1}^{K}\frac{\Gamma(\alpha_i+\delta_i)p^{\alpha(i)-1}}{\Gamma(\alpha_i)\Gamma(\delta_i)\big(1+(\beta/\zeta)sp_i\big)^{\alpha(i)+\delta(i)}}.$$

Note that τ is independent of β and ζ by transforming $z = \beta s/\zeta$. The prior parameters are $\underline{\alpha}$, $\underline{\delta}$, and γ are updated to $\underline{\alpha}+\underline{n}$, $\underline{\delta}-\underline{n}$, and $\gamma - n$ in the posterior. If $\underline{\delta}-\underline{n}-\underline{l} \in R^{+K}$ so that the expectations exist, the predictive distribution (3) is proportional to

$$(\alpha_i + n_i/\delta_i - n_i - 1)E[S^{\gamma-n-1}|\underline{\alpha}+\underline{n}+\underline{e}_i, \underline{\delta}-\underline{n}-\underline{e}_i].$$

4. The Logistic Normal Distribution

The generalized, logistic normal distribution follows the construction of section 3, but takes an interesting form that warrants a separate development. Let $\underline{Z}$

be a K-dimensional, multivariate normal random vector with mean $\underline{\mu}$ and covariance matrix $\Sigma = (\sigma(i,j))$. The lognormal random vector $\underline{V}$ is defined by $W_i = \exp\{Z_i\}$, and the lognormal family of distributions is indicated by $LN_K(\underline{\mu}, \Sigma)$. Its density is given by

$$g(\underline{w}) = (2\pi)^{-K/2} \det \Sigma^{-1/2} \prod_{i=1}^{K} w_i^{-1} \exp\{-0.5(\ln \underline{w} - \underline{\mu})'\Sigma^{-1}(\ln \underline{w} - \underline{\mu})\} \tag{4}$$

where $\underline{w} \in R^{+K}$ and $\ln \underline{w} = (\ln w_1, \ldots, \ln w_K)'$. The joint moments of $\underline{W}$ are computed from the moment generating function of $\underline{Z}$. Let $\underline{a} \in R^K$.

$$E[\prod W_i^{a(i)}] = \exp\{\underline{a}'\underline{\mu} + 0.5\underline{a}'\Sigma\underline{a}\}. \tag{5}$$

The generalized lognormal random vector $\underline{X}$, whose distribution is indicated by $LN_K(\underline{\mu}, \Sigma, \gamma)$, has a density given by

$$f(\underline{x}) = (\underline{l}'\underline{x})^\gamma g(\underline{x})/C(\gamma, \underline{\mu}) \tag{6}$$

where g is given in (4), $C(\gamma, \underline{\mu}) = E[(\underline{l}'\underline{W})^\gamma | \underline{\mu}]$, and $\underline{W}$ is $LN_K(\underline{\mu}, \Sigma, 0)$. The distribution on the simplex S_K which is deduced from normalizing the components of $\underline{X}$ is called the "logistic normal" and is indicated by $LNS_K(\underline{\mu}, \Sigma, \gamma)$. Some results about the generalized lognormal $\underline{X}$ and the logistic normal $\underline{P}$ follow.

Result 1.

For $\underline{a} \in R^K$ the joint moments of $\underline{X}$ are

$$E[\prod X_i^{a(i)}] = E[\prod W_i^{a(i)}]C(\gamma, \underline{\mu} + \Sigma\underline{a})/C(\gamma, \underline{\mu})$$

where the first factor on the right-hand side is defined in (5) and the second and third factors are defined in (6).

PROOF: Write $E[\prod X_i^{a(i)}] = E[\exp\{\underline{a}' \ln \underline{X}\}]$. Complete the quadratic form

$$\begin{aligned} &\underline{a}' \ln \underline{x} - 0.5(\ln \underline{x} - \underline{\mu})'\Sigma^{-1}(\ln \underline{x} - \underline{\mu}) \\ &= -0.5(\ln \underline{x} - \underline{\mu} - \Sigma\underline{a})'\Sigma^{-1}(\ln \underline{x} - \underline{\mu} - \Sigma\underline{a}) + \underline{a}'\underline{\mu} + 0.5\underline{a}'\Sigma\underline{a}. \end{aligned}$$

Identify the normalizing factors of the density in (6).

Result 2.

The joint density of $\underline{P}$ and $S = \underline{l}'\underline{X}$ is

$$\begin{aligned} \tau(\underline{p}, s) = &(2\pi)^{-K/2} \det \Sigma^{-1/2} C(\gamma, \underline{\mu})^{-1} \\ &\cdot s^{\gamma-1} \exp\{-0.5\underline{l}'\Sigma^{-1}\underline{l}(\ln s + (\underline{l}'\Sigma^{-1}\underline{l})^{-1}\underline{l}'\Sigma^{-1}(\ln \underline{p} - \underline{\mu}))^2\} \\ &\cdot \exp\{-0.5(\ln \underline{p} - \underline{\mu})'A(\ln \underline{p} - \underline{\mu})\} \prod p_i^{-1} \end{aligned}$$

for $\underline{p} \in S_K$ and $s \in R^+$ where $A = \Sigma^{-1} - (\Sigma^{-1} J \Sigma^{-1} / \underline{l}'\Sigma^{-1}\underline{l})$ and $J = \underline{l}\underline{l}'$. Note that A is singular since the simplex has dimension $K - 1$.

PROOF: The joint density of $\underline{P}$ and S is proportional to

$$s^{\gamma-1} \prod p_i^{-1} \exp\{-0.5(\ln s\underline{p} - \underline{\mu})'\Sigma^{-1}(\ln s\underline{p} - \underline{\mu})\}.$$

Use $\ln s\underline{p} = (\ln s)\underline{l} + \ln \underline{p}$. Factor the quadratic into two parts: one that depends on s and $\underline{p}$ and the other that only depends on $\underline{p}$. The result follows by completing the quadratic form of the first factor and then the second.

Result 3.

The marginal density of $\underline{P}$ is

$$\begin{aligned}\tau(\underline{p}) = {} & (2\pi)^{(K-1)/2}(\underline{l}'\Sigma\underline{l})^{-1/2} \det \Sigma^{-1/2} C(\gamma, \underline{\mu})^{-1} \\ & \cdot \exp\{-0.5(\ln \underline{p} - \underline{\mu})'\Sigma^{-1}(\ln \underline{p} - \underline{\mu})\} \prod p_i^{-1} \\ & \cdot \exp\{+0.5(\ln \underline{p} - \underline{\mu} - \Sigma\underline{\gamma})' B (\ln \underline{p} - \underline{\mu} - \Sigma\underline{\gamma})\}\end{aligned}$$

where $B = \Sigma^{-1} J \Sigma^{-1} / \underline{l}'\Sigma^{-1}\underline{l}$, and $\underline{\gamma} \in R^K$ such that $\underline{l}'\underline{\gamma} = \gamma$. Note that the choice of $\underline{\gamma}$, subject to the constraint, is arbitrary since

$$B\Sigma_{\underline{\gamma}} = \gamma\Sigma^{-1}\underline{l}/\underline{l}'\Sigma^{-1}\underline{l}.$$

PROOF: Integrating the joint density of result 2 over $s \in R^+$ results in the factor

$$\begin{aligned}& \int s^{\gamma-1} \exp\left\{-0.5\underline{l}'\Sigma^{-1}\underline{l}\,(\ln s + (\underline{l}'\Sigma^{-1}(\ln \underline{p} - \underline{\mu})/\underline{l}'\Sigma^{-1}\underline{l}))^2\right\} ds \\ & = (2\pi/\underline{l}'\Sigma^{-1}\underline{l})^{1/2} \exp\{(-\gamma\underline{l}'\Sigma^{-1}(\ln \underline{p} - \underline{\mu}) + 0.5\gamma^2)/\underline{l}'\Sigma^{-1}\underline{l}\}\end{aligned}$$

since the integrand is proportional to the γ moment of a lognormal density. Write $\gamma = \underline{\gamma}'\Sigma\Sigma^{-1}\underline{l}$ to obtain

$$\gamma\underline{l}'\Sigma^{-1}(\ln \underline{p} - \underline{\mu})/\underline{l}'\Sigma^{-1}\underline{l} = \underline{\gamma}\Sigma B(\ln \underline{p} - \underline{\mu}) \text{ and } 0.5\gamma^2/\underline{l}'\Sigma^{-1}\underline{l} = 0.5\underline{\gamma}'\Sigma B\Sigma_{\underline{\gamma}}.$$

Combining the exponents gives

$$\begin{aligned}& 0.5(\ln \underline{p} - \underline{\mu})' B (\ln \underline{p} - \underline{\mu}) - \underline{\gamma}'\Sigma B(\ln \underline{p} - \underline{\mu}) + 0.5\underline{\gamma}'\Sigma B\Sigma\underline{\gamma} \\ & = 0.5(\ln \underline{p} - \underline{\mu} - \Sigma\underline{\gamma})' B (\ln \underline{p} - \underline{\mu} - \Sigma\underline{\gamma}),\end{aligned}$$

and the result follows.

Result 4.

$$S|\underline{P} \text{ is } LN_1\big(-\underline{l}'\Sigma^{-1}(\ln \underline{p} - \underline{\mu})/\underline{l}'\Sigma^{-1}\underline{l}, (\underline{l}'\Sigma^{-1}\underline{l})^{-1}, \gamma\big).$$

Result 5.

For $\underline{a} \in R^K$ and $\underline{l}'\underline{a} = a$,

$$E[\prod P_i^{a(i)}] = E[\prod W_i^{a(i)}]C(\gamma - a, \underline{\mu} + \Sigma\underline{a})/C(\gamma, \underline{\mu})$$

PROOF: After completing the quadratic form in $\ln \underline{p}$,

$$\begin{aligned} E[\prod P^{a(i)}] &= E[\exp\{\underline{a}' \ln \underline{p}\}] \\ &= \int_{S_K} \text{constant}^* \prod p_i^{-1} \exp\{-0.5(\ln \underline{p} - \underline{\mu} - \Sigma\underline{a})'\Sigma^{-1}(\ln \underline{p} - \underline{\mu} - \Sigma\underline{a})\} \\ &\quad \cdot \exp\{0.5(\ln \underline{p} - \underline{\mu} - \Sigma\underline{a})'B(\ln \underline{p} - \underline{\mu} - \Sigma\underline{a})\}d\underline{p} \\ &\quad \cdot \exp\{\underline{a}'\underline{\mu} + 0.5\underline{a}'\Sigma\underline{a}\} \end{aligned}$$

Comparing factors with the marginal density of result 3 gives result 5.

Result 6.

The posterior distribution of $\underline{P}$ given the data $\underline{n}$ is $LNS_K(\underline{\mu} + \Sigma\underline{n}, \Sigma, \gamma - n)$, so that the prior parameters $(\underline{\mu}, \Sigma, \gamma)$ are updates to $(\underline{\mu} + \Sigma\underline{n}, \Sigma, \gamma - n)$.

REMARKS: The observations enter the posterior distribution through the mean vector of the normal random vector, and the observations are weighted by the covariance matrix of the normal random vector. The parameter $-\gamma$ corresponds to the "prior sample size" in conjugate priors for the natural parameter in exponential families of distributions. Viewing the analysis sequentially, $n - \gamma$ is the "prior sample size" for the $n + 1$ observation. This terminology is somewhat misleading since γ can be any real number.

The choice of Σ determines the behavior of the predictive distribution for small sample sizes where "small" is judged relative to the number of cells K and the diagonal elements of Σ. However, if Σ is positive definite so that the support of the generalized, logistic normal distribution is the $K - 1$ dimensional simplex, then the results of Freedman (1963) provide the convergence of the posterior distribution to a point mass at the "true" distribution. Consequently, for large sample sizes the specification of Σ is not critical.

PROOF: The posterior density is proportional to $\exp\{\underline{n}' \ln \underline{p}\}\tau(\underline{p})$. Complete the quadratic form $\underline{n}' \ln \underline{p} - 0.5(\ln \underline{p} - \underline{\mu})'\Sigma^{-1}(\ln \underline{p} - \underline{\mu})$ to obtain

$$\begin{aligned} \tau(\underline{p} \mid \underline{n}) = {} & \text{constant}^* \prod p_i^{-1} \exp\{-0.5(\ln \underline{p} - \underline{\mu} - \Sigma\underline{n})'\Sigma^{-1}(\ln \underline{p} - \underline{\mu} - \Sigma\underline{n})\} \\ & \cdot \exp\big\{+0.5\big(\ln \underline{p} - (\underline{\mu} + \Sigma\underline{n}) - \Sigma(\underline{\gamma} - \underline{n})\big)'B\big(\ln \underline{p} - (\underline{\mu} + \Sigma\underline{n}) - \Sigma(\underline{\gamma} - \underline{n})\big)\big\}. \end{aligned}$$

Note that $B\Sigma(\underline{\gamma} - \underline{n}) = (\gamma - n)\Sigma^{-1}\underline{l}/\underline{l}'\Sigma^{-1}\underline{l}$ which only depends on $\underline{\gamma} - \underline{n}$ through $\gamma - n$. Identify parameters with the marginal density in result 3.

Result 7.

The predictive distribution $E[P_i \mid \underline{n}]$ is proportional to

$$W_i(0)\exp\left\{\sum_{j=1}^{K} n_j\sigma(i,j)\right\} C(\gamma - n - 1, \underline{\mu} + \Sigma\underline{n} + \Sigma\underline{e}_i)$$

where $W_i(0) = E[W_i] - \exp\{\mu_i + 0.5\sigma(i,i)\}$. If $\gamma = 1$, then the prior mean is $E[P_i] = W_i(0)\Big/\sum_{k=1,K} W_k(0)$. In the following, I will take $\gamma = 1$.

PROOF: Use results 5 and 6.

The covariance Σ reflects the modeler's beliefs about the smoothness of the p's. A canonical example is when Y is a discretization of an absolutely continuous random variable, and $p_i = \int_{A_i} f(x)dx$ where f is a density and $A_1, \ldots, A_K$ partitions its support. Then the p's have a natural ordering, and if f is continuous, neighboring p's are more related than distant p's. An alternative to the Poisson or negative binomial models for the mine explosion data is to model the p's by this method where f is, say, a gamma or lognormal density, and the A's are the time intervals. Either of these choices would adequately describe the long tail of the data but would be inadequate to describe the bulge. Of course, other choices of f may be more appropriate.

In the above situation, a reasonable model of the covariance function between P_i and P_j is

$$\sigma(i,j) = \alpha\sigma(\beta|i-j|) \text{ for } \alpha > 0 \text{ and } \beta > 0$$

where $\sigma(0) = 1$, and σ is a positive definite function such as $\sigma(x) = \exp(-x^2)$ or $\sigma(x) = (1+x^2)^{-1}$. In the literature, σ is a smoothing kernel, and β is the smoothing or scale parameter. α determines the tradeoff between the data and the prior mean.

With $\gamma = 1$, the posterior mean in result 7 depends on the C-factors

$$C(-n, \underline{\mu} + \Sigma\underline{n} + \Sigma\underline{e}_i) = E[(\sum_{j=1,K} e^{\sigma(i,j)}W_j)^{-n} \mid \underline{\mu} + \Sigma\underline{n}].$$

Using the mean value theorem, there exists a function ϕ such that

$$C(-n, \underline{\mu} + \Sigma\underline{n} + \Sigma\underline{e}_i) = \exp\{-n\phi(i)\}C(-n, \underline{\mu} + \Sigma\underline{n})$$

The predictive distribution in result 7 is proportional to

$$W_i(n,\phi) = W_i(0)\exp\{\sum_{j=1,K} n_j\sigma(i,j) - n\phi(i)\}.$$

If σ is continuous, then so is ϕ. In general, ϕ is unknown but can be approximated by a cubic spline s, which has fixed nodes. The spline's knots are estimated by maximizing the predictive likelihood

$$PL(s) = \prod_{i=1}^{K} \left[W_i(n,s) \Big/ \sum_{k=1}^{K} W_k(n,s) \right]^{n_i}$$

where the spline s replaces ϕ.

If each n_i is reasonably large, then it is possible to place a node in each cell and stably estimate the knots. This case is uninteresting since the resulting probability will be the MLE n_i/n. A more interesting situation obtains when the observations are sparse so that there are cells with few or no observations. Then the number of nodes should be less than the number of cells. My experience has been that if the nodes separate the peaks and valleys of the distribution, then the estimated, predictive distribution will detect these features. Increasing the number of nodes beyond this point does not significantly affect the estimator. As a practical technique, start with a few nodes and estimate the predictive distribution. Then increase the number of nodes until the estimated spline and predictive distribution do not change by some small, positive amount.

Table 1 presents an analysis of a data set from Maguire, Pearson, and Wynn (1952). The data is the time between explosions in mines involving more than 10 men killed in Great Britain from December 8, 1875 to May 29, 1951. Leonard (1978) and Simonoff (1983) have also analyzed this data, and have discussed some possible inadequacies of the exponential model of Maguire, Pearson, and Wynn. Here, there are 55 cells, $W_i(0) = 1$, the covariance function is $\sigma(i,j) = \alpha/(1+\beta(x_i - x_j)^2)$ where x_i is the midpoint of cell i, $\gamma = 1$, $\alpha = 0.05$, and $\beta = 0.005$. From my experience, the predictive probabilities are not sensitive to the form of the covariance function, as long as it is smooth, non-negative, symmetric about 0, and decreasing in $|x|$. The cubic spline has 5 equally spaced nodes with estimated knots 0.1066, 0.3368, 0.7018, 0.7803, and 0.4867. Qualitatively, the Bayesian, approximated, predictive distribution (BAP) is similar to the results of Leonard and Simonoff.

5. Conclusion

An argument, as put forth by a frequentist, against the model of this paper is that if the n_i are moderately large, then the predictive distribution is close to the MLE and all of the hard work about priors and posteriors and expectations is unnecessary. If the data is sparse, then the predictive distribution is overly influenced

TABLE 1: Time Intervals Between Explosions in Mines

Days	Freq	MLE	BAP	Days	Freq	MLE	BAP
1- 30	18	0.1651	0.1391	841- 870	0	0.0000	0.0017
31- 60	14	0.1284	0.1265	871- 900	1	0.0092	0.0017
61- 90	9	0.0826	0.0956	901- 930	0	0.0000	0.0015
91-120	8	0.0734	0.0825	931- 960	0	0.0000	0.0014
121-150	6	0.0550	0.0685	961- 990	0	0.0000	0.0013
151-180	4	0.0367	0.0574	991-1020	0	0.0000	0.0012
181-210	6	0.0550	0.0568	1021-1050	0	0.0000	0.0012
211-240	7	0.0642	0.0524	1051-1080	0	0.0000	0.0011
241-270	1	0.0092	0.0368	1081-1110	0	0.0000	0.0011
271-300	6	0.0550	0.0399	1111-1140	0	0.0000	0.0011
301-330	7	0.0642	0.0373	1141-1170	0	0.0000	0.0011
331-360	5	0.0459	0.0296	1171-1200	0	0.0000	0.0012
361-390	5	0.0459	0.0242	1201-1230	1	0.0092	0.0012
391-420	0	0.0000	0.0163	1231-1260	0	0.0000	0.0012
421-450	0	0.0000	0.0136	1261-1290	0	0.0000	0.0013
451-480	2	0.0183	0.0125	1291-1320	1	0.0092	0.0014
481-510	1	0.0092	0.0103	1321-1350	0	0.0000	0.0015
511-540	1	0.0092	0.0087	1351-1380	1	0.0092	0.0017
541-570	1	0.0092	0.0073	1381-1410	0	0.0000	0.0018
571-600	0	0.0000	0.0059	1411-1440	0	0.0000	0.0019
601-630	0	0.0000	0.0051	1441-1470	0	0.0000	0.0022
631-660	1	0.0092	0.0045	1471-1500	0	0.0000	0.0025
661-690	0	0.0000	0.0037	1501-1530	0	0.0000	0.0028
691-720	0	0.0000	0.0032	1531-1560	0	0.0000	0.0033
721-750	1	0.0092	0.0029	1561-1590	0	0.0000	0.0039
751-780	0	0.0000	0.0025	1591-1620	1	0.0092	0.0048
781-810	0	0.0000	0.0022	1621-1650	1	0.0092	0.0058
811-840	0	0.0000	0.0019	Total	109		

by the prior distribution, which, for a frequentist, cannot be trusted. The frequentist would demand more data, not better priors.

Of course, the obvious flaw in the last statement is that additional data is usually not available, and often it is unattainable. For example, consider the estimation of the probabilities of catastrophic accidents such as nuclear power plant failures. The predictive distribution must rely on someone's judgments, preferably an expert in the problem domain.

The logistic normal model has the feature of coherently combining expert judgment and the data. The expert judgment is expressed by probability statements and enters the model through the parameters $W_i(0)$, and the covariance Σ determines the weight of the data. Lenk and Floyd (1988b) applied this model to information systems where $W_i(0)$ is supplied by an expert system and the data consists of feedback about the system's performance.

For the finite dimensional simplex and reasonably large n the MLE exists, and the data rapidly dominates priors that are not too radical. However, in nonparametric density estimation, the MLE does not exist, and the data is not informative about the density evaluated at a point. Nonparametric density estimation, conceptually, corresponds to estimating multinomial probabilities when the dimension of the simplex is large and the data is extremely sparse. Regardless of the sample size, the prior retains a dominant role.

The Dirichlet process is not directly usable because of the weak correlational structure. Lo (1984) convolved a Dirichlet process with a smoothing kernel to evaluate the predictive distribution. The computational complexity of this method is $O(2^n)$. Lenk (1988a) extended the results of this paper for the logistic normal model to density estimation. One advantage of the logistic normal is that it can be extended to two or more dimensions. Lenk (1987) computed a nonparametric, predictive regression function $E[Y \mid X,\text{data}]$ by integrating the predictive density for X and Y that is derived under the logistic normal model.

Acknowledgments

I wish to thank an editor and referees for carefully reviewing the manuscript and making many suggestions that improved the presentation. I wish to thank Bruce Hill for introducing me to the subjectivist viewpoint.

References

Aitchison, J. (1981). "Distributions on the Simplex for the Analysis of Neutrality," in *Statistical Distributions in Scientific Work, Vol. 4*, C. Tallie, G. Patil, and B. Baldessari, eds., Dordrect, Holland: D. Reidel Publishing Company, 147-156.

Aitchison, J. and S. Shen (1980). "Logistic-Normal Distribution: Some Properties and Uses," *Biometrika*, Vol. 67, 261-272.

Bishop, Y., S. Fienberg and P. Holland (1975). *Discrete Multivariate Analysis: Theory and Practice*, Cambridge, MA: MIT Press.

de Finetti, B. (1964). "Foresight: Its Logical Laws, Its Subjective Sources," in *Studies in Subjective Probability*, H. Kyburg and H. Smokler, eds., New York: Wiley, 93-158.

Freedman, D.A. (1963). "On the Asymptotic Behavior of Bayes' Estimates in the Discrete Case," *Annals of Mathematical Statistics*, Vol. 34, 1386-1402.

Good, I.J. (1965). *The Estimation of Probabilities: An Essay on Modern Bayesian Methods*. Cambridge, MA: MIT Press.

Hewitt, E. and L.J. Savage (1955). "Symmetric Measures on Cartesian Products," *Transactions of the American Mathematical Society*, Vol. 80, 470-501.

Lenk, P. (1987). "The Logistic Normal Distribution for Bayesian Nonparametric Predictive Regression," unpublished manuscript.

Lenk, P. (1988a). "The Logistic Normal Distribution for Bayesian Nonparametric Predictive Densities," *Journal of the American Statistical Society*, Vol. 83, 509-516.

Lenk, P. and B. Floyd (1988b), "Dynamically Updating Relevance Judgements in Probabilistic Information Systems Via Users' Feedback," *Management Science*, Vol. 34, No. 12.

Leonard, T. (1973). "A Bayesian Method for Histograms," *Biometrika*, Vol. 60, 297-308.

Leonard, T. (1975). "Bayesian Estimation Models for Two-Way Tables," *Journal of the Royal Statistical Society B*, Vol. 37, 23-37.

Leonard, T. (1978). "Density Estimation, Stochastic Processes, and Prior Information," *Journal of the Royal Statistical Society (B)*, Vol. 40, 113-146.

Lindley, D. (1964). "The Bayesian Analysis of Contingency Tables," *Annals of Mathematical Statistics*, Vol. 35, 1622-1643.

Lo, A.Y. (1984). "On a Class of Bayesian Nonparametric Estimates: I. Density Estimates," *Annals of Statistics*, Vol. 12, 351-357.

Maguire, B.A., E.S. Pearson, and A.H.A. Wynn (1952). "The Time Intervals Between Industrial Accidents," *Biometrika*, Vol. 39, 168-180.

Nazaret, W. (1987). "Bayesian Log Linear Estimates for Three-way Contingency Tables," *Biometrika*, 74, 401-410.

Simonoff, J. (1983). "A Penalty Function Approach to Smoothing Large Sparse Contingency Tables," *Annals of Statistics*, Vol. 11, 208-218.

Bayesian and Likelihood Methods in Statistics and Econometrics
S. Geisser, J.S. Hodges, S.J. Press and A. Zellner (Editors)

TURNING POINTS IN ECONOMIC TIME SERIES, LOSS STRUCTURES AND BAYESIAN FORECASTING

Arnold Zellner, Chansik Hong, and Gaurang Mitu Gulati
University of Chicago

Abstract

In this paper the problems of forecasting turning points in and future values of economic time series are analyzed using a Bayesian decision theoretic approach. It is shown how to obtain optimal turning point forecasts and point forecasts using symmetric and asymmetric loss structures. It is found that optimal forecasts are sensitive to the forms of loss structures. Using an autoregressive-leading indicator model for 18 countries' annual output growth rates, turning points for the period 1974-85 were forecasted. These forecasting results are presented and discussed. It is concluded that the methods used in this paper may be of practical value to applied economic forecasters.

1. Introduction

In a letter commenting on a draft of Zellner (1987a), Barnard (1987) wrote, "I very much liked your emphasis on the need for sophisticated, simple model building and testing in social science." Apparently, Barnard and many other scientists are disturbed by the complexity of many models put forward in econometrics and other social sciences. And indeed we think that they should be disturbed since not a single complicated model has worked very well in explaining past data and in predicting as yet unobserved data. In view of this fact, in Garcia-Ferrer et al., (1987) and Zellner and Hong (1989), a relatively simple, one-equation model for forecasting countries' annual output growth rates was formulated, applied, and found to produce good forecasts year by year, 1974-84 for 18 countries. This experience supports Barnard's and many others' preference for the use of sophisticatedly simple models and methods. See Zellner (1988) for further discussion of this issue.

In the present paper, we extend our previous work to consider the problem of forecasting future values and turning points of economic time series given explicit loss structures. Kling (1987, pp. 201-204) has provided a good summary of past work on forecasting turning points by Moore (1961, 1983), Zarnowitz (1967), Moore and Zarnowitz (1982), Wecker (1979), Neftci (1982), and others. In this work there is an emphasis on the importance and difficulty of forecasting turning points. Also, in our opinion, not enough attention has been given to the role of loss structures in forecasting just turning points and in forecasting turning points *and* the future values of economic variables.

The plan of our paper is as follows. In Section 2, we introduce loss structures and explain how optimal forecasts of the occurrence of turning points and future values of economic time series can be computed. Applications of our procedures are presented in Section 3 based on annual output growth rate data for eighteen countries and an autoregressive model with leading indicator variables employed in our previous work—see Garcia-Ferrer et al., (1987) and Zellner and Hong (1989). In Section 4, a summary of results and some concluding remarks are presented.

2. Turning Points, Loss Structures and Forecasting

As Wecker (1979) and Kling (1987) recognize, given a model for past observations and a definition of a turning point, probabilities relating to the occurrence of future turning points can be evaluated. In our work, we, along with Kling (1987) evaluate these probabilities using predictive probability density functions for as yet unobserved, future values of variables which take account of uncertainty regarding the values of model parameters as well as the values of future error terms. We also indicate how to take account of model uncertainty in evaluating probabilities relating to future events such as the occurrence or non-occurrence of a turning point at a future time.

2a. Forecasting Turning Points

As regards the definition of turning points in an economic time series, we recognize that there are many possible definitions and that no one definition has achieved universal acceptance. Thus in what follows, we present analysis that can be employed for whatever definition of a turning point is adopted. We shall employ one, particularly simple definition of a turning point to illustrate our analysis and then go on to use a broader definition in our applied work.

Let the given past measurements of a variable, say the growth rate of real GNP, be denoted by $\underset{\sim}{y}' = (y_1, y_2, \ldots, y_{n-1}, y_n)$ and $z \equiv y_{n+1}$ be the first future value of the series. Particular definitions of a downturn (DT) and of an upturn (UT) and their negations, based on just y_{n-1}, y_n and z are given by,

$$y_{n-1} < y_n \text{ and } \begin{cases} z < y_n & \equiv \text{Downturn } (DT_1) \\ z \geq y_n & \equiv \text{No Downturn } (NDT_1) \end{cases} \tag{1a}$$

and

$$y_{n-1} > y_n \text{ and } \begin{cases} z > y_n & \equiv \text{Upturn } (UT_1) \\ z \leq y_n & \equiv \text{No Upturn } (NUT_1) \end{cases} \tag{1b}$$

where the subscript 1 on DT, UT, etc., denotes that one past observation, y_{n-1} and one future observation, $z \equiv y_{n+1}$ have been employed in defining downturns and upturns. Below, we shall introduce broader definitions of DTs and UTs.

Given a model assumed to generate the observations, $\underset{\sim}{y}$, let $p(\underset{\sim}{y} \mid \underset{\sim}{\theta})$ be the likelihood function for the past data, where $\underset{\sim}{\theta}$ is a vector of parameters, $\pi(\underset{\sim}{\theta} \mid I)$ a prior distribution for $\underset{\sim}{\theta}$, and $p(\underset{\sim}{\theta} \mid D) \propto \pi(\theta \mid I)p(y \mid \underset{\sim}{\theta})$ the posterior distribution for $\underset{\sim}{\theta}$, where $D = (\underset{\sim}{y}, I)$ represents the past sample and prior information. Then with $p(z \mid \underset{\sim}{\theta}, D)$ representing the probability density function (pdf) for $z \equiv y_{n+1}$, given $\underset{\sim}{\theta}$ and the past data, the predictive pdf for z is given by

$$p(z \mid D) = \int_\Theta p(z \mid \underset{\sim}{\theta}, D) p(\underset{\sim}{\theta} \mid D) d\underset{\sim}{\theta} \qquad -\infty < z < \infty, \tag{2}$$

where $\underset{\sim}{\theta} \subset \Theta$, the parameter space. As is well known, $p(z \mid D)$ can be viewed as an average of the conditional pdf, $p(z \mid \underset{\sim}{\theta}, D)$ with the posterior pdf, $p(\underset{\sim}{\theta} \mid D)$ serving as the weighting function.

The predictive pdf in (2) can be employed to obtain probabilities associated with the events in (1). For example, if $y_{n-1} < y_n$, the probability of a downturn, P_{DT_1} is given by

$$P_{DT_1} = \int_{-\infty}^{y_n} p(z \mid D) dz, \tag{3}$$

while the probability of no downturn is $P_{NDT_1} = 1 - P_{DT_1}$. Note that with $y_{N-1} < y_N$, the probability of an upturn, given the definitions in (1), is zero.

With $y_{n-1} < y_n$ and the probability of a downturn P_{DT_1} as given by (3), we now wish to make a decision as to whether to forecast a downturn or no downturn. To solve this problem, consider the loss structure shown in Table 1. The two possible outcomes are DT_1 and NDT_1. If the act forecast a DT_1 is chosen, loss is scaled to be 0 if the forecast is correct and to be $c_1 > 0$ if it is incorrect. If the act forecast a NDT_1 is chosen, loss is 0 if the outcome is NDT_1 and $c_2 > 0$ if it is DT_1. In many circumstances, $c_1 \neq c_2$. From (3), probabilities associated with the outcomes DT_1 and NDT_1 are available and can be used to compute expected losses associated with the two acts shown in Table 1 as follows:

Table 1: Forecasting Loss Structure Given $y_{n-1} < y_n$[†]

Acts	Possible Outcomes	
	DT_1	NDT_1
Forecast DT_1	0	c_1
Forecast NDT_1	c_2	0

[†] c_1 and c_2 are given positive quantities.

$$EL \mid \text{Forecast } DT_1 = 0 \cdot P_{DT_1} + c_1(1 - P_{DT_1}) = c_1(1 - P_{DT_1}) \tag{4a}$$

and

$$EL \mid \text{Forecast } NDT_1 = c_2 P_{DT_1} + 0(1 - P_{DT_1}) = c_2 P_{DT_1}. \tag{4b}$$

If (4a) is less than (4b), that is $c_1(1 - P_{DT_1}) < c_2 P_{DT_1}$ or, equivalently

$$1 < \frac{c_2}{c_1}\left(\frac{P_{DT_1}}{1 - P_{DT_1}}\right) \tag{5a}$$

or

$$P_{DT_1} > \frac{c_1}{c_1 + c_2}, \tag{5b}$$

then choosing the act forecast a DT_1 will lead to lower loss than choosing the act forecast NDT_1. Note that if $c_2 = c_1$, this rule leads to a forecast of a DT_1 if $P_{DT_1} > 1/2$. On the other hand, if c_2/c_1 is much larger than 1, say $c_2/c_1 = 2$, then the condition in (5) would be satisfied for $P_{DT_1}/(1 - P_{DT_1}) > 1/2$ or $P_{DT_1} > 1/3$. This example indicates that the decision to forecast a downturn is very sensitive to the value of the ratio c_2/c_1 as well as to the value of P_{DT_1}. Thus, considering just the value of P_{DT_1} in forecasting turning points is not usually satisfactory except in the special case of symmetric loss, $c_1 = c_2$.[1]

[1]Since analysis of forecasting an upturn, UT_1, given that $y_{n-1} > y_n$ is similar to that for forecasting a downturn, it will not be presented.

2b. Point Forecasts Using Different Loss Functions

Above, we have considered the problem of forecasting turning points. Now we consider the problem of deriving point forecasts using two different loss functions, one appropriate for a downturn situation and the other for a no-downturn situation. That loss functions can be different in these two situations is an important consideration which should be taken into account in producing forecasts.

Consider forecasting the value of $z \equiv y_{n+1}$ given $y_{n-1} < y_n$ and the probability of a DT_1 from (3) employing squared error loss functions. Let

$$L_{DT_1} = k_1(\hat{z} - z)^2 \qquad k_1 > 0 \tag{6a}$$

be the loss incurred if a downturn occurs and $\hat{z}$ is used as a point forecast of z and

$$L_{NDT_1} = k_2(\hat{z} - z)^2 \qquad k_2 > 0 \tag{6b}$$

be the loss incurred if no downturn occurs and $\hat{z}$ is used as a point forecast of z. Expected loss is

$$EL = P_{DT_1} k_1 E(\hat{z} - z)^2 | DT_1 + (1 - P_{DT_1}) k_2 E(\hat{z} - z)^2 | NDT_1. \tag{7}$$

On minimizing (7) with respect to the choice of $\hat{z}$, the minimizing value for $\hat{z}$, denoted by $\hat{z}^*$, is[2]

$$\hat{z}^* = \frac{P_{DT_1} k_1 \bar{z}_{DT_1} + (1 - P_{DT_1}) k_2 \bar{z}_{NDT_1}}{P_{DT_1} k_1 + (1 - P_{DT_1}) k_2}, \tag{8}$$

a weighted average of the conditional mean of z given $z < y_n$, $\bar{z}_{DT_1}$, and the conditional mean of z given $z \geq y_n$, $\bar{z}_{NDT_1}$ with weights $P_{DT_1} k_1$ and $(1 - P_{DT_1}) \times k_2$. Note that if $k_1 = k_2$, $\hat{z}^* = \bar{z}$, the mean of the predictive pdf $p(z \mid D)$ in (2). However, if $k_1 \neq k_2$, $\hat{z}^*$ in (8) will not be equal to $\bar{z}$. For example, if $P_{DT_1} = 1/2$, (8) reduces to $\hat{z}^* = (k_1 \bar{z}_{DT_1} + k_2 \bar{z}_{NDT_1})/(k_1 + k_2)$ which differs from $\bar{z}$. Thus while $\bar{z}$ is optimal relative to an overall squared error loss function ($k_1 = k_2$), it is *not* optimal in the case that different loss functions ($k_1 \neq k_2$) are appropriate for downturn and no-downturn situations.

In (6a-b), we have allowed for the possibility that loss functions may be different for downturn and no-downturn cases. However, the use of *symmetric*, squared error loss functions may not be appropriate in all circumstances. If $L_{DT_1}(z, \hat{z})$ and

---

[2] To derive $\hat{z}^*$ in (8) express (7) as $EL = P_{DT_1} k_1 [E(z - \bar{z}_{DT_1})^2 | DT_1 + (\hat{z} - z_{NDT_1})^2] + (1 - P_{DT_1}) k_2 [E(z - \bar{z}_{NDT_1})^2 | NDT_1 + (\hat{z} - \bar{z}_{NDT_1})^2]$ and minimize with respect to $\hat{z}$.

$L_{NDT_1}(z, \hat{z})$ are general convex loss functions for DT_1 and NDT_1, respectively, then expected loss is given by

$$EL = P_{DT_1} EL_{DT_1}(z, \hat{z})|DT_1 + (1 - P_{DT_1}) EL_{NDT_1}(z, \hat{z})|NDT_1 \tag{9}$$

where the expectations on the right-side of (9) are computed using the conditional predictive pdfs $p(z \mid z < y_n, D)$ and $p(z \mid z \geq y_n, D)$. Then EL in (9) can be minimized, analytically or by computer methods to obtain the minimizing value of $\hat{z}$.

To illustrate the approach described in the previous paragraph, it may be that given a DT, over-forecasting is much more serious than under-forecasting by an equal amount. A loss function, the LINEX loss function, employed in Varian (1975) and Zellner (1986), is a convenient loss function which captures such asymmetric effects. It is given by:

$$L_{DT_1} = b_1 \left[e^{a_1(\hat{z} - z)} - a_1(\hat{z} - z) - 1 \right] \qquad \begin{matrix} b_1 > 0 \\ a_1 > 0. \end{matrix} \tag{10}$$

When $\hat{z} = z$, loss is zero and when $\hat{z} - z > 0$, a case of over-forecasting, loss rises almost exponentially with $a_1 > 0$. When $\hat{z} - z < 0$, loss rises almost linearly. Choice of the value of a_1 governs the degree of asymmetry. For example when a_1 has a small value, the loss function in (10) is close to a symmetric squared error loss function as can be seen by noting that $e^{a_1(\hat{z}-z)} \doteq 1 + a_1(\hat{z} - z) + a_1^2(\hat{z} - z)^2/2$ and substituting this expression in (10).

With NDT_1, it may be that under-forecasting is a more serious error than over-forecasting by an equal amount. The following LINEX loss function provides such asymmetric properties.

$$L_{NDT_1} = b_2 \left[e^{a_2(\hat{z} - z)} - a_2(\hat{z} - z) - 1 \right] \qquad \begin{matrix} b_2 > 0 \\ a_2 < 0. \end{matrix} \tag{11}$$

The loss functions in (10) and (11) can be inserted in (9) and an optimal point forecast can be computed. Note that the necessary condition for a minimum of (9) is

$$P_{DT_1} \frac{dEL_{DT_1} \mid DT_1}{d\hat{z}} + (1 - P_{DT_1}) \frac{dEL_{NDT_1} \mid NDT_1}{d\hat{z}} = 0. \tag{12}$$

If the derivatives in this last expression are approximated by expanding them around $\hat{z}_1$ and $\hat{z}_2$, values which set them equal to zero, respectively, the approximate value of $\hat{z}, \hat{z}^*$, which satisfies (12) is given by

$$\hat{z}^* = w\hat{z}_1 + (1 - w)\hat{z}_2, \tag{13}$$

where $w = P_{DT_1}b_1a_1^2/[P_{DT_1}b_1a_1^2 + (1 - P_{DT_1})b_2a_2^2]$. Thus it is seen that when $[P_{DT_1}/(1 - P_{DT_1})](b_1a_1^2/b_2a_2^2]$ is large, $\hat{z}^*$ is close to $\hat{z}_1$ and when it is small, $\hat{z}^*$ is close to $\hat{z}_2$, with $\hat{z}_1 = -\ell n(Ee^{-a_1z} \mid DT_1)/a_1$ and $\hat{z}_2 = -\ell n(Ee^{-a_2z} \mid NDT_1)/a_2$. Again, the optimal point forecast in (12) is sensitive not only to the values of P_{DT_1} and $1 - P_{DT_1}$, but also to the values of the loss functions' parameters, a_1, a_2, b_1, b_2.

2c. Turning Point and Point Forecasting Combined

Above, we have considered forecasting turning points and point forecasting *separately*. In this section, we provide a loss structure within the context of which it is possible to obtain a minimum expected loss solution to the joint problem of forecasting turning points and obtaining point forecasts.

The loss structure which we shall employ is shown in Table 2. It is seen that we allow for different quadratic losses, $k_1(\hat{z} - z)^2$ and $k_2(\hat{z} - z)^2$ for DT_1 and NDT_1 cases as in (6). Also, we incorporate additive elements of loss, c_1 and c_2 to reflect errors in turning point forecasts as in Table 1.

To obtain the minimum expected loss solution, we consider choice of the DT_1 forecast, derive the optimal point prediction and evaluate expected loss using it. We do the same for a NDT_1 forecast and then choose the act, forecast DT_1 or forecast NDT_1 which has the lower evaluated expected loss. As will be seen, the optimal point forecast of z has a value consistent with the selected optimal turning point forecast.

Table 2
Extended Forecasting Loss Structure*

	Possible Outcomes	
Acts	DT_1	NDT_1
Forecast DT_1	$k_1(\hat{z} - z)^2$	$k_2(\hat{z} - z)^2 + c_1$
Forecast NDT_1	$k_1(\hat{z} - z)^2 + c_2$	$k_2(\hat{z} - z)^2$

*It is assumed that $y_{n-1} < y_n$ and DT_1 and NDT_1 are defined as in (1a) of the text. The random variable $z \equiv y_{n+1}$ denotes the as yet unobserved value of y_{n+1} and $\hat{z}$ is some point forecast of z. k_1, k_2, c_1, and c_2 are given constants.

Given a choice of the DT_1 forecast, expected loss is given by:

$$EL_{DT_1} = P_{DT_1}k_1E[(\hat{z} - z)^2 \mid DT_1] + (1 - P_{DT_1})\{k_2E[(\hat{z} - z)^2 \mid NDT_1] + c_1\}. \quad (14)$$

The value of $\hat{z}$ which minimizes (14) subject to the side condition $\hat{z} < y_n$ is denoted by $\hat{z}_{DT_1}$. Then (14) is evaluated at $\hat{z} = \hat{z}_{DT_1}$. Similarly, if the NDT_1 forecast is chosen, expected loss is given by:

$$EL_{NDT_1} = P_{DT_1}\{k_1 E(\hat{z} - z)^2 \mid DT_1 + c_2\} + (1 - P_{DT_1})k_2 E(\hat{z} - z)^2 \mid NDT_1. \quad (15)$$

The value of $\hat{z}$ which minimizes (15) subject to $\hat{z} \geq y_n$ is denoted by $\hat{z}_{NDT_1}$ and it can be used to evaluate (15). Then, if $EL_{DT_1}(\hat{z} = \hat{z}_{DT_1}) < EL_{NDT_1}(\hat{z} = \hat{z}_{NDT_1})$, we forecast a DT_1 and use $\hat{z}_{DT_1}$ as our point forecast. Alternatively, if $EL_{DT_1}(\hat{z} = \hat{z}_{DT_1}) > EL_{NDT_1}(\hat{z} = \hat{z}_{NDT_1})$, we forecast a NDT_1 and use $\hat{z}_{NDT_1}$ as our point forecast.

In a special case $k_1 = k_2$ and $c_1 = c_2 = 0$, the above procedure yields the mean of the predictive pdf for z, denoted by $\bar{z}$, as the optimal point prediction and a DT_1 forecast if $\bar{z} < y_n$ or a NDT_1 forecast if $\bar{z} > y_n$.

As another special case, consider $k_1 = k_2 = 1$ and $c_1, c_2 > 0$. Then minimization of (14) with respect to $\hat{z}$ subject to $\hat{z} < y_n$, leads to

$$\hat{z}_{DT_1} = \begin{cases} \bar{z} & \text{if } \bar{z} < y_n \\ y_n - \Delta & \text{if } \bar{z} \geq y_n \end{cases} \quad (16)$$

where $\Delta > 0$ is arbitrarily small and

$$EL_{DT_1}(\hat{z} = \hat{z}_{DT_1}) = \begin{cases} (1 - P_{DT_1})c_1 + E(z - \bar{z})^2 & \text{if } \bar{z} < y_n \\ (1 - P_{DT_1})c_1 + E(z - \bar{z})^2 + (y_n - \bar{z})^2 & \text{if } \bar{z} \geq y_n. \end{cases}$$

Similarly, minimization of (15) with respect to $\hat{z}$ subject to $\hat{z} \geq y_n$ leads to:

$$\hat{z}_{NDT_1} = \begin{cases} y_n & \text{if } \bar{z} < y_n \\ \bar{z} & \text{if } \bar{z} \geq y_n \end{cases} \quad (17)$$

and

$$EL_{NDT_1})\hat{z} = \hat{z}_{NDT_1}) = \begin{cases} P_{DT_1}c_2 + E(z - \bar{z})^2 + (y_n - \bar{z})^2 & \text{if } \bar{z} < y_n \\ P_{DT_1}c_2 + E(z - \bar{z})^2 & \text{if } \bar{z} \geq y_n. \end{cases}$$

Thus if $\bar{z} < y_n$ the optimal turning point forecast is DT_1 if

$$(1 - P_{DT_1})c_1 + E(z - \bar{z})^2 < P_{DT_1}c_2 + E(z - \bar{z})^2 + (y_n - \bar{z})^2$$

or

$$1 < \frac{P_{DT_1}c_2 + (y_n - \bar{z})^2}{(1 - P_{DT_1})c_1} \quad (18a)$$

or

$$P_{DT_1} > \frac{c_1 - (y_n - \bar{z})^2}{c_1 + c_2}. \quad (18b)$$

Also, with $\bar{z} \geq y_n$ and (18) holding, $\bar{z}$ is the optimal point forecast for z. By similar considerations, if $\bar{z} < y_n$ and $P_{DT_1} < [c_1 - (y_n - \bar{z})^2]/(c_1 + c_2)$ then the optimal turning point forecast is NDT_1 and the optimal point forecast is $\hat{z}_{NDT_1} = y_n$.

With $\bar{z} \geq y_n$, then by the same line of reasoning, if

$$P_{DT_1} > \frac{c_1 + (y_n - \bar{z})^2}{c_1 + c_2}, \tag{19}$$

the optimal forecasts are DT_1 and $\hat{z}_{DT_1} = y_n - \Delta$. On the other hand, if the inequality in (19) is reversed, the optimal forecasts are NDT_1 and $\hat{z}_{NDT_1} = \bar{z}$.

The analysis leading to (18) and (19) was based on the assumption that $k_1 = k_2 = 1$. The above analysis, while tedious, can be extended to cover the case $k_1 \neq k_2$ with $k_1, k_2 > 0$ and $c_1, c_2 > 0$.

2d. Summary and Additional Considerations

In summary, it has been shown how the loss structures in Tables 1 and 2 can be used to obtain forecasts of turning points and how the probability of a DT and various loss functions can be employed to obtain optimal point forecasts. The main conclusion which emerges is that forecasts of turning points and of a future value are quite sensitive to assumptions regarding loss structures.

Above we have concentrated attention on the first future observation, $z = y_{n+1}$ given $y_{n-1} < y_n$ or given $y_{n-1} > y_n$—see (1). Given a predictive pdf for the next q future observations, $\underset{\sim}{z}' = (z_1, z_2, \ldots, z_q) \equiv (y_{n+1}, y_{n+2}, \ldots, y_{n+q})$, namely $p(\underset{\sim}{z} \mid D), -\infty < z_i < \infty$, $i = 1, 2, \ldots, q$, it is possible to calculate the probabilities associated with various inequalities involving the z_i's, using the predictive pdf, $p(\underset{\sim}{z} \mid D)$. For example, the probability that $z_1, z_2 < z_3 > z_4, z_5$ can be computed.

The definitions in (1) can be broadened to include more past and future observations, as Wecker (1979) and Kling (1987) indicate. As an example consider, with $z_1 \equiv y_{n+1}$ and $z_2 \equiv y_{n+2}$,

$$y_{n-2} < y_{n-1} < y_n \text{ and } \begin{cases} z_1 < y_n \text{ and } z_2 < z_1 \equiv DT_2 \\ \text{otherwise } \equiv NDT_2 \end{cases} \tag{20}$$

where $DT_2 \equiv$ downturn based on two previous observations and two future observations relative to the given value y_n and $NDT_2 \equiv$ no such downturn. Then given $y_{n-2} < y_{n-1} < y_n$, the probability of a DT_2, P_{DT_2}, is given by

$$P_{DT_2} = \int_{-\infty}^{y_n} \int_{-\infty}^{z_1} p(z_1, z_2 \mid D) dz_2 dz_1, \tag{21}$$

where it has been assumed that $\infty < z_i < \infty$, $i = 1, 2$.[3] This probability can be

[3]Note that $y_{n-2}, y_{n-1} < y_n > z_1, z_2$ is an alternative definition of a DT_2 and its probability of occurring is $\int_{-\infty}^{y_n} \int_{-\infty}^{y_n} p(z_1, z_2 \mid D) dz_1 dz_2$, somewhat different from (21).

employed along with the loss structure in Table 1 to choose between the acts forecast DT_2 and forecast NDT_2 so as to minimize expected loss. Also, this analysis can be combined with the problem of obtaining optimal point predictions of z_1 and z_2. Finally, using m past values and m future values, a DT_m can be defined and its probability calculated and used in forecasting.

Above, we have considered just one model for the observations, $\underset{\sim}{y}' = (y_1, y_2, \ldots, y_{n-1}, y_n)$. Often forecasters utilize several alternative models, $M_1, M_2, \ldots, M_r$, for example autoregressive models of differing orders, autoregressive models with various leading indicator variables, etc. Let $p_i(\underset{\sim}{y} \mid M_i, D_i)$ be the marginal pdf for $\underset{\sim}{y}$, based on model M_i and sample and prior information D_i.[4] Then the posterior probability associated with the ith model M_i, denoted by P_i is given by

$$P_i = \Pi_i p_i(\underset{\sim}{y} \mid M_i, D_i) \Big/ \sum_{i=1}^{r} \Pi_i p_i(\underset{\sim}{y} \mid M_i, D_i) \tag{22}$$

where Π_i, $i = 1, 2, \ldots, r$ is the prior probability associated with M_i.[5] Also, for each model the predictive pdf for $z = y_{n+1}$ can be derived and is denoted by $p_i(z \mid M_i, D_i)$. It can be used, for example, to compute the probability of a DT, P_{iDT}. The P_{iDT}'s and the P_i's in (22) allow us to form a probability mass function as shown in Fig. 1 for the case of $r = 4$. Fig. 1 reveals the effects of model uncertainty on the probability of a DT. That is, instead of having a single probability of a DT, when there is model uncertainty, we have several probabilities of a DT, $P_{1DT}, P_{2DT}, \ldots, P_{rDT}$ and their respective probabilities, $P_1, P_2, \ldots, P_r$. The usual practice of selecting one model and viewing it as "absolutely true" in deriving a single probability of a DT obviously abstracts from model uncertainty and is inappropriate when model uncertainty is present.

Formally, we have the marginal predictive pdf, $p(z \mid D)$, given by

$$p(z \mid D) = \sum_{i=1}^{r} P_i p_i(z \mid M_i, D_i), \tag{23}$$

where P_i is given in (21) and D is the union of the D_i. Then, for example, the

---

[4]As is well known, $p_i(\underset{\sim}{y} \mid M_i, D_i) = \int_{\Theta_i} p_i(\underset{\sim}{y} \mid M_i, \underset{\sim}{\theta}_i)\pi(\underset{\sim}{\theta}_i \mid I_i)d\underset{\sim}{\theta}_i$ where $p_i(\underset{\sim}{y} \mid M_i, \underset{\sim}{\theta}_i)$ is the likelihood function given model i, $\underset{\sim}{\theta}_i$ is a vector of parameters with prior pdf $\pi(\underset{\sim}{\theta}_i \mid I_i)$, where I_i denotes the prior parameters, $D_i = (\underset{\sim}{y}, I_i)$ and $\underset{\sim}{\theta}_i \subset \Theta_i$, the parameter space.

[5]Here we assume that the r models are mutually exclusive and exhaustive. It is possible to relax the assumption that the collection of models is exhaustive.

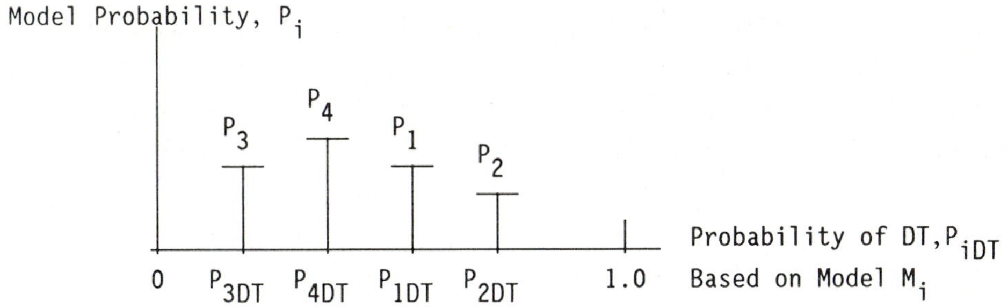

Fig. 1: Probability Mass Function for Probabilities of a DT

probability of a DT_1, P_{DT_1} is given by

$$P_{DT_1} = \int_{-\infty}^{y_n} p(z \mid D)dz \tag{24}$$

$$= \sum_{i=1}^{r} P_i P_{iDT_1},$$

where $P_{iDT_1} = \int_{-\infty}^{y_n} p_i(z \mid M_i, D_i)dz$ is the probability of DT_1 based on model M_i. It is seen from the second line of (24) that P_{DT_1} is an average of the P_{iDT_1} with the posterior model probabilities, the P_i's serving as weights. Further, various measures can be computed to characterize the dispersion and other features of the P_{iDT_1}'s. For example their variance is given by

$$\text{Var}(P_{iDT_1}) = \sum_{i=1}^{r} P_i(P_{iDT_1} - P_{DT_1})^2. \tag{25}$$

Given a loss structure such as that in Table 1, P_{DT_1} in (24) can be employed to make an optimal choice between forecast DT_1 or forecast NDT_1. Similar analysis yields results for forecasting an upturn or no upturn. For broader definitions of turning points, predictive pdfs for several future observations would replace $p_i(z \mid M_i, D_i)$ in (23) and the integral in (24) would have to be modified along the lines shown in (21).

Finally, the problem of forecasting turning points in two or more time series is of interest. Given a definition of a turning point, and a model for two or more time series, probabilities of downturns or upturns can be computed from the joint predictive pdf of future values of the several time series. Then, given a loss structure, optimal turning point forecasts can be derived. In the case of two time series, for example the rates of growth of output and of inflation, the set of possible forecasts will involve forecasting downturns for both variables, a downturn for one variable and no downturn for the other, or no downturn for both variables. Given m such possible forecasts and m possible outcomes, an $m \times m$ loss structure can be defined, an expanded version of the 2×2 case shown in Table 1. Using probabilities associated with possible outcomes, the forecast that minimizes expected loss can be determined along the lines shown in (4) for the 2×2 case. Also, the multiple models, multiple time series case can be addressed using a generalization of the multiple models, one time series case analyzed above.

3. Data and Applications

In this section some of the techniques described above will be applied in the analysis of data relating to *annual* output growth rates for 18 countries used in our previous work, Garcia-Ferrer et al., (1987) and Zellner and Hong (1987). A diskette containing the data is available on request for a small charge to cover mailing and other costs.

Shown in Fig. 2 is a boxplot of the annual rates of output growth for 18 countries over the period, 1951-85. It is seen that the annual median growth rates, given by the horizontal line in each box, appear to follow a cyclical path with peaks in 1955, 1960, 1964, 1969, 1973, 1976, 1979, and 1984 and troughs in 1952, 1958, 1963, 1966, 1971, 1975, and 1981. The average time between peaks is 4.1 years and between troughs is 4.8 years. To provide more detail, Figs. 3a and 3b provide the number of countries experiencing peaks and troughs in each year. It is seen that many countries experienced peaks in years close to or at 1955, 1960, 1964, 1969, 1973, 1976, 1979, and 1984. As regards troughs, they were encountered for many countries in years close to or at 1954, 1958, 1962, 1966-67, 1971, 1975, 1977, and 1982. These descriptive measures reflect the well known fact that economies' output growth rates, while not perfectly synchronized, tend to move up and down together, as noted by Burns and Mitchell (1946), Zarnowitz (1985) and others.

We now turn to the problem of forecasting turning points for the 18 countries' growth rates of annual output. *Here a downturn is defined to be a sequence of observations satisfying* $y_{n-1}, y_{n-2} < y_n > y_{n+1}$ *and an upturn a sequence of observations satisfying* $y_{n-2}, y_{n-1} > y_n < y_{n+1}$. While these are not the only possible definitions, they will be used since if there were a deterministic four year cycle in the data, these definitions would be exactly appropriate for identifying peaks and troughs.

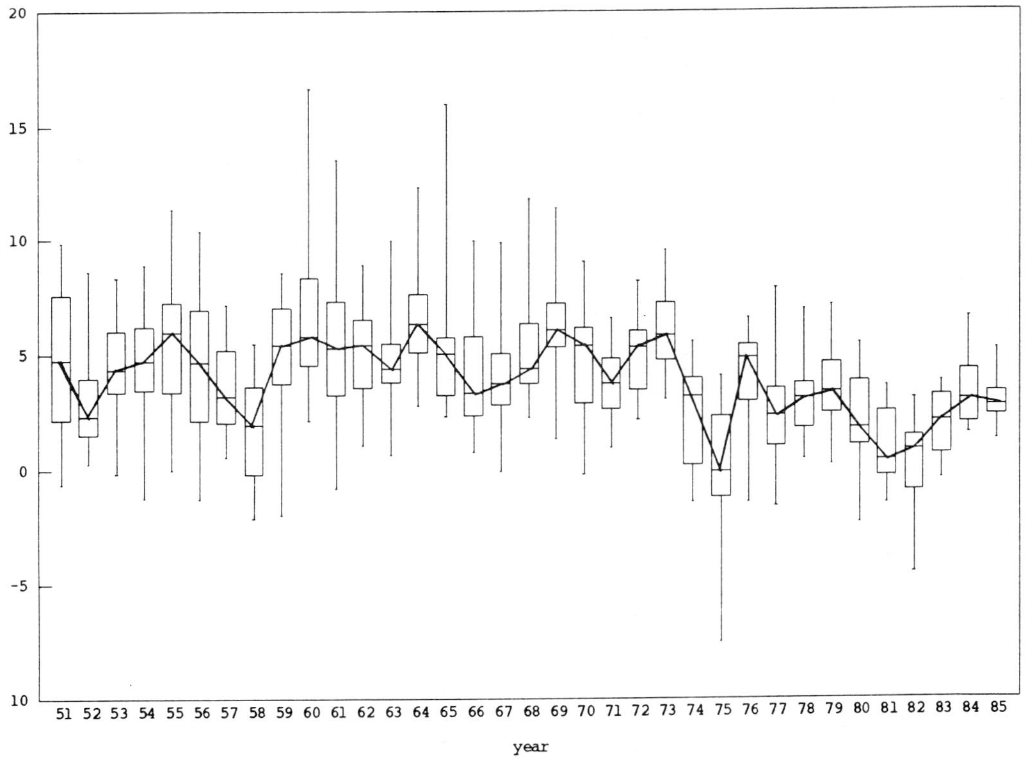

*Annual rates of growth of real GNP or GDP for the following countries have been utilized: Australia, Austria, Belgium, Canada, Denmark, Finland, France, Germany, Ireland, Italy, Japan, Netherlands, Norway, Spain, Sweden, Switzerland, United Kingdom, and United States. Data from the U. of Chicago, Graduate School of Business IMF data base.

Fig. 2: Boxplot of Annual Real Output Growth Rates for 18 Countries, 1951-1985*

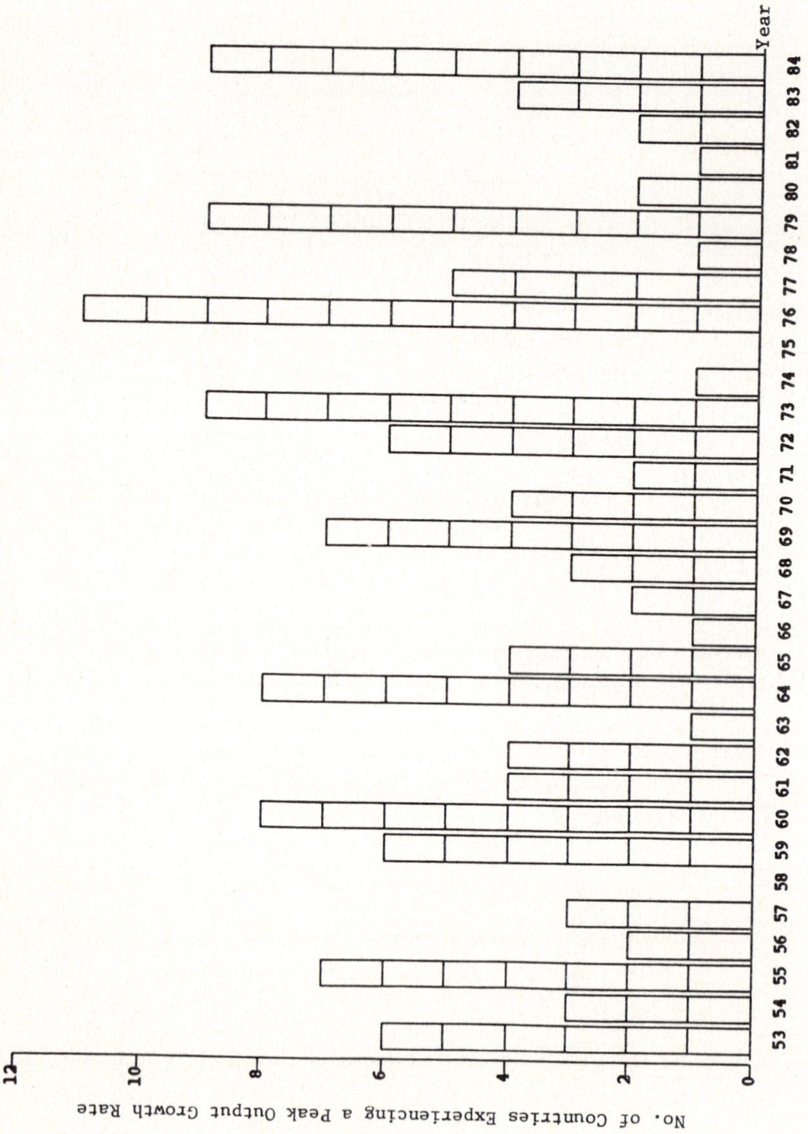

Fig. 3a: Number of Countries Experiencing a Peak Output Growth Rate by Years, 1953-84 for 18 Countries

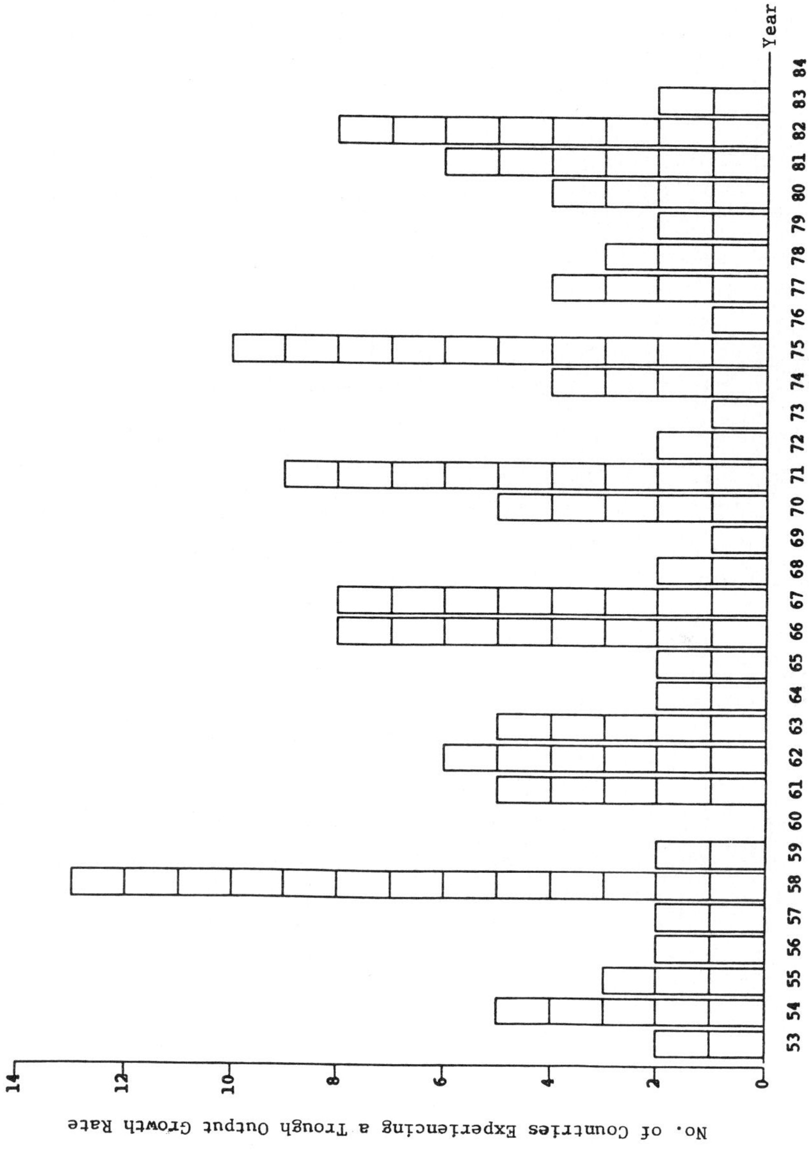

Fig. 3b: Number of Countries Experiencing a Trough Output Growth Rate by Years, 1953-84 for 18 Countries

Our forecasting model, an autoregression of order three with leading indicator variables, denoted by AR(3)LI model, used in our previous work, Garcia-Ferrer et al., (1987) and Zellner and Hong (1987) is[6]

$$\begin{aligned} y_{it} = \alpha_{i0} + \alpha_{1i}y_{it-1} + \alpha_{2i}y_{it-2} + \alpha_{3i}y_{it-3} & \qquad i = 1, 2, \ldots, 18 \\ + \beta_{1i}SR_{it-1} + \beta_{2i}SR_{it-2} + \beta_{3i}GM_{it-1} & \\ + \beta_{4i}WR_{t-1} + \epsilon_{it} & \qquad t = 1, 2, \ldots, T \end{aligned} \tag{26}$$

where the subscripts i and t denote the i'th country and t'th year, respectively and

$$\begin{aligned} y_{it} &= \text{growth rate of real output} \\ SR_{it} &= \text{growth rate of real stock prices} \\ GM_{it} &= \text{growth rate of real money} \\ WR_t &= \text{"world return," the median of the } SR_{it}\text{'s.} \\ \epsilon_{it} &= \text{error term.} \end{aligned}$$

The ϵ_{it}'s are assumed independently drawn from a normal distribution with zero mean and variance σ_i^2 for all i and t. Using annual data, 1951-1973, a diffuse prior pdf for the parameters, the countries' Student-t predictive pdfs for 1974 were computed, updated year-by-year, and used to compute probabilities of downturns and upturns. Then using the loss structure in Table 2 with $c_1 = c_2$ forecasts of downturns (DTs) and upturns (UTs) were made. In this case, a DT is forecasted when $P_{DT} > 1/2$ and NDT is forecasted when $P_{DT} < 1/2$. Also, a UT is forecasted when $P_{UT} > 1/2$ and NUT when $P_{UT} < 1/2$. For all 18 countries over the period 1974-85, under the definition two preceding observations below (or above) a current observation and the following observation below (or above) the current observation for a DT (UT), our forecasting procedure yielded the results shown in Table 3.

We see from panel A of Table 3 that 45 out of 68 or 66% of the DT/NDT forecasts are correct. Of the 40 DT forecasts 35 or 40 or 88% are correct. However, only 10 of 28 or 36% of the NDT forecasts are correct. As regards forecasts of UT's or NUT's, 60 of 82 forecasts, 75% are correct. Of the 63 UT forecasts, 43 or 68% are correct while for the NUT forecasts, 17 of 19, or 90% are correct. Thus except for the NDT forecasts, the turning point forecasts are quite good.

---

[6]See Appendix tables for a description of the statistical fits of (26) to our data and Garcia-Ferrer et al., (1987) and Zellner and Hong (1987) for the quality of point forecasts yielded by (26).

Table 3

Forecasts of Turning Points for 18 Countries' Output Growth Rates, 1974-85*

A. Downturn (DT) and No Downturn (NDT)

Forecast	Correct	Incorrect	Total
DT	35	5	40
NDT	10	18	28
Total	45	23	68
Percent	(66%)	(34%)	(100%)

B. Upturn (UT) and No Upturn (NUT)

Forecast	Correct	Incorrect	Total
UT	43	20	63
NUT	17	2	19
Total	60	22	82
Percent	(75%)	(25%)	(100%)

*With y_t denoting a growth rate in year t, the following definitions have been employed: $y_{n-2}, y_{n-1} < y_n > y_{n+1} \equiv DT$; $y_{n-2}, y_{n-1} < y_n \leq y_{n+1} \equiv NDT$; $y_{n-2}, y_{n-1} > y_n < y_{n+1} \equiv UT$; and $y_{n-2}, y_{n-1} > y_n \geq y_{n+1} \equiv NUT$.

As regards the 18 incorrect NDT forecasts, 4 occurred for calculated P_{DT}'s between .40 and .49, 8 for calculated P_{DT}'s between .30 and .39 and 4 for calculated P_{DT}'s between .20 and .29. Thus 16 of the 18 incorrect NDT forecasts had substantial probabilities of DTs associated with them. If a more conservative rule for forecasting DTs had been employed, say forecast a downturn if $P_{DT} > 0.2$, then only 2 of the NDT forecasts would be incorrect. The results of using a cut-off probability of 0.2, that is forecast a DT if $P_{DT} > .2$, which would be optimal given a cost parameter ratio $c_2/c_1 = 4$ in Table 1, are: 51 or 85% of 60 DT forecasts correct and 6 or 75% of 8 NDT forecasts correct. These figures indicate very well the sensitivity of turning point forecasting to asymmetry of loss structures.

4. Summary and Concluding Remarks

The problems of forecasting turning points and future values of economic time series were considered with the major result being that such forecasts are very sensitive to properties of loss structures. An operational procedure for forecasting turning points was formulated and applied to forecast turning points in output growth rates for 18 countries, 1974-85 utilizing an AR(3)LI model fitted with pre-1974 data and updated year-by-year in the forecast period. In general, with the exception of no-downturn forecasts, the results were encouraging, namely 66% of the DT and NDT forecasts correct and 75% of the UT and NUT forecasts correct. These results indicate that our AR(3)LI model and our forecasting techniques may be of practical value to applied economic forecasters not only in providing forecasts regarding turning points but also in computing probabilities associated with future, as yet unobserved values of economic variables.

Acknowledgment

Research financed in part by the National Science Foundation and by income from the H.G.B. Alexander Endowment Fund, Graduate School of Business, University of Chicago.

References

Barnard, G.A. (1987). *Personal Communication.*

Burns, A.F. and W.C. Mitchell (1946). *Measuring Business Cycles.* New York: National Bureau of Economic Research, Inc.

Garcia-Ferrer, A., R.A. Highfield, F. Palm, and A. Zellner (1987). "Macro-economic Forecasting Using Pooled International Data," *Journal of Business and Economic Statistics*, Vol. 5, 53-67.

Kling, J.L. (1987). "Predicting the Turning Points of Business and Economic Time Series," *Journal of Business*, Vol. 60, 201-238.

Moore, G.H. (1961). *Business Cycle Indicators*, Vol. 1. Princeton, NJ: Princeton University Press.

Moore, G.H. (1983). *Business Cycles, Inflation and Forecasting*, 2nd ed. Cambridge, MA: Ballinger.

Moore, G.H. and V. Zarnowitz (1982). "Sequential Signals of Recession and Recovery," *Journal of Business*, Vol. 55, 57-85.

Neftci, S.H. (1982). "Optimal Prediction of Cyclical Downturns," *Journal of Economic Dynamics and Control*, Vol. 4, 225-241.

Press, S.J. and A. Zellner (1978). "Posterior Distribution for the Multiple Correlation Coefficient with Fixed Regressors," *Journal of Econometrics*, Vol. 8, 307-321.

Varian, H. (1975). "A Bayesian Approach to Real Estate Assessment," in *Studies in Bayesian Econometrics and Statistics in Honor of Leonard J. Savage*, S.E. Fienberg and A. Zellner, eds., Amsterdam: North-Holland, 195-208.

Wecker, W.E. (1979). "Predicting the Turning Points of a Time Series," *Journal of Business*, Vol. 52, 35-50.

Zarnowitz, V. (1967). *An Appraisal of Short-Term Economic Forecasts*. New York: National Bureau of Economic Research, Inc.

Zarnowitz, V. (1985). "Recent Work on Business Cycles in Historical Perspective," *Journal of Economic Literature*, Vol. 23, 523-580.

Zellner, A. (1986). "Bayesian Estimation and Prediction Using Asymmetric Loss Functions," *Journal of the American Statistical Association*, Vol. 81, 446-451.

Zellner, A. (1987). "Science, Economics and Public Policy," *American Economist*, Vol. 31, 3-7.

Zellner, A. (1988). "Causality and Causal Laws in Economics," in *Causality*, D. Aigner and A. Zellner, eds., special issue of the *Journal of Econometrics*, Vol. 39, 7-21,

Zellner, A. and C. Hong (1989). "Forecasting International Growth Rates Using Bayesian Shrinkage and Other Procedures," *Journal of Econometrics*, Vol. 40, 183-202.

Appendix Table 1

Bayesian Diffuse Prior Regression Output for the AR(3)LI Model in Equation (26), 18 Countries, 1954-73[a]

Country	Const.	y_{it-1}	y_{it-2}	y_{it-3}	SR_{it-1}	SR_{it-2}	MG_{it-1}	WR_{t-1}	R^2 [c]	$E(\sigma \mid D)$ [d]	r_1 [e]
	Posterior Means and Standard Deviations[b]										
Belgium	-0.11	0.26	0.54	0.18	-0.09	-0.09	-0.06	0.22	0.55	1.68	0.08
	(1.86)	(0.25)	(0.35)	(0.26)	(0.08)	(0.04)	(0.11)	(0.11)			
Denmark	4.26	-0.31	-0.13	0.10	-0.03	0.02	0.36	0.03	0.40	2.25	-0.06
	(1.92)	(0.31)	(0.33)	(0.24)	(0.08)	(0.06)	(0.15)	(0.08)			
France	5.26	-0.10	0.02	0.13	-0.04	0.01	-0.03	0.10	0.56	0.98	-0.29
	(1.68)	(0.29)	(0.25)	(0.21)	(0.03)	(0.02)	(0.05)	(0.03)			
Germany	4.17	-0.23	-0.22	0.08	0.03	0.00	0.44	0.18	0.72	2.48	-0.11
	(2.36)	(0.27)	(0.25)	(0.20)	(0.05)	(0.04)	(0.21)	(0.11)			
Ireland	2.00	0.41	0.15	-0.13	0.05	-0.04	0.05	-0.10	0.43	2.57	-0.06
	(1.38)	(0.41)	(0.40)	(0.34)	(0.06)	(0.05)	(0.14)	(0.08)			
Italy	5.68	0.10	0.21	-0.44	0.04	-0.03	0.05	0.06	0.54	1.89	0.06
	(2.39)	(0.37)	(0.35)	(0.27)	(0.04)	(0.04)	(0.11)	(0.07)			
Netherlands	1.41	0.35	-0.02	0.10	-0.11	-0.08	0.39	0.33	0.78	1.67	0.22
	(2.05)	(0.19)	(0.20)	(0.20)	(0.07)	(0.03)	(0.12)	(0.10)			
United Kingdom	4.90	-0.53	-0.10	-0.11	0.01	0.00	0.20	0.03	0.42	1.71	-0.11
	(2.01)	(0.43)	(0.55)	(0.41)	(0.06)	(0.07)	(0.12)	(0.10)			
United States	4.98	-0.23	-0.04	-0.33	0.04	-0.03	0.67	-0.09	0.76	1.52	0.18
	(1.20)	(0.19)	(0.21)	(0.17)	(0.05)	(0.04)	(0.21)	(0.05)			
Australia	7.66	-0.43	-0.54	0.28	0.06	0.11	0.38	-0.13	0.81	1.66	-0.36
	(3.17)	(0.29)	(0.23)	(0.29)	(0.06)	(0.06)	(0.15)	(0.07)			
Austria	6.34	-0.03	-0.16	-0.25	-0.02	0.03	0.17	0.09	0.45	2.11	0.00
	(2.34)	(0.29)	(0.30)	(0.26)	(0.06)	(0.04)	(0.14)	(0.10)			
Canada	6.57	-0.29	0.08	-0.12	0.12	0.01	0.00	-0.12	0.61	1.61	-0.14
	(2.85)	(0.39)	(0.30)	(0.32)	(0.06)	(0.06)	(0.05)	(0.08)			

Appendix Table 1 (cont'd)

Posterior Means and Standard Deviations[b]

Country	Const.	y_{it-1}	y_{it-2}	y_{it-3}	SR_{it-1}	SR_{it-2}	MG_{it-1}	WR_{t-1}	R^2 [c]	$E(\sigma \mid D)$ [d]	r_1[e]
Finland	4.94	0.08	-0.27	0.11	-0.01	-0.09	0.15	0.17	0.70	2.71	0.00
	(1.91)	(0.25)	(0.20)	(0.21)	(0.07)	(0.05)	(0.13)	(0.11)			
Japan	6.89	0.29	-0.13	0.14	-0.08	-0.02	0.02	0.13	0.28	3.61	-0.04
	(5.63)	(0.38)	(0.42)	(0.40)	(0.08)	(0.06)	(0.12)	(0.16)			
Norway	3.46	0.08	0.12	-0.12	0.02	0.01	0.13	0.00	0.15	1.91	0.10
	(1.96)	(0.28)	(0.30)	(0.29)	(0.06)	(0.05)	(0.15)	(0.06)			
Spain	6.23	0.18	-0.36	0.12	0.09	-0.01	0.05	-0.03	0.33	3.89	0.05
	(5.00)	(0.52)	(0.47)	(0.43)	(0.11)	(0.10)	(0.31)	(0.20)			
Sweden	2.41	0.50	-0.10	0.00	-0.03	-0.06	-0.18	0.10	0.57	1.39	0.06
	(2.03)	(0.31)	(0.32)	(0.27)	(0.03)	(0.03)	(0.09)	(0.06)			
Switzerland	5.47	-0.25	-0.13	-0.28	-0.08	0.02	0.34	0.21	0.84	1.18	0.11
	(1.35)	(0.17)	(0.17)	(0.14)	(0.03)	(0.02)	(0.07)	(0.06)			

[a] The estimation period is 1954-73, 20 observations for all countries except for Australia, 1960-73, Canada, 1959-73, Japan, 1956-73, and Spain, 1958-73.

[b] Figures not in parentheses are posterior means which are equal to least squares estimates. Figures in parentheses, standard deviations of marginal posterior pdf's, are $\sqrt{\nu(\nu-2)}$ times the usual asymptotic least squares standard errors where $\nu = n - k$, where n = number of observations and k = number of parameters estimated, that is ν = usual degrees of freedom.

[c] R^2, the sample squared multiple correlation coefficient is approximately the mean of the posterior distribution of the squared population multiple correlation coefficient—see Press and Zellner (1978).

[d] $E(\sigma \mid D)$ is the posterior mean of the error term standard deviation, σ, namely $E(\sigma \mid D) = s(\nu/2)^{1/2} \cdot \Gamma[(\nu-1)/2]/\Gamma(\nu/2)$, where $\nu = n - k$, the degrees of freedom and s^2 = sum of squared residuals divided by ν. For $\nu = 20 - 8 = 12$, $E(\sigma \mid D)$ is equal to 1.068 times the non-Bayesian standard error of estimate, s.

[e] r_1, the sample first order residual serial correlation coefficient is an approximate posterior mean of ρ_1, the population first order serial correlation.

Appendix Table 2

Bayesian Diffuse Prior Regression Output for the AR(3)LI Model in Equation (26), 18 Countries, 1954-85[a]

Posterior Means and Standard Deviations[b]

Country	Const.	y_{it-1}	y_{it-2}	y_{it-3}	SR_{it-1}	SR_{it-2}	MG_{it-1}	WR_{t-1}	R^2 [c]	$E(\sigma \mid D)$ [d]	r_1 [e]
Belgium	-0.43	0.20	0.66	0.15	-0.03	-0.10	0.00	0.19	0.71	1.49	0.01
	(0.80)	(0.14)	(0.15)	(0.16)	(0.03)	(0.03)	(0.08)	(0.05)			
Denmark	1.71	-0.01	-0.15	0.17	0.01	-0.03	0.16	0.01	0.37	2.28	0.03
	(1.08)	(0.19)	(0.20)	(0.19)	(0.03)	(0.02)	(0.07)	(0.05)			
France	0.46	0.17	0.34	0.34	-0.03	-0.01	0.01	0.10	0.57	1.59	0.23
	(0.86)	(0.21)	(0.21)	(0.19)	(0.02)	(0.02)	(0.06)	(0.04)			
Germany	1.61	0.01	-0.05	0.14	0.05	-0.01	0.40	0.08	0.72	2.19	-0.21
	(0.89)	(0.18)	(0.17)	(0.14)	(0.03)	(0.03)	(0.10)	(0.05)			
Ireland	2.02	0.27	-0.05	0.14	0.04	-0.02	0.05	-0.04	0.28	2.43	-0.08
	(1.16)	(0.27)	(0.27)	(0.25)	(0.05)	(0.03)	(0.10)	(0.06)			
Italy	1.94	-0.05	0.23	0.04	0.01	-0.02	0.19	0.10	0.60	2.10	0.34
	(1.09)	(0.20)	(0.18)	(0.15)	(0.03)	(0.02)	(0.08)	(0.05)			
Netherlands	-0.09	0.42	0.03	0.28	-0.01	-0.05	0.33	0.14	0.76	1.71	-0.13
	(0.73)	(0.12)	(0.14)	(0.12)	(0.05)	(0.02)	(0.07)	(0.06)			
United Kingdom	1.76	-0.10	0.07	0.24	0.05	0.00	0.18	-0.02	0.44	1.79	-0.08
	(0.84)	(0.19)	(0.20)	(0.19)	(0.03)	(0.03)	(0.07)	(0.05)			
United States	3.89	-0.16	0.12	-0.27	0.09	-0.06	0.52	-0.08	0.72	1.59	-0.01
	(0.73)	(0.14)	(0.15)	(0.13)	(0.04)	(0.03)	(0.13)	(0.04)			
Australia	1.83	0.11	-0.04	0.45	0.06	-0.02	0.05	-0.02	0.43	2.22	0.09
	(1.43)	(0.22)	(0.20)	(0.21)	(0.04)	(0.03)	(0.12)	(0.05)			

Appendix Table 2 (cont'd)

Posterior Means and Standard Deviations[b]

Country	Const.	y_{it-1}	y_{it-2}	y_{it-3}	SR_{it-1}	SR_{it-2}	MG_{it-1}	WR_{t-1}	R^2 [c]	$E(\sigma \mid D)$ [d]	r_1 [e]
Austria	3.01	0.20	-0.24	0.19	0.03	0.01	0.17	0.04	0.46	2.13	-0.04
	(1.26)	(0.21)	(0.21)	(0.18)	(0.05)	(0.03)	(0.07)	(0.05)			
Canada	2.17	0.10	0.37	-0.06	0.04	-0.05	0.09	0.01	0.44	2.24	-0.14
	(1.19)	(0.23)	(0.23)	(0.21)	(0.04)	(0.03)	(0.05)	(0.05)			
Finland	3.15	0.29	-0.28	0.19	0.00	-0.07	0.12	0.10	0.57	2.60	0.01
	(1.28)	(0.20)	(0.18)	(0.17)	(0.05)	(0.03)	(0.08)	(0.07)			
Japan	2.24	0.43	-0.03	0.27	-0.05	-0.04	0.08	0.09	0.57	3.02	-0.04
	(1.59)	(0.26)	(0.27)	(0.21)	(0.06)	(0.04)	(0.08)	(0.08)			
Norway	3.02	0.23	-0.01	-0.09	-0.01	-0.03	0.14	-0.01	0.26	1.60	0.05
	(1.21)	(0.19)	(0.20)	(0.20)	(0.02)	(0.02)	(0.07)	(0.04)			
Spain	2.99	0.32	-0.23	0.22	0.01	0.02	0.15	0.08	0.55	2.67	-0.01
	(1.65)	(0.28)	(0.25)	(0.22)	(0.05)	(0.04)	(0.15)	(0.06)			
Sweden	1.95	0.81	-0.44	0.16	-0.04	-0.01	-0.18	0.05	0.48	1.66	-0.05
	(0.85)	(0.27)	(0.26)	(0.20)	(0.02)	(0.02)	(0.09)	(0.03)			
Switzerland	1.75	0.21	-0.02	0.05	-0.02	0.00	0.21	0.15	0.72	1.93	-0.21
	(0.66)	(0.16)	(0.16)	(0.13)	(0.04)	(0.03)	(0.06)	(0.05)			

[a]The estimation period is 1954-85, 32 observations for all countries except for Australia, 1960-85, Canada, 1959-85, Japan, 1956-85, Spain, 1958-84, and Sweden, 1954-83.
[b]For footnotes b, c, d, and e, see Appendix Table 1.

VII. INCOMPLETELY SPECIFIED PRIOR DISTRIBUTIONS

Bayesian and Likelihood Methods in Statistics and Econometrics
S. Geisser, J.S. Hodges, S.J. Press and A. Zellner (Editors)

INCORPORATING PARTIAL PRIOR INFORMATION: RANGES OF POSTERIOR PROBABILITIES

L. Mark Berliner and Prem K. Goel
Ohio State University

Abstract

This paper considers the incorporation of partial prior information concerning an unknown parameter in statistical inference and decision making problems. Such prior information is combined with experimental information to produce ranges of posterior probabilities concerning the parameter of interest. An example from Martz and Waller (1983) is pursued in depth.

1. Introduction

Many "textbook" presentations of Bayesian analysis appear to suggest that the approach is composed of two "automatic" steps:

(1) Choose a prior;

(2) Combine prior and experimental information, via Bayes' Theorem to obtain a posterior distribution upon which to base inferences or decisions.

While these steps form the basic tools of Bayesian analysis, an overmechanized view of their implementation is often inaccurate. Our goal is to present a simple, though "non-textbook" Bayesian argument *not* based on a single prior.

The problem of interest here is the Bayesian statistical analysis based on *partial* prior information. We assume that the experimenter is either unwilling or unable to specify a single prior distribution. The suggestion then is to replace the "automatic" steps above by:

(1)′ Construct a class of priors consistent with the available partial prior information;

(2)′ Combine each prior in the class and experimental information, via Bayes' Theorem, thereby producing *ranges* of posterior measures (e.g. probabilities).

The following motivation of this suggestion is enlightening: Consider how the subjectivist constructs a prior in practice. Though various techniques exists (cf. Kadane, et al., 1980 and Winkler, 1967), nearly all approaches begin by inputing a finite number, say k, of characteristics the prior should possess. For example, one might specify k quantiles of the prior; an alternative is to specify some moment information, e.g. mean and variance. Of course, in general these characteristics do not uniquely specify a prior, but rather a class of priors. Typically, the prior is then chosen to be that distribution in a convenient class, e.g. the conjugates, which most closely matches the k inputs. Another approach is to choose a prior, consistent with the k inputs, according to some information theoretic criteria. A major example is the maximum entropy prior discussed by Jaynes (1968). However, such choices are essentially arbitrary and only a class of priors is consistent with the experimenter's input in the prior elicitation process.

The consideration of classes of priors is by no means a new idea. An extensive literature concerning the *robust Bayesian viewpoint* and the incorporation of partial prior information is reviewed and extended in Berger (1984, 1985). Other highly related references are Berger (1979), DeRobertis (1978), Diaconis and Zabell (1982), Lambert and Duncan (1985), and Goldstein (1984).

Finally, we note that there are other approaches to decision making, without employing a single prior, based on nonprobabilistic methods. Major references include Dempster (1968) and Shafer (1976, 1982). For brevity, we remark only that we prefer probabilistic representations of uncertainty for both theoretical and practical (i.e. ease of understanding) reasons.

2. Formulation

Suppose that θ represents an unknown parameter. Let Θ denote the parameter space, the collection of all possible values of θ. Also, let $\pi(\theta)$ denote a prior distribution (or density) on Θ. Further, let X denote an observable random variable whose distribution (density) depends on θ, and let $X \sim f(x \mid \theta)$.

Given the outcome $X = x$, the posterior density of θ, denoted by $\pi(\theta \mid x)$, is given by

$$\pi(\theta \mid x) = \frac{f(x \mid \theta)\pi(\theta)}{\int_\Theta f(x \mid \theta)\pi(\theta)d\theta},$$

for θ continuous. (If θ is discrete, integrals are replaced by obvious summations.) We will consider the following prior structures: Assume that the parameter space is partitioned into k disjoint subsets,

$$\Theta = T_1 \cup \ldots \cup T_k$$

The choice of the partition is usually arbitrary, though in some multiple decision problems, the T_i's may arise naturally. The partial prior information is specified through prior probabilities

$$p_i = Pr[\theta \in T_i], \qquad i = 1, \ldots, k, \tag{2.1}$$

where p_i sum to one. Without any further specifications, (2.1) generates a class, say Γ, of priors:

$$\Gamma = \{\pi(\theta) = \sum_{i=1}^{k} p_i q_i(\theta) : q_i(\theta) \text{ is any density with support set } T_i, \quad i = 1, \ldots, k\}. \tag{2.2}$$

Next, for a fixed prior π, consider the posterior probabilities of the T_i's:

$$p_i(x) = Pr[\theta \in T_i \mid x], \qquad i = 1, 2, \ldots, k.$$

Our interest is in the ranges of $p_i(x)$ as π varies over Γ. In particular, in the next section we obtain the following bounds:

$$\ell_i(x) = \inf_{\pi \in \Gamma} p_i(x), \; u_i(x) = \sup_{\pi \in \Gamma} p_i(x) \tag{2.3}$$

3. Ranges of Posterior Probabilities

Before presenting the results, some preliminaries are required. First, for an arbitrary prior on Θ, say $q(\theta)$, the marginal density of X is given by

$$m(x \mid q) = \int_{\Theta} f(x \mid \theta) q(\theta) d\theta.$$

Next, for any prior $\pi \in \Gamma$ given by (2.2), direct application of Bayes Theorem gives the posterior distribution of θ, which is representable as a convex combination of the posteriors corresponding to the q_i, i.e.,

$$\pi(\theta \mid x) = \sum_{i=1}^{k} p_i(x) q_i(\theta \mid x).$$

Furthermore, the posterior probabilities $p_i(x)$, are given by

$$p_i(x) = \frac{p_i m(x \mid q_i)}{\sum_{j=1}^{k} p_j m(x \mid q_j)}, \qquad i = 1, \ldots, k. \tag{3.1}$$

As indicated in the following theorem, the calculation of (2.3) involves consideration of the extrema of $f(x \mid \theta)$ for $\theta \in T_i$, $i = 1, \ldots, k$.

For each $i = 1, \ldots, k$ define $\bar{f}_i(x)$ and $\underline{f}_i(x)$ as follows:

$$\bar{f}_i(x) = \sup_{\theta \in T_i} f(x \mid \theta) \text{ and } \underline{f}_i(x) = \inf_{\theta \in T_i} f(x \mid \theta).$$

THEOREM 1: *For Γ in (2.2), the extrema in (2.3) are given by*

$$\ell_i(x) = \frac{p_i \underline{f}_i(x)}{p_i[\underline{f}_i(x) - \bar{f}_i(x)] + U}, \tag{3.2}$$

with

$$U = \sum_{j=1}^{k} p_j \bar{f}_i(x),$$

and

$$u_i(x) = \frac{p_i \bar{f}_i(x)}{p_i[(\bar{f}_i(x) - \underline{f}_i(x)] + L)}, \tag{3.3}$$

with $L = \sum_{j=1}^{k} p_j \underline{f}_i(x)$, *for* $i = 1, 2, \ldots, k$.

PROOF: Simple algebra allows us to express the $p_i(x)$ as

$$p_i(x) = [1 + N_i/D_i]^{-1},$$

where $N_i = \sum_{j \neq i} p_j m(x \mid q_j)$ and $D_i = p_i m(x \mid q_i)$. Clearly $\ell_i(x)$ obtains when N_i is as large as possible while D_i is as small as possible. (The reverse is true for $u_i(x)$.) Hence, extrema of $p_i(x)$ occur at various extrema of $m(x \mid q_j)$, $j = 1, \ldots, k$.

Next, consider $m(x \mid q_i) = \int f(x \mid \theta) q_i(\theta) d\theta$. For x fixed, since q_i is constrained to have support at most T_i, $m(x \mid q_i)$ is bounded above by $\bar{f}_i(x)$. The minimization of $m(x \mid q_i)$ is analogous. ∎

3.1 Notes

Huber (1973) presented Theorem 1 for $k = 2$; also, see Berger and Berliner (1986). For $k = 2$ (i.e., hypothesis testing) this type of analysis, as well as fascinating generalizations, may be found in Edwards, Lindman, and Savage (1963), Berger and Sellke (1987), and Casella and Berger (1987). DeRobertis (1978) considered classes similar to (2.2) but for which the p_i themselves are assumed only to lie in fixed intervals, say $a_i \leq p_i \leq b_i$. Her results, however, do not seem to apply to (2.2).

3.2 Implications of the Ranges

The derivation of the ranges in Theorem 1 did not use the information that the posterior probabilities $p_i(x)$ must sum to one. The appropriate adjustment is to base inferences on $\Gamma(x)$ defined by

$$\Gamma(x) = \{\boldsymbol{v} = (v_1, \ldots, v_k)^t : \ell_i(x) \leq v_i \leq u_i(x), \quad i = 1, \ldots, k \text{ and } \sum_{i=1}^{k} v_i = 1\}. \tag{3.4}$$

Note that for $k = 2$, $\Gamma(x)$ provides no refinement of (3.2), (3.3), since it is easy to see that $\ell_1(x) = 1 - u_2(x)$ and $\ell_2(x) = 1 - u_1(x)$. The following facts provide some insight into the structure of $\Gamma(x)$ in general.

PROPOSITION 1: *Let U and L be as defined in (3.2) and (3.3). Then*

$$1 \geq \sum_{i=1}^{k} \ell_i(x) \geq L/U \tag{3.5}$$

and

$$1 \leq \sum_{i=1}^{k} u_i(x) \leq U/L \tag{3.6}$$

PROOF: The portions of the bounds involving 1 are clear by inspection. To show that $\sum_{i=1}^{k} \ell_i(x) \geq L/U$, it follows from (3.2) that for each i,

$$\ell_i(x) \geq \frac{p_i \underline{f}_i(x)}{L}.$$

Summing both sides of this inequality yields the result. The derivation of (3.6) is analogous.

PROPOSITION 2: *$\Gamma(x)$ is nonempty and convex.*

PROOF: $\Gamma(x)$ is nonempty by construction. (This is also clear from (3.5) and (3.6).) Next, let $\boldsymbol{V}^{(1)}$ and $\boldsymbol{V}^{(2)}$ be elements of $\Gamma(x)$. For any $0 \leq \alpha \leq 1$, it follows that for $i = 1, \ldots, k$,

$$\ell_i(x) \leq \alpha V_i^{(1)} + (1-\alpha)V_i^{(2)} \leq u_i(x).$$

Hence, $\alpha \boldsymbol{V}^{(1)} + (1-\alpha)\boldsymbol{V}^{(2)}$ is in $\Gamma(x)$, establishing convexity.

REMARKS: Note that as the partition of Θ gets finer, i.e., a subset is divided into several subsets, the weighted upper and lower bounds, U and L, of the likelihood function approach each other. In other words, U decreases and L increases. In this sense, the bounds become tighter. In the limit, when Γ contains only one prior distribution, $\Gamma(x)$ also contains one posterior distribution.

4. Illustration

Martz and Waller (1982) (abbreviated from now on as MW), pp. 177-184 present an example particularly well suited to the analysis given here. The unknown parameter of interest θ is the mean life (in hours) of an engine. The prior information, elicited from two experts, A and B, are in terms of the probabilities that q is in certain intervals of a partition of the positive real line. In our notation the prior input is given in Table 1.

Table 1: Prior Information in MW Example

i	T_i	p_i	
		Expert A	Expert B
1	0-1000	.01	.15
2	1000-2000	.04	.15
3	2000-3000	.20	.20
4	3000-4000	.50	.20
5	4000-5000	.15	.15
6	> 5000	.10	.15

MW also consider the results of an experiment in which two independent observations, X_1 and X_2, are taken from an exponential distribution with density

$$f(y \mid \theta) = \theta^{-1} \exp(-y/\theta), \quad y(\text{hours}) > 0.$$

To analyze the experimental results, we only need the sufficient statistic for θ; namely $X = X_1 + X_2$. Under the exponential model, X has a gamma distribution. However, since only the likelihood function for θ enters into a Bayesian analysis (i.e., any normalizing constants are irrelevant, cf. Berger, 1985, Chapters 1, 4), we need only consider

$$f(x \mid \theta) \; \alpha \; (1/\theta^2)\exp(-x/\theta), \quad x > 0.$$

The datum used here is $x = 4500$ hours, which is same as the one used by MW.

Separately for each expert, MW proceed using a *generalized* prior density for θ constructed by letting the density be the constant value p_i for all θ in T_i, $i = 1, \ldots, 6$. Of course, the resulting "prior" is not a proper probability distribution; the actual mass given to T_6 is infinite. Nevertheless, a formal application of Bayes Theorem, presented in detail by MW, does produce "posterior probabilities" of the six intervals.

For both sets of prior probabilities and the given data, we have computed the posterior bounds in (2.3). These results, as well as the "posterior probabilities", are given in Table 2.

The most obvious observation to be made from Table 2 is that the MW results do not correspond to an approximation to *any* Bayes analysis. (See Hill, 1985, for related discussion). For both experts the MW "posterior probabilities" for T_6 (.40 or .55) are considerably higher than they could be for any prior distribution consistent with the prior information. (These numbers are also intuitively quite absurd: Both experts assigned probability of at most .15 to T_6 and the datum does not "indicate" that θ is large, since the MLE for θ is 2250 hr.)

Do our results therefore form an indictment of the use of uniform-like generalized priors? We believe the answer is a qualified no. In our opinion, generalized Bayes analysis with uniform priors is a powerful tool in situations in which no prior information is available. On the other hand, whenever possible, one should check results against something like Theorem 1 to avoid anomalies. Moreover, a "mixed modes" analysis, attempting to incorporate partial prior information using uniform generalized priors (for a single parameter), does not appear to be reasonable.

It should be noted that the MW analysis gets into serious trouble because T_6 is unbounded. The same sort of analysis, if applied for a bounded parameter space, is a proper Bayes (in fact, maximum entropy) analysis.

5. Discussion

The finest form of robust Bayesian analysis arises when Γ is constructed so that (i) Γ contains all priors deemed to be plausible, but (ii) Γ does not contain unrealistic,

Table 2: Bounds for the MW Example

Expert A

T_i	MW Results	ℓ_i	u_i
0-1000	7.4×10^{-4}	.00	6.1×10^{-3}
1000-2000	2.5×10^{-2}	1.9×10^{-2}	5.7×10^{-2}
2000-3000	.16	.21	.29
3000-4000	.34	.48	.61
4000-5000	.08	.10	.16
> 5000	.40	.00	.08

Expert B

T_i	MW Results	ℓ_i	u_i
0-1000	.01	.00	.11
1000-2000	.09	8.7×10^{-2}	.26
2000-3000	.15	.24	.39
3000-4000	.13	.20	.35
4000-5000	.07	.12	.22
> 5000	.55	.00	.15

"silly" priors. In particular, in addition to the k probabilities used to generate Γ, one usually has some notions of the shape (e.g., unimodality) the prior should have. Such information has been ignored in (2.2). The by- product is the inclusion of "wild" priors. Unfortunately, from the proof of Theorem 1, we see that the extrema of posterior probabilities obtained there occurred at "nonsense" priors. Using a Γ that is too large leads to "conservative" bounds. In general the selection of Γ will involve a trade off between the degree of conservatism allowed and mathematical tractability. On the bright side, it is possible to work with realistic, though complex, classes of priors (cf. Berger and Berliner, 1986, O'Hagan and Berger, 1988) and such problems form an active area of research.

Acknowledgments

This research was supported by the Air Force Office of Scientific Research under Contract AFOSR-84-0162, the National Science Foundation under Grant DMS-8400683, and the Office of Naval Research under Contract N00014-84-K-0422. The U.S. Government is authorized to reproduce and distribute reprints for Governmental purposes notwithstanding any copyright notation thereon. We are grateful to the referees for their valuable comments.

References

Berger, J.O. (1984). "The Robust Bayesian Viewpoint" (with discussion), in *Robustness of Bayesian Analysis*, J. Kadane, ed., Amsterdam: North-Holland.

Berger, J.O. (1985). *Statistical Decision Theory and Bayesian Analysis*, 2nd ed., New York: Springer-Verlag.

Berger, J.O. and L.M. Berliner (1986). "Robust Bayes and Empirical Bayes Analysis with ϵ-Contaminated Priors," *Annals of Statistics* , Vol. 14, 461-486.

Berger, J.O. and T. Sellke (1987). "Testing of a Point Null Hypothesis: The Irreconcilability of Significance Levels and Evidence," *Journal of the American Statistical Society*, Vol. 82, 112-122.

Berger, R. (1979). "Gamma Minimax Robustness of Bayes Rules," *Communications in Statistics*, Vol. 8, 543-560.

Casella, G. and R. Berger (1987). "Reconciling Bayesian and Frequentist Evidence in the One-Sided Testing Problem," *Journal of the American Statistical Society*, Vol. 82.

Dempster, A.P. (1968). "A Generalization of Bayesian Inference," *Journal of the Royal Statistical Society, (B)*, Vol. 30, 205-248.

DeRobertis, L. (1978). "The Use of Partial Prior Knowledge in Bayesian Inference," Ph.D. Thesis, Yale University, New Haven.

Diaconis, P. and S. Zabell (1982). "Updating Subjective Probability," *Journal of the American Statistical Association*, Vol. 77, 822-830.

Edwards, W., H. Lindman, and L.J. Savage (1963). "Bayesian Statistical Inference for Psychological Research," *Psychological Review*, Vol. 70, 193-242.

Goldstein, M. (1984). "Turning Probabilities into Expectations," *Annals of Statistics*, Vol. 12, 1551-1557.

Hill, B. (1985). "Book Review: Bayesian Reliability Analysis by H. Martz and R. Waller," *Journal of the American Statistical Association*, Vol. 80, 253-254.

Huber, P.J. (1973). "The Use of Choquet Capacities in Statistics," *Bulletin of the International Statistical Institute*, Vol. 45, 181-191.

Jaynes, E.T. (1968). "Prior Probabilities," *IEEE Transactions on Systems Science and Cybernetics*, SSC-4, 227-241.

Kadane, J.B., J.M. Dickey, R.L. Winkler, W.S. Smith, and S.C. Peters (1980). "Interactive Elicitation of Opinion for a Normal Linear Model," *Journal of the American Statistical Association*, Vol. 75, 845- 854.

Lambert, D. and G.T. Duncan (1985). "Bayesian Inference Based on Partial Prior Information," Technical Report No. 209, Department of Statistics, Carnegie-Mellon University, Pittsburgh.

Martz, H.F. and R.A. Waller (1983). *Bayesian Reliability Analysis.* New York: Wiley.

O'Hagan, A. and J.O. Berger (1988). "Ranges of Posterior Probabilities for Quasi-unimodal Priors with Specified Quantiles," *Journal of the American Statistical Association* Vol. 83, 503-508.

Schafer, G. (1976). *A Mathematical Theory of Evidence.* Princeton: University Press.

Schafer, G. (1982). "Belief Functions and Parametric Models" (with discussion), *Journal of the Royal Statistical Society (B)*, Vol. 44, 322-352.

Winkler, R.L. (1967). "The Assessment of Prior Distributions in Bayesian Analysis," *Journal of the American Statistical Association*, Vol. 62, 776-800.

Bayesian and Likelihood Methods in Statistics and Econometrics
S. Geisser, J.S. Hodges, S.J. Press and A. Zellner (Editors)
Elsevier Science Publishers B.V. (North-Holland), 1990

ESTIMATION OF THE LINEAR REGRESSION MODEL UNDER CONDITIONS OF INCOMPLETE PRIOR INFORMATION

Wojciech Bijak

Central School of Planning and Statistics; Warsaw, POLAND

Abstract

The problem of joint estimation of the regression parameters and variance in the normal linear regression model is considered. We assume that incomplete prior information leads us to determine some class $\mathcal{G}$ of prior distributions. It is proved that, under different types of loss functions, the Bayes estimators, with respect to the conjugate prior distribution, are $\mathcal{G}$-minimax in the class $\mathcal{D}$ of all estimators of unknown parameters.

1. Introduction

Complete knowledge of the prior distribution of unknown parameters is assumed in the Bayesian approach to statistical and econometric problems. Such a situation as well as the other extreme of a complete absence of past experience or other prior information rarely occurs in practice. Hodges and Lehmann (1952) have paid attention to the fact that the situation occurring in practice usually lies between the two extremes given above. Statisticians often dispose of such prior information which allows only for determination of some class of prior distributions. In this situation we will say that the statisticians dispose of incomplete prior information or partial prior information.

Problems of statistical inference under conditions of incomplete prior information were considered by Repin and Tartakovskij (1977), Berger (1980, 1984) (from the robust Bayesian point of view), Truchaev (1981), and Bijak (1986), among others. Different types of criteria were proposed in the works listed above to solve statistical problems with partial prior information. The most popular $\mathcal{G}$-minimax criterion was also used by Robbins (1964), Blum and Rosenblatt (1967), Jackson, et al. (1970), Zacks (1971), Solomon (1972a, 1972b), Pilz (1983, 1984).

Posterior sensitivity analysis is another important approach to statistical inference with incomplete prior information [see e.g., Potter and Anderson (1980, 1983), Leamer (1982), Polasek (1984)]. This approach has been omitted in this paper.

2. The Model and the Main Assumptions

Suppose that $\mathcal{R}^n$ is n dimensional Euclidean space and $\mathcal{R}_+$ is the set of positive real numbers. Let $\mathcal{M}_n^>$, $\mathcal{M}_n^{\geq}$ and $\mathcal{M}_n^s$ denote the sets of symmetric positive definite matrices, symmetric non-negative definite matrices and symmetric matrices of order n, respectively. Additionally let $|\mathbf{B}|$ and $tr\mathbf{B}$ denote determinant and trace of a matrix $\mathbf{B}$, respectively. We do not introduce special symbols for random variables. The meaning of denotation emerges from a context. If we write $E(z)$ then z is a random variable, but if we write $z \in \mathcal{R}$ it means that z is a value of random variable.

Consider the normal linear regression model (NLRM) defined in the following way. Suppose that $\mathbf{y}$ is a $(n \times 1)$ vector of normally and independently distributed random variables having common variance $\sigma^2 \in \mathcal{R}$ such that

$$E(\mathbf{y} \mid \mathbf{X}, \beta, \sigma^2) = \mathbf{X}\beta$$
$$\mathrm{Cov}(\mathbf{y} \mid \mathbf{X}, \beta, \sigma^2) = \sigma^2\mathbf{I}$$

where $\mathbf{X}$ is a $(n \times k)$ matrix of known constants, β is a $(k \times 1)$ vector of unknown regression parameters and $\mathbf{I}$ is the identity matrix of order $n[\mathbf{y} \sim N(\mathbf{X}\beta, \sigma^2\mathbf{I})]$.

We are interested in estimation of the vector of parameters $\theta = (\beta', \sigma^2)'$. In our investigations we assume the following types of loss function $L(\theta, \hat{\theta})$:

1. A quadratic loss function

$$L_1(\theta, \hat{\theta}) = (\theta - \hat{\theta})'\tilde{\mathbf{Q}}(\theta - \hat{\theta})$$

where $\hat{\theta}$ means an estimate of θ, $\tilde{\mathbf{Q}} = \begin{bmatrix} \mathbf{Q} & \mathbf{q}_{12} \\ \mathbf{q}_{21} & q_{22} \end{bmatrix} \in \mathcal{M}_{k+1}^>$ and $\mathbf{Q} \in \mathcal{M}_k^>$.

2. A quadratic loss function with a relative squared error loss for variance

$$L_2(\theta, \hat{\theta}) = \left[(\beta - \hat{\beta})', \frac{\sigma^2 - \hat{\sigma}^2}{\sigma^2}\right] \tilde{\mathbf{Q}} \begin{bmatrix} \beta - \hat{\beta} \\ \dfrac{\sigma^2 - \hat{\sigma}^2}{\sigma^2} \end{bmatrix}.$$

3. A quadratic loss function with logarithmic squared error loss function for variance

$$L_3(\theta, \hat{\theta}) = [(\beta - \hat{\beta})', \log\,\sigma^2 - \log\,\hat{\sigma}^2]\tilde{\mathbf{Q}} \begin{bmatrix} \beta - \hat{\beta} \\ \log\,\sigma^2 - \log\,\hat{\sigma}^2 \end{bmatrix}.$$

Let γ be a prior distribution of the vector θ. We suppose that prior knowledge about the parameter vector θ may be represented by any prior from some set $\mathcal{G}$ of prior distributions. The set $\mathcal{G}$ reflects the sort of incomplete prior information accessible for example in repeatable statistical investigations provided by NLRM. We examine the following sets $\mathcal{G}$ of prior distributions:

$$\mathcal{G}_1 = \{\gamma : E(\beta \mid \sigma^2) = \mathbf{m},\ \mathbf{m} \in \mathcal{M} \subset \mathcal{R}^k,\ \mathrm{Cov}(\beta \mid \sigma^2) = \sigma^2\Sigma,\ \Sigma \in \mathcal{S} \subset \mathcal{M}_k^>,$$
$$\beta \sim N(\mathbf{m}, \sigma^2\Sigma),\ E(\sigma^2) = w,\ \mathrm{Var}(\sigma^2) = s,\ (w, s) \in \mathcal{V} \subset \mathcal{R}_+^2\},$$

where the sets $\mathcal{M},\ \mathcal{S}\ \mathcal{V}$ are closed and bounded,

$$\mathcal{G}_2 = \{\gamma : E(\beta \mid \sigma^2) = \mathbf{m},\ \mathbf{m} \in \mathcal{M} \subset \mathcal{R}^k,\ \mathrm{Cov}(\beta \mid \sigma^2) = \sigma^2\Sigma,\ \Sigma \in \mathcal{S} \subset \mathcal{M}_l^>,$$
$$\beta \sim N(\mathbf{m}, \sigma^2\Sigma),\ E(\sigma^2) = w,\ w \in \mathcal{W} \subset \mathcal{R}_+,\ E[(\sigma^2)^{-1}] = g,$$
$$E[(\sigma^2)^{-2}] = h,\ (g, h) \in \mathcal{H} \subset \mathcal{R}_+^2\},$$

where the sets $\mathcal{M},\ \mathcal{S},\ \mathcal{W},\ \mathcal{H}$ are closed and bounded,

$$\mathcal{G}_3 = \{\gamma : \gamma = \gamma_o(\mathbf{m}, \Sigma, \lambda, \mu),\ \mathbf{m} \in \mathcal{M} \subset \mathcal{R}^k,\ \Sigma \in \mathcal{S} \subset \mathcal{M}_k^>,$$
$$(\lambda, \mu) \in \mathcal{L},\ \lambda > 1,\ \mu > 0\}$$

where $\gamma_o(\mathbf{m}, \Sigma, \lambda, \mu)$ is a conjugate prior distribution of θ which is normal-inverted gamma 1 (NIG1) with the density function

$$p(\beta, \sigma \mid \mathbf{m}, \Sigma, \lambda, \mu) =$$
$$(2\pi)^{-k/2}|\sigma^2\Sigma|^{-1/2} \exp\left\{-\frac{1}{2\sigma^2}(\beta - \mathbf{m})'\Sigma(\beta - \mathbf{m})\right\}(\sigma^2)^{-(\lambda+1)} \exp\{-\mu(\sigma^2)^{-1}\},$$

and where the sets $\mathcal{M},\ \mathcal{S},\ \mathcal{L}$ are closed and bounded.

The set $\mathcal{G}_3$ consists only of conjugate prior distributions and it is the subset of the sets $\mathcal{G}_1,\ \mathcal{G}_2$.

Let $\mathcal{D}$ denote the class of all estimators θ, and $\mathcal{D} = \mathcal{D}_b \times \mathcal{D}_v$, where $\mathcal{D}_b,\ \mathcal{D}_v$ are the classes of all estimators of β and σ^2 respectively. For our purpose let us introduce the subclass $\mathcal{D}_{lq} = \mathcal{D}_l \times \mathcal{D}_q$, where $\mathcal{D}_l \subset \mathcal{D}_b,\ \mathcal{D}_q \subset \mathcal{D}_b$ and

$$\mathcal{D}_l = \{b(\mathbf{y}) \in \mathcal{D}_b : b(\mathbf{y}) = \mathbf{C}\mathbf{y} + \mathbf{c}\},$$

where $\mathbf{C}$ is a $(k \times n)$ matrix, $\mathbf{c}$ is a $(k \times 1)$ vector, and

$$\mathcal{D}_q = \{v(\mathbf{y}) \in \mathcal{D}_v : v(\mathbf{y}) = \mathbf{y}'\mathbf{A}\mathbf{y} + \mathbf{a}\mathbf{y} + \alpha\},$$

where $\mathbf{A} \in \mathcal{M}_n^s$, $\mathbf{a}$ is a $(1 \times n)$ vector, $\alpha \in \mathcal{R}$.

Least squares estimators and Bayes estimators with respect to the conjugate prior distributions and loss functions L_1, L_2, L_3 are in the class $\mathcal{D}_{lq}$. Easy to prove theorems can be formulated in the following way.

THEOREM 1: *For the NLRM, the loss function L_1 and $\gamma_o(\mathbf{m}, \Sigma, \lambda, \mu)$—the prior distribution of θ—the estimator $d_1^{\gamma_o} = (b_1^{\gamma_o'}, v_1^{\gamma_o})'$ is the Bayes estimator of θ, where*

$$b_1^{\gamma_o}(\mathbf{y}) = (\Sigma^{-1} + \mathbf{X'X})^{-1}(\mathbf{X'y} + \Sigma^{-1}\mathbf{m}), \tag{1}$$

$$v_1^{\gamma_o}(\mathbf{y}) = e_1(\mathbf{y'Py} - 2\mathbf{m'X'Py} + \mathbf{m'X'PXm}) + (1 - ne_1)w, \tag{2}$$

and $\mathbf{P} = (\mathbf{I} - \mathbf{X}\Sigma\mathbf{X}')^{-1}$, $e_1 = \frac{s}{2(w^2+s)+ns}$.

PROOF: It is well known that in point estimation problems, when quadratic loss function L_1 is used, posterior means are values of the Bayes point estimators. For any $\mathbf{y} \in \mathcal{R}^n$ the distribution $\gamma_o(\overline{\mathbf{m}}, \overline{\Sigma}, \overline{\lambda}, \overline{\mu})$ where $\overline{\Sigma} = (\Sigma^{-1} + \mathbf{X'X})^{-1}$, $\overline{\mathbf{m}} = \overline{\Sigma}(\mathbf{X'y} + \Sigma^{-1}\mathbf{m})$, $\overline{\lambda} = \lambda + n/2$, $\overline{\mu} = \mu + \frac{1}{2}[\mathbf{y}'(\mathbf{I} - \mathbf{X}\overline{\Sigma}\mathbf{X}')\mathbf{y} - 2\mathbf{m}'\Sigma^{-1}\overline{\Sigma}\mathbf{X'y} + \mathbf{m}'(\Sigma^{-1} - \Sigma^{-1}\overline{\Sigma}\Sigma^{-1})\mathbf{m}]$ is the posterior distribution of θ [see analogous considerations e.g., DeGroot (1970), section 11.10., Zellner (1971), section 3.2.3 and 3.2.5]. From NIG1 distribution properties we have $E(\beta \mid \mathbf{y}) = \overline{\mathbf{m}}$, $E(\sigma^2 \mid \mathbf{y}) = \overline{\mu}/(\overline{\lambda} - 1)$. Taking into consideration that $w = \mu/(\lambda - 1)$, $s = \mu^2/(\lambda - 1)^2(\lambda - 2)$ and that $\mathbf{I} - \mathbf{X}\overline{\Sigma}\mathbf{X}' = \mathbf{P}$, $\Sigma^{-1}\overline{\Sigma}\mathbf{X}' = \mathbf{X'P}$, $\Sigma^{-1} - \Sigma^{-1}\overline{\Sigma}\Sigma^{-1} = \mathbf{X'PX}$ the posterior means and consequently the Bayes estimators of β and σ^2 may be put in the form (1) and (2), respectively.

THEOREM 2: *For the NLRM, the loss function L_2 and $\gamma_o(m, \Sigma, \lambda, \mu)$—the prior distribution of θ, the estimator $d_2^{\gamma_o} = (b_2^{\gamma_o'}, v_2^{\gamma_o})'$ is the Bayes estimator of θ, where*

$$b_2^{\gamma_o}(\mathbf{y}) = b_1^{\gamma_o}(\mathbf{y}) + 2\mathbf{Q}^{-1}\mathbf{q}_{12}q_{22}e_2/k_1, \tag{3}$$

$$v_2^{\gamma_o}(\mathbf{y}) = e_2(\mathbf{y'Py} - 2\mathbf{m'X'Py} + \mathbf{m'X'PXm}) + k_1(1 - ne_2)g/k_2, \tag{4}$$

$$k_1 = q_{22} - \mathbf{q}_{21}\mathbf{Q}^{-1}\mathbf{q}_{12}, \quad k_2 = hq_{22} - g^2\mathbf{q}_{21}\mathbf{Q}^{-1}\mathbf{q}_{12}, \quad e_2 = \frac{(h - g^2)k_1}{n(h - g^2)k_1 + 2k_2}.$$

PROOF: For any $\mathbf{y} \in \mathcal{R}n$ the posterior expected loss $E[L_2(\theta, \hat{\theta}) \mid \mathbf{y}]$ is minimized by a solution of equation

$$\begin{bmatrix} \mathbf{Q} & \overline{g}\mathbf{q}_{12} \\ \overline{g}\mathbf{q}_{21} & \overline{h}\mathbf{q}_{22} \end{bmatrix} \begin{bmatrix} b(\mathbf{y}) \\ v(\mathbf{y}) \end{bmatrix} = \begin{bmatrix} \mathbf{Q}E(\beta \mid \mathbf{y}) + \mathbf{q}_{12} \\ \mathbf{q}_{21}E[\beta(\sigma^2)^{-1} \mid \mathbf{y}] + \overline{g}q_{22} \end{bmatrix},$$

where $\overline{g} = E[(\sigma^2)^{-1} \mid \mathbf{y}]$, $\overline{h} = E[(\sigma^2)^{-2} \mid \mathbf{y}]$.

Solving we obtain

$$b_2^{\gamma_o}(\mathbf{y}) = E(\beta \mid \mathbf{y}) + \mathbf{Q}^{-1}\mathbf{q}_{12}\mathbf{q}_{21}\{\overline{g}^2 E(\beta \mid \mathbf{y}) - \overline{g}E[\beta(\sigma^2)^{-1} \mid \mathbf{y}]\}/\overline{k}_2 + \mathbf{Q}^{-1}\mathbf{q}_{12}q_{22}/\overline{k}_2$$
$$v_2^{\gamma_o}(\mathbf{y}) = \mathbf{q}_{21}\{E[\beta(\sigma^2)^{-1} \mid \mathbf{y}] - \overline{g}E(\beta \mid \mathbf{y})\}/\overline{k}_2 + \overline{g}k_1/\overline{k}_2,$$

where $\overline{k}_2 = \overline{h}q_{22} - \overline{g}^2\mathbf{q}_{21}\mathbf{Q}^{-1}\mathbf{q}_{12}$.

For any $\mathbf{y} \in \mathcal{R}n$ the distribution $\gamma_o(\overline{\mathbf{m}}, \overline{\Sigma}, \overline{\lambda}, \overline{\mu})$ is the posterior distribution of θ. From NIG1 distribution properties we have $\overline{g} = \overline{\lambda}/\overline{\mu}$, $\overline{h} = \overline{\lambda}(\overline{\lambda}+1)/\overline{\mu}^2$, $E[\beta(\sigma^2)^{-1} \mid \mathbf{y}] = \overline{g}E(\beta \mid \mathbf{y})$ and for prior moments $g = \lambda/\mu$, $h = \lambda(\lambda+1)/\mu^2$. Hence we obtain the Bayes point estimate and the Bayes estimator of β and σ^2 in the form (3) and (4), respectively.

THEOREM 3: *For the NLRM, the loss function L_3 and $\gamma_o(\mathbf{m}, \Sigma, \lambda, \mu)$—the prior distribution of θ—the estimator $d_3^{\gamma_o} = (b_3^{\gamma_o'}, v_3^{\gamma_o})'$ is the Bayes estimator of θ, where*

$$b_3^{\gamma_o}(\mathbf{y}) = b_1^{\gamma_o}(\mathbf{y}), \tag{5}$$
$$v_3^{\gamma_o}(\mathbf{y}) = e_3(\mathbf{y}'\mathbf{P}\mathbf{y} - 2\mathbf{m}'\mathbf{X}'\mathbf{P}\mathbf{y} + \mathbf{m}'\mathbf{X}'\mathbf{P}\mathbf{X}\mathbf{m}) + 2e_3\mu, \tag{6}$$

$e_3 = \frac{1}{2}\exp\{-\psi(\lambda+n/2)\}$, and for $z \in \mathcal{R}_+$ $\psi(z) = \frac{d\log\Gamma(z)}{dz} = \frac{\Gamma'(z)}{\Gamma(z)}[\Gamma(\cdot)$ is the gamma function].

PROOF: For any $\mathbf{y} \in \mathcal{R}^n$ the posterior expected loss $E[L_3(\theta, \hat{\theta}) \mid \mathbf{y}]$ is minimized by a solution of equation

$$\tilde{\mathbf{Q}}\begin{bmatrix} b(\mathbf{y}) \\ \log\ v(\mathbf{y}) \end{bmatrix} = \tilde{\mathbf{Q}}\begin{bmatrix} E(\beta \mid \mathbf{y}) \\ E(\log\sigma^2 \mid \mathbf{y}) \end{bmatrix}$$

Solving we obtain $b_3^{\gamma_o}(\mathbf{y}) = E(\beta \mid \mathbf{y})$, $v_3^{\gamma_o}(\mathbf{y}) = \exp\{E(\log\sigma^2 \mid \mathbf{y})\}$. For posterior distribution $\gamma_o(\overline{\mathbf{m}}, \overline{\Sigma}, \overline{\lambda}, \overline{\mu})$ we have $E(\log\sigma^2 \mid \mathbf{y}) = \log\overline{\mu} - \psi(\overline{\lambda})$. So the Bayes estimate and consequently the Bayes estimator of β and σ^2 has the form (5) and (6) respectively.

In the next sections we show that $\mathcal{G}$-minimax estimators of θ exist and that they have the same form as the Bayes estimators $d_1^{\gamma_o}$, $d_2^{\gamma_o}$, $d_3^{\gamma_o}$.

3. $\mathcal{G}$-Minimax Estimation with Quadratic Loss

It is reasonable to use the $\mathcal{G}$-minimax criterion for choice of the "best" estimator in conditions of incomplete prior information. The "best" estimator means that the worst Bayes risk, with respect to prior distributions over $\mathcal{G}$, is as small as possible over $\mathcal{D}$.

DEFINITION: The estimator $d^* \in \mathcal{D}$ is said to be a $\mathcal{G}$-minimax estimator of θ in $\mathcal{D}$ if

$$\sup_{\gamma\in\mathcal{G}} r(\gamma, d^*) = \inf_{d\in\mathcal{D}} \sup_{\gamma\in\mathcal{G}} r(\gamma, d)$$

where $r(\gamma, d)$ denotes the Bayes risk of d with respect to γ, $r(\gamma, d) = E[R(\theta, d)]$ and $R(\theta, d) = E\{L[\theta, d(\mathbf{y})] \mid \theta\}$.

For the estimators $d \in \mathcal{D}q$ the Bayes risk can be presented explicitly.

THEOREM 4: *For the NLRM, the loss function* L_1, *any* $\gamma \in \mathcal{G}_1$ *and any* $d \in \mathcal{D}_{lq}$

$$\begin{aligned} r(\gamma, d) = wtr\tilde{\mathbf{Q}} & \begin{bmatrix} \mathbf{CC}' + (\mathbf{CX}-\mathbf{I})\Sigma(\mathbf{CX}-\mathbf{I})' & [\mathbf{C}+(\mathbf{CX}-\mathbf{X})\Sigma\mathbf{X}']\hat{\mathbf{a}}' \\ \hat{\mathbf{a}}[\mathbf{C}'+\mathbf{X}\Sigma(\mathbf{CX}-\mathbf{I})'] & \hat{\mathbf{a}}\mathbf{P}^{-1}\hat{\mathbf{a}}' \end{bmatrix} \\ & + [\hat{\mathbf{c}}', -\hat{\alpha}]\tilde{\mathbf{Q}} \begin{bmatrix} \hat{\mathbf{c}} \\ -\hat{\alpha} \end{bmatrix} \\ & + q_{22}[s(1 - tr\mathbf{AP}^{-1})^2 + 2(w^2+s)tr(\mathbf{AP}^{-1})^2] \end{aligned} \tag{7}$$

where $\hat{\mathbf{c}} = (\mathbf{CX}-\mathbf{I})\mathbf{m} + \mathbf{c}$, $\hat{\mathbf{a}} = 2\mathbf{m}'\mathbf{X}'\mathbf{A} + \mathbf{a}$, $\hat{\alpha} = w - \mathbf{m}'\mathbf{X}'\mathbf{AXm} - \mathbf{aXm} - \alpha - wtr\mathbf{AP}^{-1}$.

PROOF: For arbitrary $\gamma \in \mathcal{G}_1$ and $d \in \mathcal{D}_{lq}$

$$\begin{aligned} r(\gamma, d) &= E\left\{E[L_1(\theta, d(\mathbf{y})) \mid \theta]\right\} \\ &= E\{E[(\beta - d(\mathbf{y}))'\mathbf{Q}(\beta - d(\mathbf{y})) \mid \theta]\} \\ &\quad + 2E\{E[(\beta - b(\mathbf{y}))'\mathbf{q}_{12}(\sigma^2 - v(\mathbf{y})) \mid \theta]\} \\ &\quad + E\{E[q_{22}(\sigma^2 - v(\mathbf{y}))^2 \mid \theta]\} \end{aligned} \tag{8}$$

It is known [see e.g., Solomon (1972a), Pilz (1983)] that under more general assumptions (without assumptions of normality of random vectors $\mathbf{y}$ and β) we have

$$E\{E[(\beta - b(\mathbf{y}))'\mathbf{Q}(\beta - b(\mathbf{y})) \mid \theta]\} = wtr\mathbf{Q}[\mathbf{CC}' + (\mathbf{CX}-\mathbf{I})\Sigma(\mathbf{CX}-\mathbf{I})'] + \hat{\mathbf{c}}'\mathbf{Q}\hat{\mathbf{c}}$$

From the fact that for normally distributed random variables $\mathbf{y}$ and β, first and third central moments equal zero (after some transformations) it follows that

$$E\{E[(\beta - b(\mathbf{y}))'\mathbf{q}_{12}(\sigma^2 - v(\mathbf{y})) \mid \theta]\} = wtr[\mathbf{C}' + \mathbf{X}\Sigma(\mathbf{CX}-\mathbf{I})']\mathbf{q}_{12}\hat{\mathbf{a}} - \hat{\mathbf{c}}'\mathbf{q}_{12}\hat{\alpha}$$

To express the last component of the sum (8) we have to use the known fact [see Humak (1984) section A.2.] that for a normally distributed random vector $\mathbf{z} \sim N(\omega, \Omega)$, $\mathbf{z} \in \mathcal{R}^i$ and for arbitrary matrix $\mathbf{B}$ of order i we have

$$tr[(\mathbf{B} \otimes \mathbf{B})\Phi] = tr\mathbf{B}[(tr\mathbf{B}\Phi)\Phi + 2\Phi\mathbf{B}\Phi] = (tr\mathbf{B}\Phi)^2 + 2tr(\mathbf{B}\Phi)^2 \tag{9}$$

where $\otimes$ means Kronecker product of matrices and Φ is the matrix of fourth central moments of the random vector $\mathbf{z}$. Due to the above fact we obtain (by some transformations)

$$\begin{aligned} &E\{E[q_{22}(\sigma^2 - v(\mathbf{y}))^2 \mid \theta]\} \\ &= q_{22}[w\hat{\mathbf{a}}\mathbf{P}^{-1}\hat{\mathbf{a}}' + \hat{\alpha}^2 + s(1 - tr\mathbf{A}\mathbf{P}^{-1})^2 + 2(w^2 + s)tr(\mathbf{A}\mathbf{P}^{-1})^2] \end{aligned}$$

Finally the Bayes risk for any $\gamma \in \mathcal{G}$ and any $d \in \mathcal{D}_{lq}$ may be written as (7).

If $\mathbf{C}_1 = (\Sigma^{-1} + \mathbf{X}'\mathbf{X})^{-1}\mathbf{X}'$, $\mathbf{C} = \mathbf{C}_1 + \mathbf{C}_o$ then $\mathbf{C}\mathbf{C}' + (\mathbf{C}\mathbf{X} - \mathbf{I})\Sigma(\mathbf{C}\mathbf{X} - \mathbf{I})' = \mathbf{C}_1\mathbf{C}_1' + (\mathbf{C}_1\mathbf{X} - \mathbf{I})\Sigma(\mathbf{C}_1\mathbf{X} - \mathbf{I})' + \mathbf{C}_o\mathbf{P}^{-1}\mathbf{C}_o' = (\Sigma^{-1} + \mathbf{X}'\mathbf{X})^{-1} + \mathbf{C}_o\mathbf{P}^{-1}\mathbf{C}_o'$ and $\mathbf{C} + (\mathbf{C}\mathbf{X} - \mathbf{I})\Sigma\mathbf{X}' = \mathbf{C}_o\mathbf{P}^{-1}$ because $\mathbf{C}_1(\mathbf{C}_1\mathbf{X} - \mathbf{I})\Sigma\mathbf{X}' = 0$.

Let $\mathbf{c}_1 = (\mathbf{I} - \mathbf{C}_1\mathbf{X})\mathbf{m}$, $\mathbf{c} = \mathbf{c}_1 + \mathbf{c}_o$ then $\hat{\mathbf{c}} = \hat{\mathbf{c}}_o = \mathbf{C}_o\mathbf{X}\mathbf{m} + \mathbf{c}_o$.

If $\mathbf{A}_1 = e_1\mathbf{P}$, $e_1 = s/[2(w^2 + s) + ns]$, $\mathbf{A} = \mathbf{A}_1 + \mathbf{A}_o$ then $s(1 - tr\mathbf{A}\mathbf{P}^{-1})^2 + 2(w^2 + s)tr(\mathbf{A}\mathbf{P}^{-1})^2 = s(1 - tr\mathbf{A}_1\mathbf{P}^{-1})^2 + 2(w^2 + s)tr(\mathbf{A}_1\mathbf{P}^{-1}) + str(\mathbf{A}_o\mathbf{P}^{-1})^2 + 2(w^2 + s)tr(\mathbf{A}_o\mathbf{P}^{-1})^2$, because $-s(1 - ne_1) + 2(w^2 + s)e_1 = 0$. Let $\mathbf{a}_1 = -2\mathbf{m}'\mathbf{X}'\mathbf{A}_1$, $\mathbf{a} = \mathbf{a}_1 + \mathbf{a}_o$, $\alpha_1 = w - \mathbf{m}'\mathbf{X}'\mathbf{A}_1\mathbf{X}\mathbf{m} - \mathbf{a}_1\mathbf{X}\mathbf{m} - wtr\mathbf{A}_1\mathbf{P}^{-1}$, $\alpha = \alpha_1 + \alpha_o$ then $\hat{\mathbf{a}} = \hat{\mathbf{a}}_o = 2\mathbf{m}'\mathbf{X}\mathbf{A}_o + \mathbf{a}_o$, $\hat{\alpha} = \hat{\alpha}_o = -\mathbf{m}'\mathbf{X}'\mathbf{A}_o\mathbf{X}\mathbf{m} - \mathbf{a}_o\mathbf{X}\mathbf{m} - \alpha_o - wtr\mathbf{A}_o\mathbf{P}^{-1}$.

Let us define $d_1^\gamma = (b_1^{\gamma'}, v_1^\gamma)'$, where

$$\begin{aligned} b_1^\gamma(\mathbf{y}) &= \mathbf{C}_1\mathbf{y} + \mathbf{c}_1, \\ v_1^\gamma(\mathbf{y}) &= \mathbf{y}'\mathbf{A}_1\mathbf{y} + \mathbf{a}_1\mathbf{y} + \alpha_1. \end{aligned}$$

The Bayes risk (7) may be expressed in the following form

$$\begin{aligned} r(\gamma, d) = r(\gamma, d_1^\gamma) + wtr\tilde{\mathbf{Q}}\begin{bmatrix} \mathbf{C}_o \\ \hat{\mathbf{a}}_o \end{bmatrix}\mathbf{P}^{-1}[\mathbf{C}_o', \hat{\mathbf{a}}_o'] + [\hat{\mathbf{c}}_o', -\hat{\alpha}_o]\tilde{\mathbf{Q}}\begin{bmatrix} \hat{\mathbf{c}}_o \\ -\hat{\alpha}_o \end{bmatrix} \\ + q_{22}[s(tr\mathbf{A}_o\mathbf{P}^{-1}) + 2(w^2 + s)tr(\mathbf{A}_o\mathbf{P}^{-1})^2], \end{aligned}$$

where $r(\gamma, d_1^\gamma) = wtr\mathbf{Q}(\Sigma^{-1} + \mathbf{X}\mathbf{X})^{-1} + q_{22}(2s^2 + sw^2)/[(n + 2)s + 2w^2]$.

The following corollary is a consequence of Theorem 4.

Corollary 1: *For the NLRM, the loss function L_1 and any $d \in \mathcal{D}_{lq}$, $d \neq d_1^\gamma$*

$$\sup_{\gamma \in \mathcal{G}_1} r(\gamma, d) > \sup_{\gamma \in \mathcal{G}_1} r(\gamma, d_1^\gamma)$$

PROOF: For arbitrary $\gamma \in \mathcal{G}_1$, any $d \in \mathcal{G}_{lq}$, $d \neq d_1^\gamma$ we have $r(\gamma, d) - r(\gamma, d_1^\gamma) \geq 0$. Since $\tilde{\mathbf{Q}} \in \mathcal{M}_{k+1}^{>}$, it follows that $[\hat{\mathbf{c}}_o', -\hat{\alpha}_o]\tilde{\mathbf{Q}}\begin{bmatrix}\hat{\mathbf{c}}_o \\ -\hat{\alpha}_o\end{bmatrix} \geq 0$ and equality holds if and only if $\hat{\mathbf{c}}_o = 0$, $\hat{\alpha}_o = 0$. Matrix $\mathbf{P}^{-1}$ is a positive definite matrix, hence $\begin{bmatrix}\mathbf{C}_o \\ \hat{\mathbf{a}}_o\end{bmatrix}\mathbf{P}^{-1}[\mathbf{C}_o', \hat{\mathbf{a}}_o']$ is a non-negative definite matrix and $w > 0$, so $wtr\tilde{\mathbf{Q}}\begin{bmatrix}\mathbf{C}_o \\ \hat{\mathbf{a}}_o\end{bmatrix}\mathbf{P}^{-1}[\mathbf{C}_o', \hat{\mathbf{a}}_o'] \geq 0$. The equality holds if and only if $\mathbf{C}_o = \beta 0$ and $\hat{\mathbf{a}}_o = \beta 0$. Since $q_{22} > 0$, $s > 0$, it follows that $q_{22}[s(tr\mathbf{A}_o\mathbf{P}^{-1}) + 2(w^2+s)tr(\mathbf{A}_o\mathbf{P}^{-1})^2] \geq 0$ and the equality holds if and only if $\mathbf{A}_o = \beta 0$. Finally $r(\gamma, d) = r(\gamma, d_1^\gamma)$ if and only if $\mathbf{C}_o = \beta 0$, $\mathbf{c}_o = \beta 0$, $\mathbf{A}_o = \beta 0$, $\mathbf{a}_o = \beta 0$, $\alpha_o = 0$. For any $\gamma \in \mathcal{G}_1$ and any $d \in \mathcal{D}_{lq}$, $d \neq d_1^\gamma$ we have $r(\gamma, d) > r(\gamma, d_1^\gamma)$ and consequently for all $d \in \mathcal{D}_{lq}$, $d \neq d_1^\gamma$ we obtain

$$\sup_{\gamma \in \mathcal{G}_1} r(\gamma, d) > \sup_{\gamma \in \mathcal{G}_1} r(\gamma, d_1^\gamma).$$

In light of Corollary 1. the problem of finding a $\mathcal{G}$-minimax estimator of θ may be reduced to the problem of determining Σ, w, and s which maximize $r(\gamma, d_1^\gamma)$, viz.

$$\sup_{\Sigma \in \mathcal{S}, (w,s) \in \mathcal{V}} \{wtr\mathbf{Q}(\Sigma^{-1} + \mathbf{X}'\mathbf{X})^{-1} + q_{22}(2s^2 + sw^2)/[(n+2)s + 2w^2]\} \tag{10}$$

The mathematical programming problem (10) can be solved in two steps:

1. Find a matrix Σ_* such that

$$\Sigma_* = \arg[\sup_{\Sigma \in \mathcal{S}} tr\mathbf{Q}(\Sigma^{-1} + \mathbf{X}'\mathbf{X})^{-1}]. \tag{11}$$

2. Find w_*, s_* such that

$$\begin{aligned}(w_*, s_*) &= \arg\{\sup_{(w,s)\in\mathcal{V}} \{wtr\mathbf{Q}(\Sigma^{-1} + \mathbf{X}'\mathbf{X})^{-1} \\ &\qquad + q_{22}(2s^2 + sw^2)/[(n+2)s + 2w^2]\}\}.\end{aligned} \tag{12}$$

THEOREM 5: *Let Σ_*, w_*, s_* be an arbitrary solution of the optimization problem (10). Then for any $\mathbf{m} \in \mathcal{M}$ the estimator $d_1^{\gamma^*} = (b_1^{\gamma^*\prime}, v_1^{\gamma^*})'$ where*

$$\begin{aligned}b_1^{\gamma^*}(\mathbf{y}) &= (\Sigma_*^{-1} + \mathbf{X}'\mathbf{X})^{-1}(\mathbf{X}'\mathbf{y} + \Sigma_*^{-1}\mathbf{m}), \\ v_1^{\gamma^*}(\mathbf{y}) &= \mathbf{e}_1^*(\mathbf{y}'\mathbf{P}_*\mathbf{y} - 2\mathbf{m}'\mathbf{X}'\mathbf{P}_*\mathbf{y} + \mathbf{m}'\mathbf{X}'\mathbf{P}_*\mathbf{X}\mathbf{m}) + (1 - ne_1^*)w_*, \\ e_1^* &= s_*/[2(w_*^2 + s_*) + ns_*) + ns_*], \ \mathbf{P}_* = (\mathbf{I} - \mathbf{X}\Sigma_*\mathbf{X}')^{-1}\end{aligned}$$

is $\mathcal{G}_1$-minimax in the class $\mathcal{D}_{lq}$.

PROOF: Theorem 5 follows immediately from Definition, Corollary 1 and previous considerations.

A solution of the problem (12) is easy to obtain due to the fact that the optimization is only over two variables. Generally a solution of the problem (11) may be obtained by a numerical procedure. An explicit solution may be given in a special case. Let $\mathcal{S}$ be the set of matrices bounded from above [see Leamer (1982)] such that $\mathcal{S} = \{\Sigma \in \mathcal{M}_k^> : \Sigma \leq \hat{\Sigma},\ \hat{\Sigma} \in \mathcal{M}_k^>\}$ or $\mathcal{S} = \{\Sigma \in \mathcal{M}_k^> : \underset{\wedge}{\Sigma} \leq \Sigma \leq \hat{\Sigma},\ \underset{\wedge}{\Sigma}, \hat{\Sigma} \in \mathcal{M}_k^>\}$ where, for instance $\Sigma \leq \hat{\Sigma}$ means that $\hat{\Sigma} - \Sigma \in \mathcal{M}_k^>$. Then the matrix $\hat{\Sigma}$ is the solution of the problem (11). If $\Sigma \leq \hat{\Sigma}$ then $\hat{\Sigma}^{-1} \leq \Sigma^{-1}$. For the non-negative definite matrix $\mathbf{X}'\mathbf{X}$ we have $\hat{\Sigma}^{-1} + \mathbf{X}'\mathbf{X} \leq \Sigma^{-1} + \mathbf{X}'\mathbf{X}$. It follows that $(\hat{\Sigma}^{-1} + \mathbf{X}'\mathbf{X})^{-1} \geq (\Sigma^{-1} + \mathbf{X}'\mathbf{X})^{-1}$. For a positive definite matrix $\mathbf{Q}$ we have $tr\mathbf{Q}(\Sigma^{-1} + \mathbf{X}'\mathbf{X})^{-1} \geq tr\mathbf{Q}(\Sigma^{-1} + \mathbf{X}'\mathbf{X})^{-1}$. It follows that $\hat{S}$ is the solution of the problem (11).

If the matrix $\mathbf{X}'\mathbf{X}$ is invertible then we obtain the optimization problem

$$\inf_{\Sigma \in \mathcal{S}} tr(\mathbf{X}'\mathbf{X})^{-1}\mathbf{Q}(\mathbf{X}'\mathbf{X})^{-1}[\Sigma + (\mathbf{X}'\mathbf{X})^{-1}]^{-1}$$

equivalent to the problem (11). Since $\Sigma \in \mathcal{M}_k^> \subset \mathcal{M}_k^s$, $(\mathbf{X}'\mathbf{X})^{-1}\mathbf{Q}(\mathbf{X}'\mathbf{X})^{-1} \in \mathcal{M}_k^>$, $[\Sigma + (\mathbf{X}'\mathbf{X})^{-1}[^{-1} \in \mathcal{M}_k^>$, then the function $f(\Sigma) = tr(\mathbf{X}'\mathbf{X})^{-1}\mathbf{Q}(\mathbf{X}'\mathbf{X})^{-1}[\Sigma + (\mathbf{X}'\mathbf{X})^{-1}]^{-1}$ is the ratio of polynomials of $[k(k+1)]/2$ variables and it is continuous, differentiable and convex. The last property follows from the fact that for any $\mathbf{B},\ \mathbf{G} \in \mathcal{M}_k^>$ and any $\delta \in (0,1)$ $[\delta\mathbf{B} + (1-\delta)\mathbf{G}]^{-1} \leq \delta\mathbf{B}^{-1} + (1-\delta)\mathbf{G}^{-1}$. It may be proved also [details are available in Bijak (1986)], that the function f has no stationary points, which means that any solution of the problem (11) can not be in the interior of the set $\mathcal{S}$. The properties mentioned above of the function f imply that the problem (11) is a convex mathematical programming problem if $\mathcal{S}$ is convex; there exist numerical methods for solving it properly.

THEOREM 6: *For the NLRM with loss function L_1 and any $\mathbf{m} \in \mathcal{M}$, the estimator $d_1^{\gamma^*}$ (see Theorem 5) is $\mathcal{G}_1$-minimax in the class $\mathcal{D}$ of all estimators of θ.*

PROOF: By assumption, $\mathcal{S}$ and $\mathcal{V}$ are closed and bounded. Thus $\Sigma_* \in \mathcal{S}$ and $(w_*, s_*) \in \mathcal{V}$. For any $\mathbf{m} \in \mathcal{M}$, the prior distribution NIG1 $\gamma_o^*(\mathbf{m}, \Sigma_*, w_*/s_* + 2, w_*(w_*/s_* + 1))$ is in the class $\mathcal{G}_1$. The estimator $d_1^{\gamma_o^*}$ is the Bayes estimator with respect to γ_o^* (see Theorem 1). Hence for all $d \in \mathcal{D}$

$$r(\gamma_o^*, d_1^{\gamma_o^*}) \leq r(\gamma_o^*, d) \leq \sup_{\gamma \in \mathcal{G}_1} r(\gamma, d)$$

According to Theorem 5 the estimator $d_1^{\gamma_o^*}$ is $\mathcal{G}$-minimax in the class $\mathcal{D}_{lq}$. This means that

$$r(\gamma_o^*, d_1^{\gamma_o^*}) = \sup_{\gamma \in \mathcal{G}_1} r(\gamma, d_1^{\gamma_o^*})$$

Thus, the conclusion is that for any $d \in \mathcal{D}$

$$\sup_{\gamma \in \mathcal{G}_1} r(\gamma, d_1^{\gamma_o^*}) \leq \sup_{\gamma \in \mathcal{G}_1} r(\gamma, d)$$

which finishes the proof.

If we assume that $\mathbf{y} \sim N(\mathbf{X}\beta, \sigma^2\mathbf{V})$, where $\mathbf{V} \in \mathcal{M}_n^>$ is known, then our results can be easily generalized. Furthermore, in the proof of Theorem 4, it is only required that the third central moments equal zero and that equation (9) is satisfied. Such conditions are fulfilled e.g., for vectors of independent, symmetrical and mesokurtic random variables. We can assume the above mentioned conditions on third and fourth central moments instead of normality of the random vectors $\mathbf{y}$ and β.

4. $\mathcal{G}$-Minimax Estimation with Other Loss

Quadratic loss function is appropriate in many point estimation problems. In estimation of variance σ^2 in NLRM relative squared error loss or logarithmic squared error loss is more adequate. We assume these loss functions for variance in L_2, L_3.

A theorem similar to Theorem 4 is valid for the estimators $d \in \mathcal{D}_{lq}$, loss function L_2 and prior distributions $\gamma \in \mathcal{G}_2$.

THEOREM 7: *For the NLRM, the loss function L_2, any $\gamma \in \mathcal{G}_2$ and any $d \in \mathcal{D}_{lq}$*

$$r(\gamma, d) = tr\tilde{\mathbf{Q}} \begin{bmatrix} w\mathbf{CC}' + (\mathbf{CX} - \mathbf{I})\Sigma(\mathbf{CX} - \mathbf{I})' & [\mathbf{C} + (\mathbf{CX} - \mathbf{I})\Sigma\mathbf{X}']\hat{\mathbf{a}}' \\ \hat{\mathbf{a}}[\mathbf{C}' + \mathbf{X}\Sigma(\mathbf{CX} - \mathbf{I})'] & g\hat{\mathbf{a}}\mathbf{P}^{-1}\hat{\mathbf{a}}' \end{bmatrix}$$

$$+ [\hat{\mathbf{c}}', -\hat{\alpha}]\tilde{\mathbf{Q}} \begin{bmatrix} \hat{\mathbf{c}} \\ -\hat{\alpha} \end{bmatrix} + q_{22}(h - g^2)\tilde{\alpha}^2 + q_{22} 2tr(\mathbf{AP}^{-1})^2 \tag{13}$$

where $\hat{\mathbf{c}} = (\mathbf{CX} - \mathbf{I})\mathbf{m} + \mathbf{c}$, $\hat{\mathbf{a}} = 2\mathbf{m}'\mathbf{X}'\mathbf{A} + \mathbf{a}$, $\hat{\alpha} = 1 - g\mathbf{m}'\mathbf{X}'\mathbf{AXm} - g\mathbf{aXm} - g\alpha - tr\mathbf{AP}^{-1}$, $\tilde{\alpha} = \mathbf{m}'\mathbf{X}'\mathbf{P}^{-1}\mathbf{Xm} + \mathbf{aXm} + \alpha$.

PROOF: To prove Theorem 7 we need the same facts and similar transformations as in the proof of Theorem 4.

Let $\mathbf{C} = \mathbf{C}_2 + \mathbf{C}_o$, $\mathbf{C}_2 = (\Sigma^{-1} + \mathbf{X}'\mathbf{X})^{-1}\mathbf{X}'$, $\mathbf{c}_1 = (\mathbf{I} - \mathbf{CX})m + \mathbf{Q}^{-1}\mathbf{q}_{12}\hat{\alpha}$, $\mathbf{c} = \mathbf{c}_1 + \mathbf{c}_o$ then $\hat{\mathbf{c}} = \mathbf{Q}^{-1}\mathbf{q}_{12}\hat{\alpha} + \mathbf{c}_o$ and for any $\gamma \in \mathcal{G}_2$ and any $d \in \mathcal{D}_{lq}$

$$r(\gamma, d) = trw\mathbf{Q}(\Sigma^{-1} + \mathbf{X}'\mathbf{X})^{-1} + wtr\hat{\mathbf{Q}} \begin{bmatrix} \mathbf{C}_o \\ \hat{\mathbf{a}} \end{bmatrix} \mathbf{P}^{-1}[\mathbf{C}_o', \hat{\mathbf{a}}'] + \mathbf{c}_o'\mathbf{Q}\mathbf{c}_o$$
$$+ k_1\hat{\alpha}^2 + q_{22}(h - g^2)\tilde{\alpha}^2 + q_{22}2tr(\mathbf{AP}^{-1})^2,$$

where $\hat{\mathbf{Q}} = \begin{bmatrix} w\mathbf{Q} & \mathbf{q}_{12} \\ \mathbf{q}_{21} & gq_{22} \end{bmatrix}$.

If $\alpha_1 = k_1 g(1 - tr\mathbf{A}\mathbf{P}^{-1})/k_2 - \mathbf{m}'\mathbf{X}'\mathbf{A}\mathbf{X}\mathbf{m} - \mathbf{a}\mathbf{X}\mathbf{m}$ where $k_1 = q_{22} - \mathbf{q}_{21}\mathbf{Q}^{-1}\mathbf{q}_{12}$, $k_2 = hq_{22} - g^2\mathbf{q}_{21}\mathbf{Q}^{-1}\mathbf{q}_{12}$ then $\hat{\alpha} = \hat{\alpha}_1 = 1 - g\mathbf{m}'\mathbf{X}'\mathbf{A}\mathbf{X}\mathbf{m} - g\mathbf{a}\mathbf{X}\mathbf{m} - g\alpha_1 - tr\mathbf{A}\mathbf{P}^{-1} = (k_2 - g^2k_1)(1 - tr\mathbf{A}\mathbf{P}^{-1})/k_2$ and $\tilde{\alpha}_1 = \mathbf{m}'\mathbf{X}'\mathbf{A}\mathbf{X}\mathbf{m} + \mathbf{a}\mathbf{X}\mathbf{m} + \alpha_1 = gk_1(1 - tr\mathbf{A}\mathbf{P}^{-1})/k_2$. Let $\alpha = \alpha_1 + \alpha_o$ then $k_1\hat{\alpha}^2 + q_{22}(h - g^2)\tilde{\alpha}^2 = k_1\hat{\alpha}_1^2 + q_{22}(h - g^2)\tilde{\alpha}_1^2 + k_2\alpha_o^2$. We obtain that for any $\gamma \in \mathcal{G}_2$ and any $d \in \mathcal{D}_{lq}$

$$r(\gamma, d) = trw\mathbf{Q}(\Sigma^{-1} + \mathbf{X}'\mathbf{X})^{-1} + tr\hat{\mathbf{Q}} \begin{bmatrix} \mathbf{C}_o \\ \hat{\mathbf{a}} \end{bmatrix} \mathbf{P}^{-1}[\mathbf{C}_o', \hat{\mathbf{a}}'] + \mathbf{c}_o'\mathbf{Q}\mathbf{c}_o + k_2\alpha_o^2$$

$$+ q_{22}k_1(h - g^2)(1 - tr\mathbf{A}\mathbf{P}^{-1})^2/k_2 + q_{22}2tr(\mathbf{A}\mathbf{P}^{-1})^2.$$

Let $\mathbf{A} = \mathbf{A}_2 + \mathbf{A}_o$, where $\mathbf{A}_2 = e_2\mathbf{P}$, $e_2 = (h - g^2)k_1/[n(h - g^2)k_1 + 2k_2]$, and let $\mathbf{a}_2 = -2\mathbf{m}'\mathbf{X}\mathbf{A}_2$, $\mathbf{a} = \mathbf{a}_2 + \mathbf{a}_o$ then $\hat{\mathbf{a}} = \hat{\mathbf{a}}_o = 2\mathbf{m}'\mathbf{X}'\mathbf{A}_o + \mathbf{a}_o$, $k_1(h - g^2)/k_2 = 2e_2/(1 - ne_2)$. Furthermore $q_{22}k_1(h - g^2)(1 - tr\mathbf{A}\mathbf{P}^{-1})^2/k_2 + q_{22}2tr(\mathbf{A}\mathbf{P}^{-1})^2 = q_{22}2e_2 + q_{22}k_1(h - g^2)(tr\mathbf{A}_o\mathbf{P}^{-1})^2/k_2 + q_{22}2tr(\mathbf{A}_o\mathbf{P}^{-1})^2$.

Finally we have

$$\begin{aligned} r(\gamma, d) = {} & trw\mathbf{Q}(\Sigma^{-1} + \mathbf{X}'\mathbf{X})^{-1} + q_{22}2\mathbf{e}_2 + tr\hat{\mathbf{Q}} \begin{bmatrix} \mathbf{C}_0 \\ \hat{\mathbf{a}}_o \end{bmatrix} \mathbf{P}^{-1}[\mathbf{C}_o', \hat{\mathbf{a}}_o'] \\ & + \mathbf{c}_o'\mathbf{Q}\mathbf{c}_o + k_2\alpha_o^2 + q_{22}2e_2(tr\mathbf{A}_o\mathbf{P}^{-1})^2/(1 - ne_2) \\ & + q_{22}2tr(\mathbf{A}_o\mathbf{P}^{-1})^2 \end{aligned}$$

The above considerations enable us to define estimator d_2^γ which is optimal for any fixed prior distribution $\gamma \in \mathcal{G}_2$. Let $d_2^\gamma = (b_2^{\gamma'}, v_2^\gamma)'$, where

$$\begin{aligned} b_2^\gamma(\mathbf{y}) &= \mathbf{C}_2\mathbf{y} + \mathbf{c}_2 + \mathbf{Q}^{-1}\mathbf{q}_{12}\hat{\alpha}_2, \\ v_2^\gamma(\mathbf{y}) &= \mathbf{y}'\mathbf{A}_2\mathbf{y} + \mathbf{a}_2\mathbf{y} + \alpha_2, \end{aligned}$$

$$\begin{aligned} \mathbf{c}_2 &= (\mathbf{I} - \mathbf{C}_2\mathbf{X})\mathbf{m}, \\ \alpha_2 &= k_1g(1 - tr\mathbf{A}_2\mathbf{P}^{-1})/k_2 + \mathbf{m}'\mathbf{X}'\mathbf{A}_2\mathbf{X}\mathbf{m} - \mathbf{a}_2\mathbf{X}\mathbf{m} \\ &= k_1g(1 - ne_2)/k_2 + e_2\mathbf{m}'\mathbf{X}'\mathbf{P}\mathbf{X}\mathbf{m}, \\ \hat{\alpha}_2 &= (k_2 - g^2k_1)(1 - tr\mathbf{A}_2\mathbf{P}^{-1})/k_2 \\ &= q_{22}2(h - g^2)/[n(h - g^2)k_1 + 2k_2] = \frac{q_{22}2e_2}{k_1}. \end{aligned}$$

We can also prove the following corollary.

COROLLARY 2: *For the NLRM with the loss function L_2 and any $d \in \mathcal{D}_{lq}$, $d \neq d_2^\gamma$*

$$\sup_{\gamma \in \mathcal{G}_2'} r(\gamma, d) > \sup_{\gamma \in \mathcal{G}_2'} r(\gamma, d_2^\gamma)$$

where $\mathcal{G}_2' = \{\gamma \in \mathcal{G}_2 : g/w > \mathbf{q}_{21}\mathbf{Q}^{-1}\mathbf{q}_{12}/q_{22}\}$.

PROOF: For any $\gamma \in \mathcal{G}_2$ and any $d \in \mathcal{D}_{lq}$ the Bayes risk $r(\gamma, d)$ can be presented in the following form

$$\begin{aligned} r(\gamma, d) = r(\gamma, d_2^\gamma) + tr\hat{\mathbf{Q}} \begin{bmatrix} \mathbf{C}_o \\ \hat{\mathbf{a}}_o \end{bmatrix} \mathbf{P}^{-1}[\mathbf{C}_o', \hat{\mathbf{a}}_o'] + \mathbf{c}_o'\mathbf{Q}\mathbf{c}_o + k_2\alpha_o^2 \\ + q_{22}2e_2(tr\mathbf{A}_o\mathbf{P}^{-1})^2/(1 - ne_2) + q_{22}2tr(\mathbf{A}_o\mathbf{P}^{-1})^2 \end{aligned}$$

where $r(\gamma, d_2^\gamma) = trw\mathbf{Q}(\Sigma^{-1} + \mathbf{X}'\mathbf{X})^{-1} + 2q_{22}e_2$. By assumption, $g/w > \mathbf{q}_{21}\mathbf{Q}^{-1}\mathbf{q}_{12}/q_{22}$ and hence $\hat{\mathbf{Q}} \in \mathcal{M}_{k+1}^{>}$. Furthermore, $k_2 > 0$, $2q_{22}e_2/(1 - ne_2) = q_{22}k_1(h - g^2)/k_2 > 0$. By the same argumentation as in the proof of Corollary 1, we have that $r(\gamma, d) - r(\gamma, d_2^\gamma) \geq 0$ and $r(\gamma, d) - r(\gamma, d_2^\gamma) = 0$ if and only if $\mathbf{C}_0 = \beta 0$, $\mathbf{c}_o = \beta 0$, $\mathbf{A}_o = \beta 0$, $\mathbf{a}_o = \beta 0$, $\alpha_o = 0$. Finally for any $\gamma \in \mathcal{G}_2'$ and any $d \in \mathcal{D}_{lq}$, $d \neq d_2^\gamma$ we have $r(\gamma, d) > r(\gamma, d_2^\gamma)$ and hence thesis of Corollary 2.

Similarly as in Section 3, we may reduce the problem of finding $\mathcal{G}_2'$-minimax estimator of θ to the problem of determining Σ, g, and h which maximize $r(\gamma, d_2^\gamma)$, viz.

$$\sup_{\Sigma \in \mathcal{S}, (g,h) \in \mathcal{H}} \{wtr\mathbf{Q}(\Sigma^{-1} + \mathbf{X}'\mathbf{X})^{-1} + q_{22}2(h - g^2)k_1/[n(h - g^2)k_1 + 2k_2]\} \quad (14)$$

The problem (14) may be solved in two steps.

1. Find a matrix Σ_* such that

$$\Sigma_* = \arg[\sup_{\Sigma \in \mathcal{S}} tr\mathbf{Q}(\Sigma^{-1} + \mathbf{X}'\mathbf{X})^{-1}]. \quad (15)$$

2. Find g_*, h_* such that

$$(g_*, h_*) = \arg\{\sup_{(g,h) \in \mathcal{H}} \{wtr\mathbf{Q}(\Sigma_*^{-1} + \mathbf{X}'\mathbf{X})^{-1} + q_{22}2(h - g^2)k_1/[n(h - g^2)k_1 + 2k_2]\}\}. \quad (16)$$

Problem (15) was considered in Section 3. Problem (16) is not trivial, because w usually depends on g and h. Let us pay attention to a case when w is a continuous function of g and h. Let $\mathcal{G}_2'' = \{\gamma \in \mathcal{G}_2' : w = w(g, h)$ and $w(\cdot)$ is continuous$\}$.

THEOREM 8: *Let Σ_*, g_*, h_* be an arbitrary solution of the optimization problem (14). Then for any $\mathbf{m} \in \mathcal{M}$ the estimator $d_2^{\gamma^*} = (b_2^{\gamma^{*\prime}}, v_2^{\gamma^*})'$ is $\mathcal{G}_2''$-minimax in the class $\mathcal{D}_{lq}$, where*

$$
\begin{aligned}
b_2^{\gamma^*}(\mathbf{y}) &= (\Sigma_*^{-1} + \mathbf{X}'\mathbf{X})^{-1}(\mathbf{X}'\mathbf{y} + \Sigma_*^{-1}\mathbf{m}) + 2\mathbf{Q}^{-1}\mathbf{q}_{12}q_{22}e_2^*/k_1, \\
v_2^{\gamma^*}(\mathbf{y}) &= e_2^*(\mathbf{y}'\mathbf{P}_*\mathbf{y} - 2\mathbf{m}'\mathbf{X}'\mathbf{P}_*\mathbf{y} + \mathbf{m}'\mathbf{X}'\mathbf{P}_*\mathbf{X}\mathbf{m}) + k_1(1 - ne_2^*)g_*/k_2^*, \\
k_2^* &= h_* q_{22} - g_*^2 \mathbf{q}_{21}\mathbf{Q}^{-1}\mathbf{q}_{12}, \\
e_2^* &= \frac{(h_* - g_*^2)k_1}{n(h_* - g_*^2)k_1 + 2k_2^*}, \\
\mathbf{P}_* &= (\mathbf{I} - \mathbf{X}\Sigma_*\mathbf{X}')^{-1}.
\end{aligned}
$$

PROOF: Theorem 8 follows from Definition, Corollary 2 and above considerations.

THEOREM 9: *For the NLRM with loss function L_2 and any $\mathbf{m} \in \mathcal{M}$, the estimator $d_2^{\gamma^*}$ (see Theorem 8) is $\mathcal{G}_3$-minimax in the class $\mathcal{D}$ of all estimators of θ.*

PROOF: For a NIG1 distribution which is $\gamma_o(\mathbf{m}, \Sigma, \lambda, \mu)$ we have $w = \mu/(\lambda - 1) = g/(2g^2 - h)$, hence $\mathcal{G}_3 \subset \mathcal{G}_2''$. For any $\mathbf{m} \in \mathcal{M}$ the estimator $d_2^{\gamma^*}$ is $\mathcal{G}_3$-minimax in the class $\mathcal{D}_{lq}$. There is a prior distribution $\gamma_o^*(\mathbf{m}, \Sigma_*, g_*^2/(h_* - g_*^2), g_*/(h_* - g_*^2))$ in the set $\mathcal{G}_3$ such that the estimator $d^{\gamma_o^*}$ ($d^{\gamma_o^*} = d^{\gamma^*}$) is the Bayes estimator with respect to γ_o^*. Hence the estimator $d_2^{\gamma^*}$ is $\mathcal{G}_3$-minimax in the class $\mathcal{D}$.

Generalizations mentioned at the end of Section 3 may be done also in estimation of parameter vector θ with loss function L_2. It is an open problem to determine the largest sets of prior distributions for which Theorem 8 and Theorem 9 are true.

Let us consider loss function L_3 and the set of prior distributions $\mathcal{G}_3$. Let $d_3^\gamma = (b_3^{\gamma\prime}, v_3^\gamma)'$ where $b_3^\gamma(\mathbf{y}) = E(\beta \mid \mathbf{y})$, $v_3^\gamma(\mathbf{y}) = \exp\{E(\log\ \sigma^2 \mid \mathbf{y})\}$ hence $r(\gamma, d_3^\gamma) = r(\gamma, b_3^\gamma) + r(\gamma, v_3^\gamma)$. We have proved in Section 2 that $b_3^\gamma(\mathbf{y}) = E(\beta \mid \mathbf{y}) = (\Sigma^{-1} + \mathbf{X}'\mathbf{X})^{-1}(\mathbf{X}'\mathbf{y} + \Sigma^{-1}\mathbf{m})$, $v_3^\gamma(\mathbf{y}) = \exp\{E(\log\ \sigma^2 \mid \mathbf{y})\} = \exp\{\log\ \overline{\mu} - \psi(\overline{\lambda})\}$ and in Section 3 that $r(\gamma, b_3^\gamma) = w tr \mathbf{Q}(\Sigma^{-1} + \mathbf{X}'\mathbf{X})^{-1} = [\mu/(\lambda - 1)] tr \mathbf{Q}(\Sigma^{-1} + \mathbf{X}'\mathbf{X})^{-1}$. Since the loss function is positive we have

$$
\begin{aligned}
r(\gamma, v_3^\gamma) &= E\{E[(\log\ \sigma^2 - E(\log\ \sigma^2 \mid y))^2 \mid \theta]\} \\
&= E\{E[(\log\ \sigma^2 - E(\log\ \sigma^2 \mid y))^2 \mid y]\} \\
&= E\{[\log\ \overline{\mu} - \psi(\overline{\lambda})]^2 + \zeta(2, \overline{\lambda}) - [\log\ \mu - \psi(\lambda)]^2\} = \zeta(2, \lambda + n/2),
\end{aligned}
$$

where for $z > 1$

$$\zeta(z,q) = [\Gamma(z)]^{-1}\int_0^\infty [t^{z-1}\exp\{-qt\}/(1-\exp\{-t\})]dt$$

$$= \sum_{k=0}^{\infty}(q+k)^{-z}.$$

The Bayes risk of the estimator d_3^γ is, therefore, equal to

$$r(\gamma, d_3^\gamma) = [\mu/(\lambda-1)]tr\mathbf{Q}(\Sigma^{-1} + \mathbf{X}'\mathbf{X})^{-1} + \zeta(2, \lambda + n/2)$$

Now we can prove the analogous theorem to Theorem 9.

THEOREM 10: *For the NLRM with loss function L_3 and any $\mathbf{m} \in \mathcal{M}$, the estimator $d_3^{\gamma^*} = (b_3^{\gamma^{*'}}, v_3^{\gamma^*})'$ is $\mathcal{G}_3$-minimax in the class $\mathcal{D}$, where*

$$b_3^{\gamma^*}(\mathbf{y}) = (\Sigma_*^{-1} + \mathbf{X}'\mathbf{X})^{-1}(\mathbf{X}'\mathbf{y} + \Sigma_*^{-1}\mathbf{m}),$$
$$v_3^{\gamma^*}(\mathbf{y}) = \mathbf{e}_3^*(\mathbf{y}'\mathbf{P}_*\mathbf{y} - 2\mathbf{m}'\mathbf{X}'\mathbf{P}_*\mathbf{y} + \mathbf{m}'\mathbf{X}'\mathbf{P}_*\mathbf{X}\mathbf{m}) + 2e_3^*\mu,$$
$$\Sigma_* = \arg[\sup_{\Sigma\in\mathcal{S}} tr\mathbf{Q}(\Sigma^{-1} + \mathbf{X}'\mathbf{X})^{-1}],$$
$$(\lambda_*, \mu_*) = \arg\{\sup_{(\lambda,\mu)\in\mathcal{L}} [\mu/(\lambda-1)]tr\mathbf{Q}(\Sigma_*^{-1} + \mathbf{X}'\mathbf{X})^{-1} + \zeta(2, \lambda + n/2)]\},$$
$$e_3^* = \frac{1}{2}\exp\{-\psi(\lambda_* + n/2)\}, \quad \mathbf{P}_* = (\mathbf{I} - \mathbf{X}\Sigma_*\mathbf{X}')^{-1}.$$

PROOF: For any $\gamma \in \mathcal{G}_3$ the estimator $d_3^{\gamma^*}$ is the Bayes estimator with respect to γ^* (see Theorem 3), hence for any $d \in \mathcal{D}$ we have $r(\gamma, d) \geq r(\gamma, d_3^\gamma)$ and, further, $\sup_{\gamma\in\mathcal{G}_3} r(\gamma, d) \geq \sup_{\gamma\in\mathcal{G}_3} r(\gamma, d_3^\gamma)$. We obtain the following maximizatiom problem

$$\sup_{\Sigma\in\mathcal{S},(\lambda,\mu)\in\mathcal{L}} \{[\mu/(\lambda-1)]tr\mathbf{Q}(\Sigma^{-1} + \mathbf{X}'\mathbf{X})^{-1} + \zeta(2, \lambda + n/2)]\} \tag{17}$$

which may be solved in two steps [similar to the problems (10), (14)]:

1. Find a matrix Σ_* such that

$$\Sigma_* = \arg[\sup_{\Sigma\in\mathcal{S}} tr\mathbf{Q}(\Sigma^{-1} + \mathbf{X}'\mathbf{X})^{-1}].$$

2. Find λ_*, μ_* such that

$$(\lambda_*, \mu_*) = \arg\{\sup_{(\lambda,\mu)\in\mathcal{S}} [\mu/(\lambda-1)]tr\mathbf{Q}(\Sigma_*^{-1} + \mathbf{X}'\mathbf{X})^{-1} + \zeta(2, \lambda + n/2)]\}.$$

Let Σ_*, λ_*, μ_* be an arbitrary solution of the problem (17). Then for any $d \in \mathcal{D}$

$$\sup_{\gamma \in \mathcal{G}_3} r(\gamma, d) \geq \sup_{\gamma \in \mathcal{G}_3} r(\gamma, d_3^\gamma) = \sup_{\gamma \in \mathcal{G}_3} r(\gamma, d_3^{\gamma^*}).$$

From Definition we immediately conclude that $d_3^{\gamma^*}$ is $\mathcal{G}_3$-minimax estimator in the class $\mathcal{D}$ of all estimators of parameter vector θ.

5. Concluding Remarks

Note that $\mathcal{G}$-minimax criterion does not lead to choice of $\mathbf{m}$ [Repin and Tartakovskij (1977) paid attention to fact that $\mathcal{G}$-minimax decision rules are not often uniquely determined]. For any $\mathbf{m} \in \mathcal{M}$ the estimators $d_1^{\gamma^*}$, $d_2^{\gamma^*}$, $d_3^{\gamma^*}$ are $\mathcal{G}$-minimax and we have obtained classes of $\mathcal{G}$-minimax estimators $\mathcal{D}_1^* = \{d_1^{\gamma^*} : \mathbf{m} \in \mathcal{M}\}$, $\mathcal{D}_2^* = \{d_2^{\gamma^*} : \mathbf{m} \in \mathcal{M}\}$ and $\mathcal{D}_3^* = \{d_3^{\gamma^*} : \mathbf{m} \in \mathcal{M}\}$. Choice of $\mathbf{m}$ is arbitrary from $\mathcal{G}$-minimax criterion point of view. A study of posterior sensitivity to parameter $\mathbf{m}$ variation may be easily done. If we want to have an unique estimator we must introduce additional information about $\mathbf{m}$ (if available). The most natural approach, from the Bayesian point of view, is to introduce a prior distribution (a noninformative prior distribution) for $\mathbf{m}$ over $\mathcal{M}$ and choose e.g., $\mathbf{m}_* = E(\mathbf{m})$. Another approach is a modification of the assumption on a set $\mathcal{M}$ such that a central point of $\mathcal{M}$ will be known [see Solomon (1972a, 1972b) and Pilz (1984)].

The analysis presented in this paper enables us to determine the worst prior distributions too. We have proven that prior distributions γ^* are the least favorable. The mathematical programming problems (10), (14) and (17) show us that γ^* depend on loss functions L_1, L_2 and L_3.

Particular results may be obtained when we determine the sets $\mathcal{G}$ more precisely. Computationally attractive cases occur when the sets $\mathcal{S}$, $\mathcal{V}$, $\mathcal{W}$, $\mathcal{H}$, $\mathcal{L}$ are defined by linear inequalities.

References

Berger, J.O. (1980). *Statistical Decision Theory.* New York: Springer-Verlag.

Berger, J.O. (1984). "The Robust Bayesian Viewpoint," in *Robustness in Bayesian Analysis*, J.B. Kadane, ed., Amsterdam: North-Holland, 63-124.

Bijak, W. (1986). "Estimation of the Linear Econometric Model Under Conditions of Incomplete Prior Information," Doctoral Dissertation, Central School of Planning and Statistics, Warsaw, Poland (in Polish).

Blum, J.R. and J. Rosenblatt (1967). "On Partial A Priori Information in Statistical Inference," *The Annals of Mathematical Statistics*, Vol. 38, 1671-1678.

DeGroot, M.H. (1970). *Optimal Statistical Decisions.* New York: McGraw-Hill.

Hodges, J.L., Jr. and E.L. Lehmann (1952). "The Use of Previous Experience in Reaching Statistical Decisions," *The Annals of Mathematical Statistics*, Vol. 23, 396-407.

Humak, K.M.S. (1984). *Statistische Methoden der Modellbildung III.* Berlin: Akademie-Verlag.

Jackson, D.A., T.M. O'Donovan, W.J. Zimmer and J.J. Deely (1970). "$\mathcal{G}$-Minimax Estimators in the Exponential Family," *Biometrika*, Vol. 57, 439-443.

Leamer, E.E. (1982). "Sets of Posterior Means with Bounded Variance Priors," *Econometrica*, Vol. 50, 725-736.

Pilz, J. (1983). *Bayesian Estimation and Experimental Design in Linear Regression Model.* Leipzig: Teubner-Verlag.

Pilz, J. (1984). "Robust Bayes Estimation Under Weak Prior Knowledge," in *Robustness of Statistical Methods and Nonparametric Statistics*, D. Rasch and M.L. Tiku, eds., Berlin: Deutscher Verlag der Wissenschaften, 85-89.

Polasek, W. (1984). "Multivariate Regression Systems. Estimation and Sensitivity Analysis of Two Dimensional Data," in *Robustness in Bayesian Analysis*, J.B. Kadane, ed., Amsterdam: North-Holland, 229-309.

Potter, J.M. and B.D.O. Anderson (1980). "Partial Prior Information and Decision Making," *IEEE Transactions on Systems Man and Cybernetics* , Vol. 10, 125-133.

Potter, J.M. and B.D.O. Anderson (1983). "Statistical Inference with Partial Prior Information," *IEEE Transactions on Information Theory*, Vol. 29, 688-695.

Repin, V.G. and G.P. Tartakovskij (1977). *Statističeskij analiz pri apriornoj neopredelennosti i adaptacija informacionnych sistem*, Moskva: Sovetskoe Radio.

Robbins, H.E. (1964). "The Empirical Bayes Approach to Statistical Decision Problems," *The Annals of Mathematical Statistics*, Vol. 35, 1-20.

Solomon, D.L. (1972a). "Λ-Minimax Estimation of a Multivariate Location Parameter," *Journal of the American Statistical Association*, Vol. 67, 641-646.

Solomon, D.L. (1972b). "Λ-Minimax Estimation of a Scale Parameter," *Journal of the American Statistical Association*, Vol. 67, 647-649.

Truchaev, R.I. (1981). *Modeli Priniatia Rešenij V Uslovijach Neopredelennosti.* Moskva: Nauka.

Zacks, S. (1971). *The Theory of Statistical Inference* . New York: J. Wiley.

Zellner, A. (1971). *An Introduction to Bayesian Inference in Econometrics.* New York: J. Wiley.

Bayesian and Likelihood Methods in Statistics and Econometrics
S. Geisser, J.S. Hodges, S.J. Press and A. Zellner (Editors)

BAYESIAN ANALYSIS WITH INCOMPLETELY SPECIFIED PRIOR DISTRIBUTIONS

Albert Madansky
The University of Chicago

Abstract

This paper develops Bayesian procedures that use two kinds of incompletely specified prior distributions. One of these priors uses knowledge about some quantiles of the prior distribution, the second uses monotonicity of the prior. The methods are developed for the case of the parameter in a Bernoulli model.

1. Introduction

A curious and quite interesting statistical problem is the practical analogue of the phenomenon embodied in the question raised by Laplace "given that the sun has arisen every day in the past, what is the probability that it will arise tomorrow?". The current analogue, with slightly smaller n, can be posed in a drug-testing context "Given that we have administered a drug to n randomly-selected individuals and found the drug not to be harmful, what is the probability that the drug is not harmful?". The frequentist has no good answer to this question. Assuming that θ is the unknown parameter of the Bernoulli trials by which we model the n observations, we see that our maximum likelihood estimate of θ is $n/n = 1$, and that the usual approximate confidence interval around θ is the point 1.[1]

The Bayesian who adopts a uniform prior distribution for θ would estimate θ by $\hat{\theta} = (n+1)/(n+2)$, and the posterior distribution would be a Beta distribution

[1]This is the result of formally using the normal approximation to the binomial to develop a confidence interval. The approximate upper confidence bound is $\hat{\theta}_u = \hat{\theta} + z_\gamma\sqrt{\hat{\theta}(1-\hat{\theta})/n}$, where z_γ is an appropriate percentile of the normal distribution and $\hat{\theta}$ is the maximum likelihood estimate of θ. When $\hat{\theta} = 1$, this reduces to the point 1. An exact interval consists of those values of θ which would have been accepted as null hypotheses given our observations. If we observe n successes in n trials, our lower confidence limit is $\gamma^{1/n}$, where γ is the confidence coefficient. For $\gamma = .95$ and n = 50, for example, the limit is .9989747 and for n = 10 it is still quite large, namely .9948838.

with parameters $(n+1, n+2)$. Thus a credible interval around θ could be calculated from this posterior distribution.[2] But what if the Bayesian does not wish to stick his neck out and postulate a uniform, or any other prior distribution? What kinds of inferences can he make about θ without postulating a prior distribution?

One response to such questions is to seek out robust Bayesian procedures, such as those described in the survey paper by Berger (1984) and the more recent papers cited in O'Hagan and Berger (1988). Another response to such questions is to seek out Bayesian procedures that incorporate knowledge about some quantiles of the prior distribution, without committing oneself to a full specification of a prior density. We consider such an approach in Section 1. Yet another response to such questions is to incorporate knowledge about the shape of the prior distribution. The main part of this paper studies the implications of one bit of prior knowledge, namely that the prior density is monotonically nondecreasing, on the posterior distribution of θ.

2. Incorporating Partial Prior Information

In general, let θ be a parameter whose range is the real line. Suppose the intervals $I_1, I_2, ..., I_m$ are contiguous intervals partitioning the real line, with $I_i = (a_{i-1}, a_i)$ and let $f_i^0(\theta)$ be that part of the prior density $f_{\boldsymbol{\theta}}^0(\theta)$ defined on I_i. Let the set C be a union of intervals I_i where the i's are some subset Ω of the indices $\{1, 2, ..., m\}$. Define

$$\omega_i = \int_{a_{i-1}}^{a_i} f_i^0(\theta)\ell(\theta)d\theta,$$

where $\ell(\theta)$ is the likelihood function of a set of observations (and where we suppress these observations as arguments of ℓ in order to highlight the argument θ). Let

$$\overline{F}^1(C) = \left[1 + \frac{\sum_{i\notin\Omega} \underline{\omega}_i}{\sum_{i\in\Omega} \overline{\omega}_i}\right]^{-1}$$

and

$$\underline{F}^1(C) = \left[1 + \frac{\sum_{i\notin\Omega} \overline{\omega}_i}{\sum_{i\in\Omega} \underline{\omega}_i}\right]^{-1}$$

where

$$\overline{\omega}_i = \sup \omega_i, \ \ \underline{\omega}_i = \inf \omega_i,$$

[2] Jeffreys (see Jeffreys, 1939 and Wrinch and Jeffreys, 1921) considered this problem from a different point of view, summarized in Geisser (1980). He would estimate θ by $\hat{\theta} = (n+1)/(n+3)$, given n successes in n trials.

where the sup and inf are taken over the set of "permissible" prior densities. Then, as has been shown in O'Hagan and Berger (1988) (see also Berliner and Goel, 1986), $F^1(C)$, the posterior probability of the set C, is bounded by $\underline{F}^1(C)$ and $\overline{F}^1(C)$.

In our specific Bernoulli trials case, let $f^0_{\boldsymbol{\theta}}(\theta)$ be the prior density of $\boldsymbol{\theta}$, $f^1_{\boldsymbol{\theta}}(\theta)$ be the posterior density of $\boldsymbol{\theta}$, and suppose we have n observations with 0 failures. Then $\ell(\theta) = \theta^n$ and

$$f^1_{\boldsymbol{\theta}}(\theta) = \frac{\theta^n f^0_{\boldsymbol{\theta}}(\theta)}{\int_0^1 \theta^n f^0_{\boldsymbol{\theta}}(\theta) d\theta}$$

Perhaps the Bayesian is willing to postulate two points on the prior cumulative distribution function. Suppose he says that, for $\theta_1 < \theta_2$,

$$F^0_{\boldsymbol{\theta}}(\theta_1) \leq p_1, 1 - F^0_{\boldsymbol{\theta}}(\theta_2) \geq p_2.$$

What are the implications for $F^1_{\boldsymbol{\theta}}(\theta)$?

This corresponds to the situation in which the parameter space $[0,1]$ is divided into three intervals, $I_1 = [0,\theta_1], I_2 = (\theta_1,\theta_2)$, and $I_3 = [\theta_2, 1]$, where $\Omega = I_1, \overline{\omega}_1 = p_1\theta_1^n$, and $\underline{\omega}_3 = p_2\theta_2^n$. Since we have set no bounds on the prior probability of being in I_2, we take $\underline{\omega}_2 = 0$, so that

$$F^1_{\boldsymbol{\theta}}(\theta_1) \leq \frac{1}{1 + (\theta_2/\theta_1)^n (p_2/p_1)}$$

This bound is not very strong. Table 1 compares this bound with the value of $F^1_{\boldsymbol{\theta}}(\theta_1)$ assuming a uniform prior distribution for selected values of θ_1 and n, assuming that $\theta_2 = .9999$.

3. Monotonically Increasing Prior Density Function

Let us now consider the implication of another kind of statement that a Bayesian might make about the prior density of $\boldsymbol{\theta}$ without committing himself to an explicit statement of the functional form of $f^0_{\boldsymbol{\theta}}(\theta)$. Suppose we are willing to assume that $f^0_{\boldsymbol{\theta}}(\theta)$ is monotonically increasing in θ (or equivalently that $F^0_{\boldsymbol{\theta}}(\theta)$is a convex function of θ). Such a density reflects the subjective assignment of an ever-increasing probability to $\boldsymbol{\theta}$ equalling θ as θ gets closer to 1 in our drug-testing (or sun-rising) example. (O'Hagan and Berger, 1987 and Berger and O'Hagan, 1987 consider two other restrictions on the shape of the prior density, namely that they be unimodal and quasi-unimodal, and derive bounds for the posterior distribution. In one sense the monotone prior density case adds a greater restriction to the class of densities that they deal with, and so we should be able to obtain sharper results. But they make the added assumption

Table 1: Comparison of Posterior and Upper Bound on Posterior Distribution

$F_{\boldsymbol{\theta}}^1(\theta)$ assuming uniform prior

	θ		
n	.9	.99	.999
10	.314	.895	.989
100	.239 $\times 10^{-4}$	.362	.999
1000	0	.427 $\times 10^{-4}$	.367
10000	0	0	.451 $\times 10^{-4}$

Upper Bound on $F_{\boldsymbol{\theta}}^1(\theta_1)$

$\theta_2 = .9999,\ p_1 = \theta_1,\ p_2 = .0001$

	θ_1		
n	.9	.99	.999
10	.99968	.99989	.99990
100	.19449	.99973	.99989
1000	*	.32082	.99975
10000	*	*	.55095

* - less than 10^{-34}

that there are specified quantiles, whereas we make no such assumptions, and so our results are not directly comparable.)

Let $G_{\boldsymbol{\theta}}^1(\theta) = \int_0^\theta x^n f_{\boldsymbol{\theta}}^0(x)dx$ and $\overline{G}_{\boldsymbol{\theta}}^1(\theta) = \int_\theta^1 x^n f_{\boldsymbol{\theta}}^0(x)dx$ i.e., $G_{\boldsymbol{\theta}}^1(\theta)/G_{\boldsymbol{\theta}}^1(1)$ is the posterior probability that $\boldsymbol{\theta}$ is at most θ and $\overline{G}_{\boldsymbol{\theta}}(\theta)/G_{\boldsymbol{\theta}}^1(1)$ is the posterior probability that $\boldsymbol{\theta}$ is at least θ conditional on observing n successes in n Bernoulli trials. By monotonicity,

$$G_{\boldsymbol{\theta}}^1(\theta) \leq f_{\boldsymbol{\theta}}^0(\theta) \int_0^\theta x^n dx = f_{\boldsymbol{\theta}}^0(\theta)\theta^{n+1}/(n+1)$$

and

$$\overline{G}_{\boldsymbol{\theta}}^1(\theta) \geq f_{\boldsymbol{\theta}}^0(\theta) \int_\theta^1 x^n dx = f_{\boldsymbol{\theta}}^0(\theta)(1-\theta^{n+1})/(n+1),$$

so that the posterior odds ratio $R_{\boldsymbol{\theta}}^1(\theta) = G_{\boldsymbol{\theta}}^1(\theta)/\overline{G}_{\boldsymbol{\theta}}^1(\theta)$ is bounded by

$$R_{\boldsymbol{\theta}}^1(\theta) \leq \frac{\theta^{n+1}}{1-\theta^{n+1}}$$

But this upper bound is the odds-ratio for the uniform distribution. Thus the uniform density provides an extreme case of a monotone density, in that it produces an upper bound on the odds-ratio.

One can improve on this bound by utilizing a property of monotonic densities (actually, an implication of the convexity of the distribution function, see Theorem 25 of Eggleston (1958)), namely that if $\theta \leq \theta^*$,

$$F^0_{\boldsymbol{\theta}}(\theta)/\theta \leq F^0_{\boldsymbol{\theta}}(\theta^*)/\theta^*.$$

We use this result to obtain bounds on the posterior odds ratio $R^1_{\boldsymbol{\theta}}(\theta)$ by bounding separately the numerator and denominator of $F^1_{\boldsymbol{\theta}}(\theta)$ and combining these bounds to bound the ratio. Remembering that $f^0_{\boldsymbol{\theta}}(x)dx = dF^0_{\boldsymbol{\theta}}(x)$, we note first that integration by parts yields the expression

$$G^1_{\boldsymbol{\theta}}(\theta) = \theta^n F^0_{\boldsymbol{\theta}}(\theta) - \int_0^\theta n x^{n-1} F^0_{\boldsymbol{\theta}}(x)dx$$

whereupon, applying the aforementioned property of monotonic densities, we see that

$$\begin{aligned} G^1_{\boldsymbol{\theta}}(\theta) &\geq \theta^n F^0_{\boldsymbol{\theta}}(\theta) - n\int_0^\theta (x^n/\theta) F^0_{\boldsymbol{\theta}}(\theta)dx \\ &= \frac{\theta^n}{n+1} F^0_{\boldsymbol{\theta}}(\theta) \end{aligned}$$

due to the monotonicity assumption.

Since the monotonicity assumption also implies that, if $\theta < \theta^*$,

$$\frac{1 - F^0_{\boldsymbol{\theta}}(\theta)}{1-\theta} \leq \frac{1 - F^0_{\boldsymbol{\theta}}(\theta^*)}{1-\theta^*}$$

to bound $\overline{G}^1_{\boldsymbol{\theta}}(\theta)$, we use the same form of argument, so that

$$\begin{aligned} \overline{G}^1_{\boldsymbol{\theta}}(\theta) = \int_0^1 x^n f^0_\theta(x)dx &= \theta^n[1 - F^0_{\boldsymbol{\theta}}(\theta)] + n\int_\theta^1 x^{n-1}[1 - F^0_{\boldsymbol{\theta}}(x)]dx \\ &\geq \theta^n[1 - F^0_{\boldsymbol{\theta}}(\theta)] + n\int_\theta^1 x^{n-1}\frac{[1 - F^0_{\boldsymbol{\theta}}(\theta)]}{1-\theta}(1-x)dx \\ &= \frac{(1-\theta^{n+1})(1 - F^0_{\boldsymbol{\theta}}(\theta))}{(1-\theta)(n+1)} \end{aligned}$$

In addition we can obtain a general upper bound on $G^1_{\boldsymbol{\theta}}(\theta)$, namely $G^1_{\boldsymbol{\theta}}(\theta) \leq \theta^n F^0_{\boldsymbol{\theta}}(\theta)$. Moreover, generally,

$$\overline{G}^1_{\boldsymbol{\theta}}(\theta) \leq 1 - F^0_{\boldsymbol{\theta}}(\theta).$$

Putting these together, we obtain the following bounds on the posterior odds ratio $R^1_{\boldsymbol{\theta}}(\theta)$:

$$\frac{(n+1)(1-\theta)\theta^n}{1-\theta^{n+1}} R^0_{\boldsymbol{\theta}}(\theta) \geq R^1_{\boldsymbol{\theta}}(\theta) \geq \frac{\theta^n}{n+1} R^0_{\boldsymbol{\theta}}(\theta)$$

When the prior distribution is uniform, $R^0_{\boldsymbol{\theta}}(\theta) = \theta/(1-\theta)$, so that this upper bound turns out to be $(n+1)$ times the odds-ratio for the uniform distribution, which, as seen above, is itself an upper bound for $R^1_{\boldsymbol{\theta}}(\theta)$. But there are other prior distributions for which this bound is a better upper bound than that based on the uniform distribution, as will be seen below. Selected values of these bounds (actually, bounds on $R^1_{\boldsymbol{\theta}}(\theta)/R^0_{\boldsymbol{\theta}}(\theta)$) are given in Table 2, along with values of $R^1_{\boldsymbol{\theta}}(\theta)/R^0_{\boldsymbol{\theta}}(\theta)$) for the uniform distribution for comparison purposes.

These bounds are not the best possible bounds obtainable based on assuming monotonicity of $f^0_{\boldsymbol{\theta}}(\theta)$, since we only used this assumption to establish lower bounds on $G^1_{\boldsymbol{\theta}}(\theta)$ and $\overline{G}^1_{\boldsymbol{\theta}}(\theta)$, and used general upper bounds for these quantities. Indeed, these bounds are attained for prior densities $f^0_{\boldsymbol{\theta}}(\theta)$ which are not monotonic. This can be seen most readily when one recognizes that the lower bounds based on the assumption of monotonicity are attained by uniform prior densities for θ over the range $(\theta_0,1)$ and the upper bounds are attained by "spike" prior densities, i.e., densities whose probability mass is concentrated at a single point (either 1 for the bound on $\overline{G}^1_{\boldsymbol{\theta}}(\theta)$ or θ_0 for the bound on $G^1_{\boldsymbol{\theta}}(\theta)$). The density

$$f^0_{\boldsymbol{\theta}}(\theta) = \begin{cases} 0 & \theta < \theta_0 \\ \frac{1-\lambda}{1-\theta_0} & \theta_0 \leq \theta < 1 \\ \lambda & \theta = 1 \end{cases}$$

produces a posterior odds ratio which attains the lower bound, and the density

$$f^0_{\boldsymbol{\theta}}(\theta) = \begin{cases} 0 & \theta < \theta_0 \\ \lambda & \theta = \theta_0 \\ \frac{1-\lambda}{1-\theta_0} & \theta > \theta_0 \end{cases}$$

produces a posterior odds ratio which attains the upper bound, when we observe n out of n successes.

What are the extreme types of monotonic prior density functions, i.e., functions $f^0_{\boldsymbol{\theta}}(\theta)$ such that the resulting $R^1_{\boldsymbol{\theta}}(\theta)$ are maximized or minimized? At first blush this appears to be a problem in the calculus of variations. However, if one sets up

Table 2: Bounds Based Partially on Monotonicity of Prior Density

Lower Bound on $R^1_\theta(\theta)/R^0_\theta(\theta)$

	θ		
n	.9	.99	.999
10	.0317	.0822	.0900
100	$.2360 \times 10^{-4}$	.0036	.0090
1000	*	$.4313 \times 10^{-7}$	$.3673 \times 10^{-3}$
10000	*	*	$.4517 \times 10^{-8}$

Upper Bound on $R^1_\theta(\theta)/R^0_\theta(\theta)$

	θ		
n	.9	.99	.999
10	.5590	.9505	.9950
100	$.2689 \times 10^{-3}$	.5798	.9508
1000	*	$.4321 \times 10^{-4}$	.5818
10000	*	*	$.4517 \times 10^{-3}$

$R^1_\theta(\theta)/R^0_\theta(\theta)$ assuming uniform prior

	θ		
n	.9	.99	.999
10	.0508	.0864	.0905
100	$.2656 \times 10^{-5}$	.0057	.0094
1000	*	$.4317 \times 10^{-6}$	$.5811 \times 10^{-3}$
10000	*	*	$.4517 \times 10^{-7}$

* - less than 10^{-34}

this problem as an isoperimetric problem one finds that the Euler equation for this problem is degenerate, and so one must resort to special arguments to determine the extremal densities.

The posterior odds ratio is given by

$$R^1_\theta(\theta) = \frac{\int_0^\theta x^n f^0_\theta(x)dx}{\int_0^1 x^n f^0_\theta(x)dx - \int_0^\theta x^n f^0_\theta(x)dx}$$

and so, for given value of $\beta = \int_0^1 x^n f^0_\theta(x)dx, R^1_\theta(\theta)$ is maximized or minimized if $F^1_\theta(\theta) = \int_0^\theta x^n f^0_\theta(x)dx$ is. Moreover, since we would like to express $R^1_\theta(\theta)$ as a function of $R^0_{\boldsymbol{\theta}}(\theta)$, we constrain $F^0_{\boldsymbol{\theta}}(\theta)$ to equal α, so that $R^0_{\boldsymbol{\theta}}(\theta) = \alpha/(1-\alpha)$.

Let $g_\theta(x) = \alpha$ for $x < \theta$. Then

$$\int_0^\theta x^n g_\theta(x)dx = \alpha\frac{\theta^{n+1}}{n+1} \leq \int_0^\theta x^n f^0_{\boldsymbol{\theta}}(x)dx = G^1_{\boldsymbol{\theta}}(\theta),$$

as shown above. Therefore the minimizing $f^0_{\boldsymbol{\theta}}(x)$ must equal $g_\theta(x)$ for $x < \theta$. The odds-ratio is then

$$\frac{\alpha\theta^{n+1}/(n+1)}{\beta - \alpha\theta^{n+1}/(n+1)}$$

which is minimized if β is maximized. By monotonicity of $f^0_{\boldsymbol{\theta}}(x)$, $f^0_{\boldsymbol{\theta}}(x) \geq \alpha$ for $x \geq \theta$. Moreover, β is maximized subject to this constraint by keeping $f^0_{\boldsymbol{\theta}}(x) = \alpha$ for $x < 1$ and setting $f^0_{\boldsymbol{\theta}}(1) = \lambda$, a "spike" at $x = 1$.

Since $\int_0^1 f^0_{\boldsymbol{\theta}}(x)dx = 1$, we see that $\lambda = 1-\alpha, \beta = 1 - n\alpha/(n+1)$, and the minimum odds-ratio is

$$R^1_{\boldsymbol{\theta}}(\theta) = \frac{\alpha\theta^{n+1}}{(n+1) - n\alpha - \alpha\theta^{n+1}}.$$

But

$$R^0_{\boldsymbol{\theta}}(\theta) = \frac{\alpha\theta}{1-\alpha\theta},$$

so that α can be expressed as

$$\alpha = \frac{R^0_{\boldsymbol{\theta}}(\theta)}{\theta(1 + R^0_{\boldsymbol{\theta}}(\theta))}.$$

Thus a lower bound on $R^1_{\boldsymbol{\theta}}(\theta)$ in terms of $R^0_{\boldsymbol{\theta}}(\theta)$ is

$$R^1_{\boldsymbol{\theta}}(\theta) \geq \frac{\theta^{n+1}}{(n+1) - n - \theta^{n+1} + (n+1)\theta/R^0_{\boldsymbol{\theta}}(\theta)}$$

attainable by the monotonic density which maximized β above, $f^0_{\boldsymbol{\theta}}(x) = \alpha$ for $x < 1$ and $f^0_{\boldsymbol{\theta}}(1) = \lambda$, i.e., a density which is uniform on (0,1) with a "spike" at $x = 1$.

To obtain an upper bound for $R^1_{\boldsymbol{\theta}}(\theta)$, we first note that this is accomplished by maximizing $F^1_{\boldsymbol{\theta}}(\theta)$ subject to these constraints. Now let $g_\theta(x) = 1-\alpha$ for $\theta < x$. Then

$$\int_\theta^1 x^n g_\theta(x)dx = \frac{(1-\alpha)(1-\theta^{n+1})}{n+1} \leq \int_\theta^1 x^n f^0_{\boldsymbol{\theta}}(x)dx = \overline{G}^1_{\boldsymbol{\theta}}(x),$$

as seen above. Therefore the maximizing $f^0_{\boldsymbol{\theta}}(x)$ must equal $g_\theta(x)$ for $x > \theta$. The odds-ratio is then

$$\frac{1-(1-\alpha)(1-\theta^{n+1})/(n+1)}{\beta-(1-\alpha)(1-\theta^{n+1})/(n+1)},$$

which is maximized if β is minimized. By monotonicity of $f_\theta^0(x)$, $f_\theta^0(x) \leq 1-\alpha$ for $x \leq \theta$. Moreover, β is minimized subject to this constraint by making $f_\theta^0(x)$ as small as possible, subject to the constraint that

$$\int_0^\theta f_\theta^0(x)dx = \alpha$$

If $f_\theta^0(x)$ is to distribute its mass on the entire interval (0, θ), the minimizing $f_\theta^0(x)$ must be uniform. But $f_\theta^0(x) = \mathrm{c}$ for $x < \theta$ implies that

$$1 = \int_0^\theta f_\theta^0(x)dx + \int_\theta^1 f_\theta^0(x)dx$$

$$= c\theta + (1-\alpha)(1-\theta),$$

or

$$c = 1 - \alpha + \alpha/\theta$$

This contradicts the monotonicity constraint, $f_\theta^0(x) \leq 1-\alpha$ for $x \leq \theta$. Thus either $f_\theta^0(x)$ is uniform on (0, 1) or its mass is distributed uniformly on (λ, θ) and on (θ, 1). Then

$$1 = c(\theta - \lambda) + (1-\alpha)(1-\theta)$$

implies that

$$c = \frac{\theta + \alpha(1-\theta)}{\theta - \lambda}.$$

But $c \leq 1 - \alpha$ if and only if

$$\theta + \alpha(1-\theta) \leq (\theta-\lambda)(1-\alpha)$$

or

$$\alpha \leq \lambda(\alpha - 1),$$

which is impossible. The minimizing $f_\theta^0(x)$ is, then, a uniform density, $f_\theta^0(x) = (1-\lambda)^{-1}$ for $\lambda \leq x \leq 1$ and 0 otherwise, where $\lambda < \theta$.

Then the maximum odds-ratio is

$$R_\theta^1(\theta) = \frac{\theta^{n+1} - \lambda^{n+1}}{1 - \theta^{n+1}},$$

and, since $R_\theta^0(\theta) = (\theta - \lambda)/(1-\theta)$,

$$R_\theta^1(\theta) \leq \frac{\theta^{n+1} - [\theta - R_\theta^0(\theta)(1-\theta)]^{n+1}}{1-\theta^{n+1}}$$

Table 3: Bounds Based Completely on Monotonicity of Prior Density

Lower Bound on $R^1_{\boldsymbol{\theta}}(\theta)$ assuming $R^0_{\boldsymbol{\theta}}(\theta) = 1$

	θ		
n	.9	.99	.999
10	.0331	.0823	.0900
100	$.2922 \times 10^{-6}$	.0036	.0090
1000	*	$.4532 \times 10^{-7}$	$.3675 \times 10^{-3}$
10000	*	*	$.4522 \times 10^{-8}$

Upper Bound on $R^1_{\boldsymbol{\theta}}(\theta)$ assuming $R^0_{\boldsymbol{\theta}}(\theta) = 1$

	θ		
n	.9	.99	.999
10	.3321	.9039	.9900
100	$.2391 \times 10^{-4}$	.3645	.9047
1000	*	$.4274 \times 10^{-4}$	.3675
10000	*	*	$.4513 \times 10^{-4}$

* - less than 10^{-34}

Notice that when $R^0_{\boldsymbol{\theta}}(\theta) = \theta/(1-\theta)$, i.e., when the prior distribution of $\boldsymbol{\theta}$ is uniform, then the lower and upper bounds are equal to $\theta^{n+1}/(1-\theta^{n+1})$, the posterior odds ratio for the uniform distribution. As $R^0_{\boldsymbol{\theta}}(\theta)$ approaches 0, both the upper and lower limits approach 0. Unlike the bounds given in Table 2, these bounds are not multiplicative factors of $R^0_{\boldsymbol{\theta}}(\theta)$. We can see from an examination of the form of these bounds that they are tighter than those given by Table 2. We present in Table 3 values of the minimum and maximum posterior odds-ratio for various values of θ when $R^0_{\boldsymbol{\theta}}(\theta) = 1$, i.e., when *a priori* we would give even odds that the true value of θ is at most the value tabulated. A comparison with Table 2 shows very little tightening of the odds ratio by these bounds, except for a few cases where the upper bound for the posterior odds ratio is reduced by about 50%. Comparing Table 4, containing values of the minimum and maximum posterior odds ratio when the prior odds ratio is .1, with .1 times the values of Table 2 yields similar results.

Table 4: Bounds Based Completely on Monotonicity of Prior Density

Lower Bound on $R^1_{\boldsymbol{\theta}}(\theta)$ assuming $R^0_{\boldsymbol{\theta}}(\theta) = .1$

θ

10	.00318	.00822	.00900
100	$.2656 \times 10^{-7}$	.00036	.00090
1000	*	$.4317 \times 10^{-8}$	$.3673 \times 10^{-4}$
10000	*	*	$.4518 \times 10^{-9}$

Upper Bound on $R^1_{\boldsymbol{\theta}}(\theta)$ assuming $R^0_{\boldsymbol{\theta}}(\theta) = .1$

θ

n	.9	.99	.999
10	.05289	.09457	.09947
100	$.1617 \times 10^{-4}$	.05515	.09462
1000	*	$.2720 \times 10^{-4}$	.05537
10000	*	*	$.2855 \times 10^{-4}$

* - less than 10^{-34}

4. Generalizations

One can of course impose other conditions on $f^0_{\boldsymbol{\theta}}(\theta)$ and develop implications from these conditions and hence alternative bounds on the posterior probability of θ and/or the odds-ratio. One such is that

$$\frac{d \log f^0_{\boldsymbol{\theta}}(\theta)}{d \log \theta} \geq c \geq 0,$$

i.e., the prior density can rise rapidly but not fall rapidly. Then

$$\log f^0_{\boldsymbol{\theta}}(\theta) - \log f^0_{\boldsymbol{\theta}}(\lambda) \gtreqless c[\log \theta - \log \lambda] \text{ as } \theta \gtreqless \lambda$$

or

$$f^0_{\boldsymbol{\theta}}(\theta) \gtreqless f^0_{\boldsymbol{\theta}}(\lambda)(\theta/\lambda)^c \text{ as } \theta \gtreqless \lambda.$$

(This condition is a generalization of the monotonicity condition.) Then

$$R^1_{\boldsymbol{\theta}}(\theta) = \frac{\int_0^\theta x^n f^0_{\boldsymbol{\theta}}(x)dx}{\int_\theta^1 x^n f^0_{\boldsymbol{\theta}}(x)dx}$$

$$\leq \frac{\int_0^\theta x^n (x/\theta)^c f^0_{\boldsymbol{\theta}}(x)dx}{\int_\theta^1 x^n (x/\theta)^c f^0_{\boldsymbol{\theta}}(x)dx}$$

$$= \frac{\theta^{n+c+1}}{1-\theta^{n+c+1}}.$$

More generally, if in n trials we observe m successes, then the same argument shows that $R^1_{\boldsymbol{\theta}}(\theta)$ is bounded by the posterior odds-ratio based on assuming a $B(c+1,1)$ prior.

The argument that led to the first set of bounds generalizes to any likelihood function $\ell(y;\theta)$ on θ in [0,1], for then

$$G^1_{\boldsymbol{\theta}}(\theta) = \int_0^\theta \ell(y;x) f^0_{\boldsymbol{\theta}}(x)dx$$

$$= \ell(y;\theta)F^0_{\boldsymbol{\theta}}(\theta) - \int_0^\theta \ell'(y;x)F^0_{\boldsymbol{\theta}}(x)dx$$

$$\geq F^0_{\boldsymbol{\theta}}(\theta)[\ell(y;\theta) - \frac{1}{\theta}\int_0^\theta \ell'(y;x)x dx]$$

$$= F^0_{\boldsymbol{\theta}}(\theta)[\ell(y;\theta) - \frac{1}{\theta}\theta\ell(y;\theta) + \frac{1}{\theta}\int_0^\theta \ell(y;x)dx]$$

$$= F^0_{\boldsymbol{\theta}}(\theta)\frac{1}{\theta}\int_0^\theta \ell(y;x)dx,$$

where $\ell'(y;x) = \partial\ell(y;x)/\partial x$.

A similar approach to $\overline{G}^1_{\boldsymbol{\theta}}(\theta)$ yields

$$\overline{G}^1_{\boldsymbol{\theta}}(\theta) = \ell(y;\theta)[1 - F^0_{\boldsymbol{\theta}}(\theta)] + \int_\theta^1 \ell'(y;x)[1 - F^0_{\boldsymbol{\theta}}(x)]dx$$

$$\geq [1 - F^0_{\boldsymbol{\theta}}(\theta)]\ell(y;\theta) + \int_\theta^1 \ell'(y;x)\frac{[1 - F^0_{\boldsymbol{\theta}}(\theta)]}{1-\theta}(1-x)dx$$

$$= [1 - F^0_{\boldsymbol{\theta}}(\theta)][\ell(y;\theta) - \frac{1-\theta}{1-\theta}\ell(y;\theta) + \frac{1}{1-\theta}\int_\theta^1 \ell(y;x)dx]$$

$$= \frac{[1 - F^0_{\boldsymbol{\theta}}(\theta)]}{1-\theta}\int_\theta^1 \ell(y;x)dx.$$

The bounds $G_{\theta}^{1}(\theta) \leq \theta^{n} F_{\theta}^{0}(\theta)$ and $\overline{G}_{\theta}^{1}(\theta) \leq 1 - F_{\theta}^{0}(\theta)$ generalize in the following way. Let θ_1^* maximize ℓ(y;x) in the interval $0 \leq x \leq \theta$ and θ_2^* maximize ℓ(y;x) in the interval $\theta \leq x \leq 1$. Then

$$G_{\theta}^{1}(\theta) \leq \ell(y;\theta_1^*)F_{\theta}^{0}(\theta)$$

and

$$\overline{G}_{\theta}^{1}(\theta) \leq \ell(y;\theta_2^*)[1 - F_{\theta}^{0}(\theta)],$$

so that

$$\frac{(1-\theta)\ell(y;\theta_1^*)}{\int_\theta^1 \ell(y;x)dx} \geq \frac{R_{\theta}^{1}(\theta)}{R_{\theta}^{0}(\theta)} \geq \frac{\int_0^\theta \ell(y;x)dx}{\theta\ell(y;\theta_2^*)}$$

For example, if $\ell(y;\theta) = \binom{n}{y}\theta^y(1-\theta)^{n-y}$, then, when $\theta < y/n$, $\theta_1^* = \theta$ and $\theta_2^* = y/n$, and, when $\theta > y/n$, $\theta_1^* = y/n$ and $\theta_2^* = \theta$.

Acknowledgments

The author benefited from many discussions in 1957-1960 with L.J. Savage about this problem. Savage treated this problem in lectures given at the University of Chicago in 1959-1960, and presented the bounds leading to Tables 1 and 2 in those lectures, essentially using the same arguments. I am indebted to Harry Roberts for providing me with his notes of the Savage lectures, along with his and other participants' comments on my results as presented to the University of Chicago Econometrics and Statistics Colloquium in 1976. I am further indebted to the referees for acquainting me with recent unpublished literature on this problem area, and for helping me to make this paper clearer than was its original version. The current lack of clarity of this paper is due entirely to me.

References

Berger, J.O. (1984). "The Robust Bayesian Viewpoint," in *Robustness of Bayesian Analysis*, J.B. Kadane, ed., Amsterdam: North-Holland, 63-144.

Berger, J.O. and A. O'Hagan (1987). "Ranges of Posterior Probabilities for Unimodal Priors with Specified Quantiles," Technical Report #86-59, Department of Statistics, Purdue University. (To appear in J.M. Bernardo, M.H. DeGroot, D.V. Lindley, and A.F.M. Smith, eds., *Bayesian Statistics III*, Oxford: Oxford University Press.)

Berliner, L.M. and P. Goel (1986). "Incorporating Partial Prior Information: Ranges of Posterior Probabilities," Technical Report 357, Department of Statistics, Ohio State University.

Eggleston, H.G. (1958). *Convexity.* Cambridge: Cambridge University Press.

Geisser, S. (1980). "The Contributions of Sir Harold Jeffreys to Bayesian Inference," in A. Zellner, ed., *Bayesian Analysis in Econometrics and Statistics: Essays in Honor of Harold Jeffreys.* Amsterdam: North-Holland, 13-20.

Jeffreys, H. (1939). *Theory of Probability.* Oxford: Oxford University Press.

O'Hagan, A. and J.O. Berger (1988). "Ranges of Posterior Probabilities for Quasi-unimodal Priors with Specified Quantiles," *Journal of the American Statistical Association*, Vol. 83, 503-508.

Wrinch, D.H. and H. Jeffreys (1921). "On Certain Fundamental Principles of Scientific Inquiry," *Philosophical Magazine*, Vol. 42, 369-390.

VIII. APPROXIMATIONS AND COMPUTATIONAL METHODS

Bayesian and Likelihood Methods in Statistics and Econometrics
S. Geisser, J.S. Hodges, S.J. Press and A. Zellner (Editors)

ASPECTS OF REPARAMETRIZATION IN APPROXIMATE BAYESIAN INFERENCE

Jorge A. Achcar and Adrian F. M. Smith
Federal University of São Carlos, SP, BRAZIL and University of Nottingham, UK

Abstract

Recent proposals for approximating posterior moments and densities using Laplace and related methods are reviewed and the effects of reparametrization on the accuracy of such approximations are illustrated with selected examples.

Key Words — Bayesian inference, Laplace approximation, parametrization.

1. Introduction

Given data x and specified likelihood and prior density functions, $l(\boldsymbol{\theta} \mid x)$ and $\pi(\boldsymbol{\theta})$, for $\boldsymbol{\theta}$, Bayesian inference summaries for (possibly transformed) components of $\boldsymbol{\theta}$ typically require the evaluation of integrals of the form

$$E\big(g(\boldsymbol{\theta}) \mid x\big) = \int g(\boldsymbol{\theta})\pi(\boldsymbol{\theta} \mid x)d\boldsymbol{\theta}, \tag{1}$$

for selected real-valued functions $g(\boldsymbol{\theta})$, where

$$\pi(\boldsymbol{\theta} \mid x) = \frac{l(\boldsymbol{\theta} \mid x)\pi(\boldsymbol{\theta})}{\int l(\boldsymbol{\theta} \mid x)\pi(\boldsymbol{\theta})d\boldsymbol{\theta}}. \tag{2}$$

In general, the integrations required for (1) and (2) cannot be carried out exactly using analytic methods and some form of approximation is necessary. One possibility is to use numerical integration methods (see, for example, Naylor and Smith, 1982, and Smith *et al.*, 1987); another is to use analytic approximation techniques for integrals, or ratios of integrals. In the context of numerical integration strategies, the papers cited clearly demonstrate that careful consideration of the parametrization used in defining the likelihood function is an essential prerequisite for the successful development of numerical approximations. In the context of analytic approximation strategies, although there has been a recent upsurge of interest in parametrization within the framework of likelihood inference (see, for example, Cox and Reid, 1987), very little seems to have been done within the Bayesian framework.

This paper reports some preliminary, largely numerical, studies into the effects of reparametrization on the accuracy of some recently proposed approximation techniques for Bayesian inference. The techniques are reviewed in Section 2; illustrative analyses of approximations to posterior moments are given in Section 3 for one-parameter likelihoods and in Section 4 for two-parameter likelihoods; some summary concluding remarks follow in Section 5.

2. Approximation Techniques

2.1 Techniques Based on the Laplace Method

We note that (1) can be written in the form

$$E(g(\boldsymbol{\theta}) \mid x) = \int e^{-nh^*(\boldsymbol{\theta})} d\boldsymbol{\theta} \Big/ \int e^{-nh(\boldsymbol{\theta})} d\boldsymbol{\theta}, \tag{3}$$

where, with n-vector x fixed and all functions having argument $\boldsymbol{\theta}$, $-nh = \ln(\pi) + \mathcal{L}$, $-nh^* = \ln(g) + \ln(\pi) + \mathcal{L}$, with $\mathcal{L} = \ln(l)$ denoting the log-likelihood function and π the prior density appearing in (2).

Consider first the case of a uniparameter likelihood, with $\boldsymbol{\theta} = \theta$. If $\hat{\theta}$ maximises $-h$ (i.e., $\hat{\theta}$ is the posterior mode) and $\sigma = [h''(\hat{\theta})]^{-\frac{1}{2}}$, the Laplace method of approximation sets

$$\int e^{-nh(\theta)} d\theta \approx \sqrt{2\pi}\sigma n^{-\frac{1}{2}} e^{-nh(\hat{\theta})} \tag{4}$$

for the denominator of (3). Similarly, if $g(\theta)$ is a suitably smooth positive function, $\hat{\theta}^*$ maximises $-h^*$ and $\sigma^* = [h^{*''}(\hat{\theta}^*)]^{-\frac{1}{2}}$, the Laplace approximation for the numerator of (3) is given by $\sqrt{2\pi}\sigma^* n^{-\frac{1}{2}} \exp\big(-nh^*(\hat{\theta}^*)\big)$. It has then been shown by Tierney and Kadane (1986) that the implied approximation for (1) given by

$$\hat{E}\big(g(\theta) \mid x\big) = (\sigma^*/\sigma) \exp[-n\big(h^*(\hat{\theta}^*) - h(\hat{\theta})\big)] \tag{5}$$

satisfies $E\big(g(\theta) \mid x\big) = \hat{E}\big(g(\theta) \mid x\big)\big(1 + O(n^{-2})\big)$ and thus provides a potentially powerful general technique for approximating integrals of the form (1) (see, also, Tierney, Kass and Kadane, 1986, and Kass, Tierney and Kadane, 1987, for further underpinning and extensions of the method).

In the case of a multiparameter likelihood, with $\boldsymbol{\theta} \in \mathbb{R}^m$, the Laplace approximation to the denominator of (3) is given by

$$\int e^{-nh(\boldsymbol{\theta})} d\boldsymbol{\theta} \approx (2\pi)^{m/2} \mid n\boldsymbol{D}^2 h(\hat{\boldsymbol{\theta}}) \mid^{-\frac{1}{2}} \exp\big(-nh(\hat{\boldsymbol{\theta}})\big), \tag{6}$$

where $\hat{\boldsymbol{\theta}}$ maximises $-h(\boldsymbol{\theta})$ and $\boldsymbol{D}^2 h(\hat{\boldsymbol{\theta}})$ is the Hessian matrix of h evaluated at $\hat{\boldsymbol{\theta}}$. Writing $\sigma = |n\boldsymbol{D}^2 h(\hat{\boldsymbol{\theta}})|^{-\frac{1}{2}}$ and $\sigma^* = |nD^2 h^*(\hat{\boldsymbol{\theta}}^*)|^{-\frac{1}{2}}$ for the analogous term derived from the numerator of (3), the approximation to (3) proposed by Tierney and Kadane takes the form (5), with $\hat{\boldsymbol{\theta}}^*$ and $\hat{\boldsymbol{\theta}}$ replacing $\hat{\theta}^*$ and $\hat{\theta}$.

If $\boldsymbol{\theta} = (\psi, \boldsymbol{\lambda})$ and interest centres on the marginal posterior density for ψ, we see from (2) that applying the Laplace approximation gives

$$\begin{aligned} \pi(\psi \mid \boldsymbol{x}) &\propto \int e^{-nh_\psi(\boldsymbol{\lambda})} d\boldsymbol{\lambda} \\ &\propto |\boldsymbol{D}^2 h_\psi(\hat{\boldsymbol{\lambda}}_\psi)|^{-\frac{1}{2}} \exp\big(-nh_\psi(\hat{\boldsymbol{\lambda}}_\psi)\big), \end{aligned} \tag{7}$$

where $-nh_\psi = \ln(\pi) + \mathcal{L}$, considered as a function of $\boldsymbol{\lambda}$ for fixed ψ, and $\hat{\boldsymbol{\lambda}}_\psi$ maximises $-h_\psi$. The form (7) thus provides (up to proportionality) a pointwise approximation to the ordinates of the marginal posterior density for ψ.

2.2 Techniques Based on the Pearson Family

The Laplace method for integrals is based on a second-order Taylor approximation applied to the exponent term in the integrand and is thus clearly likely to perform best when posterior distributions are in some sense close to normality. Under fairly general conditions this might be expected to occur for very large samples, but may not be true for small or moderate samples, particularly when parameters are defined on ranges other than the real line.

One possible response to this observation is to develop approximation techniques that would be better suited to posterior forms more resembling distributions other than normal, such as gammas or betas. Such an approach has been followed in the one-parameter case by Morris (1988), who takes as his starting point the Pearson family of distributions, characterised by parameters m, μ_0 and a quadratic function Q, and specifying a density for a single parameter θ of the form

$$q_Q(\theta \mid m, \mu_0) = K_Q(m, \mu_0) p(\theta)/Q(\theta),$$

where

$$\begin{aligned} p(\theta) &= \exp\left[-m \int \frac{\theta - \mu_0}{Q(\theta)} d\theta\right], \\ K_Q^{-1}(m, \mu_0) &= \int [p(\theta)/Q(\theta)] d\theta, \\ Q(\theta) &= q_0 + q_1\theta + q_2\theta^2 \end{aligned}$$

and the range of θ is such that $0 < Q(\theta) < \infty$.

It is shown by Morris that, for a given choice of quadratic function Q, a suitable approximation to the integral of a unimodal function $f(\theta)$ is given by

$$\int f(\theta)d\theta \approx f(\hat{\theta})/q_Q(\hat{\theta} \mid \hat{m}, \hat{\theta}), \tag{9}$$

where $\hat{m} = r''(\hat{\theta})Q(\hat{\theta})$ and $\hat{\theta}$ maximises $r(\theta) = \ln[f(\theta)Q(\theta)]$. Details of the forms of K^{-1}, Q and p for familiar cases of the Pearson densities are given in Morris (1988), where it is also shown that the approximation can often be further simplified to the expression

$$\int f(\theta)d\theta \approx \frac{f(\hat{\theta})\sqrt{2\pi}}{[-r''(\hat{\theta})]^{\frac{1}{2}}}. \tag{10}$$

2.3 Parametrization Issues

The following considerations motivate a study of parametrization issues in the context of the kinds of approximate Bayesian inference procedures described in Sections 2.1 and 2.2.

First, we note that although the multiparameter version of the Laplace approximation (5) is invariant under full rank linear transformations of $\boldsymbol{\theta}$, in general the proposed approximation procedures are not invariant to changes in the (otherwise arbitrary) parametrization chosen when specifying the likelihood and prior density functions. Secondly, we note that the Pearson family approach currently only offers an alternative to the Laplace method in the one-parameter case and, in any case, requires a judicious selection of the particular form of Pearson density (or, equivalently, Q function) to be used. Thirdly, in those cases where potential marked non-normality of the posterior density causes one to be wary of applying the Laplace method, appropriate reparametrization not only offers a general alternative to the Pearson family approach, but is also available for the multiparameter case.

In this paper, we shall attempt to highlight and illustrate these issues by studying the explicit forms of approximation derived from the various possible approaches for a selection of one- and two-parameter examples.

3. One-Parameter Illustrations

3.1 The Exponential Likelihood

Suppose $x_1, \ldots, x_n$ are a random sample from an exponential distribution, with parameter $\theta = E(X_i)$, so that the likelihood function has the form $l(\theta \mid x) =$

$\theta^{-n}\exp[-\theta^{-1}n\bar{x}]$, where $n\bar{x} = (x_1 + \ldots + x_n)$. Suppose further that the prior density has the conjugate form $\pi(\theta) \propto \theta^{-\alpha-1}\exp[-\theta^{-1}\beta]$, $\alpha, \beta > 0$, so that the posterior density $\pi(\theta|x)$ has the same form, but with α and β replaced by $\alpha + n$ and $\beta + n\bar{x}$. It is then clear that by considering these latter quantities as the "equivalent" sample size and sufficient statistic, respectively, we can, without loss of generality, study the approximation and reparametrization problem by assuming throughout that the prior $\pi(\theta) \propto \theta^{-1}$ has been specified.

Combining $l(\theta \mid x)$ with this prior, it is easily verified that the exact posterior mean for θ is given by

$$E(\theta \mid x) = n\bar{x}/(n-1). \tag{11}$$

Working with the parametrization of the likelihood in terms of θ, we find, after some algebra, that the Laplace approximation (5), with $g(\theta) = \theta$, yields

$$n^{n-\frac{1}{2}}e\bar{x}/(n+1)^{n-\frac{1}{2}}. \tag{12}$$

Suppose, instead, that we were to work with the reparametrization $\phi = \ln(\theta)$, so that the functions involved in (1) become $g(\phi) = e^{\phi}$, $l(\phi \mid x) = e^{-n\phi}e^{-e^{-\phi}n\bar{x}}$, $\pi(\phi) \propto$ constant. In this new parametrization, again after some algebra, we obtain

$$(n-1)^{n-\frac{3}{2}}e\bar{x}/n^{n-\frac{3}{2}}. \tag{13}$$

Alternatively, using the original parametrization θ, we could consider the Pearson approximation obtained by applying the method of Section 2.2 separately to the integrals I_1, I_2 appearing in the numerator and denominator of $E(\theta \mid x) = I_1/I_2$, where, based on $l(\theta \mid x)$ and the prior,

$$I_1 = \int_0^\infty \theta^{-n}\exp(-\theta^{-1}n\bar{x})d\theta, \qquad I_2 = \int_0^\infty \theta^{-(n+1)}\exp(-\theta^{-1}n\bar{x})d\theta.$$

The choice of quadratic function $Q(\theta) = \theta$ corresponds to the gamma form as the choice of a Pearson density and leads to the approximation

$$(n-1)^{n-\frac{3}{2}}e\bar{x}/n^{n-\frac{3}{2}}, \tag{14}$$

which, we observe, coincides with the Laplace approximation in the reparametrization $\phi = \ln(\theta)$. The choice $Q(\theta) = \theta^2$ corresponds to the reciprocal gamma form (see Morris, 1988) and yields the approximation

$$(n-2)^{n-\frac{3}{2}}en\bar{x}/(n-1)^{n-\frac{1}{2}}. \tag{15}$$

Comparison of (12)-(15) with (11) reveals that the percentage errors of the approximations do not depend on $\bar{x}$ and can therefore be simply studied as functions of n. Table 1 presents some selected numerical comparisons.

Table 1: Exponential Likelihood:
Percentage Errors of Approximations of $E(\theta \mid x)$

	n	1	2	3	4	5	6	7	8	9
Laplace	θ (12)	26.0	11.7	6.6	4.3	3.0	2.2	1.7	1.3	1.1
	$\ln(\theta)$ (13)	3.9	1.4	0.7	0.4	0.3	0.2	0.2	0.1	0.1
Pearson	$Q(\theta) = \theta^2$ (15)		3.9	1.4	0.7	0.4	0.3	0.2	0.2	0.1

In particular, we note that the Laplace approximation based on the reparametrization $\ln(\theta)$ is a considerable improvement on the Laplace approximation in the original parametrization.

3.2 The Binomial Likelihood

Suppose x is the number of successes in a binomial experiment involving n trials. If the likelihood is parametrized in terms of the success probability θ, we can take $l(\theta \mid x) = \theta^x(1-\theta)^{n-x}$, and we shall carry out comparative studies assuming the prior $\pi(\theta) \propto \theta^{-\frac{1}{2}}(1-\theta)^{-\frac{1}{2}}$. First, we shall take $g(\theta) = \theta$, so that the posterior mean is the inference summary to be approximated.

It is easily verified that the exact posterior mean is given by

$$E(\theta \mid x) = (x + \frac{1}{2})/(n+1), \tag{16}$$

and, after some algebra, that the Laplace approximation based on the parametrization θ is given by

$$(n-1)^{n+\frac{1}{2}}(x+\frac{1}{2})^{x+1}\Big/\left[n^{n+\frac{3}{2}}(x-\frac{1}{2})^x\right]. \tag{17}$$

If, instead, we were to work with the reparametrization $\phi = \sin^{-1}\sqrt{\theta}$, the functions involved in (1) become $g(\phi) = \sin\phi^2$, $l(\phi \mid x) = (\sin^2\phi)^x(1-\sin^2\phi)^{n-x}$, $\pi(\phi) \propto$ constant and the Laplace approximation can be shown to give

$$n^{n+\frac{1}{2}}(x+1)^{x+1}/[(n+1)^{n+\frac{3}{2}}x^x]. \tag{18}$$

The Pearson approach, based on $Q(\theta) = \theta(1-\theta)$, which corresponds to the beta density, leads to the approximation

$$(n+1)^{n+\frac{1}{2}}(x+\frac{3}{2})^{x+1}/[(n+2)^{n+\frac{3}{2}}(x+\frac{1}{2})^x], \tag{19}$$

which can also be shown to be the Laplace approximation using the parametrization $\phi = \ln[\theta/(1-\theta)]$. In this case, we have illustrated, in Table 2, the comparative behaviour of these approximations for some selected values of n and x.

Table 2: Binomial Likelihood: Percentage Errors of Approximations of $E(\theta \mid x)$

	n	2	3	4	5	6	7	8
	(17)	20.4	3.2	2.7	5.5	7.0	7.9	8.5
$x = 1$	(18)	3.2	2.6	2.3	2.2	2.0	2.0	2.0
	(19)	1.5	1.8	1.9	2.0	2.0	2.1	2.1
	(17)		10.4	4.9	2.3	0.9	0.1	0.4
$x = 2$	(18)		1.4	1.1	0.9	0.9	0.8	0.8
	(19)		0.4	0.5	0.7	0.7	0.8	0.8
	(17)			6.0	3.5	2.1	1.3	0.8
$x = 3$	(18)			0.8	0.6	0.5	0.5	0.5
	(19)			0.3	0.3	0.4	0.4	0.5
	(17)				3.4	2.5	1.7	1.2
$x = 4$	(18)				0.5	0.4	0.4	0.3
	(19)				0.2	0.2	0.2	0.2

Here, we note that the (in some sense "optimal" form of Pearson approximation does best, but the Laplace approximation under the $\phi = \sin^{-1}\sqrt{\theta}$ reparametrization is very satisfactory, and is again often a considerable improvement on the approximation using the original parametrization.

Clearly, it is natural to ask whether the pattern of the performance of these various approximations carries over to inferences for other quantities of interest. Preliminary studies indicate that this is indeed the case, and we shall illustrate the kinds of results that obtain by presenting a selection of results for θ^2.

In this case, the exact posterior mean, $E(\theta^2 \mid x)$ is given by

$$(x+\frac{1}{2})[(n-x+\frac{1}{2})/(n+2)+(x+\frac{1}{2})]/(n+1)^2; \tag{20}$$

the Laplace approximation in the original parametrization, θ, is given by

$$(n-1)^{n+\frac{1}{2}}(x+\frac{3}{2})^{x+2}\Big/\left[(n+1)^{n+\frac{5}{2}}(x-\frac{1}{2})^{x}\right]; \tag{21}$$

in the parametrization $\phi = \sin^{-1}\sqrt{\theta}$ it is given by

$$(n)^{n+\frac{1}{2}}(x+2)^{x+2}/[(n+2)^{n+\frac{5}{2}}x^x], \tag{22}$$

and in the parametrization $\phi = \ln[\theta/(1-\theta)]$ it is given by

$$(n+1)^{n+\frac{1}{2}}(x+\frac{5}{2})^{x+2}/[(n+3)^{n+\frac{5}{2}}(x+\frac{1}{2})^x]. \tag{23}$$

An illustration of the comparative behaviour of the percentage errors of approximation is given in Table 3, for selected values of n and x.

Table 3: Binomial Likelihood:
Percentage Errors of Approximation of $E(\theta^2 \mid x)$

	n	2	3	4	5	6	7	8
	(21)	17.7	2.3	9.1	12.0	13.3	14.1	14.4
$x = 1$	(22)	10.1	7.1	5.2	3.9	2.9	2.2	1.8
	(23)	17.7	13.8	11.5	10.1	9.1	8.4	7.7
	(21)		9.3	2.6	0.4	1.9	2.8	3.3
$x = 2$	(22)		5.4	4.3	3.4	2.8	2.4	2.0
	(23)		8.5	7.1	6.2	5.5	5.0	4.6
	(21)			5.5	2.3	0.7	0.3	0.9
$x = 3$	(22)			3.4	2.8	2.4	2.0	1.8
	(23)			5.1	4.4	3.9	3.5	3.2
	(21)				3.5	1.8	0.8	0.2
$x = 4$	(22)				2.3	2.0	1.7	1.5
	(23)				3.4	3.0	2.7	2.5

Although there is no uniformly best choice, the Laplace approximation (22), corresponding to the parametrization $\phi = \sin^{-1}\sqrt{\theta}$, clearly achieves the best overall performance.

4. Two-Parameter Illustrations

4.1 Ratio of Exponential Parameters

Suppose that x_{ij}, $i = 1,\ 2$, $j = 1, \ldots, n$ are independent random samples from exponential distributions with $E(X_{ij}) = \theta_i$, $i = 1,\ 2$. In the parametrization $\boldsymbol{\theta} = (\theta_1, \theta_2)$, the likelihood function can be taken to be

$$l(\boldsymbol{\theta} \mid x) = (\theta_1\theta_2)^{-n} \exp(-n\bar{x}_1\theta_1^{-1} - n\bar{x}_2\theta_2^{-1})$$

and we shall adopt the prior form

$$\pi(\theta_1, \theta_2) \propto (\theta_1\theta_2)^{-1}.$$

Suppose further that $\psi = \theta_1/\theta_2$ is the parameter of interest and that we wish to approximate the posterior mean for ψ, whose exact form is easily verified to be

$$E(\psi \mid x) = n\bar{x}_1/(n-1)\bar{x}_2. \tag{24}$$

The Laplace approximation derived from the parametrization (θ_1, θ_2) is given by

$$(\bar{x}_1/\bar{x}_2)[n^{n-\frac{3}{2}}(n+2)^{n+\frac{1}{2}}/(n+1)^{2n-1}]. \tag{25}$$

The corresponding approximation derived from the parametrization (ϕ_1, ϕ_2), $\phi_i = \ln(\theta_i)$, $i = 1,\ 2$, for which the prior is $\pi(\phi_1, \phi_2) \propto$ constant, is given by

$$(\bar{x}_1/\bar{x}_2)[(n-1)^{n-\frac{3}{2}}(n+1)^{n+\frac{1}{2}}/n^{2n-1}]. \tag{26}$$

However, there are many other equally plausible reparametrizations which could be considered here. For example, we could directly use the parameter of interest, ψ, together with an arbitrary choice for the second parameter, such as $\lambda = \theta_1$. This implies a prior of the form $\pi(\psi, \lambda) \propto \psi^{-1}\lambda^{-1}$ and leads to the Laplace approximation

$$(\bar{x}_1/\bar{x}_2)[n^{n+\frac{1}{2}}(n+1)^{n-\frac{3}{2}}/(n-1)^{n-\frac{1}{2}}(n+2)^{n-\frac{1}{2}}]. \tag{27}$$

Alternatively, we could invoke the orthogonal parameter procedures recently suggested by Cox and Reid (1987). In this case, given the parameter of interest ψ, the orthogonal parameter is given by $\lambda = \theta_1^2\psi^{-1}$, which implies a prior $\pi(\psi, \lambda) \propto \psi^{-1}\lambda^{-1}$ and the Laplace approximation

$$(\bar{x}_1/\bar{x}_2)[(n+1)^{2n-1}/n^{n-\frac{1}{2}}(n+2)^{n-\frac{1}{2}}]. \tag{28}$$

Noting, from (24) and (25)-(28), that the percentage errors of approximation depend only on n and not on $\bar{x}_1/\bar{x}_2$, the comparative behaviour of these approximations is illustrated numerically in Table 4 for a selection of values of n.

Table 4: Ratio of Exponential Parameters: Percentage Errors of Approximations

n	2	3	4	5	6	7	8	9	10
(25)	16.2	5.4	2.5	1.3	0.8	0.5	0.4	0.3	0.2
(26)	2.6	0.7	0.3	0.1	0.1		... < 0.1...		
(27)	38.8	21.1	13.2	9.1	6.6	5.0	3.9	3.2	2.6
(28)	40.3	21.6	13.5	9.2	6.7	5.0	4.0	3.2	2.6

Here, again, the parametrization for which (in the language of Box and Tiao, 1973) the non-informative prior becomes locally uniform provides a far better Laplace approximation than does the original parametrization, whereas the parametrizations directly involving ψ lead to very poor approximations, with the orthogonal choice of Cox and Reid performing worst of all.

4.2 Two Variance Components

Consider the random effects model $x_{jk} = \theta + e_j + e_{jk}$, $j = 1, \ldots, J$, $k = 1, \ldots, K$, where $E(e_j) = E(e_{jk}) = 0$, $\text{var}(e_j) = \sigma_2^2$ and $\text{var}(e_{jk}) = \sigma_1^2$. Assuming the non-informative prior for θ, σ_1^2 and σ_2^2 given by

$$\pi(\theta, \sigma_1^2, \sigma_2^2) \propto \sigma_1^{-2}(\sigma_1^2 + K\sigma_2^2)^{-1},$$

see, for example, Box and Tiao (1973), it is easily verified that the corresponding joint marginal posterior density for σ_1^2 and σ_2^2 is given by

$$\pi(\sigma_1^2, \sigma_2^2 \mid x) \propto (\sigma_1^2)^{-(\frac{v_1}{2}+1)}(\sigma_1^2 + K\sigma_2^2)^{-(\frac{v_2}{2}+1)} \exp\left[-\frac{1}{2}\left(\frac{v_1 m_1}{\sigma_1^2} + \frac{v_2 m_2}{\sigma_1^2 + K\sigma_2^2}\right)\right], \tag{29}$$

where $v_1 = J(K-1)$, $v_2 = J-1$, $v_1 m_1 = S_1$, $v_2 m_2 = S_2$, with $S_1 = \sum_j \sum_k (x_{jk} - \bar{x}_{j\cdot})^2$, $S_2 = K\sum_j (\bar{x}_{j\cdot} - \bar{x}_{\cdot\cdot})^2$. Writing $\sigma_{12}^2 = \sigma_1^2 + K\sigma_2^2$, from (29), it can be verified that the exact posterior means for σ_{12}^2/σ_1^2, $\sigma_1^2 + K\sigma_2^2$, σ_1^2 and σ_2^2 are given by

$$E\{\sigma_{12}^2/\sigma_1^2 \mid x\} = \frac{v_2 m_2 I_\chi\left(\frac{v_2}{2}-1, \frac{v_1}{2}+1\right)}{(v_2-2) m_1 I_\chi\left(\frac{v_2}{2}, \frac{v_1}{2}\right)} \tag{30a}$$

$$E\{\sigma_1^2 \mid x\} = \frac{v_1 m_1 I_\chi\left(\frac{v_2}{2}, \frac{v_1}{2}-1\right)}{2\left(\frac{v_1}{2}-1\right) I_\chi\left(\frac{v_2}{2}, \frac{v_1}{2}\right)} \tag{30b}$$

$$E\{\sigma_1^2 + K\sigma_2^2 \mid x\} = \frac{v_2 m_2 I_\chi\left(\frac{v_2}{2}-1, \frac{v_1}{2}\right)}{2\left(\frac{v_2}{2}-1\right) I_\chi\left(\frac{v_2}{2}, \frac{v_1}{2}\right)} \tag{30c}$$

$$E\{\sigma_2^2 \mid x\} = \frac{(v_2 m_2) I_\chi\left(\frac{v_2}{2}-1, \frac{v_1}{2}\right)}{2K\left(\frac{v_2}{2}-1\right) I_\chi\left(\frac{v_2}{2}, \frac{v_1}{2}\right)} - \frac{(v_1 m_1) I_\chi\left(\frac{v_2}{2}, \frac{v_1}{2}-1\right)}{2K\left(\frac{v_1}{2}-1\right) I_\chi\left(\frac{v_2}{2}, \frac{v_1}{2}\right)}, \tag{30d}$$

where $\chi = v_2 m_2/(v_1 m_1 + v_2 m_2)$ and $I_\chi(p, q)$ is the incomplete Beta function.

After considerable algebra, Laplace approximations based on the original parametrization (σ_1^2, σ_2^2) can be shown to be

$$\hat{E}\{\sigma_{12}^2/\sigma_1^2 \mid x\} \approx \frac{(v_2 m_2)\left(\frac{v_1}{2}+2\right)^{\frac{(v_1+1)}{2}} \left(\frac{v_2}{2}\right)^{\frac{(v_2-3)}{2}}}{(v_1 m_1)\left(\frac{v_1}{2}+1\right)^{\frac{(v_1-1)}{2}} \left(\frac{v_2}{2}+1\right)^{\frac{(v_2-1)}{2}}} \tag{31a}$$

$$\hat{E}\{\sigma_1^2 + K\sigma_2^2 \mid x\} \approx \frac{e(v_2 m_2)\left(\frac{v_2}{2}\right)^{\frac{(v_2-3)}{2}}}{2\left(\frac{v_2}{2}+1\right)^{\frac{(v_2-1)}{2}}} \tag{31b}$$

$$\hat{E}\{\sigma_1^2 \mid x\} \approx \frac{e(v_1 m_1)\left(\frac{v_1}{2}\right)^{\frac{(v_1-3)}{2}}}{2\left(\frac{v_1}{2}+1\right)^{\frac{(v_1-1)}{2}}} \tag{31c}$$

$$\hat{E}\{\sigma_2^2 \mid x\} \approx \frac{(v_1 m_1)^{v_1/2} (v_2 m_2)^{v_2/2} \exp\left\{\frac{(v_1+v_2)}{2}+2\right\} e^{-nh(\hat{x},\hat{y})}}{2^{(v_1+v_2)/2}\left(\frac{v_1}{2}+1\right)^{(v_1-1)/2}\left(\frac{v_2}{2}+1\right)^{(v_2-1)/2}\{\det n\boldsymbol{D}^2 h(\hat{x},\hat{y})\}^{\frac{1}{2}}}, \tag{31d}$$

where $\hat{x}$ and $\hat{y}$ maximize $-nh(x,y) = -a_1 \ln x + \ln y - a_2 \ln(x + Ky) - \frac{1}{2}(p_1/x + p_2/(x+Ky))$, $a_1 = \frac{1}{2}v_1 + 1$, $a_2 = \frac{1}{2}v_2 + 1$, $p_1 = v_1 m_1$, $p_2 = v_2 m_2$ and $\boldsymbol{D}^2 h(\hat{x}, \hat{y})$ is the Hessian of h evaluated at $\hat{x}$ and $\hat{y}$.

In the parametrization $\xi_1 = \ln(\sigma_1^2)$ and $\xi_2 = K^{-1}\ln(\sigma_1^2 + K\sigma_2^2)$, which has a prior $\pi(\xi_1, \xi_2) \propto$ constant, we find:

$$\hat{E}\{\sigma_{12}^2/\sigma_1^2 \mid x\} \approx \frac{(v_2 m_2)\left(\frac{v_1}{2}+1\right)^{\frac{(v_1+1)}{2}}\left(\frac{v_2}{2}-1\right)^{\frac{(v_2-3)}{2}}}{(v_1 m_1)\left(\frac{v_1}{2}\right)^{\frac{(v_1-1)}{2}}\left(\frac{v_2}{2}\right)^{\frac{(v_2-1)}{2}}} \tag{32a}$$

$$\hat{E}\{\sigma_1^2 + K\sigma_2^2 \mid x\} \approx \frac{e(v_2 m_2)\left(\frac{v_2}{2}-1\right)^{(v_2-3)/2}}{2\left(\frac{v_2}{2}\right)^{(v_2-1)/2}} \tag{32b}$$

$$\hat{E}\{\sigma_1^2 \mid x\} \approx \frac{e(m_1 v_1)\left(\frac{v_1}{2}-1\right)^{\frac{(v_1-3)}{2}}}{2\left(\frac{v_1}{2}\right)^{(v_1-1)/2}} \tag{32c}$$

$$\hat{E}\{\sigma_2^2 \mid x\} \approx \frac{K(v_1 m_1)^{v_1/2}(v_2 m_2)^{v_2/2} e^{(v_1+v_2)/2} e^{-nh(\hat{\xi}_1,\hat{\xi}_2)}}{2^{(v_1+v_2)/2}\left(\frac{v_1}{2}\right)^{(v_1-1)/2}\left(\frac{v_2}{2}\right)^{(v_2-1)/2}\{\det^n \boldsymbol{D}^2 h(\hat{\xi}_1,\hat{\xi}_2)\}^{\frac{1}{2}}}, \tag{32d}$$

where $\hat{\xi}_1$ and $\hat{\xi}_2$ maximises

$$-nh(\xi_1,\xi_2) = -\ln(K) + \ln(e^{K\xi_2} - e^{\xi_1}) - v_1\xi_1/2 - Kv_2\xi_2/2 - [v_1 m_1 e^{-\xi_1} + v_2 m_2 e^{-K\xi_2}]/2$$

and $\boldsymbol{D}_2 h(\hat{\xi}_1, \hat{\xi}_2)$ is the Hessian of h evaluated at $\hat{\xi}_1$ and $\hat{\xi}_2$.

In Table 5, we consider a numerical comparison of these approximations applied to the well-known dyestuff data given by Box and Tiao (1973, Section 5.1.2), for which $K = 5$; $S_1 = 58830.00$; $v_1 = 24$; $m_1 = 2451.25$; $S_2 = 56357.00$; $v_2 = 5$ and $m_2 = 11271.50$. Once again, we observe the (often dramatic) increase in accuracy obtained by using a parametrization which implies a locally uniform non-informative prior.

5. Concluding Remarks

The preliminary studies reported here serve to demonstrate that the accuracy of Laplace-type procedures for approximating posterior moments and densities can be very sensitive to the parametrization adopted in defining the likelihood. Selected illustrative examples suggest that reparametrizations which lead to locally uniform non-informative prior representations typically improve the accuracy of approximations, often very dramatically. Work in progress is seeking to provide a theoretical underpinning for these findings by detailed examination of the error terms of the Laplace approximation, and by studying the closely related integrand decomposition techniques suggested by Mosteller and Wallace (1984).

Table 5: Two Variance Components (Box and Tiao Example) (Percentage Errors in Parentheses)

		Laplace Approximation	
	Exact (30)	(σ_1^2, σ_2^2) (31)	$(\ln\ \sigma_1^2, K^{-1}\ln(\sigma_1^2 + K\sigma_2^2))$ (32)
$E(\sigma_{12}^2/\sigma_1^2 \mid x)$	7.69 (0%)	6.42 (19.80%)	7.50 (2.53%)
$E(\sigma_1^2 + K\sigma_2^2 \mid x)$	18850.36 (0%)	15632.20 (20.59%)	18383.47 (2.54%)
$E(\sigma_1^2 \mid x)$	2665.25 (0%)	2672.40 (0.27%)	2672.40 (0.27%)
$E(\sigma_2^2 \mid x)$	3237.02 (0%)	530.02 (510.73%)	3243.40 (0.20%)

Acknowledgments

The first author acknowledges support from the Brazilian Institution CAPES-MEC for a visit to the University of Nottingham; the work of the second author forms part of a programme of research supported under the UK Science and Engineering Research Council's Complex Stochastic Systems Initiative.

References

Box, G.E.P. and G.C. Tiao (1973). *Bayesian Inference in Statistical Analysis*, Addison-Wesley.

Cox, D.R. and N. Reid (1987). "Parameter Orthogonality and Approximate Conditional Inference," *J. R. Statist. Soc., B*, Vol.49, No.1, 1-39.

Kass, R.E., L. Tierney, and J.B. Kadane (1987). "The Validity of Posterior Expansions Based on Laplace's Method," Technical Report, Department of Statistics, Carnegie-Mellon University.

Morris, C.N. (1988). "Approximating Posterior Distributions and Posterior Moments," to appear in *Bayesian Statistics 3*, J.M. Bernardo. M.H. DeGroot, D.V. Lindley, and A.F.M. Smith, eds., Oxford University Press.

Mosteller, F. and D.L. Wallace (1984). *Applied Bayesian and Classical Inference: The Case of the Federalist Papers.* New York: Springer.

Naylor, J.C. and A.F.M. Smith (1982). "Applications of a Method for the Efficient Computation of Posterior Distributions," *Applied Statist.*, Vol. 31, 214-225.

Smith, A.F.M., A.M. Skene, J.E.H. Shaw, and J.C. Naylor (1987). "Progress with Numerical and Graphical Methods for Practical Bayesian Statistics," *The Statistician*, Vol. 36, 75-82.

Tierney, L. and J.B. Kadane (1986). "Accurate Approximations for Posterior Moments and Marginal Densities," *J. Amer. Statist. Assn.*, Vol. 81, 82-86.

Tierney, L., R.E. Kass, and J.B. Kadane (1986). "Approximations of Posterior Expectations and Variances Using Laplace's Method," Technical Report, Department of Statistics, Carnegie-Mellon University.

Bayesian and Likelihood Methods in Statistics and Econometrics
S. Geisser, J.S. Hodges, S.J. Press and A. Zellner (Editors)

A METHOD FOR MAXIMIZING LIKELIHOOD FUNCTIONS

Jill H. Larkin and Joseph B. Kadane
Carnegie-Mellon University

Abstract

We find maximum likelihood estimates (or maximum posterior estimates) by iteratively fitting unrestricted quadratics in the parameter space to the logarithm of the likelihood function. This technique depends on and makes use of the normal shape of likelihoods for large enough samples. Conventional maximization methods either use local information (hill climbing, quasi-Newton algorithm) or compare individual evaluations at many points (grid searches). Unlike local gradient methods, ours is unlikely to be caught by a local maximum with a small "catchment area." Unlike grid searches, it uses the relations among the calculated values to estimate the entire function. The procedure is demonstrated with a real problem of interest in developmental psychology. We believe that these ideas can provide a useful tool for studying complex likelihoods in multidimensional spaces.

Key Words – Asymptotic normality, iterative estimation, maximum likelihood, regression.

1. Introduction

This paper concerns maximization of likelihood functions when the first derivative equations cannot be solved explicitly. In such cases, many commonly used programs employ techniques (e.g., grid search) suitable for maximization of any function no matter how badly behaved. This paper suggests a more aggressive numerical strategy, using the typical normal shape for the likelihood functions of large samples of identically distributed and conditionally independent observations. Since the method relies on this restrictive shape, it might be expected to perform relatively well when those assumptions apply, and relatively poorly when they do not. We believe, however, that this method may play a useful role among those available to statisticians seeking to maximize likelihood functions.

The paper first (Section 2) describes the procedure. Section 3 presents a heuristically useful graphical techniques. Section 4 applies the procedure to an example. Section 5 compares the method with other maximize-likelihood methods. Section 6 describes some implementation problems and their solutions.

The work reported here concerns a problem that is common to the likelihood approach favored by George Barnard, and to the Bayesian approach. To adherents of the latter, we remark that the prior can be considered formally as a factor in the likelihood. Maximization of the posterior density is important to Bayesians; for example, to implement Laplacean approximations (see Tierney and Kadane (1986)).

2. Description of the Method

Consider a model with N parameters. Each parameter may take a range of values. The set of all possible values of the parameters form a region (possibly unbounded) in a N-dimensional space. Let θ be a vector in that space corresponding to a possible set of values of the N parameters of the model. We define the function $\Lambda(\theta)$ as -2 times the logarithm of the likelihood $\mathcal{L}$. The goal is to find the value of θ corresponding to the maximum of the likelihood $\mathcal{L}$ or to the minimum of Λ.

If the likelihood that has the structure of identically distributed random variables that are independent (conditional on the parameters), Λ is quadratic over the whole of the interesting range for a very large sample (see Wald (1949), LeCam (1953), Walker (1969) and Perlman (1983)). Additionally, Perlman (Theorem 4.3) shows that the loglikelihood is ultimately convex on every fixed neighborhood.

Then assuming that the likelihood is approximately normal with mean θ_0 and variance matrix V,

$$\begin{aligned}\Lambda(\theta) &= -2ln\{K\exp(-1/2)[(\theta-\theta_0)'V^{-1}(\theta-\theta_0)]\}\\ &= -2ln(K)+[(\theta-\theta_0)'V^{-1}(\theta-\theta_0)]\\ &= \theta'V^{-1}\theta-2\theta_0'V^{-1}\theta+K',\end{aligned} \tag{1}$$

where $K'=-2\ell_n(K)+\theta_0'V^{-1}\theta_0$.

The goal is to find the value θ_0 of θ for which $\mathcal{L}$ is maximal and correspondingly Λ is minimal. To find the value we generate a number of points $\{\theta_i\}, i=1,\dots n$. For convenience, and because it will correspond to the notion use in regression process, we denote $\Lambda(\theta_i)$ by y_i. We also express in components the vector θ_i as $(\theta_{i,1},\theta_{i,2},\dots,\theta_{i,n})$

and the matrix V^{-1} as $[V^{r,s}]$. Then we can rewrite (1) as:

$$y_i = \sum_{r=1}^{N}\sum_{s=1}^{N} \theta_{i,r}V^{r,s}\theta_{i,s} - 2\sum_{r=1}^{N}\sum_{s=1}^{N} \theta_{0,r}V^{r,s}\theta_{i,s} + K'$$
$$= \sum_{r=1}^{N} \theta_{i,r}^2 V^{r,r} + 2\sum_{r=1}^{N}\sum_{s=r+1}^{N} \theta_{i,r}V^{r,s}\theta_{i,s} - 2\sum_{r=1}^{N}\sum_{s=1}^{N} \theta_{0,r}V^{r,s}\theta_{i,s} + K' \quad (2)$$

The right side of (2) is an approximation of the left side. To minimize the error, we choose the parameters V, θ, and K' so as to minimize the sum of the squared differences between the left and the right sides. This is the equivalent to running a regression with the dependent variable y_i and independent variables:

$$\theta_{i,r}^2 (1 \le r \le N),$$
$$\theta_{i,r}\theta_{i,s} (1 \le r < s \le N),$$
$$\theta_{i,r} (1 \le r \le N). \quad (3)$$

The regression equation is:

$$y_i = \sum_{r=1}^{N} \gamma_r \theta_{i,r}^2 + \sum_{r=1}^{N}\sum_{s=r+1}^{N} \alpha_{r,s}\theta_{i,r}\theta_{i,s} + \sum_{r=1}^{N} \beta_r \theta_{i,r} + \delta + e_i \quad (4)$$

where the α's, β's, γ's and δ are regression parameters, and $\{e_i\}$ are errors associated with each y_i. The number of independent variables is N γ's plus $N(N-1)/2$ α's plus N β's plus one δ or, $(N+1)(N+2)/2$ independent variables. Suppose X is the $(N+1)(N+2)/2$ by N matrix of independent variables given in expression (3). Then using the least squares method, estimating the α's, β's, γ's and δ is possible if the information matrix can be inverted.

But what we really want to know is the estimate of θ_0 and its covariance matrix V. These are our estimates of the parameters appearing in the original model. They can be found from the estimates of gamma's alpha's, beta's and delta as follows:

$$\hat{V}^{i,i} = \hat{\gamma}_i \quad (5a)$$
$$\hat{V}^{i,j} = .5\hat{\alpha}_{\max(i,j),\,\min(i,j)} \quad (i \ne j) \quad (5b)$$
$$\hat{\theta}_0 = -.5V\hat{\beta} \quad (5c)$$

The form of (5b) guarantees that the estimate of $\hat{V}$ is the symmetric,but does not necessarily guarantee that it is positive-definite. (We discuss in Section 6 what can be done if $\hat{V}$ is not positive definite.) Assuming that $\hat{V}^{-1}$ is positive-definite, then it can be inverted to solve (5c) for θ_0.

Now we have the estimates for the parameters θ_0 of interest and of their covariance matrix V. However, the validity of these estimates depends on the function Λ being well approximated by a quadratic over the region considered. This is guaranteed to be the case only if the sample size is very large. However, by Taylor's Theorem, Λ is quadratic in any small region. Therefore, we can successively improve our estimates by the following iterative procedure:

1. Estimate θ_0 and V as described above, initially using points $\{\theta_i\}$ distributed uniformly through the region of legal values of θ.
2. Generate a new set of points $\{\theta_i\}$ that are normally distributed with the mean at the previous estimate for θ_0 and covariance matrix equal to cV where c is a constant which can be adjusted, and V is the previous estimate of the covariance matrix.
3. Repeat step 1, using the new set $\{\theta_i\}$, and then step 2 until the values of θ_0 and V do not change appreciably with further iterations.

In our work, we have gotten stable estimates by starting with $c = 3$. The set of points generated then have mean location θ_0 and the variance in location V^{-1}. We then reduce c by .2 in subsequent iteration cycles. If, however, c is too small, the curvature of lambda is too small to allow a stable inversion of V^{-1}. When this happens, we discard that iteration, and replace it with a cycle using $c + .2$. Our final stable values are almost all obtained with value of $c = 1$.

Judgment is also required to choose the number of points to be generated. We have used 3^N, where N is the number of dimensions in the parameter space. For large N, however, or for computationally difficult likelihood functions, this number might make computation time infeasible, and s smaller number might be tried.

3. Techniques for Graphing Multi-dimensional Functions

The procedure described above depends on global properties of the likelihood function considered. A graphical presentation of the likelihood evaluation allows visual inspection of these global properties. Such graphs are heuristically useful in visually estimating maximum likelihood points (as a check on the numerical results) and in diagnosing problems such as violently non-quadratic behavior of Λ and non-positive eigenvalues of V. The following techniques provide interpretable graphs of Λ. The first produces graphs of Λ as a function of a grid of points distributed through a region of the parameter space. The second technique shows cross sections on Λ as a function of one parameter.

3a. Graphing the Likelihood for a Grid of Parameter Values

Consider, for example, generating four values (.2,.4,.6,.8) for each of three dimensions to produce a three-dimensional grid of 4^3=64 points $\{\underset{\sim}{\theta}_i\}$. These points are generated by starting with $\underset{\sim}{\theta}_1$=(.2,.2,.2), the smallest values for all coordinates, then incrementing the last coordinate through its sequence of values, then incrementing the third coordinate to its next value and again cycling through the values of the last coordinate and so on. The result is the list of points $\underset{\sim}{\theta}_1$ listed below.

$$\underset{\sim}{\theta}_1, \underset{\sim}{\theta}_2, \underset{\sim}{\theta}_3, \ldots, \underset{\sim}{\theta}_i, \ldots, \underset{\sim}{\theta}_{63}, \underset{\sim}{\theta}_{64} =$$
$$(.2,.2,.2),(.2,.2,.4),(.2,.2,.6),(.2,.2,.8),(.2,.4,.2),(.2,.4,.4),\ldots,(.8,.8,.6),(.8,.8,.8)$$

Each of these points has a value of Λ associated with it. We graph Λ as a function of the index associated with the points $\underset{\sim}{\theta}_i$. The result looks like Figure 1 constructed from data similar to that described later in Section 4. Calculated values of Λ, indicated by points in this figure, have been connected in groups of 4 by smooth curves. In this application there are 5 parameters each sampled in cycles of 4 for a total of 4^5=1024 points $\{\underset{\sim}{\theta}_i\}$. Each parameter corresponds to one coordinate of the vector $\underset{\sim}{\theta}_i$. Figure 1 show values of the index i between $384 = 1 \times 4^4 + 2 \times 4^3$ and 448 $= 1 \times 4^4 + 3 \times 4^3$. In this portion of the graph, therefore, the first parameter takes its lowest value (.2) and the second parameter takes the second lowest value (.4). Within this portion of the graph, the third parameter takes the values (.2,.4,.6,.8), each of these values corresponding to the sequence of 16 values of the index i. The fourth and fifth parameters also each take the values (.2,.4,.6,.8). Each of the curves in Figure 1 connects values of Λ corresponding to a single value of the fourth parameter, and within these curves the value of the fifth parameter take the values (.2,.4,.6,.8).

In Figure 1, one can readily see that within the values of i shown, the minimum value of Λ, is at

$$i = 407 = 1 \times 4^4 + 2 \times 4^3 + 1 \times 4^2 + 1 \times 4^1 + 3.$$

At this point (as in the entire graph) the first and the second parameters take their first and second values (.2 and .4). The third, fourth, and fifth parameters take respectively the values .4, .4, and .6, at this maximum-likelihood point. This kind of graph gives a quick estimate of where the maximum-likelihood points lie. The procedure can be repeated within smaller regions to give more detailed information.

Figure 1 also gives information about how much each of the parameters affects Λ. The cycle of 64 in i , shows the effect of the third parameter Λ. There is only a moderate effect over the region shown. The four cycles of 16 show the effect of the fourth parameter, and the 16 cycles of 4 show the effect on the fifth parameter.

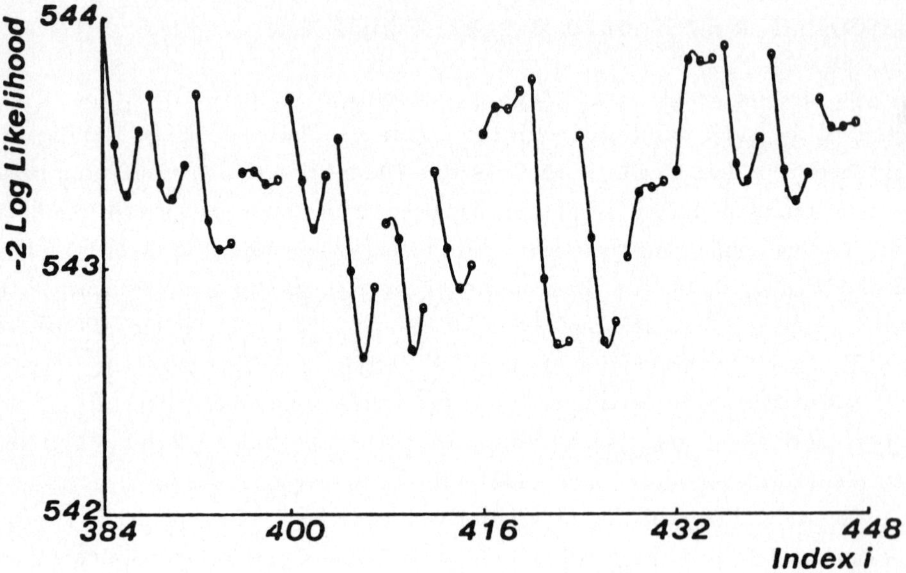

Figure 1: Graph of Λ as a function of i, the position of its parameter-vector argument in a cycling list.

Notice the very large effect of the fifth parameter. In the cycle of four surrounding the minimum value of Λ, all other values are notably higher. Near the minimum, all of the nested cycles show a roughly quadratic shape, supporting the applicability of our quadratic regression procedure. Also there is a pronounced upward curve for all cycles in the region of the minimum, suggesting no problems with non-positive eigenvalues of V.

A more formal analysis might be based on this graphical technique by using spectral analysis on the "time series" of loglikelihoods. In Figure 1 each parameter has a characteristic period (here 4, 16, 64 ...). In the spectrum of this graph, high spikes corresponding to the period of a parameter would indicate a parameter that influences the loglikelihood greatly, suggesting small standard errors (or high precision) for that parameter. This analysis corresponds to our informal observation that in Figure 1 the fifth parameter with period 4 has the largest effect on Λ. We have not, however, pursued this more formal approach.

3b. Graphing Cross Sections of the Likelihood

Once a maximum-likelihood point has been found, one can graph Λ (-2 times the logarithm of the likelihood) as a function of any single parameter while holding the others fixed at their maximum likelihood values. The result is a set of cross-sections of the multi-dimensional function Λ. These graphs first verify that a maximum-likelihood (minimum Λ) point has been found. They can also be used to check the assumption that Λ is globally quadratic. Figure 2 shows the 5 parameter cross sections for data similar to that analyzed in the next section. Each cross section passes through the maximum-likelihood point.

For each of the five parameters, the minimum on this graph indicates the maximum likelihood estimate of that parameter. Two of the parameters (labeled 2 and 3) have estimates of zero. Parameters 1, 4, 5 have the indicated estimates .237, .667, and .881.

4. Example

We have used the procedure described above to relate psychology data (subjects' responses) to a set of mental rules hypothesized to underly these responses. This analysis goes beyond mere description of the nature of the responses; it gives direct estimates of the psychologically interesting quantities, the extent to which various mental rules are used (Sophian, Larkin and Kadane, 1985)

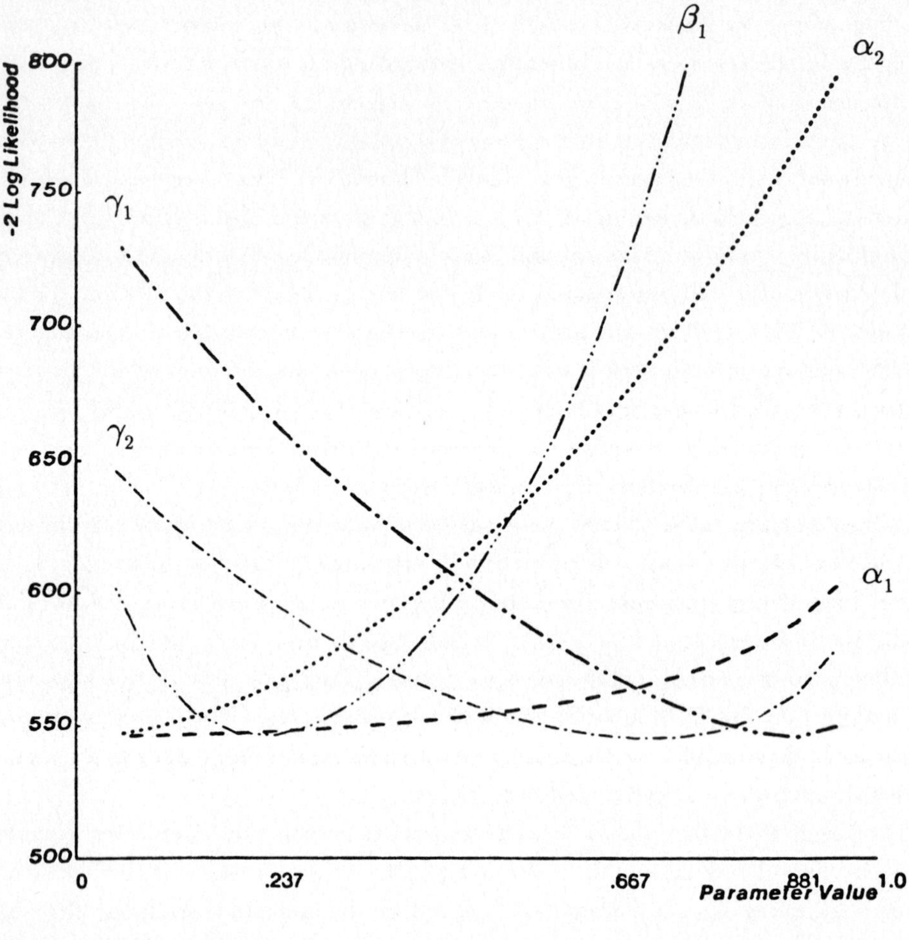

Figure 2: Five parametric cross-sectional graphs of Λ through the maximum-likelihood point. The five curves show values of the five parameters described in Section 4, fit to a different data set.

Children aged 20, 30, and 42 months watched individually while a small toy was hidden in one of the three covered cups. After two of the cups were interchanged, the child searched for the toy. Either the cup with the toy was exchanged with another cup, a relevant transposition, or the two other cups were exchanged, an irrelevant transposition. The chosen cup is the child's response.

Very young children search using an initial location rule, i.e., they look in the cup at the place where they saw the toy disappear. Older children, of course, use a correct rule. A possible intermediate rule focuses on the transposed cups, and search at random in one of them (a moved cup rule). We assume that one of these rules represents each child's understanding of the situation and that throughout the experiment the child applies just one rule. We call these rules task-adjusted because they use information from the task. Let α_i be the *a priori* probability that rule i is a given child's rule.

A young child does not always apply his or her task-adapted rule. The child may be distracted and not encode information from the task, or for some other reason may not use this information. In this case, in our psychological model the child uses one of two fallback rules: (1) Choose a "favorite" location or (2) choose a location at random. We let γ_v be the probability that a child uses his task-adapted rule on each of the two problem varieties v (relevant and irrelevant transpositions), and we let β_j be the conditional probabilities that if a child is not using his task-adapted rule he will use the fallback rule j. Finally we introduce probabilities δ_r, that each location r is a child's favorite location.

If we consider a child with a task-adapted rule i and a favorite location r solving a problem of variety v, the likelihood of this child's response x is:

$$\gamma_\nu p(x \mid i, r, v, t) + (1 - \gamma_v) \sum_{j=1}^{2} \beta_j q(x \mid j, r, v, t) \tag{6}$$

where $p(x \mid i, r, \nu, t)$ and $q(x \mid j, r, \nu, t)$ are the probabilities of response x for trial t on problem variety v on the basis of favorite location r and task-adapted rule i or fallback rule j respectively.

If we assume independence over trials t and over children c the complete likelihood function is then:

$$\prod_{c=1}^{C} \sum_{i=1}^{3} \alpha_i \sum_{r=1}^{3} \delta_r \sum_{\nu=1}^{2} \prod_{i=1}^{T(\nu)} [\gamma_\nu p(x \mid i, r, \nu, t) + (1 - \gamma_\nu) \sum_{j=1}^{2} \beta_j q(x \mid j, r, \nu, t)] \tag{7}$$

where C is the number of children and $T(v)$ is the number of trials on problem variety v.

There are seven independent parameters since the α_i, β_j, and δ_r sum to one. The following reparameterization gives a vector$\underset{\sim}{\theta}$ with coordinates all bounded by 0 and 1.

$$
\begin{aligned}
\theta_1 &= \alpha_1 \\
\theta_2 &= (1-\alpha_1)\alpha_2 \\
\theta_3 &= \beta_1 \\
\theta_4 &= \gamma_1 \\
\theta_5 &= \gamma_2 \\
\theta_6 &= \delta_1 \\
\theta_7 &= (1-\delta_1)\delta_2
\end{aligned}
$$

In our initial efforts to fit Λ by the equation (1), V^{-1} was not positive definite. It had zero, or slightly negative eigenvalues with eigenvectors roughly spanning the space of θ_6 and θ_7. These parameters indicate how likely each location is to be a child's favorite location. Since children do not often use the favorite-location rule, very little data bears on these parameters. Therefore the likelihood doesn't change significantly for changes in θ_6 and θ_7, resulting in corresponding near-zero eigenvalues in V^{-1}. We therefore removed θ_6 and θ_7 from the model, and postulated $\delta_1 = \delta_2 = \delta_3 = 1/3$. Because Λ is insensitive to these parameters, any choice would probably have done about as well.

For the two younger groups, the estimates of α_2 were consistently negative, and graphs (like that in Figure 2) supported this negativity. We therefore replace the original model with the simpler model in which α_2, the probability of using the "moved box" rule, was set to zero. For the oldest children, not surprisingly, the estimate of the probability α_1 of using the primitive "initial location" rule was consistently negative and was set equal to zero. In this case the value of Λ decreased slightly from 543.5 to 543.3. This happens occasionally probably because the quadratic model can be fit more accurately when all the parameters have zeros in the region being sampled. More commonly, of course, Λ increases slightly reflecting a slight decrease in the maximum likelihood model is simplified.

For the oldest children, one further simplification was made setting equal to zero the parameters γ_1 describing the conditional probability of using the favorite-box fallback rule. These simplifications increased Λ by less than .5 for each group. Thus simplifying the model did not appreciably change the likelihood of the data based model. Table 1 shows the complete set of parameter estimates for these data.

Table 1: Parameter values, estimated by our technique indicating children's use of various rules for searching. Standard error estimates are given in parentheses.

	Age in Months		
	20	30	42
Probability of each task-adapted rule being a child's rule			
Initial Location	.955 (.274)	.474 (.122)	0
Moved Box	0	0	.170 (.022)
Combination (correct)	.045 (.274)	.526 (.122)	.830 (.022)
Probability of using some task-adapted rule			
For relevant-transposition problems	.345 (.076)	.463 (.094)	.656 (.050)
For irrelevant-transposition problems	.492 (.144)	.621 (.044)	.796 (.051)
Conditional probability of fallback rules			
Favorite box	.283 (.064)	.267 (.051)	0
Random	.717 (.064)	.733 (.051)	1
-2 times log likelihood (Λ)	725.0	621.4	543.2

5. Comparison with Other Methods

Maximization methods conventionally either use information from single points in the parameter space (hill climbing, quasi-Newton algorithm) or compare individual evaluations at many points (grid searches). While often effective, these procedures encounter difficulties such as local maxima and rates of change that are very different in different directions. This paper describes an alternative procedure for maximum likelihood estimation of parameters and their standard deviations. This procedure begins with a grid-like search at randomly distributed points, but rather than individual likelihood evaluations, it uses all of the evaluations to estimate the global shape of the likelihood function. This procedure therefore uses more information than does a conventional grid search. The initial estimate of the likelihood function locates a region in which the maximum likelihood point is likely to be found and further search is concentrated there. The procedure is iterated, using more closely spaced points with the region, to obtain a new estimate of the shape of the likelihood function and a new region for more intense search.

Unlike local gradient methods, the quadratic regression method is unlikely to be caught by a local maximum with a small "catchment area." Unlike grid searches, it uses the relations among the circulated values to estimate the entire function. This process gives an estimate of the maximum likelihood point which need not be one of the actual points sampled in the search. Although the procedure may be misled in cases where the likelihood function is non-normal, we believe it is generally less fragile than the available alternatives. Given the general difficulty of reliable maximization, it offers a useful additional approach.

The computation is extensive but feasible. In our experience a model with five parameters requires from three to five iterations of the fitting algorithm. Each iteration requires roughly 3^N likelihood evaluations, where N is the number of parameters.

5a. Comparison With Other Maximization Methods

Hill climbing methods rely on general ideas of maximization applicable to arbitrary functions (with sufficient smoothness, etc.), and hence do not take advantage of the special structure that likelihood functions tend to have. Some of them, especially Newton's method and the method of scoring (Kass, 1983), do use a local quadratic approximation. However, "if x_0 (the starting value) is not close enough to x (the value sought), the Newton iteration may diverge hopelessly" (Gill, Murray and Wright, 1981, p. 85). Thus Morgan (1978) finds that at one of the eight starting points for the location/scale Cauchy example, the Newton-Ralphson method diverged. The pictures in Morgan (1983) show that this likelihood has a surface with many ridges, although Copas (1975) shows it is unimodal. We have explored Morgan's Cauchy example using the quadratic regression technique. Not surprisingly, it does not perform well, probably because the likelihood is multimodal in the location parameter, given the scale parameter.

There are reasons, then, for believing that the methods described in this paper are more robust than the standard hill-climbing techniques, and less likely to diverge or be distracted by local maxima that are not global. We believe that further comparisons are needed to explore the strengths and weaknesses in each approach. Prominent among the alternatives that should be considered are the David-Fletcher-

Powell quasi-Newton methods (see Dennis and Schnabel, 1983, and Chambers, 1973), the Nelder-Mead (1965) simplex method, and Newton methods. Additionally the idea of stable parameters (Ross, 1970, 1976) may benefit all the methods by providing a good reparameterization. To facilitate this comparision we offer upon request our algorithm implemented in Fortran using IMSL subroutines.

5b. Comparison with a Numerical Method for Finding the Standard Errors

We suspect that such a direct and crude method as ours can fail to find reasonable values for the standard errors of parameters, even when it finds maximum likelihood values reasonably accurately. To find out, we compared our methods with Kass' (1984) derivative-free algorithm on the models reported in Section 4. The results are given in Table 2. The standard errors in Table 2 correspond to the parameters in Table 1, but in Table 2 we have shown only the standard errors, and, of those, only one element of each set in which constraints on the parameters require the standard errors to be equal.

Since Kass' algorithm is especially designed for finding observed information at the maximum, and since it works in double-precision, we believe his results to be more accurate than ours. Against that standard, our algorithm did reasonably well.

5c. Other Use of Regression Techniques

We have been preceded in using regression techniques to find standard errors of maximum likelihood estimates in non-linear models by work of Schim van der Loeff and Harkema (1976), following the suggestion of Kloek. They used the technique to evaluate standard errors at the presumed maximum of the likelihood. Expanding the likelihood in the Taylor series around the maximum, and using the fact that the first derivatives are zero at the maximum, they find an analog to our equation (2) without linear terms in θ. They apparently used other techniques to find the maximum likelihood estimates.

Recently an interesting method of Veall (1987, 1988) adapts work of Clough (1969) to finding a confidence level that a global maximum has been found.

6. Difficulties with the Method

This section describes problems encountered in implementing the likelihood maximization method described here, and the solutions we have developed.

Table 2: Standard errors estimated by the quadratic regression method and those given by the numerical method of Kass (1984). Both sets are based on the model in Section 4.

	Estimation Quad. Reg.	**Method** Numerical	**Age in Months**
Probability of each task-adapted rule being a child's rule			
Initial Location	.274	.087	20
	.122	.526	30
Moved Box	.022	.089	42
Probability of using some task-adapted rule			
For relevant-transposition problems	.076	.049	20
	.094	.047	30
	.050	.044	42
For irrelevant-transposition problems	.144	.062	20
	.044	.059	30
	.051	.049	42
Conditional probability of fallback rules			
Favorite box	.064	.070	20
	.051	.089	30

6a. Globally Non-Quadratic Functions

There is one major caveat in using this procedure. From the first step our procedure fits a quadratic function to evaluations of Λ at points sampled globally from throughout the parameter space. The danger is being misled by the violently non-quadratic behavior of Λ so that subsequent search is confined to an inappropriate region. There are, however, some safegaurds. First, non-positive eigenvalue(s) of the matrix V may signal globally non-quadratic behavior along the direction(s) spanned by the associate eigenvector(s). Second, the two graphical techniques described in Section 3 allow visual inspection of the evaluations on Λ and may aid in identifying anomalies.

If global non-quadratic behavior seems likely, the remedy is first to use cues to this global behavior, eigenvalues and graphs, to identify regions and directions that seem to have the lowest values of lambda. Generate a large number of points in these regions. If this process identifies a quadratic region, with values lower than those in the rest of the space, the method can be applied in that region. If it does not, then the procedure described here is probably not appropriate.

6b. Excessive Comparison Time for Maximization

Complicated Likelihoods in High Dimensional Space

Because likelihoods are given in this paper by arbitrary subroutines, they may be arbitrarily costly to evaluate. Additionally, the number of parameters, N, may be arbitrarily large. Hence, doing evaluations at 3^N parameter points may be quite costly indeed.

The structure of (5) is a standard quadratic regression, and the nature of our algorithm allows for choice of the points $\underset{\sim}{\theta}_i$ at which likelihood is to be evaluated. Hence experimental design ideas might be used to reduce the number of points required. Since the likelihood functions we have worked with to date have not been so costly, our work has not required us to explore how this would be done, and how great the savings might be.

6c. Reducing the Space of Numerical Estimation

There are likelihood functions of many parameters that have the property that part of it can be maximized explicitly, conditional on the rest, and the conditional covariance matrix can be evaluated or approximated directly. In these cases one can work solely with the reduced likelihood function, which is then to be maximized numerically.

6d. Unbounded Initial Space

If one or more parameters do not initially have finite legal ranges, then, as in all maximization techniques, one must start with a guess of the mean value and the variance of that parameter. These values are then used to generate the first set of points.

6e. Parameter Estimates Out of Range

The minimum of the quadratic fit to Λ can have one or more coordinates that lie outside of the bounds on the parameters $\underset{\sim}{\theta}$. For example, in the search data presented as an illustration for the 42-month-old children parameters associated with the initial- and favorite-box rules are effectively zero. Their estimates are negative. The cross sections shown in Figure 2 come from a similar model, and estimates for parameters 2 and 3 are clearly negative. There are three approaches in this situation.

First, one can simply proceed with the next iteration. Subsequent iterations of this kind may produce parameter estimates within the bound region.

Second, if a single parameter has an estimate consistently out of bounds, one can replace this parameter with a fixed value equal to the approximate bound, thus producing a simpler model with one fewer parameter. If the maximum likelihood for this simpler model is close to that for the more complex model, then this simplification is justified.

If, however, several parameters have estimates consistently out of range, the problem is more difficult. Replacing any one of the parameters by the appropriate boundary value might (or might not) cause other estimates to change substantially. If several parameters are involved, there may be a large number of possible orders in which parameters can be removed. Although we have not encountered this situation, we present the following general method for finding the maximum likelihood estimates subject to the constraint. Suppose the constraints are of the interval type, $a_i \leq \theta_i \leq b_i, i = 1, 2, \dots N$. Suppose without loss of generality that $\underset{\sim}{\theta}_0 = (\underset{\sim}{\theta}_{01}, \underset{\sim}{\theta}_{02})$ where $\underset{\sim}{\theta}_{0,1}$ satisfies the constraints but each element of $\theta_{0,2}$ does not. Finally suppose that $\theta_{0,2}$ has K elements. We make use of the following lemma, a simple consequence of well known facts about marginal and conditional normal distributions.

LEMMA: *Let*

$$Q(y,z) = [(y - \underset{\sim y}{\mu}), (z - \underset{\sim z}{\mu})] \begin{bmatrix} \Sigma^{11} & \Sigma^{12} \\ \Sigma^{2} & \Sigma^{22} \end{bmatrix} \begin{bmatrix} (y - \underset{\sim y}{\mu}) \\ (z - \underset{\sim 2}{\mu}) \end{bmatrix}$$

be a quadratic form in y *and* z *where*

$$\begin{bmatrix} \Sigma^{11} & \Sigma^{12} \\ \Sigma^{21} & \Sigma^{22} \end{bmatrix}$$

is symmetric and positive definite. Then

$$Q(y,z) = (z - d)'\Sigma^{22}(z - d) - (y - \underset{\sim y}{\mu})'\Sigma_{11}^{-1}(y - \underset{\sim y}{\mu}),$$

where

$$d = \underset{\sim z}{\mu} - (\Sigma^{22})^{-1}\Sigma^{21}(y - \underset{\sim y}{\mu})$$

and

$$\Sigma_{11}^{-1} = \Sigma^{11} - \Sigma^{12}(\Sigma^{22})^{-1}\Sigma^{21}.$$

The minimum, over z, *of* $Q(y,z)$ *subject to the constraint that* $y = y_0$ *is found at*

$$z = \mu_z - (\Sigma^{22})^{-1}\Sigma^{21}(y_0 - \mu_y) \tag{8}$$

and the value of Q at that point is

$$(y_0 - \mu_y)'(\Sigma_{11})^{-1}(y_0 - \mu_y), \tag{9}$$

which may be interpreted as a "penalty," the amount that the minimum of Q(y,z) is made larger for having to satisfy the constraint $y = y_0$.

We can apply equation (8) by setting y equal to every proper subset of the model parameters $\underset{\sim}{\theta}_0$, and each coordinate of $\underset{\sim}{\mu}$ equal to a boundary (a_i,b_i) for the corresponding coordinate of $\underset{\sim}{\theta}_0$. The vector z contains the remaining coordinates of $\underset{\sim}{\theta}_0$. We solve each of these equations, and check that each d_i satisfies the constraints $a_i \leq d_i \leq b_i$. The likelihood is maximum (Q is minimum) for $\underset{\sim}{\theta}_0$ such that d_i satisfied the preceding constraints with minimum penalty from equation (9). The maximum-likelihood estimate for $\underset{\sim}{\theta}_0$ is then the corresponding values of y (boundary values) and z (freely adjusted values).

6f. Estimated V Not Positive Definite

The symmetry of the estimate of V^{-1} given equations (5a) and (5b) assures that V^{-1} has real eigenvalues. If these eigenvalues are all positive, then the estimate of V is positive definite, and hence a possible covariance matrix $\underset{\sim}{\theta}$. However, some eigenvalues may be zero (or very close to zero) or may be negative. We deal with these case separately.

If the estimated V^{-1} has nearly zero eigenvalues, then Λ is nearly flat in the direction(s) spanned by the associated eigenvectors. The data could be uninformative about the parameter(s) in question. For example, in the search data, this situation consistently occurred for parameters δ_1 and δ_2 describing the probability of each location being a child's favorite location.

In the first case the solution is simply to remove these parameters from the model and replace them by some plausible values. Since the likelihood function depends so little on these parameters, the exact choice of value has little effect on the model. If the second case is suspended, a finer grid of points should be generated. This may yield a quadratic fit in Λ has a downward curvature in the directions spanned by the associated eigenvectors. Thus the maximum-likelihood parameter estimates must occur at the boundary of the allowable space. The following modification of the cross-sectional graphs shown in Figure 2 should clarify the problem. Figure 2 shows variation of Λ along the direction of the parameters in the model. A modified graph could show variation in Λ along the directions on the eigenvectors of the model, thus indicating the shape in Λ in the direction were the curvature is negative. We have, however, never encountered this situation.

7. Conclusion

Familiar numerical methods for maximizing likelihood functions make little use of the global structure of these functions. While conventional hill climbing methods such as Newton-like algorithms also make use of the locally quadratic nature of the surface, they typically make use of many fewer points to make the next estimate. An exactly quadratic surface will be maximized in a single step by such an algorithm, but deviations from an exactly quadratic shape, such as local modes, could cause difficulties. An alternate strategy, a grid search, uses no structural information about the surfaces.

By contrast we use many points to fit a quadratic surface in an entire region and iterate this fitting. The procedure is demonstrated on a real problem of interest in developmental psychology (Sophian, Larkin and Kadane, 1985). We believe that these ideas can provide a valuable tool for studying complex likelihoods in multidimensional spaces.

The question remains of when to use which method. We think this is a question best addressed by several researchers each working on their own problems of interest. However, we do think that the method described here is an alternative that should be considered. Possibly the ultimate algorithm will be a hybrid that will use diagnostics to move among methods intelligently.

Acknowledgments

This work was supported in part by NSF, by NICHHD, by the Defense Advanced Research Projects Agency (DOD), and by the Office of Naval Research. We thank Rob Kass and Dorota Dabrowska for their helpful comments.

References

Chambers, J.M. (1973). "Fitting Nonlinear Models: Numerical Techniques," *Biometrika*, Vol. 60, 1-13.

Clough, D.L. (1969). "An Asymptotic Extreme-Value Sampling Theory for Estimation of a Global Maximum," *Canadian Operations Research Society*, Vol. 7, 102-115.

Copas, J.B. (1975). "On the Unimodality of the Likelihood for the Cauchy Distribution," *Biometrika*, Vol. 62, 701-704.

Dennis, J.E. and R.B. Schnabel (1983). *Numerical Methods for Unconstrained Optimization and Nonlinear Equations*, Englewood Cliffs, NJ: Prentice-Hall.

Gill, P.E., W. Murray, and M.H. Wright (1981). *Practical Optimization*, London: Academic Press.

Kass, R.E. (1983). "The Rate of Convergence of the Fisher Scoring and Gauss-Newton Algorithms," Technical Report #284), Department of Statistics, Carnegie-Mellon.

Kass, R.E. (1984). "A Program to Compute Observed Information Without Analytic Derivatives," Technical Report #342), Department of Statistics, Carnegie-Mellon.

LeCam, L. (1953). "On Some Asymptotic Properties of Maximum Likelihood Estimates and Related Bayes Estimates," *University of California Publications in Statistics*, Vol. 1, 277-328.

Morgan, B.J.T. (1978). "A Simple Comparison of Newton-Ralphson and the Method of Scoring," *Int. J. Math Educ. in Sci. & Tech.*, Vol. 9, No. 3, 343-348.

Morgan, B.J.T. (1983). "Illustration of Three Dimensional Surfaces," *BIAS*, Vol. 10, No. 2, 158-166.

Nelder, J.A. and R. Mead (1965). "A Simplex Method for Function Maximization," *Computer Journal*, Vol. 7, 308-313.

Perlman, M.D. (1983). "The Limiting Behavior of Multiple Roots of the Likelihood Equation," in *Recent Advances in Statistics*, M.H. Rizvi, J. Rustagi and D. Siegmund, eds., New York: Academic Press.

Ross, G.J.S. (1970). "The Efficient Use of Functional Minimization of Non-Linear Maximum-Likelihood Estimation," *Applied Statistics*, Vol. 19, 205-222.

Ross, G.J.S. (1975). "Simple Non-Linear Modeling for General Use," *Bull. I.S.I.*, Vol. 46, No. 2, 585-591.

Schim van der Loeff, S. and R. Harkema (1976). "Three Models of Firm Behavior: Theory and Estimation with an Application to the Dutch Manufacturing Sector," *The Review of Economics and Statistics*, Vol. 58, 13-21.

Sophian, C., J.H. Larkin, and J.B. Kadane (1985). "A Developmental Model of Search: Stochastic Estimation of Children's Rule Use," in *Children's Searching: The Development of Search Skill and Spatial Representation*, H.M. Wellman, ed., Hillsdale, NJ: Lawrence Erlbaum Associates.

Tierney, L.J. and J.B. Kadane (1986). "Accurate Approximations for Posterior Moments and Marginal Densities," *JASA*, Vol. 81, 82-86.

Veall, M.R. (1987). "Testing for a Global Maximum," Unpublished Technical Report, Program for Quantitative Studies in Economics and Population, McMaster University.

Veall, M.R. (1988). "Using the Clough Method to Test for a Global Maximum: Performance on Some Test Functions," Unpublished Technical Report, Department of Economics, McMaster University.

Wald, A. (1949). "Note on the Consistency of the Maximum Likelihood Estimate," *Ann. Math Statis.*, Vol. 20, 595-601.

Walker, A.M.F. (1969), "On the Asymptotic Behavior of Posterior Distributions," *Journal of the Royal Statistical Society, Series B*, Vol. 31, 80-88.

Bayesian and Likelihood Methods in Statistics and Econometrics
S. Geisser, J.S. Hodges, S.J. Press and A. Zellner (Editors)

THE VALIDITY OF POSTERIOR EXPANSIONS BASED ON LAPLACE'S METHOD

Robert E. Kass, Luke Tierney, and Joseph B. Kadane
Carnegie-Mellon University, University of Minnesota, and Carnegie-Mellon University

Abstract

We present methods for justifying heuristic derivations of asymptotic expansions for predictive densities, odds factors, marginal posterior densities, and posterior moments given by various authors. In applications, an observed sample is imbedded in an infinite sequence to which an expansion is applied. We therefore begin by specifying what constitutes a well-behaved sequence using an approach similar to that of Chen (1985). We then go on to give conditions under which sequences are well-behaved with probability one. The latter results are closely related to those of Johnson (1970).

Key Words and Phrases: Bayesian inference, second-order theory, asymptotics.

1. Introduction

Laplace's method for the asymptotic evaluation of integrals (Laplace, 1847) is a basic technique of mathematical analysis (described, for instance, in Erdelyi, 1956), which has been used frequently in statistical theory. In Bayesian inference Laplace's method has been used by various authors to derive approximate posterior expectations and marginal densities, predictive densities, and Bayes factors. The purpose of the present paper is to present methods for justifying these approximations. Our approach is similar to that used by Chen (1985) in discussing asymptotic posterior Normality, and our results are closely related to those of Johnson (1970). For additional references and discussion, with emphasis on the role of asymptotic expansions in interactive data analysis, see Kass, Tierney, and Kadane (1988).

The primary problem we concern ourselves with is that of approximating the expectation of a function $g(\theta)$, where θ is a parameter vector having a posterior probability density based on a sample of n observations. The posterior density will be proportional to a likelihood function $L = L_n$ and a prior density π so that the expectation we seek to evaluate asymptotically will have the form

$$E(g(\theta)) = \frac{\int g(\theta)L(\theta)\pi(\theta)d\theta}{\int L(\theta)\pi(\theta)d\theta}. \tag{1.1}$$

If the likelihood function is well-behaved, it will have a dominant peak at its maximum. Laplace's method will then be suitable for application to both the numerator and denominator of (1.1): each may be written in the form

$$\mathcal{I} = \int b(\theta)\exp[-nh_n(\theta)]d\theta \tag{1.2}$$

where $-h_n$ has a maximum at a point $\hat{\theta}$ and b and $h = h_n$ may be expanded in a Taylor series about $\hat{\theta}$. The method consists of approximating the integrand by the product of a low-order polynomial and the factor

$$\exp[-\frac{n}{2}\Sigma h_{ij}(\theta_i - \hat{\theta}_i)(\theta_j - \hat{\theta}_j)]$$

where h_{ij} is the (i,j)-component of the Hessian of h at $\hat{\theta}$; the resulting integral is then evaluated in terms of the moments of a multivariate Normal distribution.

Note that there are infinitely many choices for the functions b and h in any particular application. For example, in writing the denominator of (1.1) in the form of (1.2), two obvious choices for h are $h_n(\theta) = n^{-1}\log[L(\theta)]$ and $h_n(\theta) = n^{-1}\log[L(\theta)\pi(\theta)]$, which determine $\hat{\theta}$ to be, respectively, the MLE and the posterior mode. Tierney and Kadane (1986) used the latter in the denominator and then, assuming $g(\theta)$ was positive, took $h_n(\theta) = n^{-1}\log[g(\theta)L(\theta)\pi(\theta)]$ in the numerator. They showed that for this pair of choices, which we will call *fully exponential*, first-order approximations in each of the numerator and denominator of (1.1), having error of order $O(n^{-1})$, lead to a second-order approximation of the ratio, having error of order $O(n^{-2})$. This may be contrasted with the methods of Lindley (1961, 1980) and Mosteller and Wallace (1964), in which h was the same function in both numerator and denominator of (1.1), and which involved third derivative terms in the second-order approximations.

In a recent paper, Tierney, Kass, and Kadane (1989b) obtained a second-order expansion of the expectation of a general function $g(\theta)$ (which need not be positive), by applying the fully exponential method to approximate the moment generating function $E(\exp[sg(\theta)])$ and then differentiating. They showed that this expansion is formally equivalent to the one used by Lindley and Mosteller and Wallace, while maintaining the computational simplicity of the fully exponential method for positive functions.

The arguments given by many authors who have applied Laplace's method to the calculation of features of posterior distributions have usually been heuristic. They have essentially relied on two presumptions: first, the well-known conditions for the rigorous justification of Laplace's method in producing *analytical* expansions (see Erdelyi, 1956) may be replaced by suitable extensions to cover the cases of interest in statistics; that is, for well-behaved sequences of data, the posterior distribution will be sufficiently well-behaved that Laplace's method will apply. Second, under certain general conditions, these well-behaved data sequences will be the rule rather than the exception in a *stochastic* sense. Using arguments related to those of Le Cam (1953), Johnson (1967, 1970) effectively showed that in certain cases these presumptions are correct, by providing a set of conditions under which the expansions will be valid with probability one. Johnson treated the i.i.d. case in detail, and noted how his results could be extended to cover the Markov case. (See also Johnson and Ladalla, 1979, who treated multiparameter problems, and Ghosh, Sinha, and Joshi, 1982, Joshi, 1984, and Bickel, Götze, and van Zwet (1985) who provided rigorous treatments of uniform expansions.)

Our approach is somewhat different than Johnson's in that we explicitly separate the analytical and stochastic parts of the problem. From the Bayesian point of view this seems desirable since, in applications, the use of Laplace's method effectively imbeds the *given* set of observations into an infinite sequence to which the method applies. We state sufficient conditions on the behavior of the sequence of loglikelihood functions for validity of analytical expansions, and call a sequence of loglikelihood functions satisfying these conditions *Laplace regular*. In linear regression, Laplace regularity becomes an assumption on the sequence of design matrices and residual vectors: the eigenvalues of X^TX/n must be bounded and bounded away from zero, and $\hat{\sigma}_n$ must be bounded away from zero. No probability statements are required. We then go on to consider the family of sampling densities constituting the model and provide conditions to guarantee that Laplace regularity of the loglikelihood functions will hold with probability one. When a model satisfies these conditions we refer to it as a *Laplace regular model.* In linear regression, the model is Laplace regular if the eigenvalues of X^TX/n are bounded and bounded away from zero. It is apparent from an examination of Johnson's arguments that demonstrating Laplace regularity of a model is essentially what is needed in extending his results. We have not tried to establish a very general theorem guaranteeing Laplace regularity of stochastic process models, but instead intend Laplace regularity to be verified for interesting special families as the need arises in practice.

We treat second-order approximations, that is, expansions that incorporate the first correction term beyond the MLE or modal approximation for expectations and the Normal approximation for marginal densities. Higher-order theory is analogous. Our analytical results are presented in section 2, and our results relating Laplace regularity of models to Laplace regularity of sequences of loglikelihood functions are given in section 3.

We should emphasize the technical nature of this work on two accounts. First, although our point of view is Bayesian, our results may be useful in non-Bayesian theory as well. Second, to say that an expansion is valid asymptotically is not to say that it provides a good approximation in any particular instance. Although these expansions have been shown to be quite useful (in the references cited by Kass, Tierney, and Kadane, 1988), it is not hard to construct finite-sample examples in which the approximations perform poorly. Numerical techniques including Gaussian quadrature and Monte Carlo remain essential alternatives to asymptotic methods in practical work. See Naylor and Smith (1982) and Smith et al. (1985) for discussion of these. The choice of parameterization in which Laplace's method is applied is also an important practical concern, as shown by the examples in Tierney, Kass, and Kadane (1988), and Achcar and Smith (1989). Alternative expansions, based on Pearson family kernels that replace the Normal kernel of Laplace's method, have been shown by Morris (1988) to offer improvements in many one-parameter problems.

2. Analytical Expansions

Suppose Θ is an open subset of R^m, $\{h_n : n = 1, 2, \ldots\}$ is a sequence of six-times continuously differentiable real functions on Θ, having local minima $\{\hat{\theta} = \hat{\theta}_n : n = 1, 2, \ldots\}$, and b is a four-times continuously differentiable real function on Θ. We use $B_\delta(\theta)$ to denote the open ball of radius δ centered at θ. The Hessian of $h = h_n$ at θ will be denoted by $D^2h[\theta]$, and its (i,j)-component will be written either as $\partial_{ij}h[\theta]$ or h_{ij} while the components of its inverse will be written h^{ij}.

The pair $(\{h_n\}, b)$ will be said to satisfy the *analytical assumptions for Laplace's method* if there exist positive numbers ε, M and η, and an integer n_0 such that $n \geq n_0$ implies

(i) for all $\theta \in B_\varepsilon(\hat{\theta})$ and all $1 \leq j_1, \ldots, j_d \leq m$ with $0 \leq d \leq 6$, $| \partial_{j_1 \cdots j_d} h_n[\theta] | < M$;

(ii) $\det(D^2 h_n[\hat{\theta}]) > \eta$;

and the integral in (1.2) exists and is finite, and

(iii) for all δ for which $0 < \delta < \varepsilon$, $B_\delta(\hat{\theta}) \subseteq \Theta$ and

$$[\det(nD^2 h_n[\hat{\theta}])]^{1/2} \cdot \int_{\Theta - B_\delta(\hat{\theta})} b(\theta) \exp[-n(h_n(\theta) - h_n(\hat{\theta}))] d\theta = O(n^{-2}).$$

THEOREM 1: *If* $(\{h_n\}, b)$ *satisfy the analytical assumptions for Laplace's method then*

$$\int_{\Theta} b(\theta)\exp[-nh(\theta)]d\theta = (2\pi)^{m/2}[\det(nD^2h)]^{-1/2}\exp[-nh(\hat{\theta})]$$

$$\cdot\left(b(\hat{\theta}) + \frac{1}{n}\left\{\frac{1}{2}\Sigma h^{ij}b_{ij} - \frac{1}{6}\Sigma h_{ijk}b_s\mu^4_{ijks}\right.\right.$$

$$+\frac{1}{72}b(\hat{\theta})\Sigma h_{ijk}h_{qrs}\mu^6_{ijkqrs} \tag{2.1}$$

$$\left.\left.-\frac{1}{24}b(\hat{\theta})\Sigma h_{ijks}\mu^4_{ijks}\right\} + O(n^{-2})\right)$$

where μ^4_{ijks} *and* μ^6_{ijkqrs} *are the fourth and sixth central moments of a multivariate Normal distribution having covariance matrix* $(D^2h)^{-1}$, *i.e.,*

$$\begin{aligned}
\mu^4_{ijks} &= \sigma_{ij}\sigma_{ks} + \sigma_{ik}\sigma_{js} + \sigma_{is}\sigma_{jk}\\
\mu^6_{ijkqrs} &= \sigma_{ij}\sigma_{kq}\sigma_{rs} + \sigma_{ij}\sigma_{kr}\sigma_{qs} + \sigma_{ij}\sigma_{ks}\sigma_{qr}\\
&\quad + \sigma_{ik}\sigma_{jq}\sigma_{rs} + \sigma_{ik}\sigma_{jr}\sigma_{qs} + \sigma_{ik}\sigma_{js}\sigma_{qr}\\
&\quad + \sigma_{iq}\sigma_{jk}\sigma_{rs} + \sigma_{iq}\sigma_{jr}\sigma_{ks} + \sigma_{iq}\sigma_{js}\sigma_{kr}\\
&\quad + \sigma_{ir}\sigma_{jh}\sigma_{qs} + \sigma_{ir}\sigma_{jq}\sigma_{ks} + \sigma_{ir}\sigma_{js}\sigma_{kq}\\
&\quad + \sigma_{is}\sigma_{jk}\sigma_{qr} + \sigma_{is}\sigma_{jq}\sigma_{kr} + \sigma_{is}\sigma_{jr}\sigma_{kq}
\end{aligned}$$

where $\sigma_{ij} = h^{ij}$ *and all derivatives* b_s, *etc., are evaluated at* $\hat{\theta}$.

PROOF: For simplicity, we treat the case $m = 1$, the general case involving obvious modifications. Let $\hat{\theta}$ be the maximum of $-h_n$, let $u \equiv n^{1/2}(\theta - \hat{\theta})$ so that for a given u, $(\theta - \hat{\theta})^k = (n^{-1/2}u)^k = 0(n^{-k/2})$, and consider the expansion of the integrand $b(\theta)\exp[-nh_n(\theta)]$ about $\hat{\theta}$ that includes terms of order greater than $O(n^{-2})$. These terms involve the cubic expansion of b, the quadratic expansion of $\exp[-x]$, and the quintic expansion of h_n, as follows. We have

$$nh(\theta) = \{nh(\hat{\theta})\} + \left\{\frac{1}{2}h''(\hat{\theta})u^2\right\} + \left\{\sum_{k=3}^{5}(n^{(k-2)/2}k!)^{-1}h^{(k)}(\hat{\theta})u^k\right\} + r_n(u)$$

where $r_n(u)$ is bounded over $B_\varepsilon(\hat{\theta})$ by a polynomial in u having coefficients that are of order $O(n^{-2})$. Applying the expansion of $\exp[-x]$ to the third bracketed term above, and expanding b, we get

$$b(\theta)\exp[-nh(\theta)] = \{\exp[-nh(\hat{\theta})]\}\left\{\exp[-\frac{1}{2}h''(\hat{\theta})u^2]\right\}$$

$$\cdot\left\{1 - \frac{1}{6}n^{-1/2}h^{(3)}(\hat{\theta})u^3 + \frac{1}{72}n^{-1}[h^{(3)}(\hat{\theta})]^2u^6 - \frac{1}{24}n^{-1}h^{(4)}(\hat{\theta})u^4\right.$$

$$\left.-\frac{1}{120}n^{-3/2}h^{(5)}(\hat{\theta})u^5 + \frac{1}{72}n^{-3/2}h^{(3)}(\hat{\theta})h^{(4)}(\hat{\theta})u^7 + R_{1n}(\theta,\hat{\theta})\right\}$$

$$\cdot\left\{b(\hat{\theta}) + n^{-1/2}b'(\hat{\theta})u + \frac{1}{2}n^{-1}b''(\hat{\theta})u^2\right.$$

$$\left.+\frac{1}{6}n^{-3/2}b^{(3)}(\hat{\theta})u^3 + R_{2n}(\theta,\hat{\theta})\right\} \tag{2.2}$$

where $R_{jn}(\theta,\hat{\theta}) = O(n^{-2})$ uniformly on $B_\varepsilon(\hat{\theta})$, j=1,2. A heuristic argument would conclude by using $d\theta = n^{-1/2}du$, and noting that the odd moments of a Normal distribution vanish so that formal integration of (2.2) would yield (2.1).

Let us rewrite (2.2) as

$$b(\theta)\exp[-nh(\theta)] = \{\exp[-nh(\hat{\theta})]\}\left\{\exp[-\frac{1}{2}h''(\hat{\theta})u^2]\}\{I_n(\theta,\hat{\theta}) + R_n(\theta,\hat{\theta})\right\}$$

where $R_n(\theta,\hat{\theta})$ is the sum of all terms involving R_{1n}, R_{2n}, or u^s with $s \geq 7$. This expansion holds (for $n \geq n_0$) on a neighborhood $B_\delta(\hat{\theta})$ with $\delta < \varepsilon$. Condition (iii) reduces consideration to the integral of $b(\theta)\exp[-nh(\theta)]$ over $B_\delta(\hat{\theta})$, on which region expansion (2.2) holds; there are thus two integrals to evaluate, one involving I_n and one involving R_n.

The integral of $\{\exp[-\frac{1}{2}h''(\hat{\theta})u^2]\}I_n(\theta,\hat{\theta})$ over $B_\delta(\hat{\theta})$ is evaluated by first noting that on changing variables from θ to u the domain of integration becomes $u(B_\delta(\hat{\theta})) = B_{\delta(n)}(0)$ where $\delta(n) = n^{1/2}\delta$. Since this domain is expanding at the rate $O(n^{1/2})$, and I_n is a polynomial in u, which is being integrated against a Normal density in u, the replacement of this domain by the whole real line incurs an error of exponentially decreasing order. Using $d\theta = n^{-1/2}du$, writing $\mu^4 = 3[h''(\hat{\theta})]^{-2}$ and $\mu_6 = 15[h''(\hat{\theta})]^{-3}$ as the fourth and sixth central moments of a Normal distribution having precision $h''(\hat{\theta})$, and noting that the odd moments vanish then yields the right-hand side of (2.1) as the integral of $\{\exp[-\frac{1}{2}h(\hat{\theta})]\} \cdot \{\exp[-\frac{1}{2}h''(\hat{\theta})u^2]\} \cdot I_n(\theta,\hat{\theta})$ over $B_\delta(\hat{\theta})$.

The terms comprising R_n may be represented explicitly using the mean value form of the remainders in terms of higher derivatives of b and h evaluated at points between $\hat{\theta}$ and θ, e.g., one such term is $(1/120)n^{-2}h^{(6)}(\xi)u^6$, where $\xi = \xi(u)$ is between $\hat{\theta}$ and θ; it is one piece of the error term appearing as $R_{1n}(\theta,\hat{\theta})$. Now $|h^{(6)}(\theta)| \leq M$ on $B_\delta(\hat{\theta})$ implies $|\int \exp\{-\frac{1}{2}h''(\hat{\theta})u^2\}n^{-2}h^{(6)}(\xi)u^6 du| \leq Mn^{-2}\int \exp\{-\frac{1}{2}h''(\hat{\theta})u^2\}u^6 du = O(n^{-2})$ (the integrals being taken over $B_{\delta(n)}(0)$ where $\delta(n) = n^{1/2}\delta$). The other terms are similar. Thus, the integral of $\{\exp[-nh(\hat{\theta})]\}\cdot\{\exp[-\frac{1}{2}h''(\hat{\theta})u^2]\}\cdot R_n(\theta,\hat{\theta})$ over $B_\delta(\hat{\theta})$ is of order $O(n^{-2})$. □

Now suppose $-nh_n$ is the log of a posterior density. Our conditions (i)-(iii) are closely related to those of Chen (1985). Chen assumes $\Theta = R^m$. Then (iii) with $b = 1$ and the right-hand side being $o(1)$ becomes Chen's condition (C3). It follows from Chen's Theorem 2.1 that under (i) and (ii) this condition is necessary and sufficient for asymptotic posterior Normality. To obtain an expansion of the expectation (1.1) we need only assume that both $(\{h_n\},\pi)$ and $(\{h_n\},g\pi)$ satisfy the analytical assumptions for Laplace's method, where $-nh_n$ is the loglikelihood function. In applications, however, it is often easier to verify an alternative condition, which is stronger than condition (iii) when $b = 1$:

(iii′) for all δ for which $0 < \delta < \varepsilon$, $B_\delta(\hat{\theta}) \subseteq \Theta$ and

$$\limsup_{n\to\infty}\sup_{\theta}\{h_n(\hat{\theta}) - h_n(\theta) : \theta \in \Theta - B_\delta(\hat{\theta})\} < 0.$$

An additional convenience of this condition is that we will be able to treat at once the posterior expectations of all four-times continuously differentiable functions g based on four-times continuously differentiable priors π. When a sequence of loglikelihood functions $-nh_n(\theta) = \ell_n(\theta) = \log L_n(\theta)$ satisfies conditons (i), (ii), and (iii′) we will say it is *Laplace regular.*

EXAMPLE: *Linear regression.* The Normal linear model is $Y_i = x_i^T\beta + \varepsilon_i$ where ε_i are i.i.d. Normal$(0,\sigma^2)$, β is an unknown vector in R^p and x_i is a given vector in R^p, for i = 1, 2, ..., n. When the vectors x_1^T, ..., x_n^T are collected into a matrix X_n, the loglikelihood function based on $y = (y_1, \ldots, y_n)$ becomes

$$\ell(\beta,\sigma) - \ell(\hat{\beta},\hat{\sigma}) = -\frac{1}{2\sigma^2}\parallel X_n(\beta-\hat{\beta})\parallel^2 - \frac{n}{2}\left(\frac{\hat{\sigma}^2}{\sigma^2} - 1 - \log\left(\frac{\hat{\sigma}^2}{\sigma^2}\right)\right)$$

with $\hat{\beta}$ the least-squares estimator and $\hat{\sigma}^2 = n^{-1}\parallel y - X_n\hat{\beta}\parallel^2$. Let λ_{n1} and λ_{nm} be the largest and smallest eigenvalues of $n^{-1}X_n^TX_n$. From the inequality $\log u \leq u-1$, which holds for all $u \geq 0$, condition (iii′) of Laplace regularity holds if there exists $A > 0$ such that for all sufficiently large n, $A < \hat{\sigma}_n$ and for all $\delta > 0$

$$\sup_{\theta}\{-\frac{1}{n}\parallel X_n(\beta-\hat{\beta})\parallel^2 : \beta \in B_\delta(\hat{\beta})^c\} < 0.$$

The latter condition, in turn, is satisfied if there exists $a > 0$ such that for all sufficiently large n, $a < \lambda_{nm}$. (In the nonsingular case, the condition $(X_n^T X_n)^{-1} \to 0$ is necessary and sufficient for consistency of $\hat{\beta}$ under the much weaker assumption that ε_i are i.i.d. with mean zero and variance $\sigma^2 < \infty$; see Lai, Robbins, and Wei, 1978.) Condition (i) requires, in addition, that there exist $b < \infty$ such that, for all sufficiently large n, $\lambda_{n1} < b$. Condition (ii) is immediate. □

Before stating our theorems for Laplace-regular sequences, we first present a lemma which will be needed and a corollary which is of some interest on its own. We let $y^{(n)} = (y_1, \ldots, y_n)$ denote the sample on which the loglikelihood ℓ_n is based.

LEMMA 2: *Suppose $\{\ell_n\}$ is a Laplace-regular sequence of loglikelihood functions, π is a four-times continuously differentiable positive real function on Θ, and for some nonnegative integer n_0 the posterior $P\{\cdot \mid y^{(n_0)}\}$ based on $y^{(n_0)}$ and the prior π exists (i.e., the integral of $\exp\{\ell_{n_0}\}\pi$ is finite); then, with ε given in the definition of Laplace regularity, for any δ for which $0 < \delta < \varepsilon$, there exists a positive number c_1 such that for all sufficiently large n,*

$$\int_{\Theta - B_\delta(\hat{\theta})} \exp\{\ell_n(\theta) - \ell_n(\hat{\theta})\}\pi(\theta)d\theta < \exp\{-nc_1\}. \tag{2.3}$$

COROLLARY 3: *Under the conditions of Lemma 2, for any δ for which $0 < \delta < \varepsilon$, there exists a positive number c_2 such that for all sufficiently large n,*

$$P(\Theta - B_\delta(\hat{\theta}) \mid y^{(n)}) < \exp\{-nc_2\}. \tag{2.4}$$

PROOFS: Suppose first that π is a proper prior (i.e., it integrates to one); then, by condition (iii′), there exists $c_1 > 0$ such that (2.3) holds. Now note that Theorem 1 may be applied with $h_n = -n^{-1}\ell_n$ and $b = \pi$. This provides the normalizing constant for the posterior: retaining only the first term in the expansion, we have

$$\int_\Theta \exp\{\ell_n(\theta) - \ell_n(\hat{\theta})\}\pi(\theta)d\theta = (2\pi)^{m/2}[\det(-D^2\ell_n(\hat{\theta}))]^{-1/2}\pi(\hat{\theta}) \cdot \{1 + O(n^{-1})\}.$$

Combining this with (2.3), we obtain (2.4) for the case in which π is proper.

The general case is treated by applying the above argument with the prior π replaced by the posterior density based on $y^{(n_0)}$ (and π). This replacement is possible because condition (iii′) of Laplace regularity also holds when applied to the "loglikelihood" $\ell_{n,n_0}(\theta) = \log p(y_{n_0+1}, \ldots, y_n \mid y^{(n_0)}, \theta)$. This is so because $\limsup_{n\to\infty} \sup_\theta\{n^{-1}[\ell_{n_0}(\theta) - \ell_{n_0}(\hat{\theta})] : \theta \in \Theta - B_\delta(\hat{\theta})\} = 0$. □

We now present the main results of this section, establishing second-order approximations to the posterior expectation of a real function g on Θ: first in a standard form, then in a *fully exponential* form used by Tierney and Kadane (1986); finally we give the basic approximation for a marginal density.

Given a prior π, loglikelihood ℓ_n, positive function γ, and real function g we define h_n and b by $-nh_n = \ell_n + \log\gamma$ and $b = \pi/\gamma$. We then have

$$E(g(\theta)) = \frac{\int g(\theta)b(\theta)\exp[-nh_n(\theta)]d\theta}{\int b(\theta)\exp[-nh_n(\theta)]d\theta} \tag{2.5}$$

where the integrals are taken over Θ. The approximation we justify, in the notation of Theorem 1, is

$$E(g(\theta)) = g(\hat{\theta}) + \frac{1}{n}\Sigma g_i h^{ij}\left\{(b_j/b(\hat{\theta})) - \frac{1}{2}\Sigma h^{rs}h_{rsj}\right\} + \frac{1}{2n}\Sigma h^{ij}g_{ij} + O(n^{-2}). \tag{2.6}$$

Approximation (2.6) was used and discussed by Mosteller and Wallace (1964). The special case $\gamma = 1$ makes $b = \pi$ while $\hat{\theta}$ becomes the MLE, and the choice $\gamma = \pi$ makes $b = 1$ with $\hat{\theta}$ becoming the posterior mode. The latter was used by Lindley (1961, 1980).

THEOREM 4: *Under the conditions of Lemma 2 if, in addition, γ is a six-times and g is a four-times continuously differentiable real function on θ, γ is positive, and the posterior expectation of g, based on $y^{(n_0)}$ and π, is finite, then the posterior expectation of g has the expansion (2.6).*

PROOF: Letting $p(\theta \mid y^{(n_0)})$ be the posterior density based on $y^{(n_0)}$ and π, using (2.3) and arguing as in the proof of Lemma 2, for any $\delta > 0$ there is a positive number c such that

$$\int_{\Theta - B_\delta(\hat{\theta})} g(\theta)\exp\{\ell_n(\theta) - \ell_n(\hat{\theta})\}\pi(\theta)d\theta < \exp\{-nc\}.$$

Thus, the order of error will be unaffected by replacing Θ with $B_\delta(\hat{\theta})$ in the numerator and denominator of (2.5) and Theorem 1 may be applied to both integrals. □

When g is positive, an alternative to approximating the expectation in (2.5) using (2.6) is to define $-nh_n = \ell_n + \log\pi$ and $-nh_n^* = -nh_n + \log g$ so that

$$E(g(\theta)) = \frac{\int \exp[-nh_n^*(\theta)]d\theta}{\int \exp[-nh_n(\theta)]d\theta}\,. \tag{2.7}$$

Tierney and Kadane (1986) showed that first-order approximations in the numerator and denominator of (2.7) yield a second-order approximation

$$E(g(\theta)) = \frac{\det(\Sigma^*)^{1/2}\exp[-nh_n^*(\theta^*)]}{\det(\tilde{\Sigma})^{1/2}\exp[-nh_n(\tilde{\theta})]} + O(n^{-2}) \tag{2.8}$$

where $\tilde{\Sigma} = (nD^2h[\tilde{\theta}])^{-1}$ and $\Sigma^* = (nD^2h^*[\theta^*])^{-1}$, with $\tilde{\theta}$ and θ^* maximizing $-h_n$ and $-h_n^*$.

THEOREM 5: Approximation of expectations based on the fully exponential method. *Suppose the sequence of loglikelihood functions $\{\ell_n\}$ is Laplace-regular, π and g are six times continuously differentiable positive real functions on Θ, and for some nonnegative integer n_0 the posterior based on $y^{(n_0)}$ and π exists and the posterior expectation of g is finite; then the posterior expectation of g has the expansion (2.8).*

PROOF: The proof of Theorem 5 is essentially the same as that of Theorem 4. □

Suppose now that $\Theta = \Theta_1 \times \Theta_2$, with Θ_1 and Θ_2 being of dimension m_1 and m_2, γ is positive, and h_n and b are defined by $-nh_n = \ell_n + \log\gamma$ and $b = \pi/\gamma$. The marginal posterior density of θ_1 becomes

$$p(\theta_1 \mid y^{(n)}) = \frac{\int_{\Theta_2} b(\theta_1,\theta_2)\exp[-nh_n(\theta_1,\theta_2)]d\theta_2}{\int_{\Theta} b(\theta_1,\theta_2)\exp[-nh_n(\theta_1,\theta_2)]d\theta_1 d\theta_2} \tag{2.9}$$

and a first-order approximation is

$$p(\theta_1 \mid y^{(n)}) = (2\pi)^{-m_2/2} \cdot \frac{\det(\Sigma(\theta_1))^{1/2}\exp[-nh_n(\theta_1,\hat{\theta}_2(\theta_1))]b(\theta_1,\hat{\theta}_2(\theta_1))}{\det(\hat{\Sigma})^{1/2}\exp[-nh_n(\hat{\theta})]b(\hat{\theta})} \cdot \{1+0(n^{-1})\} \tag{2.10}$$

where $\hat{\Sigma}$ and $\Sigma(\theta_1)$ are the inverses of the Hessian matrices of nh_n and $nh_n(\theta_1,\cdot)$ at $\hat{\theta}$ and $(\theta_1,\hat{\theta}_2(\theta_1))$ with $\hat{\theta}$ and $\hat{\theta}_2(\theta_1)$ maximizing $-h_n$ and $-h_n(\theta_1\,,\cdot)$. We write the marginal posterior mode, i.e., the maximum of $p(\theta_1 \mid y^{(n)})$, as $\hat{\theta}_1$. As noted by Tierney and Kadane (1986), approximation (2.10) holds uniformly on bounded regions about $\hat{\theta}_1$.

THEOREM 6: Approximation of marginal densities. *Suppose the sequence of log-likelihood functions $\{\ell_n\}$ is Laplace-regular, $\Theta = \Theta_1 \times \Theta_2$ with Θ_1 and Θ_2 being open subsets of R^{m_1} and R^{m_2} , π and γ are six times continuously differentiable positive* *real functions on Θ, and for some n_0 the posterior based on $y^{(n_0)}$ and π exists; then, with ε given in the definition of Laplace regularity, for any δ for which $0 < \delta < \varepsilon$, the marginal posterior density of θ_1 has the expansion (2.10) uniformly for all θ_1 in $B_\delta(\hat{\theta}_1)$,*

PROOF: As in the proof of Lemma 2, the denominator of (2.9) may be approximated by the denominator of (2.10) (multiplied by $(2\pi)^{m/2}$), with multiplicative error of order $O(n^{-1})$, which does not depend on θ_1. For any $\delta < \varepsilon$, Theorem 1 may be applied to approximate the numerator in (2.9) for each $\theta_1 \in B_\delta(\hat{\theta}_1)$. To obtain uniformity we return to the proof of Theorem 1. Repeating the argument following (2.2), we pick $\delta^* < (\varepsilon^2 - \delta^2)^{1/2}$ and then note that Θ_2 may be replaced by $B_{\delta^*}(\hat{\theta}_2(\theta_1)) \subseteq \Theta_2$, and each error incurred in the subsequent steps may be replaced by its supremum over $\theta_1 \in B_\delta(\hat{\theta}_1)$. □

A more general form of approximation (2.10) is presented by Tierney, Kass, and Kadane (1989a).

3. Stochastic Expansions

In this section we list assumptions on a family of distributions that guarantee the sequence of loglikelihood functions will be Laplace regular with probability one. We then briefly relate our approach to that of Johnson (1970) and to the problem of guaranteeing consistency of the MLE.

Suppose Θ is an open subset of R^m, $(\Omega, \mathcal{A})$ is a measurable space, $\mathcal{P} = \{P_\theta : \theta \in \Theta\}$ is a family of probability distributions on $(\Omega, \mathcal{A})$, and $\{Y_i : i = 1, 2, \ldots\}$ is a stochastic process on $(\Omega, \mathcal{A})$ with the Y_i's taking values in $(\mathcal{Y}, \mathcal{B})$ where $\mathcal{Y}$ is a subset of R^k and $\mathcal{B}$ is the class of Borel subsets of $\mathcal{Y}$. Letting $Y^{(n)} = (Y_1, \ldots, Y_n)$, we will assume that for all n the n-dimensional distributions of $Y^{(n)}$ are dominated by a σ-finite measure, and we will denote a density of $Y^{(n)}$ under P_θ by $p(y^{(n)} \mid \theta)$; we will denote the loglikelihood function by ℓ_n, i.e., $\ell_n(\theta) = \log p(y^{(n)} \mid \theta)$.

The family $\mathcal{P}$ will be called *Laplace-regular* if there exist densities $p(y^{(n)} \mid \theta)$ such that

(i) for all $y^{(n)}$ and θ, $p(y^{(n)} \mid \theta) > 0$ and for all $y^{(n)}$, ℓ_n is six times continuously differentiable;

(ii) for all $\theta_0 \in \Theta$ there exist $\varepsilon > 0$ and $M < \infty$ such that $B_\varepsilon(\theta_0) \subseteq \Theta$ and for all $1 \le j_1, \ldots, j_d \le m$ with $d \le 6$,

$$\limsup_{n\to\infty} \sup_{\theta}\{n^{-1} \mid \partial_{j_1}\cdots_{j_d} \ell_n[\theta] \mid : \theta \in B_\varepsilon(\theta_0)\} < M,$$

with P_{θ_0}-probability one;

(iii) for all $\theta_0 \in \Theta$ and some $\varepsilon > 0$,

$$\limsup_{n\to\infty} \sup_{\theta}\{n^{-1} \det(D^2\ell_n[\theta]) : \theta \in B_\varepsilon(\theta_0)\} < 0$$

with P_{θ_0}-probability one;

(iv) for all $\theta_0 \in \Theta$ and all $\delta > 0$,

$$\limsup_{n\to\infty} \sup_{\theta}\{n^{-1}[\ell_n(\theta) - \ell_n(\theta_0)] : \theta \in \Theta - B_\delta(\theta_0)\} < 0$$

with P_{θ_0}-probability one.

THEOREM 7: *If a family $\mathcal{P}$ is Laplace regular its sequence of loglikelihood functions is Laplace regular with P_{θ_0}-probability one, for all θ_0.*

PROOF: This is essentially immediate. We need only note that (iv) is a familiar consistency condition for the MLE, and it entails condition (iii′) of Laplace regularity for ℓ_n. □

EXAMPLE: *Linear and nonlinear regression.* Consider again the linear regression setting given in the previous section. The condition that the eigenvalues of $n^{-1}X_n^T X_n$ be bounded and bounded away from zero is precisely what is needed for Laplace regularity of the family.

Suppose instead that $Y_i = f(x_i, \beta) + \varepsilon_i$, with ε_i again being i.i.d. Normal$(0, \sigma^2)$ and f being a smooth function of β. To check condition (iii′) of Laplace regularity of the sequence of loglikelihood functions we could use

$$\begin{aligned}\ell(\beta,\sigma) - \ell(\hat\beta,\hat\sigma) &= -\frac{1}{\sigma^2}\Sigma(y_i - f(x_i,\hat\beta))(f(x_i,\hat\beta) - f(x_i,\beta)) \\ &\quad - \frac{1}{\sigma^2}\Sigma(f(x_i,\hat\beta) - f(x_i,\beta))^2 - \frac{n}{2}\left(\frac{\hat\sigma^2}{\sigma^2} - 1 - \log\left(\frac{\hat\sigma^2}{\sigma^2}\right)\right) \\ &\le -\frac{1}{\sigma^2}\Sigma(y_i - f(x_i,\hat\beta))(f(x_i,\hat\beta) - f(x_i,\beta)) \\ &\quad - \frac{1}{2\sigma^2}\Sigma(f(x_i,\hat\beta) - f(x_i,\beta))^2\end{aligned}$$

the inequality following from $u - 1 \ge \log u$, as in the linear regression case.

Alternatively, Laplace regularity of the family may be checked as follows. Suppose, in addition, that β lies in a compact subset B of R^p, $D_n(\beta,\beta') = \sum_{i=1}^n (f(x_i,\beta) - f(x_i,\beta'))^2$ satisfies

(A) $n^{-1}D_n(\beta,\beta')$ converges uniformly to a continuous function $D(\beta,\beta')$ and $D(\beta,\beta') = 0$ if and only if $\beta = \beta'$,

and $D_\beta f = (\partial_{\beta_1} f[x_i,\beta],\ldots,\partial_{\beta_p} f[x_i,\beta])$ satisfies

(B) $n^{-1}\sum_{i=1}^n (D_\beta f)^T(D_\beta f)$ converges uniformly to a matrix $A(\beta)$, which is continuous in β and positive definite,

while for all $1 \leq j_1, \ldots, j_d \leq p$ with $1 \leq d \leq 6$

(C) $n^{-1}\sum_{i=1}^n (\partial_{j_1} f)\ldots(\partial_{j_d} f)$ converges uniformly.

Condition (A) corresponds to conditions (a) and (b) of Jennrich (1969) while (B) and (C) together correspond to his conditions (c) and (d) when $d = 1$ or 2. Assuming these when ε_i are i.i.d. with mean zero and variance $\sigma^2 < \infty$, Jennrich derived the asymptotic consistency and Normality of the nonlinear least-squares estimator. From the proof of Jennrich's Theorem 6 it may be seen that conditions (A)-(C) imply Laplace regularity for Normal nonlinear regression models. Alternative conditions for asymptotic consistency and Normality are given by Wu (1981). □

We conclude by relating our approach to that of Johnson (1970). We have given Laplace regularity conditions on the family of densities, and have derived asymptotic expectations and marginal densities. Our approach easily yields an asymptotic approximation to the distribution function as well: the first term gives asymptotic Normality, as in the work of LeCam (1953), Walker (1969), and Heyde and Johnstone (1979). Compare also the discussion of Chen (1985). Johnson (1970) provides additional conditions primarily to guarantee the requisite tail behavior of our regularity condition (iv); they are essentially consistency conditions for the MLE. We now present a result that relates the problem of checking condition (iv) to that of checking consistency of the MLE.

THEOREM 8: (Laplace regularity in the i.i.d. case.) *Suppose $\mathcal{P}^{(1)} = \{P_\theta^{(1)} : \theta \in \Theta\}$ is a family of probability measures dominated by a σ-finite measure on a measurable space $(\mathcal{Y},\mathcal{B})$ with $\mathcal{Y}$ in R^k, $\mathcal{B}$ the Borel sets in $\mathcal{Y}$ and Θ an open subset of R^m, and let $\{Y_i : i = 1,2,\ldots\}$ be a sequence of i.i.d. observations from $P_\theta^{(1)}$ with density $p(y \mid \theta)$. The family $\mathcal{P} = \mathcal{P}^{(\infty)}$ of infinite product measures on $(\mathcal{Y}^{(\infty)}, \mathcal{B}^{(\infty)})$ is Laplace-regular if the following conditions are satisfied:*

(i) $\theta \neq \theta'$ implies $P_\theta^{(1)} \neq P_{\theta'}^{(1)}$;

(ii) for all y and $\theta, p(y \mid \theta) > 0$ and for all y, $p(y \mid \cdot)$ is six times continuously differentiable;

(iii) *for all θ_0, there exist a neighborhood $N_1(\theta_0)$ of θ_0, a positive integer r, and a measurable real function M_1 on the r-fold product space $(\mathcal{Y}^{(r)}, \mathcal{B}^{(r)})$ such for all $\theta \in N_1(\theta_0)$,*

$$r^{-1}\sum_{i=1}^{r}[\log p(y_i \mid \theta) - \log p(y_i \mid \theta_0)] < M_1(y_1, \ldots, y_r)$$

and, under $P_{\theta_0}, EM_1(Y_1, \ldots, Y_r) < \infty$;

(iv) *for all θ_0, there exist a neighborhood $N_2(\theta_0)$ of θ_0 and a measurable real function M_2 on such that for all $\theta \in N_2(\theta_0)$ and all $1 \leq j_1, \ldots, j_d \leq m$, with $d \leq 6$,*

$$| \partial_{j_1 \cdots j_d} \log p(y \mid \theta) | < M_2(y)$$

and, under $P_{\theta_0}, EM_2(Y) < \infty$;

(v) *for all $\theta_0, \det(D^2 f[\theta_0]) > 0$, where $f(\theta) = E[\log p(y \mid \theta_0) - \log p(y \mid \theta)]$ with the expectation taken under P_{θ_0} ;*

(vi) *for all θ_0 every sequence of maxima of the loglikelihood function ℓ_n is strongly consistent.*

PROOF: This follows from the strong law of large numbers and Theorem 4.1 of Perlman (1972), together with a version of Huzurbazar's Theorem given in Proposition 4.2 in Perlman (1983), and the accompanying discussion on page 352 of Perlman (1983; the result used here being applicable in the multiparameter case). □

REMARKS:

(1) When the integral of $p(y \mid \theta)$ may be differentiated twice under the integral sign, condition (v) becomes positive-definiteness of the Fisher information matrix.

(2) Conditions for consistency of the MLE are discussed at length in Perlman (1972).

(3) It is straightforward to verify Laplace regularity for exponential and curved exponential families. See Kass and Fu (1988). Crawford (1988) has also verified Laplace regularity for certain mixture models.

Acknowledgments

This work was supported by the National Science Foundation under Grant Nos. NSF/DMS-850319 and NSF/DMS-8705646.

We thank Yves Thibaudeau, Arieh Epstein and the referees for their comments on the manuscript.

References

Achcar, J.A. and A.F.M. Smith (1989). "Aspects of Reparameterization in Approximate Bayesian Inference," (this volume).

Bickel, P.J., F. Götze, and W.R. van Zwet, (1985). "A Simple Analysis of Third-Order Efficiency of Estimates," in *Proceedings of the Berkeley Conference in Honor of Jerzy Neyman and Jack Kiefer*, Vol. II, L. LeCam and R.A. Olshen, eds. Belmont, CA: Wadsworth.

Chen, C.F. (1985). "On Asymptotic Normality of Limiting Density Functions with Bayesian Implications," *J. Roy. Statist. Soc., B*, Vol. 47, 540-546.

Crawford, S. (1988). "An Approximate Bayesian Analysis of Finite Mixture Distributions," Ph.D. Thesis, Department of Statistics, Carnegie Mellon University.

Erdelyi, A. (1956). *Asymptotic Expansions*. New York: Dover Publications.

Ghosh, J.K., B.K. Sinha, and S.N. Joshi (1982). "Expansions for Posterior Probability and Integrated Bayes Risk," *Statistical Decision Theory and Related Topics, III*, Vol. 1, Shanti S. Gupta and James O. Berger, eds. New York: Academic Press, 403-456.

Heyde, C.C. and I.M. Johnstone (1979). "On Asymptotic Posterior Normality for Stochastic Processes," *J. Roy. Statist. Soc., B*, Vol. 41, 184-189.

Jennrich, R.I. (1969). "Asymptotic Properties of Nonlinear Least Squares Estimators," *Ann. Math. Statist.*, Vol. 40, 633-643.

Johnson, R.A. (1967). "An Asymptotic Expansion for Posterior Distributions," *Ann. Math. Statist.*, Vol. 38, 1899-1907.

Johnson, R.A. (1970). "Asymptotic Expansions Associated with Posterior Distributions," *Ann. Math. Statist.*, Vol. 41, 851-864.

Johnson, R.A. and J.N. Ladalla (1979). "The Large Sample Behavior of Posterior Distributions When Sampling from Multiparameter Exponential Family Models, and Allied Results," *Sankyha, B* Vol. 41, 196-215.

Joshi, S.N. (1984). "Expansion of Bayes Risk in the Case of Double Exponential Family," *Sankhya, A*, Vol. 46, 64-74.

Kass, R.E., and J.C. Fu (1988). "Posterior Large Deviations," Department of Statistics Technical Report, Carnegie Mellon University.

Kass, R.E., L. Tierney, and J.B. Kadane (1988). "Asymptotics in Bayesian Computation," in *Bayesian Statistics 3*, J.M. Bernardo, M.H. DeGroot, D.V. Lindley, and A.F.M. Smith, eds. Oxford: Oxford University Press, 261-278.

Lai, T.L., H. Robbins, and C.Z. Wei (1978). "Strong Consistency of Least Squares Estimates in Multiple Regression," *Proc. Nat. Acad. Sci.* Vol. 75, 3034-3036.

Laplace, P.S. De (1847), *Oeuvres*, Vol. 7, Paris: Imprimerie Royale.

LeCam, L. (1953). "On Some Asymptotic Properties of Maximum Likelihood Estimates and Related Bayes Estimates," *Univ. Calif. Publ. Statist.*, Vol. 1, 277-330.

Leonard, T. (1982). "Comment: Simple Predictive Density Function," *J. Amer. Statist. Assoc.*, Vol. 77, 657-658.

Lindley, D.V. (1961). "The Use of Prior Probability Distributions in Statistical Inference and Decisions," *Proc. 4th Berkeley Symp.* Vol. 1, 453-468.

Lindley, D.V. (1980). "Approximate Bayesian Methods," *Bayesian Statistics*, J.M. Bernardo, M.H. DeGroot, D.V. Lindley, A.M.F. Smith, eds. Valencia, Spain: University Press.

Morris, C.N. (1988). "Approximating Posterior Distributions and Posterior Moments," in *Bayesian Statistics 3*, J.M. Bernardo, M.H. DeGroot, D.V. Lindley, and A.F.M. Smith, eds. Oxford: Oxford University Press, 327-344.

Mosteller, F. and D.L. Wallace (1964). *Inference and Disputed Authorship: The Federalist Papers.* Reading, Mass.: Addison-Wesley.

Naylor, J.C. and A.F.M. Smith (1982). "Applications of a Method for the Efficient Computation of Posterior Distributions," *App. Statist.*, Vol. 31, 214-225.

Perlman, M.D. (1972). "On the Strong Consistency of Approximate Maximum Likelihood Estimators," *Proceedings of the Sixth Berkeley Symposium on Mathematical Statistics and Probability*, Vol. 1, 263-282.

Perlman, M.D. (1983). "The Limiting Behavior of Multiple Roots of the Likelihood Equation," in *Recent Advances in Statistics: Papers in Honor of Herman Chernoff*, M.H. Rizvi, J. Rustagi, and D. Siegmund, eds. New York: Academic Press.

Smith, A.F.M. A.M. Skene, J.E.H. Shaw, J.C. Naylor, and M. Dransfield (1984). "The Implementation of the Bayesian Paradigm," *Commun. Statist. Theor. Meth.*, Vol. 14, No. 5, 1079-1102.

Tierney, L. and J.B. Kadane (1986). "Accurate Approximations for Posterior Moments and Marginal Densities," *J. Amer. Statist. Assoc.*, Vol. 81, 82-86.

Tierney, L., R.E. Kass, and J.B. Kadane (1989a). "Approximate Marginal Densities of Nonlinear Functions," *Biometrika*, to appear.

Tierney, L., R.E. Kass, and J.B. Kadane (1989b). "Fully Exponential Laplace Approximations to Expectations and Variances of Non-Positive Functions," *J. Amer. Statist. Assoc.*, to appear.

Walker, A.M. (1969). "On the Asymptotic Behavior of Posterior Distributions," *J. Roy. Statist. Soc., B*, Vol. 31, 80-88.

Wu, C.F. (1981). "Asymptotic Theory of Nonlinear Least Squares Estimation," *Ann. Statist.*, Vol. 9, 501-513.

AUTHOR INDEX

SUBJECT INDEX